Eastern African rain forests are remarkable in their high level of endemism. Miocene uplift of the central African plateau separated these montane and coastal forests from the main Guineo–Congolian forest of west and central Africa. Since then, stable Indian Ocean temperatures maintained a region of high rainfall throughout Pleistocene droughts that devastated forest elsewhere on the continent. Relics of the former Pan-African rain forest survived here, the study of which provides a unique insight into tropical evolutionary processes. This book brings together research on the animals, plants and geography of this intriguing residual forest, and highlights the need for effective management practices to conserve its exceptional biodiversity in the face of increasing pressure for land for cultivation.

Biogeography and ecology of the rain forests of eastern Africa

Biogeography and ecology of the rain forests of eastern Africa

EDITED BY

Jon C. Lovett

*Missouri Botanical Garden, St Louis, Missouri, USA
and Department of Botany, University of Dar es Salaam, Tanzania*

AND

Samuel K. Wasser

*Conservation and Research Center, National Zoological Park, Smithsonian
Institution, Front Royal, Virginia, USA*

CAMBRIDGE
UNIVERSITY PRESS

CAMBRIDGE UNIVERSITY PRESS
Cambridge, New York, Melbourne, Madrid, Cape Town, Singapore, São Paulo

Cambridge University Press
The Edinburgh Building, Cambridge CB2 8RU, UK

Published in the United States of America by Cambridge University Press, New York

www.cambridge.org
Information on this title: www.cambridge.org/9780521430838

First published 1993
This digitally printed version 2008

A catalogue record for this publication is available from the British Library

Library of Congress Cataloguing in Publication data
Biogeography and ecology of the rain forests of eastern Africa/
edited by Jon C. Lovett and Samuel K. Wasser.
p.   cm.
ISBN 0 521 43083 6 (hc)
1. Rain forest ecology – Africa, Eastern. 2. Rain forest fauna – Africa, Eastern. 3. Rain
forest plants – Africa, Eastern. 4. Rain forests – Africa, Eastern. 5. Rain forest fauna –
Africa, Eastern – Geographical distribution. 6. Rain forest plants – Africa, Eastern –
Geographical distribution. I. Lovett, Jon. II. Wasser, Samuel K.
QH195.A23B56 1993
574.5′2642′09676–dc20   92–36758   CIP

ISBN 978-0-521-43083-8 hardback
ISBN 978-0-521-06898-7 paperback

Contents

Contributors

S. BRØGGER-JENSEN Ornis Consult ApS. Vesterbrogade 140, DK-1620 Copenhagen, Denmark

T. C. E. CONGDON Brooke Bond Tanzania Ltd., P.O. Box 4955, Dar es Salaam, Tanzania

R. de JONG National Natuurhistorisch Museum, P.O. Box 9517, 2300 Ra Leiden, The Netherlands

C. J. GRIFFITHS 5 Lissadell Drive, Magherafelt, Co. Londonderry BT45 5AR

W. D. HAWTHORNE 7 Poplar Road, Oxford OX2 9LA

R. L. HOFFMAN Virginia Museum of Natural History, Martinsville, VA 24112, USA

K. M. HOWELL Department of Zoology and Marine Biology, University of Dar es Salaam, P.O. Box 35064, Dar es Salaam, Tanzania

F. P. JENSEN Department of Ornithology, Zoologisk Museum, Universitetsparken 15, DK-2100 Copenhagen, Denmark

J. KINGDON The Elms, Islip, Oxon OX5 2SD

J. C. LOVETT Botanisk Museum, DK–1123 Copenhagen, Denmark

R. I. MILLER c/o S. N. Stuart, Species Survival Commission, Avenue du Mont-Blanc, CH-1196 Gland, Switzerland

F. A. MTURI Department of Zoology and Marine Biology, University of Dar es Salaam, P.O. Box 35064, Dar es Salaam, Tanzania

W. A. RODGERS 4 Bannold Road, Waterbeach, Cambridge CB5 9LQ

N. SCHARFF Department of Entomology, Zoologisk Museum, Universitetsparken 15, DK-2100 Copenhagen, Denmark

S. N. STUART Species Survival Commission, IUCN, Avenue du Mont-Blanc, CH-1196 Gland, Switzerland

S. K. WASSER Conservation and Research Center, National Zoological Park, Smithsonian Institution, Front Royal, VA 22630, USA

PART I
Introduction

1 Introduction to the biogeography and ecology of the rain forests of eastern Africa

SAMUEL K. WASSER and JON C. LOVETT

The tropical rain forests of Africa are divided by a corridor of arid land that runs from the Horn of Africa, through Kenya, Tanzania, Zambia, Zimbabwe and Botswana to the Namib Desert of Namibia (Werger, 1978). The arid corridor reaches the sea on the east coast of Africa in Mozambique, where Madagascar casts a rain shadow, and in the north where the corridor covers much of Somalia. A narrow strip of relatively high rainfall lies between the deserts of the Somalia coast in the north, the Madagascar rain shadow in the south and the woodland of the central African plateau to the west. It is this humid area of eastern Tanzania, Kenya and southern Somalia that we define here as eastern Africa (Figures 1.1 and 1.2).

The area of tropical rain forest in eastern Africa is not large; it is approximately $10\,000\,km^2$, a mere 0.1% of the estimated 10 million km^2 of tropical rain forest in the world (Mabberley, 1983). Unlike the vast west and central African forests, the forests of eastern Africa are highly fragmented – discrete islands associated with localised areas of high rainfall, surrounded by a sea of comparatively arid woodland. Contrasts between these wet and dry areas are pronounced. For example, on the eastern slopes of the Uluguru Mountains of Tanzania there is a per-humid climate where more than 100 mm of rain falls each month of the year and the annual rainfall exceeds 3000 mm. A high diversity of slender boled, narrow crowned trees reach heights greater than 60 m in a closed, epiphyte-rich canopy. Streams cascade down steep slopes with banks of ferns, begonias and *Impatiens*. A few kilometres to the south is the highly seasonal climate of the Selous Game Reserve, where large populations of game animals roam wooded grasslands. In the mountain's rain shadow to the west, there are sisal estates in semi-arid land that receive less than 600 mm of rain per year.

As described by Griffiths (Chapter 2) and Lovett (Chapter 3), it is the geological and climatological histories of eastern Africa that make its forests unique. Climatic vicissitudes throughout the Pleistocene are thought to have caused major extinctions by substantially reducing the total area of African rain forest. However, the forest patches of eastern Africa appear to have escaped these changes owing to the remarkable stability of the Indian Ocean currents that bring moisture to the tropical East African coast (Figure 1.2; Hamilton, 1982). There probably has been tropical rain forest in eastern Africa since approximately 30 million years ago, when Africa drifted to its present position at the end of the Oligocene (Axelrod & Raven, 1978). The continental divide between East and West Africa was accentuated by the Miocene uplift of the central African plateau (Griffiths, Chapter 2). The great age, isolation and fragmented nature of the eastern African forests have combined to produce remarkably high levels of endemism and diversity. About one third of the more than 2000 moist forest plant species are endemic (found nowhere else; Lovett, 1988) with over 20 endemic genera (Iversen, 1991). The

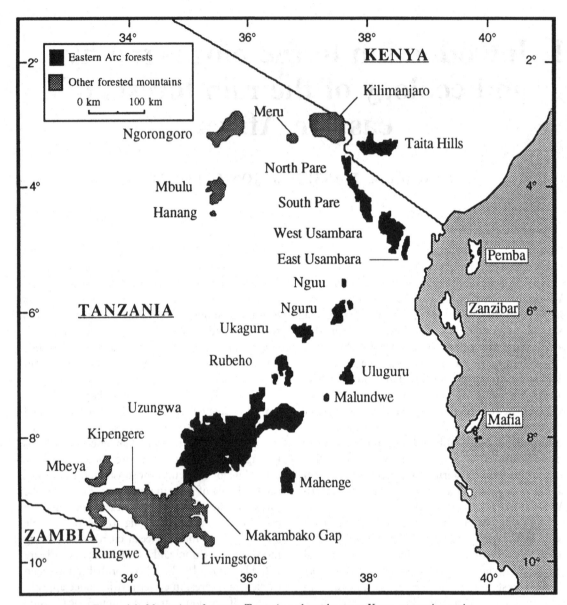

Figure 1.1. Mountains of eastern Tanzania and southeastern Kenya supporting moist forest. Eastern Arc mountain forests are shown in black.

Figure 1.2. Map of eastern Africa showing oceanic currents and trade winds (from IUCN/UNEP, 1982). Inset is a map of the arid corridor along with other arid regions throughout Africa (hatched areas, from Werger, 1978), with the wetter part of eastern Africa marked in black.

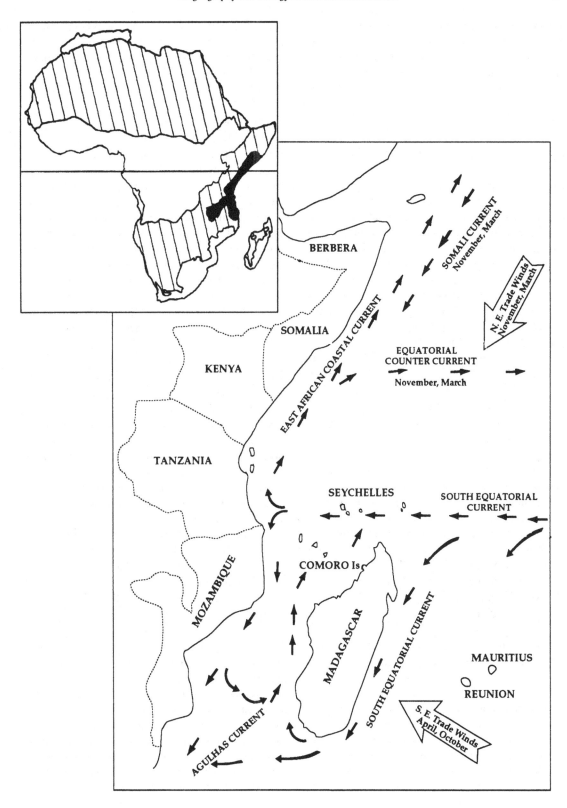

percentage of endemism is even greater in slow-moving, forest-dependent invertebrates such as millipedes (Hoffman, Chapter 6) or moisture-dependent amphibians (Howell, Chapter 9).

The eastern African forests have an altitudinal range from near sea level on the coast to almost 3000 m on Mount Kilimanjaro. A continuous altitudinal range from 400 to 2000 m is found on the Uzungwa and Nguru mountains. The eastern African forests are of three general geographical types: the Eastern Arc forests on ancient crystalline mountains; the forests of more recent origin on volcanic mountains; and coastal forests on overlying sedimentary rocks between the Eastern Arc mountains and the coast, and on the offshore islands of Pemba, Zanzibar and Mafia. Each of these forests is an effective island, whether surrounded by arid land or sea. This book is about the results of physical and biogeographical events, including isolation and speciation, that have occurred at each of these localities.

The book begins with the geological and climatological history (Griffiths, Chapter 2; Lovett, Chapter 3). The remaining chapters describe how these physical events have affected the biogeography, ecology and conservation biology of the eastern Africa forest fauna and flora. The vegetation chapters (Lovett, Chapter 4; Hawthorne, Chapter 9) describe ways in which climatological and vegetational changes reinforce one another. This sets the stage for the section describing biogeography and patterns of endemism throughout the animal kingdom, including millipedes (Hoffman, Chapter 6), spiders (Scharff, Chapter 7) butterflies (de Jong & Congdon, Chapter 8), amphibians and reptiles (Howell, Chapter 9), birds (Stuart *et al.*, Chapter 10) and mammals (Kingdon & Howell, Chapter 11). What is remarkable about these chapters is the degree of consistency in biogeographical patterns found across taxonomic groups, despite phylogenetic constraints in mobility or resistance to desiccation.

The montane forests of eastern Africa have been isolated and comparatively stable for a relatively long period of evolutionary time. This isolation has created both palaeoendemics (relicts of the formerly widespread distribution of the pan-African rain forest) and neoendemics (more recent immigrants from other habitats that were able to survive and radiate in areas of high rainfall). These patterns are compared with those in the more recent forests of the eastern African volcanoes where considerably fewer endemics are found, most of which are neoendemics. Compared with the more continuously distributed western African forests, the highly fragmented forests of eastern Africa appear to have fewer species overall but a relatively high proportion of endemics.

Movement of forest species into, and out of, the forests of eastern Africa is demonstrated throughout the book, using both northwestern and southwestern routes from the Guineo–Congolian Forest to the Eastern Arc. Most important among the northwestern routes were those through the Kenyan highlands and down the easterly, riverine drainages to the more northerly mountains of the Arc. Most important among the southwestern routes were the riverine forests passing through the Tanganyika Rift, along mountains and valleys of the Rukwa rifts and continuing into the climatic system created by Lake Nyasa. Several chapters provide evidence of these routes, based on north–south differences in species composition within the Eastern Arc montane forests. But exceptional patterns also are apparent, owing in part to interactions between phylogeny, mobility and ecology; most notable are mobility and ecological constraints on the distribution of invertebrates (see Chapters 6, 7 and 8 by Hoffman, Scharff, and de Jong & Congdon). Some species also appear to have adapted to the forest from the surrounding open woodland, and vice versa (Chapters 11 and 8 by Kingdon & Howell, and de Jong & Congdon).

The remaining chapters of the book focus on behavioural ecology and conservation biology. Two chapters (12 and 13) on forest primates, by Mturi, and Wasser, illustrate the impact that biogeographical events described throughout the book can have on the behavioural ecology of a species. Wasser's chapter suggests ways in which such behavioural and ecological changes might have facilitated processes that led to speciation (e.g. niche shifts, character displacement and hybridisation). The concluding chapter 14, by Rodgers, focuses on the conservation of the forests of eastern Africa, both historically and into the future. This chapter and others emphasise that the highly fragmented, relatively small forest patches of eastern Africa place them at considerably greater conservation risk than the larger forest blocks of western Africa (see also Newmark, 1991). Rodgers argues that forests were once much more extensive in Tanzania and Kenya (20% and 10%, respectively) but were destroyed by clearing and burning for cultivation. He draws upon biogeographical theory to develop management strategies for forest conservation in eastern Africa.

The unique nature and conservation importance of the rain forests of eastern Africa have been recognised for many years (Polhill, 1968; Lucas, 1968; Rodgers & Homewood, 1982). These points are highlighted throughout this book. Collectively, the chapters provide the readers with a comprehensive view of how evolutionary events have shaped the biogeography and ecology of these forests. It is our hope that these studies will contribute to the development and implementation of effective conservation strategies for the eastern African forests into the future.

References

AXELROD, D. I. & RAVEN, P. H. (1978). Late Cretaceous and Tertiary vegetation history of Africa. In *Biogeography and Ecology of Southern Africa*, ed. M. J. A. Werger, pp. 77–130, The Hague: Junk.

HAMILTON, A. C. (1982). *Environmental History of Africa: A Study of the Quarternary*. London: Academic Press.

IUCN/UNEP (1982). Conservation of coastal and marine ecosystems and living resources of the East African Region. *United Nations Environment Programme Regional Seas Reports and Studies* No. 11.

IVERSEN, S. T. (1991). The Usambara mountains, NE Tanzania: Phytogeography of the vascular plant flora. *Symbolae Botanicae Upsaliensis* 24, 1–234.

LOVETT, J. C. (1988). Endemism and affinities of the Tanzanian montane forest flora. In *Systematic Studies in African Botany*, ed. P. Goldblatt and P. P. Lowry, pp. 591–8. St Louis: Missouri Botanical Garden.

LUCAS, G. L. (1968). Kenya. In *Conservation of Vegetation in Africa South of the Sahara*, ed. I. Hedberg and O. Hedberg, pp. 152–63. Acta Phytogeographica Suecica, Vol. 54.

MABBERLEY, D. (1983). *Tropical Rain Forest Ecology*. London: Blackie.

NEWMARK, W. D. (1991). Tropical forest fragmentation and the local extinction of understory birds in the Eastern Usambara Mountains, Tanzania. *Conservation Biology* 5, 67–78.

POLHILL, R. M. (1968). Tanzania. In *Conservation of Vegetation in Africa South of the Sahara*, ed. I. Hedberg and O. Hedberg, pp. 166–78. Acta Phytogeographica Suecica, vol. 54.

RODGERS, W. A. & HOMEWOOD, K. M. (1982). Species richness and endemism in the Usumbara mountain forests, Tanzania. *Biological Journal of the Linnean Society* 18, 197–242.

WERGER, M. J. A. (1978). The Karoo–Namib Region. In *Biogeography and Ecology of Southern Africa*, ed. M. J. A. Werger, pp. 231–99. The Hague: Junk.

2 The geological evolution of East Africa

C. JANE GRIFFITHS

Abstract

This chapter examines the geological and geomorphological processes that have shaped the present-day landscape of the eastern part of East Africa. The underlying rock types, their age, environment of formation and interrelationships are described. The structural evolution of the landscape is traced from the Karroo *c.* 300 million years before present (myr BP), encompassing the Karroo Rifting, the breakup of Gondwanaland, the relative ages of the block mountains, volcanic mountains, plains and plateau, the East African Rift Valley system, and the vertical movements affecting the continental margin of East Africa. Reference is made to the general geology and topography of West Africa for comparison and the debate about the pre-drift position of Madagascar is briefly reviewed. Variations in soil types in relation to parent rock, age, climate, organic material and relief are considered.

Introduction

A wide variety of rock types are found in East Africa representing all three of the major groups: igneous, metamorphic, and sedimentary. They range in age from over 2000 myr BP to unconsolidated sediments accumulating at present, but represent two geological time spans: the Precambrian (>2000–570 myr BP) and the Karroo–Recent (290–0.01 myr BP) (Figure 2.1 and Table 2.1), separated by a major break in the geological record. These two groups of rocks were formed under entirely different geological conditions, resulting in distinct properties.

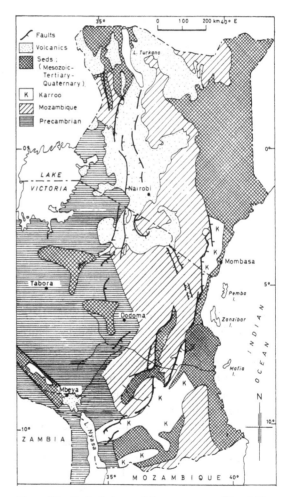

Figure 2.1. Geological map of the eastern part of East Africa.

The Precambrian rocks are highly deformed, high grade metasedimentary, with minor meta-igneous, crystalline rocks, including granulites,

Table 2.1. *Geological record for East Africa*

Era	Period/Tectonic unit (epoch)		Absolute age (myr BP)	Brief description of rock types and environment of formation
Quaternary	Recent		0.01	Volcanics associated with rifting in Arusha–Kilimanjaro and Mbeya regions
	Pleistocene		2.5	Continued marine sedimentation along the coast with non-marine inland
Tertiary	Neogene	Pliocene	7	Rifting and volcanism in northern and southern areas
		Miocene	26	Dominantly marine sediments including limestones
		Oligocene	38	and marls in coastal areas with non-marine alluvium
	Palaeogene	Eocene	54	and eluvial soils inland throughout Tertiary
		Palaeocene	65	
Mesozoic	Cretaceous		135	Marine sediments along part of the coastal belt with non-marine inland
	Jurassic		180	Dominantly marine sediments east of the faulted Mozambique Belt accumulated in the East African sedimentary basin
	Karroo		290	Continental sediments including sandstones, shales, coal measures, lagoonal evaporites and minor marine incursions near top of sequence
Palaeozoic				Major break in sequence without geological record
			438	
Precambrian	Mozambique Belt		2100	Highly deformed N–S trending metamorphic belt consisting of largely reworked older rocks with less deformed cover including marbles

gneisses and schists. The Karroo–Recent sediments include conglomerates, sandstones, limestones, shales, marls and evaporites. Volcanic rocks, which occur in association with the East African Rift Valley were formed in the Plio–Pleistocene–present day during the last 7 myr, and consist of alkaline basalts.

Following their uplift in the early Palaeozoic (438 myr BP), the Precambrian rocks have been land areas subjected to the effects of sub-aerial weathering and erosion, which has reduced large areas to plains. However, since Karroo time the region has been intermittently affected by periods of vertical movements which have formed rift valleys (grabens) and fault block mountains (horsts) on the otherwise monotonous plains. These two opposing processes of levelling and dislocation,

together with volcanic activity, are responsible for the often spectacular landscape found in East Africa today. Block faulting and volcanicity produced most of the mountain ranges where the moist forests are found. These vertical movements contrast with the lateral movements which affected other parts of the ancient supercontinent of Gondwanaland comprising Africa, South America, Madagascar, India, Antarctica and Australia (Figure 2.2). The lateral movements also began with rifting but continued with a widening of the rift by basaltic intrusions followed by invasion by seawater. The continents then drifted apart by sea-floor spreading, taking them to their present-day locations. The coastal forests are found on the Mesozoic–Quaternary marine, fluviatile and lacustrine sediments (Figure 2.1).

Figure 2.2. Reconstruction of Gondwanaland in Karroo times (from Windley, 1977).

The Precambrian rocks

The Precambrian rocks of eastern Africa belong to the Mozambique Belt which runs north–south and can be traced from Ethiopia to Mozambique. West to east, it occurs from the Gregory Rift in the eastern arm of the Rift Valley to where it is faulted beneath the younger sediments of Karroo–Recent age (Figure 2.3). Rocks of the Mozambique Belt also occur in the Seychelles and over a large area of central and eastern Madagascar.

Granitoid gneisses (metamorphosed granites), hornblende–pyroxene granulites (high grade metamorphosed rocks probably derived from sediments of sandstone and shale), marbles (metamorphosed limestones) and graphite schists (metamorphosed shales) are all found.

Radiometric dating of rocks from the Mozambique Belt has produced a range of ages extending from c. 2000 to 450 myr BP with the highest frequency occurring in the 650–450 myr BP range (Cahen & Snelling, 1966; Cahen et al., 1984). This latter span is thought to coincide with the Mozambiquian orogeny (a period of mountain building involving sedimentation, deformation, metamorphism and uplift), which led to the creation of a north–south mountain chain. Now only the deeper root zones remain, the upper layers having been removed by the processes of weathering and erosion.

Since the marbles and schists are less deformed and metamorphosed than the granulites it is believed by some authors (Sampson & Wright,

1964; Cahen et al., 1984) that they represent the sedimentary cover sequence of Mozambiquian age. These sediments were deposited on an older (granulite) basement which suffered further reworking by deformation and metamorphism during the Mozambiquian orogeny. The reheating of older rocks resets their radiometric clocks, giving a wide range of ages dependent on the degree of reworking. This relationship would account for the spread of absolute ages generated by the rocks of this belt.

The Mozambique Belt is only a part of a much more extensive orogenic belt, known as the Pan-African Belt, which affected and reworked older rocks over large areas of Africa. Figure 2.4 shows the continental extent of the Pan-African Belt. The underlying rocks of much of East and West Africa are rather uniform and consist of ancient Precambrian rocks. As will be seen in later sections, younger sedimentary or volcanic rocks overlie the ancient crystalline rocks in some places.

Karroo–Recent sediments

In Tanzania, Karroo sediments are exposed principally in the southeast, together with a block near Tanga which broadens northwards in Kenya (Figure 2.1). In addition to these surface deposits, great thicknesses of Karroo sediments lie buried beneath younger ones in the coastal zone, both onshore and offshore (Figure 2.5). Most of the Karroo sediments, consisting of sandstones, shales, conglomerates, red and grey mudstones, occasional limestones and coal measures, were deposited in a terrestrial environment under riverine or lake conditions in down-faulted basins. The Tanga beds include some brackish water sediments. These, together with great thicknesses (3000 m) of evaporites that accumulated in a graben at Mandawa in the Triassic (Kent et al., 1971), herald the initiation of the true marine conditions of the future Indian Ocean and the breakup of Gondwanaland.

The onset of marine conditions during the mid-Jurassic coincided with renewed faulting of the Mozambique Belt along the inland edge of the Somali–Kenya–Tanzania sedimentary basin

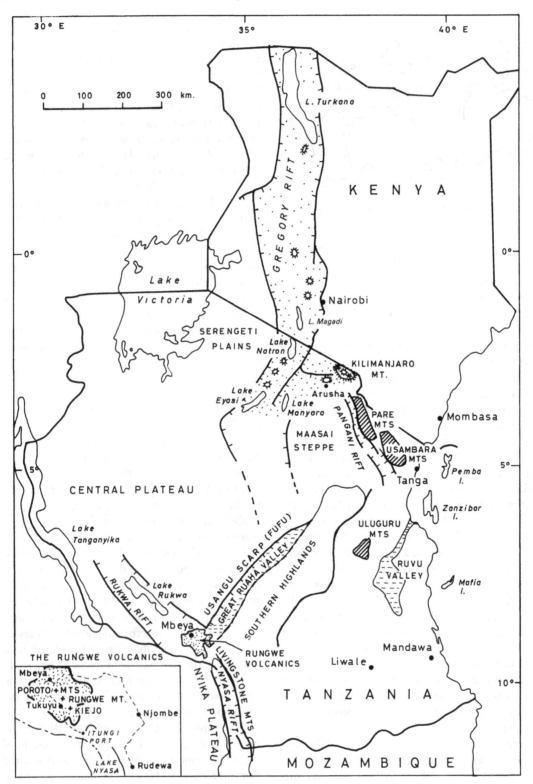

Figure 2.3. Location map.

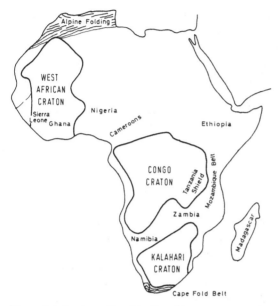

Figure 2.4. Regions affected by Pan-African orogenesis distinguished from older cratons and younger fold belts (from Clifford, 1970).

caused by sub-aerial erosion. These changes in environment have resulted in an often rapid variation of rock types in both horizontal and vertical directions. The sediments include sandstones, limestones, marls, shales and mudstones with a complete range of transitional types such as calcareous sandstones and shaley limestones. These lithological variations are too small to be mapped so that the rocks are classified on the basis of age. Greater detail is shown on Tanzanian Geological Survey 1:125 000 maps, where these exist.

Sedimentological evidence indicates that some present-day deep sea areas were land during the Neogene (26–2.5 myr BP), while areas now land were deep sedimentary basins of the same age (Kajoto, 1982). These changes of level resulted from vertical movements on reactivated Karroo faults. The offshore islands of Zanzibar, Pemba and Mafia are thought to be fault blocks (Kent, Hunt & Johnstone, 1971). The inland fault blocks of, for example, the Ulugurus and Usambaras were also blocked out along reactivated Karroo faults, another indication of the continental scale of the movements responsible for the fragmentation of Gondwanaland.

While marine and non-marine sediments were accumulating along the coast east of the Mozambique Belt in the newly forming Indian Ocean, inland, terrestrial conditions prevailed and sedimentation occurred in low lying areas such as crustal depressions, rift valley floors created by the Karroo faulting and in river valleys.

which has been developing in the Indian Ocean ever since it was born with the breakup of Gondwanaland. This basin extends offshore beyond the islands of Zanzibar and Pemba for at least 150 km (Kajato, 1982).

Sedimentation has continued in this basin since the Jurassic, although the environmental conditions have oscillated between marine and terrestrial, with occasional gaps in the sequence

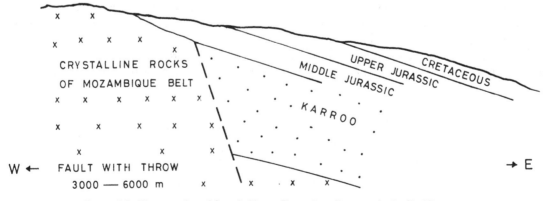

Figure 2.5. Western edge of Somali–Kenya–Tanzania sedimentary basin. Faultline concealed by mid-Jurassic marine sediments which transgressed on to Basement (based on Kent *et al.*, 1971).

Superficial deposits and soils

Superficial deposits and soils found mantling vast areas of the region's rocks, in all but high ground and in stream sections where erosion rates are high, are shown in the geological maps as being of Neogene–Recent age. The soil characteristics depend on the complex interaction of rainfall, time, topography, organic material and parent rock. Space does not permit a detailed description of the soils found in East Africa, but the following general comments may help the reader assess the possible soil type variations likely to be found.

Working in East Africa on Precambrian basement areas in the 1930s, Milne (1935) observed that topography greatly influences the soil properties. He termed the regular repetition of the particular sequences of soil profiles in association with particular topographies, a catena. In simple terms, soils fall into two main groups: the heavy, black clay-rich *mbuga* or black cotton soils occupy valleys and areas of impeded drainage; while the other group includes reddish-brown sandy, ferruginous residual soils of the hillsides from which the fine clays and silts have been washed out. The first group are alluvial deposits, and the second are eluvial.

Chemical weathering of rock proceeds rapidly in the tropics because of the high temperatures and humidity. At the prevailing high temperatures, the rate of chemical weathering is positively correlated with the amount of rainfall. This in turn favours the release of nutrients from the primary rock minerals. The chemical composition of the parent rock influences the fertility of the soil by determining the amount of nutrients that are released on weathering. Lavas and rocks which are poor in quartz give rise to fertile fine-textured soils, whereas rocks of the basement complex are generally rich in quartz and weather to coarse-textured sandy soils of low fertility. These latter are highly porous, well-drained soils and so less drought resistant than those derived from volcanic rocks.

The soils of the mountainous areas of eastern Africa have weakly developed profiles. They are formed on recently exposed geologically young surfaces, on either volcanics of Plio–Pleistocene age or steep slopes on Basement where erosion by gravity quickly removes the weathered products. However, the low potential for fertility of the parent rocks is offset in the mountains by higher than average rainfall, which enhances the rate of chemical weathering and sustains a prolific vegetation cover characteristic of the humid forests. The resulting increase in organic matter in the soil compensates for the lower nutrient status of the parent material. This fertility is rapidly lost when the vegetation is removed and the area cultivated.

Structure and evolution of the landscape

The present-day landscape owes its origin to a combination of geological processes essentially operating in opposition. On the one hand, weathering, erosion and deposition have operated continuously since the uplift of the Mozambique Belt, levelling the land to extensive plains. On the other, sporadic faulting has elevated some portions and lowered others (Figure 2.6). In some places this faulting has been accompanied by volcanism.

Across the entire face of Africa, there are wide, level plains sometimes stretching for hundreds of

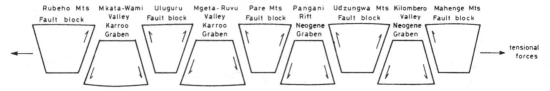

Figure 2.6. Schematic diagram to illustrate the consequences of vertical tectonics on the Precambrian rocks of Tanzania. Subsequent reactivation of the Karroo faults has maintained the landforms to the present day.

kilometres. Rising above the plains are occasional isolated hills called inselbergs. Eastern Africa has its share of such monotonous landscapes, although the Neogene–Recent rift tectonics and volcanism have so profoundly changed its face that some spectacular scenery is found in countries such as Burundi, Rwanda, Uganda, Ethiopia, Kenya, Tanzania and Malawi. Structural upheavals of this order (Figure 2.6) have not affected Central and West Africa as recently or to the same extent, so that the plainlands predominate there.

Karroo rifting

During Karroo times rift faulting prevailed and produced horsts and grabens over much of the country. Sediments accumulated in elongated basins largely under terrestrial conditions, although openings to the sea took place to the northeast in the Ruvu Valley basin area and to the southeast in the Mandawa area (Kent et al., 1971; Kajato, 1982).

The orientation of the rift faulting is often concordant with pre-existing Precambrian structural directions, a coincidence widely recognised in East Africa (King, 1970). For example, in western Tanzania, Karroo basins followed the trend of the Precambrian northwest–southeast Ubendian orogenic belt and in turn were the forerunners of the Western Rift troughs of Lakes Tanganyika and Rukwa. In contrast, in eastern Tanzania Karroo basins are generally elongated in a meridional direction concordant with the dominant trend of the Mozambique Belt. For example, Karroo sediments flank the Uluguru Mountains to the east and west.

The magnitude of the Karroo rifting is highlighted by several authors. Dixey (1956) shows that the Lake Nyasa basin started as a rift in Karroo times and continued developing during the Mesozoic. Kent et al. (1971) identified 3000 m of Karroo evaporites which accumulated in a graben largely controlled by reactivated basement faulting (Kent, 1972). At the end of Karroo times there was further down-faulting which accounts for the preservation of Karroo sediments in many places (Quennel, McKinlay & Aitken, 1956).

Kent (1972) asserts that the greatest crustal adjustments of the East African mainland coast took place during the Karroo when the modern structure of the fault-controlled coast was blocked out. No later faulting has approached its magnitude.

This period of rifting in East Africa occurred at the same time as the separation of Africa from the Antarctic (Sowerbutts, 1972), and the huge outpourings of basalt in southern Africa.

Age of block faulting

Opinions regarding the age of the various block-fault mountains in eastern Tanzania vary. Quennel et al. (1956) regard the faulting in eastern and central Tanzania which produced the fault blocks of the Pare, Uluguru and Usambara mountains as being of Neogene age. However, east of the Uluguru Mountains, the faulted edge of the Karroo sedimentary basin coincides with the boundary fault of the Ulugurus, and the mineralogical composition of the coarse-grained sediments indicates that the source area lay to the west (Sampson & Wright, 1964). This implies that the Uluguru Mountains fault block had started to form a distinctive unit as far back as the Karroo. It seems reasonable to suggest that the other fault blocks and the Southern Highlands developed over tens and even hundreds of millions of years. Probably short periods of intense uplift associated with major periods of vertical movement on a continental scale along ancient Precambrian faultlines were separated by long periods of quiescence. The blocks were probably initiated in the Karroo period (290–180 myr BP), contemporaneous with the initial breakup of Gondwanaland. Then, very long erosive phases, interrupted by widespread but less significant vertical movements, planed off the landscape to form the African and younger erosion surfaces (King, 1967). Finally, during the development of the East African Rift System the faults were again reactivated (7–0 myr BP) and the modern mountain blocks were the result.

The breakup of Gondwanaland

Fragmentation of the supercontinent took place over a protracted period and there is controversy over the exact dates at which separation of the constituent continents took place. Sowerbutts (1972) proposes that Antarctica and Africa separated during the Karroo, while South America is believed to have moved away from Africa in the Lower Cretaceous, 125 myr BP (Sclater & Tapscott, 1979) or 120 myr BP (McConnell, 1977).

On the basis of diverging pole positions for India, Antarctica, and Africa for the mid-Cretaceous, McElhinny (1970) shows that the Indian–Madagascar–Antarctic block broke away from Africa between the mid-Jurassic and mid-Cretaceous, opening up the Indian Ocean for the first time. This fits with the geological evidence from the East African coast already cited. Sowerbutts (1972) views the Jurassic–Cretaceous rifting in East Africa as being associated with the breakup of Madagascar and Africa. Rabinowitz, Coffin & Falvey (1983) are more precise and state that the Africa–Madagascar separation began 165 myr BP and ended 121 myr BP. They note that the Africa–Madagascar separation thus began at about the same time as the initial breakup of Gondwanaland, with the separation of North America from Africa (Sclater & Tapscott, 1979). Subsequently, Antarctica parted from the Indian–Madagascar block 100 myr BP and the Madagascar–Indian subcontinents separated from each other about the time of the Mesozoic–Tertiary boundary 65 myr BP (McElhinny, 1970).

The palaeoposition of Madagascar in the Gondwana supercontinent has been a subject of intense debate over the last few decades, with three possibilities proposed (Figure 2.7).

1. A northerly position adjacent to Somalia, Kenya and Tanzania (du Toit, 1937; Smith & Hallam, 1970; Dietz & Holden, 1970; Embleton & McElhinny, 1975; Rabinowitz et al., 1983).
2. A southerly position adjacent to Mozambique (Wegener, 1929; Wellington, 1954; Flores, 1970, 1984).

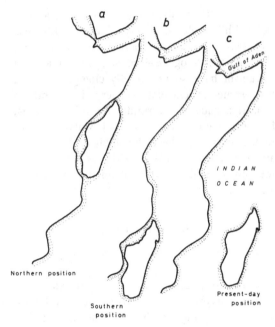

Figure 2.7. Three alternative palaeopositions of Madagascar.

3. Its present position with respect to Africa (Kent, 1972; Kutina, 1975; Kamen-Kaye, 1982).

The geological evidence supporting all of these positions is equivocal; however, physiographic, palaeomagnetic, and other geophysical evidence favours the northerly position (Rabinowitz 1971; Embleton & McElhinny, 1975, 1982; Rabinowitz et al., 1983).

Plainlands of East Africa

In eastern Africa a series of plains is found, varying from the remnant bevelled summits of the Livingstone and Uluguru mountains to the extensive areas of the Central Plateau in Tanzania, and the smooth uplifted surfaces bordering on the Rift Valley in Kenya. Down-faulting and down-warping of the coastal region accompanied the inland uplifts. Marine transgression (an extension of the seawater over a former land area) marked the breakup of Gondwanaland, establishing a stable continental margin facing the developing Indian Ocean (Kent, 1974) (Figure 2.8).

Because the vertical movements characteristic

Basins
Swells
Fold Mts. regions

Figure 2.8. The major basins and swells of Africa (after Holmes, 1965: p. 1054).

of eastern African tectonics have lowered some areas and raised others, dating the plains is not a simple correlation of levels (Figures 2.6 and 2.9a, b). Accordingly, authors do not always agree about the absolute or relative ages of particular plains. In King's view (King, 1967), the Masai steppe and Serengeti Plains are preserved as representatives of the early Tertiary (African) surface, while only remnant summits of this surface remain

standing above the younger plain of the Central Plateau of Tanzania. On the other hand, Quennel et al. (1956) regard the greater part of the Central Plateau to be the African surface which they describe as being of mid-Tertiary age. However, these authors state that the Central Plateau region was uplifted on two occasions in post-Jurassic time with the second phase, which probably continued through the Neogene, disrupting the African surface.

Recognising the vertical movements, King (1967) observes that the same topography associated with the late Tertiary surface has been down-faulted east of the Fufu scarp to form the floor of the Great Ruaha Valley (Figure 2.9b). In contrast, the up-faulted Southern Highlands that bound the valley to the southeast bear the early Tertiary planation (African Surface), and above it stand residuals of the Gondwana surface of Jurassic age. Eastwards of the fault-bound Southern Highlands to Liwale, the African surface is dominant. Towards the coast there is a convergence of these different erosional surfaces as they pass beneath marine sediments of different ages, demonstrating a coastal monoclinal flexure, or down-warp.

Westwards of the high areas of Uganda, Rwanda and Burundi, plains are again the dominant landform. Some of the plains have clearly been formed by the deposition of sediments, especially in some of the large structural

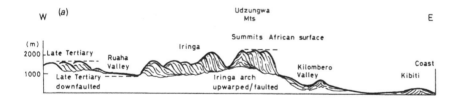

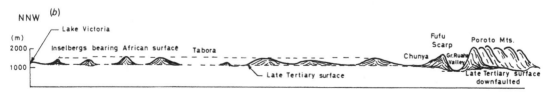

Figure 2.9(a). Generalised section across the Southern Highlands of Tanzania. (b) Generalised section across the Central Plateau of Tanzania showing erosion surfaces.

basins like the Congo. A larger proportion, though, are of erosional origin like those described from East Africa. These plains have been eroded across a wide range of different rock types and structures ranging from the Precambrian crystalline rocks of Tanzania to late Cretaceous and Eocene sands of Nigeria and the Ivory Coast which were planed during the African early Tertiary cycle of erosion (King, 1967).

An explanation of the structural pattern in East Africa

Basin and swell structures together with rift faulting are both widely recognised on a large scale in Africa today (Figures 2.8 and 2.10). These are apparently recurrent structural patterns and within East Africa are repeated on a smaller scale. It has been proposed that the structures are related through a common cause (Burke & Dewey, 1974; McConnell, 1977).

The basin and swell structure develops when a continental plate is stationary over rising mantle plumes or hot spots causing melting in the upper mantle and lower crust. The consequent increase in volume results in a swell and stretches the outer brittle skin of the crust, which in some cases is relieved by rifting and possibly volcanism (Figure 2.10). Basins occupy the areas between the swells. The Lake Victoria basin, for example, lies between the western and eastern arms of the Rift Valley. Burke & Dewey (1974) and McConnell (1977) have noted that Africa has been stationary for the past 25–30 myr during which time swells and basins, rifting and magnetism have all developed.

This mechanism, as the cause of the East African Rift Valley, differs from the commonly held notion that the Rift Valley is part of the global system associated with continental breakup (Windley, 1977).

Formation of the East African rift system

The age of the rifting and associated volcanism in the Gregory Rift in southern Kenya has been determined by Crossley (1979). There activity commenced in the mid-Miocene (15–12 myr BP), significantly later than in the northern section of the rift (30–20 myr BP), and earlier than in the Tanzanian deformation to the south which is attributed to the Pliocene (Downie & Wilkinson, 1972). Major rejuvenations in southern Kenya (5–2.5 and 0.6–0 myr BP) do not correspond with the 2–1.5 myr BP movements recorded in northern Kenya (Crossley, 1979); however, they appear to be contemporaneous with the Tanzanian rifting. This last Pleistocene phase created much of the present day rift topography resulting in the

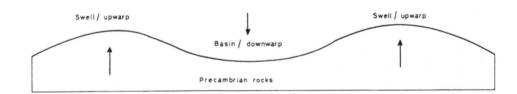

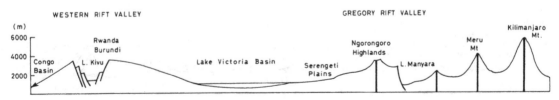

Figure 2.10. Generalised section across East Africa showing the effect of warping and subsequent tensional faulting.

Magadi–Natron and Manyara basins (Pickering, 1968; Crossley, 1979).

The central highlands of Africa, comprising the mountains of eastern Zaïre, western Uganda, Rwanda and Burundi, lie along the line of the western arm of the Rift Valley and were formed by a combination of block-faulting and volcanism. Uplift of the giant block of Precambrian rocks of the Rwenzori Range on the Zaïre–Uganda border occurred mainly in the Pliocene and Quaternary. Nyamlagira and Nyirogongo are still active volcanoes.

Evidence has already been cited to indicate that the basin of Lake Nyasa is a very old structure. Le Fournier *et al.* (1985) note that the basin of Lake Tanganyika has evolved over a period of 20 myr. However, ages of the lakes appear to be much younger with Lake Tanganyika estimated to be mid-Pliocene (*c.* 7 myr BP) and Lake Nyasa of early Pleistocene (2.0 myr BP) age. Lake Kivu formed during the late Pleistocene through the damming of river valleys by lava flows.

In comparison with the Rift Valley lakes, the Lake Victoria basin lies in a depression formed by crustal sagging between the two arms of the Rift Valley. The basin started developing contemporaneously with the rifting to the west and east and there is evidence of an earlier lake or system of lakes than the modern Lake Victoria, which is believed to be a mere 30 000 years old (Beadle, 1974).

Volcanics of Tertiary to Recent age

In the early Pleistocene the rift system extended eastwards by a system of oblique faults and down-warped basins from the Natron–Manyara trough to the Kilimanjaro region. Here it encountered the north–south faults delimiting the Pangani Rift. The formation of this east–west trough, the Kilimanjaro Depression, was the trigger for the onset of the Kilimanjaro volcanism, the oldest lavas of which are believed to be 1.0 myr old (Downie & Wilkinson, 1972).

Initially volcanic activity was widespread over the whole area now occupied by the mountain but, with time, eruptions were concentrated first to the Shira, then Mawenzi and finally to the Kibo

centres. By the beginning of the Upper Pleistocene (*c.* 120 000 yrs BP), volcanic activity became increasingly sporadic with smaller volumes erupted in each episode. Shira and Mawenzi are now extinct while the Kibo centre, which has been dormant during the Holocene or last 10 000 years, shows only residual fumarolic activity (Downie & Wilkinson, 1972). These authors conclude that there has been no eruptive activity within the last 200 years.

The lahars or volcanic mudflows which resulted from the collapse of the original craters and their lakes that occurred at late stages in the development of both Mount Meru and Mawenzi are likely to have been far more devasting to the forests on their slopes than the younger lava flows because of their much greater volume and extent.

The Mount Meru lahar spread eastwards across the Sanya Plains almost as far as Moshi and, according to Guest & Leeedal (1953), occurred in prehistoric times, while they estimate that the last eruption from Meru occurred around 1877. Since then activity has been confined to fumaroles. Earthquakes still occur from time to time, thus Meru is still classified as being an active volcano whereas Kilimanjaro is dormant.

The Mawenzi lahar occurred in the Lower Pleistocene (*c.* 500 000 BP) (Downie & Wilkinson, 1972) and spread northeastwards over an area of 1166 km^2 between Laitokitok and Chyulu Range.

Kibo has a much more restricted lahar associated with it, mainly found between 670 and 1330 m west of Moshi. Downie & Wilkinson (1972) believe it resulted from a series of landslides prior to the last Inner Crater eruptions during the Upper Pleistocene (11 000 BP).

In southern Tanzania, the isolated province of the Rungwe Volcanics comprising the Poroto Mountains, Tukuyu, Kiejo and Rungwe Mountains, lies at the Y-shaped junction of the Rukwa, Ruaha and Nyasa troughs. These lavas are similar in composition to the alkali basalts of the north. The earliest lavas erupted in the late Pliocene (5–2.5 myr BP) (Harkin, 1960), after the main rift movement. Like Kilimanjaro, the major activity occurred in the Pleistocene, the Rungwe volcano being the youngest, and the extrusives of the Poroto Mountains the oldest. The last eruption from Kiejo is reckoned to have taken place about AD 1800 (Harkin, 1960).

Conclusion

The geological history of the eastern part of East Africa is long and interesting. Palaeomagnetic evidence from different African cratons (rigid crust no longer affected by deformation) indicates that the bulk of Africa has behaved as an integral unit during the period 2300–400 myr BP (Piper, 1976). During the Mesozoic era, Karroo faulting profoundly affected the face of eastern Africa, down-throwing grabens in which vast thicknesses of sediments and evaporites accumulated, and up-lifting the forerunners of the present-day block mountains. The fragmentation of Gondwanaland commenced in the Jurassic with the widening and flooding of some of these coastal grabens, creating marine conditions along the present-day coast-line. Madagascar split from eastern Africa at an early stage in the fragmentation and from India much later.

The normal processes of weathering and erosion operated at the surface throughout the period. Mountains and hills were levelled to plains in the course of the erosional cycle while tectonic activity disrupted and dislocated these areas of low relief, rejuvenating the rivers and renewing the cycle, creating younger plains at lower levels. The younger (Neogene) surfaces are still covered by weathered material at various stages of soil formation

During the Neogene the East African Rift Valley developed, possibly in another attempt at continental breakup, but more likely in response to upper mantle and lower crustal melting. Tensional faulting affected many parts of the region, reactivating old zones of weakness, some of which were initiated in the Karroo, others in the Precambrian. Volcanic activity accompanied this rifting in some areas, and together these gave rise to mountainous landscapes contrasting strongly with the plainlands. It is the varying relief of eastern Africa rather than its underlying rock types and soils that accords it much of its uniqueness.

References

BEADLE, L. C. (1974). *The Inland Waters of Tropical Africa*. London: Longman.

BUCKLE, C. (1978). *Landforms in Africa*. London: Longman.

BURKE, K. & DEWEY, J. F. (1974). Two plates in Africa during the Cretaceous. *Nature* 249, 313–16.

CAHEN, L. & SNELLING, N. J. (1966). *The Geochronology of Equatorial Africa*. Amsterdam: North Holland Publishing Co.

CAHEN, L., SNELLING, N. J., DELHAL, J. & VAIL, J. R. (1984). *The Geochronology of Africa*. Oxford: Oxford University Press.

CLIFFORD, T. N. (1970). The structural framework of Africa. In *African Magmatism and Tectonics*, ed. T. N. Clifford and I. G. Gass, pp. 1–26. London: Oliver & Boyd.

CROSSLEY, R. (1979). The Cenozoic stratigraphy and structure of the western part of the Rift Valley in southern Kenya. *Journal of the Geological Society* 136, 393–406.

DIETZ, R. S. & HOLDEN, J. C. (1970). Reconstruction of Pangaea: breakup and dispersion of continents, Permian to Present. *Journal of Geophysical Research* 75, 4939–56.

DIXEY, F. (1956). *The East African Rift System*. Overseas Geological Mineral Resources Bulletin Supplement No. 1. London: HMSO.

DOWNIE, C. & WILKINSON, P. (1972). *The Geology of Kilimanjaro*. Sheffield: Geological Survey of Tanzania and Department of Geology, Sheffield University.

DU TOIT, A. L. (1937). *Our Wandering Continents*. Edinburgh: Oliver and Boyd.

EMBLETON, B. J. J. & MCELHINNY, M. W. (1975). The palaeoposition of Madagascar: palaeomagnetic evidence from the Isalo Group. *Earth and Planetary Science Letters* 27, 329–41.

EMBLETON, B. J. J. & MCELHINNY, M. W. (1982). Marine magnetic anomalies, paleomagnetism and the drift history of Gondwanaland. *Earth and Planetary Science Letters* 58, 141–50.

FLORES, G. (1970). Suggested origin of the Mozambique Channel. *Transactions of the Geological Society of South Africa* 73, 1–16.

FLORES, G. (1984). The S. E. Africa triple junction and the drift of Madagascar. *Journal of Petroleum Geology* 7, 403–18.

GUEST, N. J. & LEEDAL, G. P. (1953). Volcanic activity of Mt. Meru. *Records of the Geological Survey of Tanganyika* 3, 40–6.

HARKIN, D. A. (1960). The Rungwe Volcanics at the northern end of Lake Nyasa. *Geological Survey of Tanganyika Memoir II*. Dar es Salaam: Government Printer.

HOLMES, A. (1965). *Principles of Physical Geology*. London: Nelson.

KAJATO, H. J. (1982). Gas strike spurs search for oil in Tanzania. *Oil and Gas Journal* **125**, 123–31.

KAMEN-KAYE, M. (1982). Mozambique–Madagascar geosyncline, I: deposition and architecture. *Journal of Petroleum Geology* **5**, 3–30.

KENT, P. E. (1972). Mesozoic history of the East Coast of Africa. *Nature* **238**, 147–8.

KENT, P. E. (1974). Continental margin of East Africa – a region of vertical movements. In *The Geology of Continental Margins*, ed. C. A. Burk & C. L. Drake, pp. 313–20. New York: Springer-Verlag.

KENT, P. E., HUNT, J. A. & JOHNSTONE, D. W. (1971). The geology and geophysics of coastal Tanzania. *Institute of Geological Sciences, Geophysical Paper* 6.

KING, B. C. (1970). Vulcanicity and rift tectonics in East Africa. In *African Magmatism and Tectonics*, ed. T. H. Clifford and I. G. Gass, pp. 263–83. Edinburgh: Oliver and Boyd.

KING, L. C. (1967). *The Morphology of the Earth*, 2nd edn. London: Oliver and Boyd.

KUTINA, J. (1975). Tectonic development and metallogeny of Madagascar with reference to the fracture pattern of the Indian Ocean. *Geological Society of America Bulletin* **86**, 582–92.

LE FOURNIER, J., CHOROWICZ, J., THOUIN, C., BALZER, F., CHENET, P.-Y., HENRIET, J.-P., MASSON, D., MONDEGUER, A., ROSENDAHL, B., ANDERSON, F.-L. S. & TIERCELIN, J.-J. (1985). Le bassin du lac Tanganyika: évolution tectonique et sédimentaire. *Comptes rendus de l'Academie des Sciences, Paris* **301**, II, 14.

MCCONNELL, R. B. (1977). East African Rift System dynamics in view of Mesozoic apparent polar wander. *Journal of the Geological Society* **134**, 34–40.

MCELHINNY, M. W. (1970). Formation of the Indian Ocean. *Nature* **228**, 977–9.

MILNE, G. (1935). Composite units for the mapping of complex soil associations. *Transactions of the 3rd International Congress of Soil Scientists* **1**, 345–7.

PICKERING, R. (1968). Ngorongoro's geological history. *Ngorongoro Conservation Area Booklet* No. 2. Dar es Salaam: East African Literature Bureau.

PIPER, J. D. A. (1976). Palaeomagnetic evidence for a Proterozoic supercontinent. *Philosophical Transactions of the Royal Society of London, Ser. A* **280**, 469–90.

QUENNEL, A. M., MCKINLAY, A. C. M. & AITKEN, W. G. (1956). Summary of the Geology of Tanganyika, Part I. Introduction and Stratigraphy. Geological Survey of Tanganyika Memoir 1. Dar es Salaam: Government Printer.

RABINOWITZ, P. D. (1971). Gravity anomalies across the East African continental margin. *Journal of Geophysical Research* **76**, 7107–17.

RABINOWITZ, P. D., COFFIN, M. F. & FALVEY, D. (1983). The separation of Madagascar and Africa. *Science* **220**, 67–9.

SAMPSON, D. N. & WRIGHT, A. E. (1964). *The Geology of the Uluguru Mountains*. Geological Survey of Tanzania Bulletin 37. Dar es Salaam: Government Printer.

SCLATER, J. G. & TAPSCOTT, C. (1979). The History of the Atlantic. *Scientific American* **240**, 120–32.

SMITH, A. G. & HALLAM, A. (1970). The fit of the southern continents. *Nature* **225**, 139–44.

SOWERBUTTS, W. T. C. (1972). Rifting in Eastern Africa and the fragmentation of Gondwanaland. *Nature* **235**, 435–6.

WEGENER, A. (1929). *The Origin of Continents and Oceans*, 4th German edn., translation. London: Methuen.

WELLINGTON, J. H. (1954). The significance of the middle courses of the Zambesi and Limpopo Rivers and their relations with the Mozambique Plain. *South African Journal of Science* **50**, 182–5.

WINDLEY, B. F. (1977). *The Evolving Continents*. Chichester: Wiley.

3 Climatic history and forest distribution in eastern Africa

JON C. LOVETT

Introduction

The African climate has been far from stable. Equatorial rainfall is created by oceanic solar heating generating the intertropical convergence zone (ITCZ), so rainfall and hence forest growth is in part dependent on the relative position of the solar equator. The continent was south of its present position at the end of the Cretaceous with the equator running through the present-day Sahara, suggesting that the rainfall was high in what is today a desert. The amount of rain carried inland by the ITCZ is dependent on oceanic currents which are in turn modified by positions of continents; for example, Antarctica creates the cold Benguela current which brings aridity to southwestern Africa, and Madagascar casts a monsoonal rainshadow on southeast Africa. Ocean currents are also affected by periodic variations in the Earth's orbit, the Milankovitch cycle, which is thought to be responsible for repeated Pleistocene global cooling and warming (Imbrie & Imbrie, 1980). Finally, rainfall on the continent is affected by positions of mountains and lakes which create their own weather systems within the overall ITCZ and oceanic current climates.

Climatic fluctuations are partly responsible for eastern African moist forest biota distribution patterns. The three main patterns to elucidate are: apparently ancient links to the Guineo–Congolian forests of western and central Africa; apparently recent links to the western Guineo–Congolian forests; and high degree of endemism within the eastern African forests. The climatic history of eastern Africa can be conveniently divided into three time periods: (i) the breakup of Gondwanaland about 100 Myr BP (million years before present) and the tectonic events following it; (ii) uplift of the central African plateau during the Miocene; and (iii) Pleistocene global climatic fluctuations during the last 2.3 Myr. Reviews of the relationship between biological distribution patterns and climatic and geological events are given by Bonnefille (1984); Flenley (1979); Hamilton (1976, 1982, 1988) and Wickens (1976). The geological events are reviewed by Griffiths (Chapter 2) and present-day climate is described by Griffiths (1972) and Kingdon (1971) with specific details on eastern Africa in Hamilton (1989) and Pocs (1974, 1976).

Breakup of Gondwanaland and northward drift of Africa

In the Triassic to early Cretaceous there were two main land masses, a southern Gondwanaland and a northern Laurasia. They were separated by the Tethys sea and possessed markedly different biotas. Evidence of this comes from plant fossils (Roche, 1974) and the north–south split of conifer genera (Sneath, 1967): African examples are *Widdringtonia* and *Podocarpus* to the south and *Juniperus* in the north. Africa was south of its present position in the centre of Gondwanaland flanked by South America to the west and India and Antarctica to the east. At this time much of central and eastern Africa would have been dry as it lay in the centre of Gondwanaland and it is unlikely that it had an extensive moist forest cover.

From the mid-Cretaceous onwards Gond-wanaland began to drift apart so that by the beginning of the Tertiary (65 Myr BP) oceanic currents could flow around Africa. At the end of the Cretaceous, Africa was about 15–18° south of its present position putting the ITCZ over what is now the Sahara. Moist air would have been coming from the Tethys Sea to the north and the newly formed seas (Atlantic and Indian) on the western and eastern edge of the continent. Fossil evidence from the late Cretaceous and early Tertiary indicates the presence of moist forest in present-day North Africa and it has been suggested that a lowland equatorial rain forest covered the area in the Palaeocene 60 Myr BP (Axelrod & Raven, 1978).

At the time of the breakup of Gondwanaland there was rifting taking place in east and southeast Africa (Sowerbutts, 1972). This suggests that the progenitors of the Eastern Arc and eastern rift mountains were formed about 100 Myr BP. They were probably in existence before northward movement of Africa closed the Tethys sea. This northward movement limited northern oceanic currents, reducing rainfall and beginning fragmentation of the pan-African rain forest. From the mid-Cretaceous to the mid-Tertiary there is no evidence of major tectonic disturbance in Africa and most of southern and central Africa was eroded in the 'great African planation' (King, 1978). Progenitors of the Eastern Arc mountains would have been eroded during this period, and it is unlikely that they were major mountains prior to the mid-Tertiary uplift.

For 30 million years following the development of the pan-African rain forest, Africa moved northwards. The Sahara still possessed forest or woodland in the Eocene about 40 Myr BP (Aubreville, 1970) but aridity was spreading during the Oligocene. By the time Africa completed its union with Eurasia in the mid-Miocene 17–18 Myr BP, the pan-African forest would have begun to be fragmented into a western area under the Atlantic Ocean climate and an eastern area under the Indian Ocean climate. The Eastern Arc mountains would have been attracting orographic precipitation with a corresponding rain shadow to their west, though its extent would have depended

on how high they were at this time following the long period of erosion.

Closure of the Tethys sea stopped the flow of equatorial oceanic currents through it, and the northern high pressure areas of the ITCZ which brought rain into northern Africa during the early Tertiary were now largely over land and so carried less moisture into the equatorial region. Lower Miocene fossil evidence (24 Myr BP) from Bugishu near Mount Elgon (Chaney, 1933) has an abundance of small leaves suggesting a seasonal climate supporting dry woodland, as do the Ethiopian sites of the same period (Axelrod & Raven, 1978). The mid-Miocene flora of Rusinga Island in Lake Victoria (Chesters, 1957) has been interpreted as riverine thicket in savanna woodland (Axelrod & Raven, 1978; Kortlandt, 1983), and a comparison of mammal remains in five East African fossil sites (Andrews, Lord & Nesbit Evans, 1979, reviewed by Kortlandt, 1983) suggests a mixed woodland savanna in the area. However, the Ruisenga Island fossils are those of evergreen forest and semi-evergreen species (A. C. Hamilton, personal communication) and so Guineo–Congolian moist forests still existed in central Africa at that time.

The eastern African coast, and what mountains there were, would have received increasing amounts of moisture from the Indian Ocean as the breakup of Gondwanaland proceeded, becoming forested in the Miocene if not before. At this time it has been suggested that connection between the western Guineo–Congolian forests and eastern African coastal and Eastern Arc forests would have been by gallery forests along rivers extending to eastern Africa from western Africa across a flat, eroded landscape (Axelrod & Raven, 1978). However, fossil evidence from the early Miocene indicates that there was easterly drainage from the area around Lake Rudolf in Kenya and westerly drainage from Lake Victoria (Andrews & van Couvering, 1975). This suggests that the early Miocene east–west watershed was slightly east of its present position, but still some distance from the Eastern Arc mountains. Direct contact by gallery forest would have been limited at that time, though the watershed may have been low enough to permit migration of forest organisms (Andrews & van Couvering, 1975). The central highlands of Africa were not formed until the Pliocene and Quaternary and so the Atlantic rainfall system would have penetrated further eastward than at present. For example, a late Miocene (12.2 Myr BP) leaf flora from the Kenyan rift valley, in what is now dry *Acacia* wooded grassland, contains fossils of the moist forest grasses *Humbertochloa* and *Leptaspis* (Jacobs & Kabuye, 1987). At the end of the Miocene major topographic changes in East and Central Africa dramatically altered drainage patterns and so limited dispersal along rivers.

Uplift of the Central African Plateau

During the mid-Tertiary the ancient peneplaned surface of the central African plateau was flexed and warped (Burke & Wilson, 1972). The East African hinterland swelled upwards and the Congo sank into a basin. These topographic changes would have altered drainage patterns and accentuated the continental divide. At the end of the Miocene, *c.* 10 Myr BP, further lifting raised the central plateau to its present level and major rifts developed. Rivers were rejuvenated and the rise of the east–west continental divide limited the western Guineo–Congolian climatic and biotic influence, separating it from the eastern forests by a north–south band of dry seasonal country called the arid corridor (reviewed by Werger, 1978). The arid corridor is today seen in those areas where, for at least three consecutive months a year, there is less than 10 mm of rain a month. It connects northeastern and southwestern African arid areas, and comes east to the tropical Indian Ocean in northern Mozambique west of Madagascar. Called a corridor because of north–south disjunctions of arid land species, it presents a formidable barrier to east–west migration or dispersal of moist forest organisms. The remaining possible migration routes between the western and eastern forests would have either been northwards over the Kenyan highlands and down the rivers draining eastwards; or southwards via the Lake Tanganyika rift, along the mountains and valleys of the southern Rukwa rift, into the climatic system created by Lake Nyasa and across

the forest gap at Makambako in the southern Uzungwa Mountains.

From the early Pliocene, climatic history becomes clearer following analysis of deep-sea sediments and the development of models based on changes in the earth's orbit and orientation (Hays, Imbrie & Shackleton, 1976; Imbrie & Imbrie, 1980). The evidence and proposed climatic changes are reviewed by Hamilton (1988). From the beginning of the Pliocene (*c.* 9 Myr BP) to the mid-Pliocene (6.4 Myr BP) the climate was warm and humid, an observation supported by fossil evidence from northwest Ethiopia, suggesting a moist forest with western Guineo–Congolian affinities around 8 Myr BP (Yemane, Bonnefille & Faure, 1985). Off the southwest African coast deep-sea drilling indicates that strong, persistent upwelling of cold waters (the present Benguela system) began in the early late Miocene about 10 Myr BP (Siesser, 1980). After 6.4 Myr BP, southward movement of Antarctica enhanced the cold Benguela current. After 5 Myr BP a major ice sheet developed on Antarctica (Shackleton & Kennett, 1975), strengthening the cold current, bringing further aridity to the southwest coast of Africa and compressing the southern Guineo–Congolian forests. Kalahari sands, thought to be primarily of aeolian origin, dating from the Tertiary or early Quaternary, have been found as far north as the River Zaïre and as far east as Zimbabwe. Subtropical rain forest at the Cape containing taxa such as *Microcachrys*, Winteraceae, Sarcolaenaceae and Chloranthaceae which are no longer represented in Africa, was lost at the close of the Tertiary (Coetzee & Muller, 1984). Between 4.6 and 2.43 Myr BP the climate was again warm and wet. Lake Turkana in northern Kenya possessed a much richer riverine forest than at present, with western Guineo–Congolian elements (Williamson, 1985; Dechamps & Maes, 1985). During the warm, wet periods of 9–6.4 Myr BP and 4.6–2.43 Myr BP migration across the arid corridor via the northern and, more particularly, the southern migration route would have been possible for readily dispersed plants tolerant of a range of moist forest types.

Pleistocene climatic fluctuations

From the evidence of deep-sea cores it has been suggested that there have been 21 glacials or near glacial periods since 2.3 Myr BP (van Donk, 1976), with the earliest fossil evidence in Africa (from northeast Ethiopia) for a cooling and consequent depression of vegetation belts being about 2.51–2.35 Myr BP (Bonnefille, 1983). Climatic changes resulting from the glacial periods in East Africa have been reviewed by Hamilton (1982, 1988). It is the most recent world glacial maximum, which ended in equatorial Africa c. 14 000–12 000 yr BP, that has provided the most evidence about effects of the glacial periods on the East African climate. This evidence has come from two main sources: geomorphological features, in particular the glacial moraines on Mount Kenya, Kilimanjaro, Rwenzori, Elgon, the Aberdares and Meru; and fossil evidence, in the form of changes in pollen levels and species composition in pollen cores and plant and animal macrofossils.

Before the advent of carbon-14 dating methods it was thought that remains of high lake levels indicated that the East African climate was cooler and wetter during the glacial maximum, giving rise to the 'pluvial theory' (Nilsson, 1931). Now it is beyond doubt that the last ice age in East Africa represented a period of dry cold climate. Lake Victoria was almost non-existent at around 14 370 BP (Kendall, 1969; Livingstone, 1980), so convectional rainfall associated with it would have been very much reduced. A drier climate also caused water levels in Lakes Nyasa and Tanganyika to be about 250–500 m lower before about 25 000 BP, and the water level was low for tens of thousands of years (Scholz & Rosendahl, 1988). Levels were also low before about 12 000 BP (Scholz & Rosendahl, 1988).

During the last glaciation equatorial Atlantic surface temperatures decreased by 4–5 °C with greater reduction off northwest and southwest Africa (CLIMAP, 1976; van Zinderen Bakker, 1982; Hamilton, 1988). This would have resulted in decreased moisture coming into West and Central Africa. In contrast, the Indian Ocean surface temperatures were lowered by only 1–2 °C and the area off the eastern African coast did not decrease in temperature (Prell et al., 1980). However, a seabed core from near the mouth of the Zambezi River showed a maximum temperature difference of 3–4 °C between glacial and interglacial periods during the last 135 000 years (van Campo et al., 1990). Although the southerly Indian Ocean monsoon was weaker (Hamilton, 1988), coastal Tanzania and southeastern Kenya would have still been relatively wet owing to rain originating from the still warm Indian Ocean.

The temperature depression at about 18 000 BP calculated from altitudes of glacial moraines on high East African mountains is 6.7–9.5 °C and pollen diagrams show a lowering of vegetation zones by about 1000 m (Hamilton, 1988). Bonnefille, Roeland & Guiot (1990) suggest a temperature decrease of 4 ± 2 °C with a mean annual rainfall decrease of 30%, and date the maximum last glacial advance to 21 500 BP when the climate was cold but relatively moist. In Burundi, at a site which is currently tropical forest, they deduce that presence of the tree-line and alpine grassland at 2240 m indicates a lowering of vegetation belts by 1000–1500 m. Predictions of temperature decreases during past glacial stages in tropical Africa, from biotic and geomorphological evidence, are prone to error, as the climate was generally also much drier at the time. Rather than there being a simple lowering of vegetation belts, it is more likely that there was a general restriction in the distribution of lowland species and an expansion in the range of species tolerant of cooler and drier conditions (D. Taylor, in litt.).

If there had been a major depression of vegetation types, or a much drier climate, in coastal eastern Africa, then lower altitude wetter vegetation would have been eliminated. As the lower altitude forest vegetation is rich in species of restricted distribution and isolated taxonomic position, then the biological evidence suggests that Indian Ocean temperatures and monsoonal rainfall remained stable and that the lower altitude forest zone in eastern Africa was not substantially affected by the climatic changes observed on central African high mountains. It is clear that climatic changes during the last glacial maximum have been very different in different parts of Africa.

In contrast to glacial periods, the interglacials were warm and wet. Lake levels rose between 12 500 and 10 000 BP and pollen evidence shows that forest expanded at around the same time (Hamilton, 1982). The wettest post-glacial period was 9000–8000 BP when the north African deserts were dotted with swamps and lakes (Kutzbach & Street-Perrott, 1985) with rainfall 125–135% higher than today (Gasse *et al.*, 1990). After 5000 BP the climate became drier again. Assuming that wetter conditions occurred throughout Africa, moist forests would have been more widespread during this period. Under present climatic conditions the southern migration route along the Rukwa Rift contains patches of montane forest on mountain tops with lowland species at lower altitudes along forested stream courses running through fire climax Zambezian woodland. Present-day fires reduce the distribution of moist forest on the Mbeya range, and fire has been used in southwest Tanzania for at least 60 000 years (Hamilton & Faden, 1974). Under wetter conditions, and with less frequent fires, these forest patches would be more frequent and so would facilitate migration of readily dispersed forest organisms from west to east. Once through the southern migration route lowland species would still need to cross the Poroto and Kipengere mountains to reach the Eastern Arc and coastal forests.

The pattern of cold dry glacial periods with warm wet interglacials was probably true of other glacial periods, though two periods may have been cold and wet at 176 000 and 220 000 BP (Rossignol-Strick, 1983). This may be of importance as a cooler high rainfall period would allow a greater spread of forest vegetation than a hot wet period.

Summary

Before the breakup of Gondwanaland eastern Africa would have been relatively arid, being inland from Madagascar and India. It would have also been south of its present position with the equator passing through the present-day Sahara Desert. Fossil and oceanic current evidence suggests that tropical moist forest covered much of modern North Africa in the early Tertiary. By the mid-Tertiary Africa had reached its present latitudinal position. Towards the end of the Miocene, uplift of the central plateau separated eastern and western Africa. At the same time upwelling of the cold Benguela current brought aridity to southwest Africa. Ancient western Guineo–Congolian links in the Eastern Arc forests predate Miocene uplift of the central African plateau. Uplift of the Eastern Arc and Nyasa rift mountains 7 Myr BP would have accentuated rainfall gradients, with the eastern side of the mountains receiving a high rainfall and the western leeward side being in a rain shadow. This would have increased the barrier presented by the arid corridor to migration and dispersal of moist forest organisms from west to east, though migration and dispersal should have been possible during wet periods.

During the last 2.3 Myr Pleistocene climatic fluctuations caused considerable changes in central and north tropical African forest vegetation, but the Indian Ocean ameliorated the climate in tropical eastern Africa. Consequently the forests did not markedly reduce in area or undergo the altitudinal depression proposed for other forests on the Western Rift thereby allowing the survival of western Guineo–Congolian relict endemic species. More recent links could have occurred during Pleistocene interglacial periods by dispersal between forest patches either through the Kenyan Rift, or more likely along the southern migration route of the Rukwa Rift.

References

ANDREWS, P. & VAN COUVERING, J. A. H. (1975). Palaeoenvironments in the East African Miocene. *Approaches to Primate Palaeobiology* 5, 62–103.

ANDREWS, P., LORD, J. M. & NESBIT EVANS, F. M. (1979). Patterns of ecological diversity in fossil and modern mammalian faunas. *Biological Journal of the Linnean Society* 11, 177–205.

AUBREVILLE, A. (1970). Le flore tropical Tertiaire du Sahara. *Adansonia* 10, 9–14.

AXELROD, D. I. & RAVEN, P. H. (1978). Late Cretaceous and Tertiary vegetation history of Africa. In *Biogeography and Ecology of Southern Africa*, ed. M. J. A. Werger, pp. 77–130. The Hague: Junk.

BONNEFILLE, R. (1983). Evidence for a cooler and drier climate in the Ethiopian uplands towards 2.5 Myr ago. *Nature* 303, 487–91.

BONNEFILLE, R. (1984). The evolution of the East African environment. In *The Evolution of the East Asian Environment*, Vol. II, ed. R. O. Whyte, pp. 579–612. University of Hong Kong: Centre for Asian Studies.

BONNEFILLE, R., ROELAND, J. C. & GUIOT, J. (1990). Temperature and rainfall estimates for the past 40 000 years in equatorial Africa. *Nature* 346, 347–9.

BURKE, K. & WILSON, J. T. (1972). Is the African plate stationary? *Nature* 239, 287–9.

CHANEY, R. W. (1933). A Tertiary flora from Uganda. *Journal of Geology* 41, 702–9.

CHESTERS, K. I. M. (1957). The Miocene flora of Rusinga Island, Lake Victoria, Kenya. *Paleontographica* 101B, 30–71.

CLIMAP (1976). The surface of the Ice-Age Earth. *Science* 191, 1131–6.

COETZEE, J. A. & MULLER, J. (1984). The phytogeographic significance of some extinct Gondwana pollen types from the Tertiary of the southwestern Cape (South Africa). *Annals of the Missouri Botanical Garden* 71, 1088–99.

DECHAMPS, R. & MAES, F. (1985). Essai de reconstitution des climats et des végétation de la basse vallée de l'Omo au Plio-Pleistocene à l'aide de bois fossiles. In *L'environment des Hominides au Plio–Pleistocene*, ed. M. M. Beden *et al.*, pp. 175–222. Paris: Masson.

FLENLEY, J. R. (1979). *The Equatorial Rain Forest: A Geological History*. London: Butterworth.

GASSE, F. R., TEHET, A., DURAND, A., GILBERT, E. & FONTES, J.-C. (1990). The arid–humid transition in the Sahara and Sahel during the last glaciation. *Nature* 346, 141–6.

GRIFFITHS, J. F. (1972). Climate. In *East Africa: Its Peoples and Resources*, 2nd edition, ed. W. T. W. Morgan, pp. 107–17. Nairobi: Oxford University Press.

HAMILTON, A. C. (1976). The significance of patterns of distribution shown by forest plants and animals in tropical Africa for the reconstruction of

Upper Pleistocene palaeoenvironments: a review. *Palaeoecology of Africa, the Surrounding Islands and Antarctica* 9, 63–97.

HAMILTON, A. C. (1982). *Environmental History of East Africa: A Study of the Quaternary*. London: Academic Press.

HAMILTON, A. C. (1988). Guenon evolution and forest history. In *A Primate Radiation: Evolutionary Biology of the African Guenons*, ed. A. Gautier-Hion, F. Bourlière, J. P. Gautier and J. Kingdon, pp. 13–34. Cambridge: Cambridge University Press.

HAMILTON, A. C. (1989). The climate of the East Usambaras. In *Forest Conservation in the East Usambara Mountains, Tanzania*, ed. A. C. Hamilton and R. Bensted-Smith, pp. 97–102. Gland, Switzerland: IUCN.

HAMILTON, A. C. & FADEN, R. B. (1974). The history of the vegetation. In *East African Vegetation*, ed. E. M. Lind and M. E. S. Morrison, pp. 188–209. London: Longman.

HAYS, J. D., IMBRIE, J. & SHACKLETON, N. J. (1976). Variations in the earth's orbit: pacemaker of the ice ages. *Science* 194, 1121–32.

IMBRIE, J. & IMBRIE, J. Z. (1980). Modeling the climatic response to orbital variations. *Science* 207, 943–53.

JACOBS, B. F. & KABUYE, C. H. S. (1987). A middle Miocene (12.2 Ma) forest in the East African rift valley, Kenya. *Journal of Human Evolution* 16, 147–55.

KENDALL, R. L. (1969). An ecological history of the Lake Victoria basin. *Ecological Monographs* 39, 121–76.

KING, L. (1978). The geomorphology of central and southern Africa. In *Biogeography and Ecology of Southern Africa*, ed. M. J. A. Werger, pp. 1–17. The Hague: Junk.

KINGDON, J. (1971). *East African Mammals: An Atlas of Evolution in Africa*, Vol. I. London: Academic Press.

KORTLANDT, A. (1983). Facts and fallacies concerning Miocene ape habitats. In *New Interpretations of Ape and Human Ancestry*, ed. R. L. Ciochon and R. S. Corrucini, pp. 465–514. New York: Plenum Press.

KUTZBACH, J. E. & STREET-PERROTT, F. A. (1985). Milankovitch forcing of fluctuations in the level of tropical lakes from 18 to 0 kyr BP. *Nature* 317, 130–4.

LIVINGSTONE, D. A. (1980). Environmental changes in the Nile Headwaters. In *The Sahara and the Nile*, ed. N. A. J. Williams and H. Faure, pp. 339–59. Rotterdam: Balkema.

NILSSON, E. (1931). Quaternary glaciations and pluvial lakes in British East Africa. *Geografiska Annaler* **13**, 249–349.

PRELL, W. L., HUTSON, W. H., WILLIAMS, D. F., BÉ, A. W. H., GEITZENAUER, K. & MOLFINO, B. (1980). Surface circulation of the Indian Ocean during the Last Glacial Maximum, approximately 18 000 yr BP. *Quaternary Research* **14**, 309–36.

POCS, T. (1974). Bioclimatic studies in the Uluguru mountains (Tanzania, East Africa). I. *Acta Botanica Academiae Scientiarum Hungaricae* **22** (1–2), 115–35.

POCS, T. (1976). Bioclimatic studies in the Uluguru mountains (Tanzania, East Africa). II. Correlations between orography, climate and vegetation. *Acta Botanica Academiae Scientiarum Hungaricae* **22** (1–2), 163–83.

ROCHE, E. (1974). Paleobotanique, paleoclimatologie et dérive des continents. *Sciences Geologiques (Université Louis Pasteur de Strasburg) Bulletin* **27**, 9–24.

ROSSIGNOL-STRICK, M. (1983). African monsoons, an immediate climatic response to orbital insolation. *Nature* **304**, 46–9.

SCHOLZ, C. A. & ROSENDAHL, B. R. (1988). Low lake stands in Lakes Malawi and Tanganyika, East Africa, delineated with multifold seismic data. *Science* **240**, 1645–8.

SHACKLETON, N. J. & KENNETT, J. P. (1975). Palaeotemperature history of the Cenozoic and initiation of Antarctic glaciation: oxygen and carbon isotope analyses in DSDP sites 277, 279, and 281. *Initial Reports of the Deep Sea Drilling Project* **29**, 743–55.

SIESSER, W. G. (1980). Late Miocene origin of the Benguela upswelling system off northern Nambia. *Science* **208**, 283–5.

SNEATH, P. H. A. (1967). Conifer distributions and continental drift. *Nature* **215**, 467–70.

SOWERBUTTS, W. T. C. (1972). Rifting in Eastern Africa and the fragmentation of Gondwanaland. *Nature* **235**, 435–6.

VAN CAMPO, E., DUPLESSY, J. C., PRELL, W. L., BARRATT, N. & SABATIER, R. (1990). Comparison of terrestrial and marine temperature estimates for the past 135 kyr off southeast Africa: a test for GCM simulations of palaeoclimate. *Nature* **348**, 209–12.

VAN DONK, J. (1976). An [18]O record of the Atlantic Ocean for the entire Pleistocene. *Memoirs of the Geological Society of America Memoirs* **145**, 147–64.

VAN ZINDEREN BAKKER, E. M. (1982). African palaeoclimates 18 000 yrs BP. *Palaeoecology of Africa, the Surrounding Islands and Antarctica* **15**, 77–99.

WERGER, M. J. A. (1978). The Karoo–Namib Region. In *Biogeography and Ecology of Southern Africa*, ed. M. J. A. Werger, pp. 231–99. The Hague: Junk.

WICKENS, G. E. (1976). The flora of Jebel Marra (Sudan Republic) and its geographical affinities. *Kew Bulletin Additional Series* **5**, 1–368.

WILLIAMSON, P. G. (1985). Evidence for an early Plio–Pleistocene rainforest expansion in East Africa. *Nature* **315**, 487–9.

YEMANE, K., BONNEFILLE, R. & FAURE, H. (1985). Palaeoclimatic and tectonic implications of Neogene microflora from the north-western Ethiopian highlands. *Nature* **318**, 653–6.

PART II

Forest flora of eastern Africa

4 Eastern Arc moist forest flora

JON C. LOVETT

Introduction

In his classic paper on the vegetation of what is now Tanzania, the geographer Clement Gillman (1949) recognised the importance of the disjunct arc of mountains in the east of the country as condensers of moisture brought inland from the Indian Ocean. At this they are remarkably efficient: on the eastern side of the mountains rainfall can be well over 2000 mm per year; yet in the rain shadow only a few tens of kilometres to the west, it can be below 500 mm per year. In his list of Indian Ocean condensers, Gillman included the northern volcanic mountains of Meru and Kilimanjaro, and then ran the arc southwards along the disjunct crystalline block-faulted mountains to the Kipengere range above Lake Nyasa. Here he pointed out that high rainfall in the great amphitheatre of volcanic and crystalline mountains surrounding the northern end of Lake Nyasa was largely attributable to convection from the lake surface.

The volcanoes of Meru and Kilimanjaro are geologically recent, having been formed within the last million years. However, the crystalline block-faulted mountains of the Eastern Arc are very old, with initiation of faulting dating from 290–180 Myr BP (million years before present) and reactivation of the faults creating the modern mountains during the last 7 Myr BP (Griffiths, Chapter 2). Indications are that the Indian Ocean climate was comparatively stable during Pleistocene climatic fluctuations (Lovett, Chapter 3). It can be supposed, then, that moist forests on the eastern slopes of block-faulted mountains acting as Indian Ocean condensers have been under a stable high rainfall through the Pleistocene, and possibly since before the end of the Miocene.

Direct contact between the moist forests of West and Central Africa and those in the east would have been completely severed in the Miocene following uplift of the central African plateau, and were probably limited before then. The age and isolation of these mountains is reflected in the unique nature of their flora, and it is this which led them to be classified together as the Eastern Arc mountains (Lovett, 1985, 1988, 1990a; Beentje, 1988a). From north to south the main mountain blocks are: the Taita Hills of Kenya, and the Pare, Usambara, Uluguru, Nguru, Rubeho (Usagara), Uzungwa, and Mahenge mountains of Tanzania. The flora of these mountains is characterised by a high level of species and generic endemism in comparison with forests associated with mountains in Kenya, Ethiopia, Malawi, Zimbabwe and other parts of Tanzania: 25–30% of the c. 2000 Eastern Arc plant species are endemic (Lovett, 1988) with 16 genera endemic or near endemic to the forest and forest edge.

Eastern Arc forests have a wide altitudinal range, from 300–400 m a.s.l. at the base of the mountains to over 2000 m at the top. Low and medium altitude forests can be very tall with emergents to 60 m in height, whereas high altitude ridge top forests may have a closed canopy only 2–3 m tall. Because of local climate or clearance for cultivation not all the mountains have a continuous altitudinal range of forest from base to peak. Some forests are obviously secondary, with signs of cultivation and habitation under a currently closed canopy. Past changes that may be related to historical human disturbance are also indicated by moribund individuals of pioneer species with no regeneration in otherwise mature forest.

Forests of the Eastern Arc have been collected botanically for over 100 years (Gillett, 1961). Hildebrandt visited the Taita Hills in 1877, and later Holst, Engler, Buchwald, Heinse, Goetze, Kummer, Albers, Eick and Scheffler collected in the Tanzanian mountains between 1890 and 1900. Subsequent important collectors in the forests between 1920 and 1960 included Schlieben, Greenway, Bruce, Drummond and Hemsley, Semsei, and Proctor. A number of types from the early German collections were destroyed in 1943 when the Berlin–Dahlem Botanical Museum was bombed (Hiepko, 1987), but fortunately most had duplicates in other herbaria. Recently, there has been renewed interest in the forests with major collections being made in the Taita, Usambara, Nguru, and Uzungwa mountains (Iversen, 1987; Beentje, 1988a; Lovett, Bridson & Thomas, 1988; Pocs, Temu & Minja, 1990).

In this chapter a brief description is given of the Eastern Arc environment drawing attention to the cooling effect of the Indian Ocean and cool forest microclimate. Some of the classification systems that have been proposed for, or are applicable to, the forests are then reviewed and a system proposed with descriptions of the various forest types. Finally, the affinities and origin of the Eastern Arc flora are discussed and related to geological and climatic history.

Environment

Rainfall and dry season length are the main climatic factors determining forest limits in the Eastern Arc, with the exception of frost which determines the upper altitudinal limit. The forests occur on east-facing slopes where rainfall is over 1000–1500 mm per year. In the northern part of the range there are two peaks in rainfall with the short rains during October to December and the heavier long rains from March to May. Some areas (notably parts of the Usambara and Uluguru mountains) have an almost per-humid climate with most months receiving more than 100 mm of rain. Further south there is a marked dry season of several months (June to September) and a single rainy season (November to May, peaking in April). These different rainfall regimes are reflec-

ted in the forest by changes in species dominance and altitudinal ranges. Mist and cloud occur throughout the year and are an important source of dry season moisture in higher altitude forests where a dense epiphytic bryophyte cover (Pocs, 1976a) or microphyllous leaves (Kerfoot, 1968) extract atmospheric moisture through occult precipitation.

Dry season is the coolest time of year, with frosts occurring above 1800 m during July and August. The upper limit of forest is determined by regular occurrence of frosts, which is 2100–2450 m in the Uluguru (Pocs, 1976b) and just over 2000 m in the Usambara (Moreau, 1935). Frosts have been recorded as low as 1500 m in the Usambara Mountains (Moreau, 1935); frost damage in forests has been observed at 1800 m in the Uluguru (Pocs, 1976b); and in the Uzungwa at Mufindi a particular thicket type is associated with frost pockets at 1800 m (T. C. E. Congdon, personal communication). The lapse rate is relatively high: Pocs (1976b) gives a figure of 0.6 °C per 100 m in altitude. Higher figures have been recorded in the cool dry season in the northern Uzungwa forests with a daytime lapse rate of 1 °C per 100 m (Huddersfield New College, unpublished data). Moreau (1935) gives a hot season temperature maximum lapse rate of 1.7 °C per 100 m for the Amani–Sigi scarp in the East Usambara and 1 °C per 100 m for the West Usambara scarp. Proximity to the Indian Ocean has a cooling effect and temperatures in the East Usambara are 4–5 °C lower at 700 m and 2–3 °C lower at 500 m than those in other parts of Tanzania at similar altitudes (Hamilton, 1989a).

The forest itself creates a remarkably cool microclimate. Pocs (1974) showed that temperatures within a forest on Bondwa peak in the Uluguru Mountains differ markedly from those outside at the hottest time of day. The soil surface temperature in submontane forest at 1430 m was 30 °C less than a maize field at the same altitude; and upper montane forest at 2020 m was 20 °C cooler than open summit at 2125 m.

The forest soils are mostly weakly acidic humic ferrisols becoming mor-like at high altitudes. Soils in the East Usambara Mountains have been described by Hamilton (1989b). Basic soils occur on metamorphosed limestone at Kimboza east of the Uluguru and on the Mahenge Mountains. Forest species grow on the basic soils of termite mounds in grassland in the southern Uzungwa outside normal forest limits (Lovett & Gereau, 1990).

Classification of the Eastern Arc forests

A number of different classification systems and terminologies have been proposed for the forests of eastern Africa. Moreau (1935) divided the Usambara mountain forests into lowland, intermediate and highland. Pitt-Schenkel (1938), working in the West Usambara Magamba forest, refered to Moreau's 'Highland Evergreen Forest' as 'Warm Temperate Rain Forest' and listed 12 other synonyms for the same forest type. The term 'Temperate' to describe tropical African montane forests has also been used more recently by Donald & Theron (1983). Hedberg (1951) in a study of high altitude vegetation on the highest mountains of East and Central Africa described a Montane forest belt above 1700–3000 m, below a heath-like Ericaceous belt. Hedberg's montane forest belt had drier and wetter types, and zones of bamboo and *Hagenia*. At the CCTA/CSA (1956) meeting at Yangambi a set of terms for African closed forest formations was proposed. This divided moist forests into drier and wetter types at low to medium and higher altitudes; the higher altitude forests were called montane. Greenway (1973) published a vegetation classification for East Africa which was originally proposed in 1943 but was not available at the Yangambi meeting. In this classification lowland forest occurred between sea level and 1200 m; above this the forest was described as upland forest. Both lowland and upland forest had wetter and drier types. For Malawi, Chapman & White (1970) divided the forests into lowland (up to 1370 m), submontane (1370–2290 m) and montane (mostly above 1980 m). The term submontane forest was also used by Pocs (1976b, c) who divided the Uluguru forests into lowland (below 500 m), submontane (800–1500 m), montane (1500–1850 m), montane mossy or upper montane forest (1850–2400 m) and subalpine elfin forest (2100–2400 m). Wetter and drier types of lowland and submontane forest were recognised.

Classification terminology of Eastern Arc forests becomes further confused when phytogeographic systems are used. Monod (1957) originally proposed a straightforward system whereby the forests of eastern Africa were considered an eastern domain of the Guineo–Congolian forests of West and Central Africa. This included both montane and lowland forests below the high altitude Afroalpine region of Hauman (1955), which was composed of grassland with giant *Lobelia* and *Senecio*. Troupin (1966) combined Hauman's Afroalpine region with Hedberg's Ericaceous belt, included *Hagenia* and bamboo forest, and put them together in an African montane region; forests below 2000 m remained in the Guineo–Congolian region. White (1970) proposed an Afromontane region with an upper limit corresponding to Hedberg's Alpine belt, and a lower limit of 1065–1525 m depending on local conditions. Forests below White's Afromontane region were still considered part of the Guineo–Congolian region, though they were regarded as an eastern 'Usambara–Zululand' domain. White (1978a) divided his Afromontane region into seven regional mountain systems on the basis of distances between mountains, which breaks up the Eastern Arc, into a southern Uluguru–Mulanje and a northern Usambara–Imatong group; he also emphasised that the Afroalpine region was not very distinct from the Afromontane region. This approach was followed by Denys (1980), who found support for the north–south Eastern Arc divisions in a study of species distribution maps. White (1983) truncated the Guineo–Congolian region at Lake Victoria and regarded eastern African forests below the Afromontane region as part of an eastern coastal Zanzibar–Inhambane region.

The main problems with these various terminologies are as follows.

1. The term 'temperate' is not appropriate for the tropical African montane forests as they have primarily tropical affinities and origins. This is in contrast to the heath and grassland vegetation above the forest

limits, which does have temperate affinities, a fact that was recognised in the last century for West Africa (Hooker, 1861, 1864) and confirmed in East Africa by Hedberg (1961).

2. Owing to relatively lower temperatures in the East Usambara, equivalent forests are at higher altitudes elsewhere in Africa. The term 'lowland forest' to cover forest in the Eastern Arc up to 1200 m means that when comparing the East Usambara forest types with those in Uganda the upper boundary of Ugandan lowland forest would be at the relatively high altitude of 1700 m (Hamilton *et al.*, 1989).

3. Although there are some differences between northern and southern parts of the Eastern Arc, the similarities and high degree of species and generic endemism within it suggest that it is a distinct floristic unit and should not be divided into a southern Uluguru–Mulanje and a northern Usambara–Imatong group.

4. The Afromontane region of White (1983) is a mixture of temperate elements in vegetation above the forest limits and tropical elements in the forests, and so is phytogeographically heterogeneous. Similarly, his Zanzibar–Inhambane region is a mixture of Zambezian elements in woodlands and Guineo–Congolian elements in forests.

5. Moist forest tree community associations vary continuously with altitude (Hamilton, 1975; Hamilton *et al.*, 1989; Lovett, 1990b) and there is no justification for dividing them into upper Afromontane and lower Zanzibar–Inhambane types. Change in community associations with altitude is very marked, probably attributable to the high lapse rate, with almost complete replacement of species between upper montane and lowland forests. For practical purposes it is useful to divide this moist forest altitudinal continuum into a series of types. There is a discontinuity between moist forest and

the driest types of montane forest (Lovett, 1990b).

6. Distribution patterns of moist forest tree species throughout the altitude range of the forests are similar, indicating that both upper and lower altitude forests should be in the same phytogeographic region. In the Eastern Arc there are strong affinities to tropical west and central African forests, and a high degree of endemism.

The continuous nature of change in the forests is related to a continuous change in environmental variables: temperature, which is a function of altitude and distance from the ocean; and rainfall, which is related to topography and latitudinal variation in rainy seasons. A rigid system of vegetation classification cannot be applied, but for practical purposes the marked altitudinal variation in moist forests needs to be divided into a number of convenient types. A terminology similar to that used by Pocs (1976b, c) is used here in which the forest is divided into upper montane, montane, submontane, and lowland, with drier and wetter types. Phytogeographically all the forests are regarded as Guineo–Congolian, as there is no discontinuity between higher and lower altitude forest in either community associations or distribution patterns, and there are strong affinities to West and Central Africa. The high degree of endemism and occurrence of a number of species with distributions restricted to eastern Africa, suggests that eastern and western sections of the Guineo–Congolian region, with specific centres of endemism, should be recognised. This follows the approach of Monod (1957). Vegetation above the forest altitudinal limits has different phytogeographical affinities and is here called subalpine and alpine.

Terminology, together with suggested altitudinal and climatic limits, structural characteristics and a limited synonymy of the forest types recognised in the Eastern Arc, is given below. Data for stem densities (dbh = diameter at breast height) and basal areas are approximate and are derived from Hamilton *et al.* (1989) and Lovett (1992).

Dry montane forest

Altitude: >1500 m. Rainfall: 1000–1200 mm/year.

Synonyms: Montane forest belt (Hedberg, 1951). Dry montane forest (CCTA/CSA, 1956). Montane forest (White, 1970). Upland dry evergreen forest (Greenway, 1973). Drier types of undifferentiated Afromontane forest (White, 1983). Dry transitional montane forest (White, 1983).

Canopy height: 10–20 m with emergents to 30 m. Basal area: 20–40 m²/ha. Stem sizes: Few large trees >100 cm dbh, most trees <40 cm dbh. Stem density: 240 stems >20 cm dbh/ha.

Dry montane forest as defined here is the most problematical of the forest types as it is extremely variable and changes with altitude. Dry forest trees can have wide altitudinal ranges and extend into a variety of vegetation types, for example: *Apodytes dimidiata*, *Margaritaria discoidea*, *Parinari excelsa* and *Rapanea melanophloeos* can occur in both high altitude and low altitude dry forest. The classification of forest with *Juniperus procera* or *Hagenia abyssinica* also presents a problem. Both these species are light demanders; *Juniperus* regenerates following fire (Hall, 1984) and it is thought that the fire sensitive *Hagenia* may replace moister forest types as a pioneer following fire (Greenway, 1973). The West Usambara *Juniperus* forest is clearly distinct from other West Usambara moist forest types (Lovett, 1990b), and on Luhombero in the northern Uzungwa, *Hagenia* occurs as pure stands (J. B. Hall, unpublished data). However, both species also occur at low densities in mixed forest: *Juniperus* in the West Usambara Shume–Magamba forest and *Hagenia* on Selebu Mountain in the northern Uzungwa. Greenway (1973) included forest with *Juniperus* in upland dry evergreen forest and put *Hagenia* into upland evergreen bushlands together with other degraded or early successional stages of dry montane forest. The most practical course is to include them in dry montane forest when they are in mixed stands, and qualify the forest type when they are dominant. This would include the Sin-gle-Dominant Afromontane Forest of White (1983) into the definition of dry montane forest. At Mufindi in the southern Uzungwa Mountains, dry montane forest species occur in small forest patches on termite mounds in grassland on highly leached soils with a rainfall of 850 mm per year (Lovett & Gereau, 1990). Edaphic factors thus appear to determine the lower rainfall limit under which forest can grow.

Occasionally woodland species such as *Erythrina abyssinica* occur. Late successional stages include fire resistant trees such as *Agauria salicifolia* and *Myrica salicifolia* which grow in high altitude grassland or on podzolised soil resulting from forest clearance. They also occur in late succession in upper montane and montane forest. Southern Eastern Arc dry montane forest has edge species such as *Zeyherella magalismontana*, *Dais cotonifolia*, *Diospyros whyteana* and *Garcinia kingaensis* reaching their northern limits.

Other trees in dry upper montane forest include: *Albizia gummifera*, *Aphloia theiformis*, *Apodytes dimidiata*, *Bersama abyssinica*, *Cassine aethiopica*, *Cassipourea malosana*, *Catha edulis*, *Croton macrostachyus*, *Cussonia spicata*, *Ekebergia capensis*, *Euclea divinorum*, *Halleria lucida*, *Margaritaria discoidea*, *Nuxia congesta*, *Olea capensis*, *Olinia rochetiana*, *Parinari excelsa*, *Prunus africana*, *Rapanea melanophloeos*, *Rawsonia lucida*, *Schrebera alata*, *Vepris stolzii*, *Trichocladus ellipticus*.

Trees in lower altitude dry montane forest include: *Albizia gummifera*, *Bersama abyssinica*, *Cassine aethiopica*, *Cola greenwayi*, *Craibia brevicaudata*, *Drypetes gerrardii*, *Manilkara discolor*, *Margaritaria discoidea*, *Ochna holstii*, *Strychnos mitis*.

Upper montane forest

Altitude: >1800 m. Rainfall: >1200 mm/year.

Synonyms: Montane forest belt (Hedberg, 1951). Bamboo zone (Hedberg, 1951). Moist montane forest (CCTA/CSA, 1956). Higher altitude types of submontane moist forest (White, 1970). Wetter low canopy types of montane forests (White, 1970). Higher altitude types of upland rain forest (Greenway,

1973). Montane mossy forest (Pocs, 1976c). Upper montane forest (Pocs, 1976b). Wetter types of undifferentiated Afromontane forest (White, 1983). Afromontane bamboo (White, 1983). Subalpine elfin forest (Pocs, 1976c; included in Afromontane evergreen bushland and thicket by White, 1983).

Canopy height: 10–20 m with emergents to 25 m. Basal area: 30–70 m²/ha. Stem sizes: Few large trees >100 cm dbh, most trees <40 cm dbh. Stem density: 330 stems >20 cm dbh/ha.

Upper montane forest occurs at higher altitudes than montane forest, and under a greater rainfall than dry montane forest. It has a greater basal area than dry montane forest, contributed by large trees such as *Ocotea usambarensis* and a higher stem density. Often growing on exposed ridge tops it has a lower canopy than montane forest, and in this habitat can have very high densities of small stems. In areas of frequent mist, epiphytic bryophytes form thick coats on tree trunks and branches. Rubiaceae, Acanthaceae and tree ferns are common in the shrub layer. In areas with sufficient rainfall the upper altitudinal limit is determined by regular occurrence of frost or a high degree of exposure on ridge tops. Forest then gives way to subalpine Ericaceous heath, as in the West Usambara Mountains, or grassland with tree clumps as in the southern Uluguru and northern Uzungwa mountains. The grassland forest edge is often maintained by fire.

Bamboo thickets appear to be a successional stage (Hamilton & Perrott, 1981; Lovett, 1990a), so although they have a distinctive physiognomy, it is more practical to regard them as a type of upper montane forest. They also occur in montane forest (Pocs, 1976b). Another species typical of an early successional stage is *Macaranga kilimandscharica*. Differences between the northern and southern Eastern Arc are that (i) *Balthasaria schliebenii* and *Symphonia globulifera* have not been recorded from the Uzungwa, and (ii) *Bridelia brideliifolia*, *Ternstroemia polypetala* and the endemic *Allanblackia ulugurensis* have not been recorded from the Usambara. In the southern

Eastern Arc species such as *Cassipourea gummiflua* occur in upper montane forest whereas in the north they are found in montane forest. Other endemics include *Croton dictyophlebodes*, *Cussonia lukwangulensis*, *Pittosporum goetzei*, *Polyscias stuhlmannii*, *Syzygium sclerophyllum*. The latter two species occur at lower altitudes in the Taita Hills (R. B. Faden, personal communication).

Other trees include: *Aningeria adolfi-friedericii*, *Aphloia theiformis*, *Cassipourea malosana*, *Cryptocarya liebertiana*, *Dombeya torrida*, *Drypetes gerrardii*, *Ficalhoa laurifolia*, *Garcinia volkensii*, *Ilex mitis*, *Ochna holstii*, *Ocotea usambarensis*, *Olea capensis*, *Maesa lanceolata*, *Maytenus acuminata*, *Nuxia congesta*, *Podocarpus latifolius*, *Rapanea melanophloeos*, *Syzygium cordatum* (on wet ridge tops), *Trichocladus ellipticus*, *Xymalos monospora*.

Montane forest

Altitude: 1200–1800 m. Rainfall: >1200 mm/year.

Synonyms: Montane forest zone (Hedberg, 1951). Moist montane forest (CCTA/CSA, 1956). Middle altitude submontane moist forest (White, 1970). Taller types of montane forest (White, 1970). Lower altitude types of upland rain forest (Greenway, 1973). Montane evergreen forest (Pocs, 1976c). Afromontane rain forest (White, 1983).

Canopy: 25–40 m with emergents to 50 m. Basal area: 30–120 m²/ha. Stem sizes: Many large trees >50 cm dbh with a relatively high proportion >100 cm dbh. Stem density: 250 stems >20 cm dbh/ha.

Montane forest occurs at a lower altitude than upper montane forest, and is much taller with bigger trees though the basal areas are similar owing to upper montane forest having a higher stem density. Rubiaceae and Acanthaceae are common in the shrub layer. With decreasing rainfall it grades into dry montane forest. In the Uzungwa Mountains a longer dry season is reflected by drier forest species, such as *Albizia gummifera*, *Aphloia theiformis*, *Zeyherella magalismontana* and *Rauvolfia caffra*, growing with moist forest species. Another effect of the marked southern dry season is dominance of *Parinari excelsa*, and it may account for the apparent absence of *Entandrophragma excelsum* from the Uzungwa.

Secondary species include: *Anthocleista grandiflora*, *Macaranga kilimandscharica*, *Neoboutonia macrocalyx* and *Polyscias fulva*. Eastern Arc endemics include: *Allanblackia stuhlmannii*, *Beilschmiedia kweo*, *Drypetes usambarica* (with a subspecies on the Kenya coast), *Isoberlinia scheffleri* and *Polyceratocarpus scheffleri*. Endemics in the north include *Mammea usambarensis*. Other trees include: *Aningeria adolfi-friederici*, *Bersama abyssinica*, *Canthium oligocarpum* subsp. *captum*, *Chrysophyllum gorungosanum*, *Cleistanthus polystachyus*, *Cola greenwayi*, *Cornus volkensii*, *Craibia brevicaudata*, *Cryptocarya liebertiana*, *Drypetes gerrardii*, *Ficalhoa laurifolia*, *Hallea rubrostipulata*, *Myrianthus holstii*, *Newtonia buchananii*, *Ochna holstii*, *Ocotea usambarensis*, *Podocarpus falcatus*, *Strombosia scheffleri*, *Syzygium guineense* subsp. *afromontanum*, *Xymalos monospora*.

Submontane forest

Altitude: 800–1400 m. Rainfall: >1500 mm/year.

Synonyms: Moist forest (CCTA/CSA, 1956). Intermediate Forest (Moreau, 1935). Lower altitude types of submontane moist forest (White, 1970). Lowland rain forest (Greenway, 1973). Submontane rain forest (Pocs, 1976c). Zanzibar–Inhambane transitional rain forest (White, 1983).

Canopy: 25–40 m with emergents to 50 m. Basal area: 30–70 m²/ha. Stem sizes: Many large trees >50 cm dbh with a relatively high proportion >100 cm dbh. Stem densities: 170 stems >20 cm dbh/ha.

Submontane forest differs from montane forest in that it contains lowland species such as *Afrosersalisia cerasifera*, *Milicia excelsa* and *Parkia filicoidea*. Otherwise these forest types overlap in floristic composition and are similar in structure. As with montane forest, the long dry season in the

Uzungwa Mountains is reflected in single-species canopy dominance of *Parinari excelsa*, in contrast to multi-species dominance of the Usambara forests; and in the more frequent occurrence of deciduous species of *Albizia*. Typical secondary species include *Harungana madagascariense*, *Macaranga capensis*, *Trema guineense* with *Cylicomorpha parviflora* at higher altitudes, and *Sapium ellipticum* at lower altitudes. The introduced *Maesopsis eminii* has become a successful pioneer in the much disturbed East Usambara forests. Eastern Arc endemics include *Anisophylla obtusifolia* and *Cephalosphaera usambarensis*. Usambara endemics include *Englerodendron usambarense*; southern endemics include *Octoknema orientalis* and *Sibangea pleioneura*.

Other trees include: *Allanblackia stuhlmannii*, *Aningeria pseudoracemosa*, *Antidesma membranaceum*, *Isoberlinia scheffleri*, *Leptonychia usambarensis*, *Myrianthus holstii*, *Newtonia buchananii*, *Odyendea zimmermannii*, *Pachystela msolo*, *Strombosia scheffleri*, *Trichilia dregeana*, *Trilepisium madagascariense*.

In addition to dominance by *Parinari excelsa*, drier types of submontane forest would be indicated by species characteristic of dry montane forest, dry lowland forest and woodland. This type of forest is restricted in extent because there are few areas where rain shadows occur at this altitude.

Lowland forest

Altitude: <800 m. Rainfall: >1500 mm/year.

Synonyms: Moist forest (CCTA/CSA, 1956). Lowland forest (Moreau, 1935). Lowland moist forest (White, 1970). Lowland rain forest (Greenway, 1973). Lowland semi-evergreen and evergreen rain forest (Pocs, 1976c). Zanzibar–Inhambane lowland rain forest (White, 1983).

Canopy: 25–40 m with emergents to 50 m. Basal area: 20–50 m²/ha. Stem sizes: Many large trees >50 cm dbh with a relatively high proportion >100 cm. Stem densities: 140 stems >20 cm dbh/ha.

Lowland forest differs from submontane forest

in the absence of montane forest species, and tends to be more deciduous with species of *Albizia*, *Bombax rhodognaphalon* and *Parkia filicoidea* losing their leaves in the dry season and flushing again before the start of the rains. There are few ferns compared with the higher altitude forest types, and disturbed areas are dominated by the grass *Olyra latifolia*. The screw-pine, *Pandanus*, occurs in groundwater forests. Eastern Arc lowland forest shares many species with coastal forests. In the south the long dry season is reflected in dominance of *Parinari excelsa*. Eastern Arc and coastal forest endemics include: *Angylocalyx braunii*, *Aningeria pseudoracemosa*, *Lettowianthus stellatus*, *Newtonia paucijuga*.

Other trees include: *Afrosersalisia cerasifera*, *Anthocleista grandiflora*, *Antiaris toxicaria*, *Bequaertiodendron natalense*, *Dialium holtzii*, *Erythrophleum suaveolens*, *Filicium decipiens*, *Funtumia africana*, *Garcinia buchananii*, *Khaya nyasica*, *Macaranga capensis*, *Malacantha alnifolia*, *Milicia excelsa*, *Pachystela msolo*, *Ricinodendron heudelotii*, *Sapium ellipticum*, *Sorindeia madagascariensis*, *Tabernaemontana pachysiphon*, *Terminalia sambesiaca*, *Treculia africana*, *Trilepisium madagascariense*, *Zanha golungensis*.

Dry lowland forest

Altitude: <800 m. Rainfall: 1000–1500 mm/year.

Synonyms: Dry evergreen forest (CCTA/CSA, 1956). Lowland dry evergreen forest (White, 1970). Lowland dry evergreen forest (Greenway, 1973). Zanzibar–Inhambane undifferentiated forest (White, 1983).

Canopy: 15–20 m with emergents to 35 m.

In the Eastern Arc dry lowland forest occurs on the fringes of lowland forest and is characterised by the occurrence of woodland species such as *Pteleopsis myrtifolia*, and the riverine *Sterculia appendiculata*. It has been cleared for cultivation in many areas, though still occurs on the lower slope of Kanga Mountain in the northern Nguru and on the hills of Dindili and Kitanglangu near Morogoro.

Distribution and affinities

The Eastern Arc flora has affinities with forests to the south, west, north and overseas. In order to assess the degree of similarity between the Eastern Arc forests and those elsewhere the distribution patterns of 274 Eastern Arc moist forest tree species greater than 10 m tall or 20 cm diameter at breast height were analysed. In addition to the Eastern Arc being considered as a whole, each forest type as defined above and the species it contains was treated separately. The results of this analysis are given below, together with examples of distribution patterns of other plants.

Affinities with the west

The Eastern Arc flora has strong western affinities. Of the 274 tree species analysed, 169 (62%) also occurred to the west. This is 82% of species occurring outside the Eastern Arc. Of the species with westerly distributions 60% occur west of the western rift in the main part of the Guineo–Congolian region, of which 76% also occur to the south. All of the different forest types have species with westerly distributions (Table 4.1).

The Guineo–Congolian affinities of the Eastern Arc flora can be seen at a number of levels (Hamilton & Faden, 1974). There are widespread species such as *Parinari excelsa* and *Parkia filicoidea* which have southern tropical distributions and do not occur south of the Zambezi River (White, 1981). Both these species are found in a wide range of forest types and riverine vegetation, and are used as evidence for the southern migratory track between the western Guineo–Congolian region and eastern Africa east of the arid corridor (White, 1981; Hopkins & White, 1984). Other species which occur throughout the southern migration route include *Cleistanthus polystachyus*, *Erythrophleum suaveolens* and *Uapaca paludosa*. Some species are exceptionally widespread, for example: *Blighia unijugata*, *Funtumia africana*, *Milicia excelsa*, *Pachystela brevipes*, *Tapura fischeri* and *Trilepisium madagascariense* in lowland forests; and *Bersama abyssinica*, *Ilex mitis*, *Maesa*

Table 4.1. *Numbers of species occurring in different Eastern Arc forest types with westerly distributions. 274 species were included in the analysis, of which 169 or 62% have westerly distributions*

Forest type	Number of species[a]		
	1	2a	2b
Dry lowland	27	10	34
Lowland	39	10	42
Submontane	43	14	48
Montane	43	37	78
Upper montane	19	10	28
Dry montane	27	22	49
Total species	102	67	155

Note: [a]1. West of the eastern Zaïre highlands.

2a. West of the arid corridor but not extending west of the eastern Zaïre highlands.

2b. West of the arid corridor.

lanceolata, Newtonia buchananii, Ritchiea albersii and *Xymalos monospora* in montane forests. In contrast to the widespread species, there are a number of trees with remarkably disjunct distributions between West and eastern Africa. These include *Mesogyne insignis, Paramacrolobium coeruleum* and *Schefflerodendron usambarense* in lowland forest; and *Ternstroemia polypetala* in upper montane forest.

Other species have eastern distributions but are replaced by closely related species in the western Guineo–Congolian region. For example, *Aningeria adolfi-friedericii* occurs from the Horn of Africa to south of the Zambezi River and in the Kivu–Rwenzori highlands, but is replaced in the main part of the Guineo–Congolian region by *A. altissima* and *A. robusta*. Similarly, *Myrianthus holstii* occurs from south of the Zambezi River to the Kenyan and Kivu–Rwenzori highlands, but is replaced further west by six other species of *Myrianthus* (de Ruiter, 1976) one of which, *M. preusii* subsp. *seretii*, also extends to eastern Africa (Berg & Hijman, 1989). *Neoboutonia macrocalyx* occurs from southern Malawi to the Kenyan and Kivu–Rwenzori highlands, but is replaced by *N. mannii* and the very widespread *N. melleri* to the west. *Trichoscypha ulugurensis* subsp. *ulugurensis*

occurs from south of the Zambezi River to the central Eastern Arc; it is replaced in the Kivu–Rwenzori highlands by subsp. *montana*, and by 70 other *Trichoscypha* species further west. Some of the replacement species are markedly disjunct with their relatives. *Cylicomorpha parviflora* occurs from northern Malawi to the Kenyan highlands; the only other member of the genus is *C. solmsii* from Cameroun. *Balthasaria schliebenii* occurs in the northern and central Eastern Arc and the Kivu–Rwenzori highlands, the only other member of the genus *B. mannii* is found in São Tome.

Many Eastern Arc endemics are in otherwise exclusively western Guineo–Congolian genera, for example *Enantia, Cincinnobotrys, Polyceratocarpus* and *Octoknema*. At the family level, the monotypic Eastern Arc genus *Cephalosphaera*, in the Myristicaceae, is the only member of the family occurring east of the arid corridor and represented in the western Guineo–Congolian region by *Coelocaryon, Pycnanthus, Scyphocephalium* and *Staudtia*.

In contrast to southern moist forests, where there are relatively few species not found in the Eastern Arc (with the exception of the unique Cape forest edge flora: Phillipson & Russell, 1988), there are many widespread western Guineo–Congolian species that do not reach the Eastern Arc. Most of these, for example *Anthonotha noldeae, Klaindoxa gabonensis* and *Maesopsis eminii*, only extend as far east as the great lakes of Tanganyika and Victoria and so do not cross the arid corridor of low rainfall which runs from northeast Africa through central Tanzania to the Namib Desert (Werger, 1978). However, *Trichilia prieuriena* reaches Mbeya Mountain and *Neoboutonia melleri* reaches northern Lake Nyasa. *Maesopsis eminii* has been introduced into the East Usambara Mountains, where it is readily dispersed by hornbills and has rapidly become naturalised through invasion of gaps in indigenous forest. This suggests a historical, rather than climatic or edaphic, reason for the absence of these widespread western Guineo–Congolian species.

The strong affinities between the Eastern Arc and western Guineo–Congolian region must largely pre-date mid-Tertiary uplift of the central

Africa plateau. The early Miocene east–west watershed was slightly east of its present position (Andrews & van Couvering, 1975) and the central highlands of the western rift were not formed until the Pliocene (Griffiths, Chapter 2). The strong tropical Atlantic rain-bearing climate must have penetrated across the present-day arid corridor over a relatively flat, eroded mid-Tertiary land surface. Even if rain forest did not extend continuously from west to east, biological contact could have been through gallery forests. At the end of the Miocene, after 10 Myr BP, the central plateau was raised to its present level and major rifts developed. After this time, access to the Eastern Arc from the western Guineo–Congolian forests would require lowland species to disperse over mountains and montane species to disperse across valleys. During the Pliocene there were warm wet periods (Lovett, Chapter 3) which would have at least facilitated movement of readily dispersed moist forest plants tolerant of a range of forest types. Thus, although the main phytogeographic pattern was established by mid-Tertiary geomorphological changes, it still could have been supplemented by dispersal during suitable climatic conditions.

Affinities with the south

A high proportion of Eastern Arc species have southerly distributions. Of the 274 tree species considered, 167 (61%) occur to the south, representing 80% of species occurring outside the Eastern Arc (Table 4.2). Of the southern species, 76% occur south of the Songwe River and north of 14° S; 71% occur south to the Limpopo River; but only 44% extend south of the Limpopo. Many southern species also occur west of the Eastern Arc with 47% occurring west of the central rift and 34% reaching the Kivu–Rwenzori mountains. Only 19% are confined east of the arid corridor. Of those species which occur south of the Limpopo River, 60% also occur west of the central rift.

Few Eastern Arc tree species appear to have originated in southern Africa. Ancient southern links are found in the predominately southern hemisphere genus *Podocarpus*, which is represen-

Table 4.2. *Numbers of species occurring in different Eastern Arc forest types with southerly distributions. 274 species were included in the analysis, of which 167 or 61% have southerly distributions*

Forest type	Number of species[a]						
	1	2a	2b	3a	3b	4a	4b
Dry lowland	22	12	32	3	27	2	15
Lowland	17	17	33	8	32	3	21
Submontane	19	23	42	5	53	11	40
Montane	33	27	60	11	49	14	78
Upper montane	19	5	23	6	29	5	34
Dry montane	36	9	44	4	44	5	49
Total	74	49	120	20	120	24	127

Note: [a]1. South of the Limpopo River.
2a. South of the Zambezi River, but not south of the Limpopo River.
2b. South of the Zambezi River, and also south of the Limpopo River.
3a. South of 14° S, but not south of the Zambezi River.
3b. South of 14° S and also south of the Zambezi River and Limpopo River.
4a. South of the Songwe River, but not south of 14° S.
4b. South of the Songwe River, and also south of 14° S, the Zambezi River and the Limpopo River.

ted by three species in montane and upper montane forests. More modern southern affinities are predominately in the dry edges of upper montane forest: *Dais cotinifolia*, *Diospyros whyteana* and *Schefflera umbellifera* reach their northern limits in the Uzungwa; *Ptaeroxylon obliquum* has its northern limit on the West Usambara; *Kiggelaria africana* has its northern limit on Kilimanjaro; *Aphloia theiformis* reaches Kenya; and *Halleria lucida* reaches the Horn of Africa. All of these species occur south of the Limpopo River and do not occur west of the arid corridor. Some of these species have Madagasacan connections, for example the polymorphic monotypic genus *Aphloia* also occurs on Madagascar. The genus *Dais* has two species, one southern and southeastern African, the other in Madagascar. *Ptaeroxylon* is placed in the family Ptaeroxylaceae by Pennington & Styles (1975), the only other genus of which is the Malagasy *Cedrelopsis*. It may well be that some Eastern Arc species with southerly distributions, notably those of drier forest types, have their origins in Madagascar rather than a specifically Cape flora.

The Cape flora itself has tropical affinities. For example, Proteaceae, which are well represented in southern Africa and occur in upper montane and dry montane Eastern Arc moist forests as *Faurea*, probably evolved from tropical African and Madagascan ancestors (Goldblatt, 1978). Another dry upper montane forest genus, *Olinia*, has its centre of diversity in southern Africa with four species: the widespread *O. rochetiana* occurs in the Eastern Arc, and the other three are found south of the Zambezi River with two of these only found south of the Limpopo River. *Olinia* may be a tropical African genus which has diversified in the Cape (Goldblatt, 1978), giving rise to the southern endemic family Penaeaceae. Similarly, Goldblatt suggests that typical Cape elements in the Rutaceae–Diosmeae and Thymeleaceae could have evolved from more primitive broadleaved tropical ancestors such as *Calodendrum* (Rutaceae) and *Peddiea* (Thymelaceae), both of which are found in dry montane forest. An herbaceous genus commonly found in moist forests, *Streptocarpus*, which has diversified in Natal, is also thought to have a tropical origin (Hilliard & Burtt, 1971).

The lowland Eastern Arc and eastern African coastal forest flora which extends southwards is primarily western Guineo–Congolian and Madagascan (or further eastern) in origin and affinity rather than southern African. It has many endemic species in Guineo–Congolian genera and many widespread Guineo–Congolian species (Moll & White, 1978). Thus both lowland and upland Eastern Arc forests do not have a strong southern affinity or origin, and the southern forests themselves are predominately of tropical origin. Southern temperate moist forests suffered severely during the Miocene desiccation of the Cape. For example, fossil records exist from there of subtropical forest families such as Winteraceae and Sarcolaenaceae (Coetzee & Muller, 1984) which are no longer represented on continental Africa. The upwelling of cold waters, that constitute the present Benguela current and which bring aridity to southwest Africa, began in the early late Miocene about 10 Myr BP (Siesser, 1980). This was before the rejuvenation of the Eastern Arc faults about 7 Myr BP (Griffiths, Chapter 2) and creation of high mountains on the eroded African land surface. As there were no high mountains, southern subtropical species were unable to survive the Miocene desiccation in tropical montane refugia with temperate climates. However, some montane tropical taxa subsequently dispersed south into the temperate zone and radiated into the remarkable Cape flora.

Affinities with the north

Affinities with the north are not as strong as those to the west and south with 52% of Eastern Arc tree species occurring to the north compared with 61% to the south and 62% to the west (Table 4.3). Of the 86 species which occur in the Horn of

Table 4.3. *Numbers of species occurring in different Eastern Arc forest types with northerly distributions. 274 species were included in the analysis, of which 143 or 52% have northerly distributions*

Forest type	Number of species[a]				
	1	2a	2b	3a	3b
Dry lowland	22	8	23	4	22
Lowland	23	9	21	5	23
Submontane	20	17	35	5	27
Montane	45	24	66	8	64
Upper montane	19	6	25	1	21
Dry montane	36	11	47	2	39
Total	86	42	115	15	92

Note: [a]1. Horn of Africa (including southwest Ethiopia).
2a. Central Kenyan highlands, but not the Horn of Africa.
2b. Central Kenyan highlands, and also the Horn of Africa.
3a. Northern Tanzanian volcanics, but not the central Kenyan highlands and Horn of Africa.
3b. Northern Tanzanian volcanics, and also the central Kenyan highlands and Horn of Africa.

Africa and southwest Ethiopia, most are wide-spread: 41 also occur south of the Limpopo, 31 occur in the Kivu–Rwenzori mountains and 49 occur in the western Guineo–Congolian region. Fewer tree species reach Kilimanjaro (92) than the more distant Kenyan highlands (115). There are a number of possible explanations for this: Kilimanjaro is of relatively recent origin, with the oldest lavas one million years old and major volcanic mudflows occurring in the lower and upper Pleistocene (Griffiths, Chapter 2). Alternatively, Eastern Arc species cannot grow on volcanic soils, or the Kilimanjaro forests may be secondary, following cultivation on the fertile, well-watered slopes.

As with southerly affinities, although there are a great many Eastern Arc species occurring to the north, it appears that few of them originated there. Herbaceous genera such as *Alchemilla* and *Viola*, which have predominately temperate distributions, are represented in the Eastern Arc flora by *Alchemilla cryptantha*, *A. ellenbeckii*, *A. johnstonii*, *A. kiwuensis*, *Viola abyssinica* and *V. eminii*. All of these species are Afroalpine plants (Hedberg, 1957), occurring in damp upland moor and extending only into moist forest in bamboo thicket and disturbed areas. The Alfroalpine region has strong northern and southern temperate links (Hedberg, 1951, 1961, 1965; Killick, 1978) in contrast to the largely tropical nature of the moist forests.

Other northern elements occur in dry forests and thickets. For example, *Juniperus* is a north temperate genus of about 50 species which survived Pliocene and Pleistocene climatic fluctuations in the eastern Mediterranean (Kerfoot, 1966, 1975) and which is represented by one species, *Juniperus procera*, in eastern and southern tropical Africa (Hall, 1984; Kerfoot, 1961, 1964, 1966). Another dry montane forest tree with northern affinities, *Prunus africana*, is widespread throughout Africa south of the Sahara. The genus *Prunus* contains about 400 species which are mostly north temperate. The two African species (one doubtfully distinct) occupy a transitional position between *Prunus* and *Pygeum* with *Prunus africana* being closely related to the Asiatic *Prunus pygeoides* (Kalkman, 1965). Another montane dry

forest temperate link is the tree *Hagenia abyssinica*, a monotypic genus restricted to the East African highlands in the subgenus *Poterieae* which is predominately temperate and herbaceous (Kalkman, 1988).

An ancient northern link is found in *Cornus volkensii* (formerly in the monotypic African *Afrocrania*). *Cornus* is a north temperate genus of about 65 species. *C. volkensii* and four related species are though to be derived from the fossil *Dunstania* based on fruits from England's early Tertiary (Eyde, 1988). It differs from the other species of *Cornus* in being the only one with spiny pollen, the only dioecious *Cornus*, and belonging to the group with red fruits, though having blue-purple fruits. The most closely related species is *C. sessilis* from California, and the split from other members of *Cornus* probably dates from the Palaeocene or early Eocene 50–60 Myr BP.

Barriers to northern moist forest temperate influence in the Eastern Arc flora have been enormous. Africa did not complete its northward migration and union with Eurasia until the middle Miocene 17–18 Myr BP (Axelrod & Raven, 1978). Closure of the Tethys Sea prevented the flow of oceanic currents around northern Africa, bringing aridity to the Sahara and the Arabian Peninsula. This is in marked contrast to both the Neotropics and the Far East, where moist tropical regions are more closely connected to moist north temperate regions by areas of high rainfall. For example, Far Eastern montane forests have floristic affinity at family and genus level with the warm temperate and subtropical forests of south China and southern Japan (Whitmore, 1984).

Origins from overseas

Surprisingly few Eastern Arc tree species occur outside continental Africa. Of the 274 trees analysed only 24 species reach Madagascar and nearby islands, eight extend to the Far East and three occur in the Neotropics. The weak affinities with Madagascar contrast with the strong affinities to the western Guineo–Congolian region. This demonstrates that the Mozambique Channel is a formidable barrier to dispersal.

The affinities that do exist between Madagascar and eastern and southern African coastal vegetation led Leroy (1978) to propose that the coastal strip should be included in a broadly defined Madagascan region. Examples of links between the Eastern Arc forests and Madagascar include: *Aphloia* with one polymorphic species in eastern and southern Africa and Madagascar; *Ludia* with one eastern African species and about 20 species in Madagascar and the Mascarenes; *Dais* with one eastern and southern African species and one Madagascan species; and *Lepidiotrichilia* with one species in East Africa and three in Madagascar. Species with Madagascan affinities also occur in the western Guineo–Congolian region or have links there. For example, *Agauria salicifolia* which, like *Aphloia*, is a variable monotypic genus occurring on Madagascar, is very widespread; *Necepsia castaneifolia* has two subspecies in eastern Africa, two subspecies in Madagascar and the other two species in the genus are West African (Bouchat & Leonard, 1986); and *Ocotea* which was once thought to be exclusively eastern and southern African, Madagascan and Neotropical (Leroy, 1978; Chapman & White, 1970), now has a West African species *O. gabonensis*.

Eastern Arc endemics with Madagascan affinities are *Saintpaulia* and *Zimmermannia*. The Eastern Arc and eastern coastal moist forest lithophytic and epiphytic herbaceous endemic genus *Saintpaulia* appears to be derived from *Saintpaulia*-like *Streptocarpus* from Madagascar (Hilliard & Burtt, 1971). Similarly, the shrubby moist forest and forest edge *Zimmermannia* has six species in the Eastern Arc, and a seventh *Z. decaryi* in Madagascar which appears to be intermediate between the Eastern Arc *Zimmermannia* and the more widespread *Meineckia* (Poole, 1981; Webster, 1965). Another link is in the small saprophytic herb *Seychellaria* which has three species: *S. africana* from the southern Eastern Arc, *S. thomassettii* from the Seychelles, and *S. madagascariensis* from Madagascar (Vollesen, 1982).

Eastern Arc–Madagascar links can be explained either by long-distance dispersal, or direct overland contact prior to the breakup of Gondwanaland. The latter explanation is used by Vollesen (1982) to explain the distribution of

Seychellaria, and by Pocs (1975, 1985; Bizot, Pocs & Sharp, 1985) to explain the distribution of bryophytes such as the Rutenbergiaceae. The Africa–Madagascar separation began 165 Myr BP and ended 121 Myr BP (Griffiths, Chapter 2). It is possible that such ancient links exist in primitive plants like bryophytes. *Seychellaria* also belongs to a potentially primitive family, the Triuridaceae, from the evidence of its free carpels (Mabberley, 1987). The alternative to relictual Gondwanic distribution is long-distance dispersal. None of the plants linking Madagascar with the Eastern Arc has particularly large seeds or fruits and could have arrived by bird or wind dispersal. Genera such as *Saintpaulia* and *Zimmermannia* almost certainly evolved after the breakup of Gondwanaland. The age and climatic stability of the Eastern Arc has resulted in a buildup of Madagascan connections through long-distance dispersal over time. More may be revealed as knowledge of Madagascan taxonomy improves; for example, the species *Adenoplusia ulugurensis* was once thought to be endemic to upper montane forest edges in the Uluguru but is now known to be the Madagascan *Buddleja axillaris* (Leeuwenberg, 1979).

Links with the Far East and Neotropics are also a mixture of long-distance dispersal and ancient Gondwanic distribution patterns (Ehrendorfer, 1988; Raven & Axelrod, 1974), though again it is difficult to distinguish between an ancient long-distance dispersal event and a truly relictual distribution. It has been suggested that long-distance dispersal is an adequate explanation of all the floristic links between tropical America and Africa (Ayensu, 1973; Brenan 1978; Thorne, 1973). In the hemiparasitic Loranthaceae all African genera derive from Asia, and the three primitive genera *Helixanthera*, *Dendrophthoe* and *Taxillus* extend from Asia to Africa. Both *Helixanthera* and *Dendrophthoe* occur in the Eastern Arc (Polhill, 1989). The large Eastern Arc montane forest tree *Balthasaria schliebenii* is closely related to the Asiatic genus *Adinandra* (White, 1981); and also in the Theaceae the upper montane forest tree *Ternstromia polypetala* is one of two African species in a genus of 80 or so Asian and Neotropical species. More recent long-distance dispersal from overseas appears to be represented by widespread tree species such as *Antiaris toxicaria, Celtis wightii, Filicium decipiens* and *Gyrocarpus americanus* subsp. *americanus.* Amongst herbs, the bird dispersed epiphytic *Rhipsalis baccifera* is the only naturally occurring member of the American family Cactaceae on continental Africa, and also occurs in the Neotropics.

Lack of overseas input at species level reflects the isolation of continental Africa's moist forest flora. Many overseas family and generic links with the Neotropics and Far East may date from the breakup of Gondawanaland. At that time Africa was at the centre of the supercontinent, some distance from oceanic influence, and so much of it would have been relatively dry (Raven & Axelrod, 1974), though inland lakes could have provided areas of high rainfall as they do today. As South America and India drifted away, moist forest plants would have dispersed onto an increasingly wetter African continent. The relatively poor palm flora of Africa compared with the Neotropics and Far East may be explained by difficulty in dispersing across the widening Atlantic and Indian oceans. Present-day differences between continental African and Madagascan floras demonstrate the barrier to long-distance dispersal presented by the Mozambique Channel, though it does appear to have been crossed by *Chrysalidocarpus,* a Madagascan palm genus with a Pemba Island endemic. An alternative explanation for the poor continental palm flora is that droughts in Africa long after the breakup of Gondwanaland so restricted the area of rain forest that many species were driven to extinction (Raven & Axelrod, 1974; Dransfield, 1988).

Eastern Arc endemics

Over 27% of the 274 Eastern Arc tree species analysed are widespread throughout continental Africa, but 25% of the trees are endemic to the Eastern Arc. Of the endemics 45% occur in both northern (Usambara, Pare, Taita) and southern (Uzungwa, Mahenge) parts of the Arc. Endemic tree species occur in all forest types, though there are few in dry montane forest (only one, plus some edge and thicket endemics) and most in submontane forests (55% of endemics) with

Table 4.4. *Numbers of species occurring in different Eastern Arc forest types endemic to the Eastern Arc. 274 species were included in the analysis, of which 67 or 25% are endemic*

Forest type	Number of species[a]					
	1	2	3	4	5	6
Dry lowland	5	1	0	0	0	0
Lowland	15	3	0	2	2	0
Submontane	19	3	1	11	1	2
Montane	12	1	1	6	2	2
Upper montane	2	1	1	1	1	0
Dry montane	0	0	0	1	0	0
Total	32	7	2	18	4	4

Note: [a]1. In both north and south Eastern Arc.
2. In both north and central Eastern Arc.
3. In both central and south Eastern Arc.
4. In only northern Eastern Arc.
5. In only central Eastern Arc.
6. In only southern Eastern Arc.

similar quantities in the montane (36% of endemics) and lowland forests (33% of endemics). The northern part of the Arc is richest in endemic trees, containing 50% of all endemics of which 27% are restricted to it (Table 4.4). Endemic richness of the northern sections is concentrated in the Usambara Mountains. There are a number of possible reasons for this: the Usambara are better known botanically than other Eastern Arc mountains; there may have been extinctions on the other mountains; there are more habitats on the Usambara; or the environment on the Usambara may give rise to a greater number of endemics. The endemic richness of the Usambara is probably a combination of these factors. Recent exploration of other Eastern Arc mountains has produced a number of new taxa and distributions (Lovett *et al.,* 1988) and there are undoubtably more to find. Extinctions on other mountains may have occurred during dry periods as the southern Eastern Arc has a much more marked dry season than in the north and an increase in the length of this dry season could have resulted in the loss of moist forest taxa. Extinctions could also have resulted from relatively recent human disturbance as upper montane forests on the Uzungwa and Rubeho mountains are secondary, these having been cleared for cultivation within the last few hundred years (Greenway, 1973; Lovett & Congdon, 1989; Lovett & Minja, 1990). The Usambara Mountains are also topographically varied and so contain more habitats than the steep, forested eastern faces of other Eastern Arc mountains. There is (or was) a relatively large area of the most endemic-rich forest types in the Usambara with undulating plateaux at the altitude of montane forest in the West Usambara and at the altitude of submontane forest in the East Usambara. There are more canopy tree niches available in the Usambara as the forest canopy is diverse with more tree species co-dominant than in the northern Uzungwa mountains (Table 4.5). This can be attributed to the bimodal rainy season covering the northern Eastern Arc, whereas in the south the widespread dry season tolerant tree *Parinari excelsa* is the most dominant species. The other Eastern Arc mountains with a high rainfall and short dry season are the Uluguru. Although they are only partially explored, especially in the southeast, and much of the lowland and submontane forest on the wetter eastern side has been cleared for cultivation, a large number of endemic species have been recorded, especially in genera like *Impatiens*, *Lasianthus* and *Polystachya*. The Uluguru also have a number of endemic genera, not all of which are moist forest species (Table 4.6).

Eastern Arc endemic species can be divided into palaeoendemics and neoendemics. Palaeoendemics are taxa which have survived in a limited portion of their past territory, are relicts, and rarely have close relations in the same or adajacent regions. Neoendemics are taxa in rapidly evolving species complexes, which often have their most closely related taxa in the same or nearby regions (Rodgers & Homewood, 1982). The origin of palaeoendemics pre-dates uplift of the central African plateau and geomorphological separation of eastern and western biota, whereas the origin of neoendemics post-dates rejuvenation of the Eastern Arc faults and uplift of the mountains. Neoendemics are predominately shrubs and

herbs; there are few trees. For example, each Eastern Arc mountain can have different endemic species of *Impatiens*, *Saintpaulia*, *Polystachya* and *Zimmermannia*, whereas palaeoendemic trees such as *Allanblackia stuhlmannii*, *Cephalosphaera usambarensis*, *Lettowianthus stellatus* and *Polyceratocarpus scheffleri* occur throughout the Arc.

Palaeoendemics with western affinity are represented in many different habits and habitats; examples are given in Table 4.7. Despite the long period of separation between east and west, many eastern endemic species are so similar to their western relatives that taxonomically they would be better placed as subspecies or varieties. Examples include *Diospyros amaniensis* which is closely related to the widespread western Guineo–Congolian *D. gabunensis*; *D. occulta* which is hardly distinguishable from *D. troupinii*; and *Mammea usambarensis* which is closely related to *M. africana*. This has already been done for the Eastern Arc endemic subspecies of the western Guineo–Congolian species *Pterocarpus mildbraedii* subsp. *usambarensis* and *Greenwayodendron suaveolens* subsp. *usambaricum*.

Neoendemics are represented by taxonomically difficult species complexes in a number of genera which occur in a wide range of Eastern Arc forest types (Table 4.8). They appear to have originated from a number of different areas: *Impatiens* entered the Eastern Arc from the western Guineo–Congolian region by both the northern and southern migration routes (Grey-Wilson, 1980), whereas *Saintpaulia* and *Zimmermannia* appear to have originated in Madagascar and then speciated in the Eastern Arc. Each Eastern Arc mountain has its own neoendemic flora, and differentiation has been described at a number of taxonomic levels. For example, the Eastern Arc endemic tree *Drypetes usambarica* occurs from southeast Kenya to the southern Uzungwa, and has five different varieties on different mountains and in lowland and montane forests. The relatively widespread moist forest shrubby tree *Heinsenia diervilleoides* has an Uzungwa endemic subspecies *mufindiensis*; and much of the variation in *Impatiens* and *Saintpaulia* is described at species level.

Hybridisation may play a role in the taxonomic

Table 4.5. *The ten species which are the most common, the most frequent, and greatest contributors to basal area in 20 tree variable area plots (Lovett, 1992) for submontane forest in the West Usambara at Ambangulu (20 plots) and northern Uzungwa in the Mwanihana Forest Reserve (35 plots). The most common species has the greatest number of individuals. The most frequent species occurs in the greatest number of plots. The species which contributes the greatest basal area has the greatest cumulative basal area. The percentage contribution of each species is given in parentheses following the species name. The percentage contribution of the ten species is given at the head of each section*

Ambangulu

Most common

68.6% of individuals. 21% of species.
Allanblackia stuhlmannii (12), *Newtonia buchananii* (11), *Myrianthus holstii* (9), *Parinari excelsa* (7.3), *Leptonychia usambarensis* (7), *Isoberlinia scheffleri* (6.3), *Afrosersalisia cerasifera* (5.5), *Celtis africana* (3.5), *Drypetes gerrardii* (3.5), *Englerodendron usambarense* (3.5).

Most frequent

Newtonia buchananii (85), *Allanblackia stuhlmannii* (75), *Parinari excelsa* (55), *Afrosersalisia cerasifera* (55), *Myrianthus holstii* (50), *Isoberlinia scheffleri* (35), *Celtis africana* (35), *Chrysophyllum gorungosanum* (35), *Drypetes gerrardii* (35), *Englerodendron usambarense* (35).

Greatest basal area

84.5% of basal area. 21% of species.
Newtonia buchananii (24.8), *Allanblackia stuhlmannii* (18.8), *Parinari excelsa* (12.8), *Isoberlinia scheffleri* (9.2), *Myrianthus holstii* (4.5), *Afrosersalisia cerasifera* (3.7), *Cleistanthus polystachyus* (3.0), *Sapium ellipticum* (2.8), *Celtis africana* (2.6), *Chrysophyllum gorungosanum* (2.3).

Mwanihana Forest

Most common

59.8% of individuals. 11.8% of species.
Parinari excelsa (20.4), *Cephalosphaera usambarensis* (11.0), *Trichoscypha ulugurensis* (7.7), *Leptonychia usambarensis* (3.8), *Newtonia buchananii* (3.6), *Tabernaemontana pachysiphon* (3.1), *Strombosia scheffleri* (2.9), *Syzygium guineense* (2.7), *Sapium ellipticum* (2.6), *Myrianthus holstii* (2.0).

Most frequent

Parinari excelsa (91.4), *Newtonia buchananii* (45.7), *Cephalosphaera usambarensis* (42.9), *Leptonychia usambarensis* (40.0), *Trichoscypha ulugurensis* (37.1), *Syzygium guineense* (34.3), *Odyendea zimmermannii* (31.4), *Chrysophyllum gorungosanum* (28.6), *Cassipourea gummiflua* (25.7), *Tabernaemontana pachysiphon* (25.7).

Greatest basal area

78.6% of basal area. 11.8% of species.
Parinari excelsa (46.9), *Cephalosphaera usambarensis* (8.5), *Newtonia buchananii* (4.9), *Beilschmeidia kweo* (3.5), *Sapium ellipticum* (3.4), *Uapaca paludosa* (3.4), *Syzygium guineense* (2.4), *Albizia gummifera* (2.0), *Trichoscypha ulugurensis* (2.0), *Ficalhoa laurifolia* (1.6).

Table 4.6. *Endemic or near endemic genera of the Eastern Arc with distribution, habitat and habit notes*

Aerisilvaea A.R.-Sm. (Euphorbiaceae) – 1 species, Uluguru lowland forest on limestone. Shrub.

Cephalosphaera Warb. (Myrsticaceae) – 1 species throughout the Eastern Arc in submontane forests. Large canopy tree.

Dionychastrum A. & R. Fernandes (Melastomataceae) – 1 species, Uluguru, upper montane forest edge, subalpine scrub. Related to the Madagascan *Dionycha* Naud. Small shrub.

Dolichometra K. Schum. (Rubiaceae) – 1 species, East Usambara submontane forest. Prostrate herb.

Englerodendron Harms (Leguminosae: Caesalpinioideae) – 1 species, East and West Usambara submontane forest. Midstorey tree.

Linnaeopsis Engl. (Gesneriaceae) – 4 species, Uluguru montane forest. Herbs.

Neobenthamia Rolfe (Orchidaceae) – 1 species, Uluguru and Nguru forest edge. Herb.

Platypterocarpus Dunkley & Brenan (Celastraceae) – 1 species, West Usambara dry upper montane forest. Tree.

Pseudonesohedyotis Tennant (Rubiaceae) – 1 species, Uluguru montane and upper montane forest. Small scrub.

Rhipidiantha Bremek. (Rubiaceae) – 1 species, Uluguru montane forest. Shrub.

Saintpaulia H. Wendl. (Gesneriaceae) – 20 species throughout the Eastern Arc and eastern coastal moist forests from lowland to upper montane forests. Related to some Madagascan *Streptocarpus*. Herbs.

Thulinia Cribb (Orchidaceae) – 1 species, Nguru subalpine grassland. Herb.

Urogentias Gilg & C. Benedict (Gentianaceae) – 1 species, Uluguru and Nguru subalpine scrub. Herb.

Zimmermannia Pax (Euphorbiaceae) – 7 species, 6 throughout the Eastern Arc (1 also upper moist coastal forest), 1 on Madagascar, submontane, montane and dry montane forest. Shrubs to small trees.

Zimmermanniopsis A.R.-Sm. (Euphorbiaceae) – 1 species, Uzungwa montane forest. Shrub.

Gen. nov. aff. *Sideroxylon sensu* FTEA (Sapotaceae) – 1 species, Usambara and Nguru montane forest. Midstorey to canopy tree.

Table 4.7. *Examples of genera with endemic species in the Eastern Arc which appear to be palaeoendemics*

Genus	Number of Eastern Arc species	Number of Eastern Arc endemics	Number of western species
Allanblackia	2	2	8
Angylocalyx	1	1	5
Enantia	1	1	10
Mammea	1	1	1
Octoknema	1	1	4
Placodiscus	1	1	14
Polyceratocarpus	1	1	6
Sibangea	1	1	2
Uvariopsis	1	1	11
Zenkerella	4	4	1

complexity and origin of some neoendemics. For example, Grey-Wilson (1980) cites a number of present-day hybrid swarms of *Impatiens* at forest edges or disturbed areas in forest, and suggests that apparently stable species such as the Uluguru endemics *I. barbulata* and *I. cinnabrina* may have arisen through hybridisation. Hybridisation between the forest tree *Syzygium guineense* subsp. *afromontanum* and the riparian or swamp tree *S. cordatum* may give rise to *S. masukuense* (White, 1978b; Lovett & Congdon, 1992), and may also account for the Eastern Arc endemic *S. sclerophyllum*. The origin of stable species through hybridisation has been suggested for *Erythrina* in Mexico (Neill, 1988), and *Streptocarpus* in Natal (Hilliard & Burtt, 1971).

Other endemics may have originated from

Table 4.8. *Examples of genera with endemic species in the Eastern Arc which appear to be neoendemics*

Genus	Number of Eastern Arc species	Number of Eastern Arc endemics	Number of species worldwide
Memecylon	19	17	320
Stolzia	10	7	15
Polystachya	36	17	200
Pavetta	29	20	400
Coffea	9	8	90
Psychotria	42	28	500
Lasianthus	11	10	150
Impatiens	38	27	900
Saintpaulia	20	20	21
Streptocarpus	18	13	130
Zimmermannia	6	6	7

Source: From Bridson & Verdcourt, 1988; Cribb, 1984; Grey-Wilson, 1980; Hilliard & Burtt, 1971; Johansson, 1978; Smith, 1987; Verdcourt, 1976; Wickens, 1975.

more widespread forest edge or woodland species. For example the small pink-flowered southern Uzungwa shrub *Bersama rosea* probably arose from the widespread, variable montane forest and forest edge tree *B. abyssinica* (Verdcourt, 1958). Similarly, the small pink-flowered Nguru shrub *Sorindeia calantha* presumably originated from the widespread lowland, submontane and montane forest tree *S. madagascariensis*. The very variable *Mitriostigma* with two West African species and three southern and eastern African species (Bridson, 1979; Bridson & Verdcourt, 1988; Verdcourt, 1987) has a montane forest Eastern Arc endemic, *M. usambarensis* with robust pink-purple flowers, which may have been derived from the widespread and closely related white-flowered *Oxyanthus*.

In addition to pink flowers, longer peduncles or pedicels is a character of Eastern Arc forest endemics originating from more widespread species. For example, the Uluguru and Uzungwa endemic *Craterispermum longipedunculatum* has longer peduncles than the more widespread *C. schweinfurthii*; the montane Uzungwa endemic *Drypetes gerrardinioides* has longer pedicels than the related East and West African lowland *D.*

parvifolia; and the montane and lowland forest Eastern Arc endemic *Isoberlinia scheffleri* has longer pedicels than the widespread and variable woodland *I. angolensis*. *I. scheffleri* is one of the few examples of a woodland genus also occurring in the moist forests. Otherwise the forests are floristically very distinct from the surrounding woodlands.

Discussion

In comparison to moist forests in Malawi to the south and central Kenya to the north, Eastern Arc forests are exceptionally rich in endemic species. For example, Malawi has only ten endemic moist forest tree species (Dowsett-Lemaire, 1989, 1990) and central Kenya only eight (Beentje, 1988b) in comparison to 67 Eastern Arc endemics. The South African Cape forests, unlike other Cape vegetation types, have few species or genera of restricted distribution. Other parts of Africa where there are many moist forest species of restricted distribution are the high rainfall areas of Cameroun, Gabon and Zaïre; and the eastern side of the Kivu–Rwenzori mountains. The richness of these centres of endemism has been ascribed to moist forest survival in areas of high rainfall during proposed Pleistocene droughts (Hamilton, 1976, 1982). Certainly the Eastern Arc cannot have suffered a major loss of forest or a substantial lowering in altitude of forest types in the last 30 000 years, or the range of endemics in different forest types and of different affinities would not be present today. This biological evidence is supported by evidence from deep-sea drilling which indicates that Indian Ocean surface temperatures off the coast of tropical eastern Africa did not decrease substantially during the last glacial maximum (Prell *et al.*, 1980; van Campo *et al.*, 1990) and hence that rainfall would have been similar to that at present. Although evidence for dramatic climatic changes during the Pleistocene is strong (Hamilton, 1982) it is possible that they were limited to certain areas. For example, southeast Zaïre, Angola, northwestern Zambia and southwestern Tanzania make up an area of exceptional species richness and endemism in a woodland and riverine forest vegetation (Brenan, 1978). If there had been widespread, severe Pleistocene droughts this south–central African area might be expected to be poorer in species than it actually is. However, they may be responsible for the relative species poverty of forests in Malawi and Kenya.

Richness of the Eastern Arc moist forest flora within the context of African vegetation can be contrasted with its poverty in comparison to some other tropical forests. There are only two palms, *Phoenix reclinata* and *Elaeis guineensis*; and four laurels, *Beilschmiedia kweo*, *Cryptocarya liebertiana*, *Ocotea usambarensis* and *O. kenyensis*. Both of these groups are well represented in Madagascar, the Neotropics and Far East. To some extent, Miocene loss of the southern temperate moist forest flora, and lack of a north temperate moist forest input, explains poverty in the montane forests. However, survival of a substantial western Guineo–Congolian floristic influence indicates that the Eastern Arc has been under a moist climate since before the creation of the east–west continental divide. Either there was a loss of species before this and they were not replaced by subsequent speciation, or there never were many species. Certainly the rate of speciation varies enormously between different plants: compare the explosive speciation in *Impatiens* with hardly any morphological change in western Guineo–Congolian taxa isolated for perhaps more than 7 million years. Whatever factors created, and are creating, species in the Neotropics and Far East are largely absent from the Eastern Arc. The high number of endemics in the Usambara and Uluguru mountains, which are under a per-humid or almost per-humid climate, perhaps gives an indication that at least one of these factors is rainfall. A similar conclusion was reached by Gentry (1989) for Neotropical forests as an alternative to speciation resulting from the isolating mechanism of postulated Pleistocene refugia (Prance, 1987).

Whatever created or destroyed species in the past, present areas of secondary forest resulting from cultivation are species-poor in comparison with undisturbed forest. Montane forests on the Uzungwa plateau, Rubeho and Ukaguru mountains, which show signs of habitation in the form of pot-shards or cultivation ridges, are composed of widespread species with no or few Eastern Arc endemics. Historical records indicate that these forests were in existence at the time of the early German administration in Tanzania, nearly 100 years ago, and presumably they were there some time before then. The drastic changes in microclimate that follow forest clearance must inhibit regeneration of moist forest dependent endemics and it appears that the forests take a very long time to recover.

Summary

The Eastern Arc mountains act as condensers of moisture originating in the Indian Ocean, and have been doing so since their initial formation following the breakup of Gondwanaland. The geological and climatological definition of the Eastern Arc as ancient crystalline mountains under the direct climatic influence of the Indian Ocean is mirrored by a high level of species and generic endemism. The affinities of the flora are primarily with the Guineo–Congolian region of West and Central Africa, though direct contact westwards would have been severed following uplift of the central African plateau at the end of the Miocene. The most recent uplift of the mountains in the early Pliocene came after the late Miocene climatic changes in southern Africa which resulted in extinction of the southern temperate forests. Consequently there was not a temperate climatic refugium in the tropics for southern temperate forest species, so the Eastern Arc forests do not have a significant southern origin. Species with southern distributions often occur on the forest edge and have relatives in Madagascar. Geographical separation from north temperate forests has limited northern affinities of the Eastern Arc. Further inputs to the flora have arrived by long-distance dispersal from overseas, by hybridisation with woodland species, or by adaptation of woodland species.

Forest community associations vary continuously with altitude, topographic and latitudinal variations in rainfall; division of this continuous variation into forest types is arbitrary. Geographical distribution of species is similar within each of the different forest types as defined here. Thus there is no evidence to support eco-logical or phytogeographic divisions of the forest into upper and lower types. The strong affinities of the flora to the Guineo–Congolian region, combined with a high degree of endemism, sug-gest that the Eastern Arc moist forests are best classified as an eastern outlier of the more exten-sive Guineo–Congolian forests.

References

ANDREWS, P. & VAN COUVERING, J. A. H. (1975). Palaeoenvironments in the East African Miocene. *Approaches to Primate Palaeobiology* 5, 62–103.

AXELROD, D. I. & RAVEN, P. H. (1978). Late Cretaceous and Tertiary vegetation history of Africa. In *Biogeography and Ecology of Southern Africa*, ed. M. J. A. Werger, pp. 77–130. The Hague: Junk.

AYENSU, E. S. (1973). Phytogeography and evolution of the Velloziaceae. In *Tropical Forest Ecosystems in Africa and South America: A Comparative Review*, ed. B. J. Meggers, A. S. Ayensu and W. D. Duckworth, pp. 105–19. Washington, DC: Smithsonian Institution Press.

BEENTJE, H. J. (1988a). An ecological and floristic study of the forests of the Taita Hills, Kenya. *Utafiti (Occasional Papers of the National Museums of Kenya)* 1(2), 23–66.

BEENTJE, H. J. (1988b). Atlas of the rare trees of Kenya. *Utafiti* 1(3), 71–123.

BERG, C. C. & HIJMAN, M. E. E. (1989). Moraceae. In *Flora of Tropical East Africa*. Rotterdam: Balkema.

BIZOT, M., POCS, T. & SHARP, A. J. (1985). Results of a bryogeographical expedition to East Africa in 1968, III. *The Bryologist* 88(2), 135–42.

BOUCHET, A. & LEONARD, J. (1986). Revision du genre *Necepsia* Prain (Euphorbiacée africano-malgache). *Bulletin du Jardin Botanique National de Belgique* 56, 179–94.

BRENAN, J. P. M. (1978). Some aspects of the phytogeography of tropical Africa. *Annals of the Missouri Botanical Garden* 65, 437–78.

BRIDSON, D. M. (1979). Studies in *Oxyanthus* and *Mitriostigma* (Rubiaceae – Cinchonoideae) for part 2 of the *Flora of Tropical East Africa: Rubiaceae. Kew Bulletin* 34(1), 113–30.

BRIDSON, D. M. & VERDCOURT, B. (1988). Rubiaceae (Part 2). In *Flora of Tropical East Africa*. Rotterdam: Balkema.

CCTA/SCIENTIFIC COUNCIL FOR AFRICA (1956). *Phytogeography*. Publication No. 22. London: Commission for Technical Cooperation in Africa south of the Sahara.

CHAPMAN, J. D. & WHITE, F. (ed.) (1970). *The Evergreen Forests of Malawi*. Oxford: Commonwealth Forestry Institute.

COETZEE, J. A. & MULLER, J. (1984). The phytogeographic significance of some extinct Gondwana pollen types from the Tertiary of the southwestern Cape (South Africa). *Annals of the Missouri Botanical Garden* 71, 1088–99.

CRIBB, P. (1984). Orchidaceae (Part 2). In *Flora of Tropical East Africa*. Rotterdam: Balkema.

DENYS, E. (1980). A tentative phytogeographical division of tropical Africa based on a mathematical analysis of distribution maps. *Bulletin du Jardin Botanique National de Belgique* 50, 465–504.

DE RUITER, G. (1976). Revision of the genera *Myrianthus* and *Musanga* (Moraceae). *Bulletin du Jardin Botanique National de Belgique* 46, 471–510.

DONALD, D. G. M. & THERON, J. M. (1983). Temperate broad-leaved evergreen forests of Africa south of the Sahara. In *Ecosystems of the World*, Vol. 10. *Temperate Broad-leaved Evergreen Forests*, ed. J. D. Ovington, pp. 135–68. Amsterdam: Elsevier.

DOWSETT-LEMAIRE, F. (1989). The flora and phytogeography of the evergreen forests of Malawi. I. Afromontane and mid-altitude forests. *Bulletin du Jardin Botanique National de Belgique* 59, 3–131.

DOWSETT-LEMAIRE, F. (1990). The flora and phytogeography of the evergreen forests of Malawi. II: Lowland forests. *Bulletin du Jardin Botanique National de Belgique* 60, 9–71.

DRANSFIELD, J. (1988). The palms of Africa and their relationships. In *Modern Systematic Studies in African Botany*, ed. P. Goldblatt and P. P. Lowry, pp. 95–103. St Louis: Missouri Botanical Garden.

EHRENDORFER, F. (1988). Affinities of the African dendroflora: suggestions from karyo- and chemosystematics. In *Modern Systematic Studies in African Botany*, ed. P. Goldblatt and P. P. Lowry, pp. 105–27. St Louis: Missouri Botanical Garden.

EYDE, R. H. (1988). Comprehending *Cornus*: puzzles and progress in the systematics of Dogwoods. *The Botanical Review* 54, 233–51.

GENTRY, A. H. (1989). Speciation in tropical forests. In *Tropical Forests: Botanical Dynamics, Speciation and Diversity*, ed. L. B. Holm-Nielsen, J. C. Nielsen and H. Balslev, pp. 113–34. London: Academic Press.

GILLETT, J. B. (1961). History of the botanical exploration of the area of the *Flora of Tropical East Africa* (Uganda, Kenya, Tanganyika, and Zanzibar). In *Comptes rendus de la IV Réunion pleneire de l'AETFAT*, ed. A. Fernandes, pp. 205–29. Lisbon: Association pour l'Etude Taxonomique de la Flore d'Afrique Tropicale.

GILLMAN, C. (1949). A vegetation-types map of Tanganyika Territory. *The Geographical Review* 39, 7–37.

GOLDBLATT, P. (1978). An analysis of the flora of southern Africa: its characteristics, relationships, and origins. *Annals of the Missouri Botanical Garden* 65, 369–436.

GREENWAY, P. J. (1973). A classification of the vegetation of East Africa. *Kirkia* 9, 1–68.

GREY-WILSON, C. (1980). *Impatiens of Africa*. Rotterdam: Balkema.

HALL, J. B. (1984). *Juniperus excelsa* in Africa: a biogeographical study of an Afromontane tree. *Journal of Biogeography* 11, 47–61.

HAMILTON, A. C. (1975). A quantitative analysis of altitudinal zonation in Uganda forests. *Vegetatio* 30, 99–106.

HAMILTON, A. C. (1976). The significance of patterns of distribution shown by forest plants and animals in tropical Africa for the reconstruction of Upper Pleistocene palaeoenvironments: a review. *Palaeoecology of Africa* 9, 63–97.

HAMILTON, A. C. (1982). *Environmental History of East Africa*. New York: Academic Press.

HAMILTON, A. C. (1989a). The climate of the East Usambaras. In *Forest Conservation in the East Usambara Mountains, Tanzania*, ed. A. C. Hamilton and R. Bensted-Smith, pp. 97–102. Gland, Switzerland: International Union for Conservation of Nature and Natural Resources.

HAMILTON, A. C. (1989b). Soils. In *Forest Conservation in the East Usambara Mountains Tanzania*, ed. A. C. Hamilton and R. Bensted-Smith, pp. 87–95. Gland: IUCN.

HAMILTON, A. C. & FADEN, R. B. (1974). The history of the vegetation. In *East African Vegetation*, ed. E. M. Lind and M. E. S. Morrison, pp. 188–209. London: Longman.

HAMILTON, A. C. & PERROTT, R. A. (1981). A study of altitudinal zonation in the montane forest belt of Mt. Elgon, Kenya/Uganda. *Vegetatio* 45, 107–25.

HAMILTON, A. C., RUFFO, C. K., MWASHA, I. V., MMARI, C. & LOVETT, J. C. (1989). A survey of forest types on the East Usambara using the variable-area tree plot method. In *Forest Conservation in the East Usambara Mountains Tanzania*, ed. A. C. Hamilton and R. Bensted-Smith, pp. 213–25. Gland: IUCN.

HAUMAN, L. (1955). La 'région Afroalpine' en phytogeographie centro-Africaine. *Webbia* 11, 467–69.

HEDBERG, O. (1951). Vegetation belts of East African mountains. *Svensk Botanisk Tidskrift* 45, 140–202.

HEDBERG, O. (1957). Afroalpine vascular plants. A taxonomic revision. *Symbolae Botanicae Upsaliensis* 15, 1–411.

HEDBERG, O. (1961). The phytogeographical position of the afroalpine flora. *Recent Advances in Botany* 1, 914–19.

HEDBERG, O. (1965). Afroalpine flora elements. *Webbia* 19, 519–29.

HIEPKO, P. (1987). The collections of the Botanical Museum Berlin–Dahlem (B) and their history. *Englera* 7, 219–52.

HILLIARD, O. M. & BURTT, B. L. (1971). *Streptocarpus: an African Plant Study*. Pietermaritzburg: University of Natal Press.

HOOKER, J. D. (1861). On the vegetation of Clarance Peak, Fernando Po. *Journal of the Linnean Society, Botany* 6, 1–23.

HOOKER, J. D. (1864). On the plants of the temperate regions of the Cameroons mountains and islands in the Bight of Benin; collected by Mr. Gustav Mann, Government botanist. *Journal of the Linnean Society, Botany* 7, 171–240.

HOPKINS, H. C. & WHITE, F. (1984). The ecology and chorology of *Parkia* in Africa. *Bulletin du Jardin Botanique National de Belgique* 54, 235–66.

IVERSEN, S. T. (1987). Check-list, vascular plants of the Usambara mountains. Appendix 1. In *The SAREC-supported Integrated Usambara Rain Forest Project. Report for the Period 1983–1987*, ed. I. Hedberg and O. Hedberg, pp. 1–93. Uppsala: Department of Systematic Botany.

JOHANSSON, D. R. (1978). *Saintpaulias* in their natural environment with notes on their present status in Tanzania and Kenya. *Biological Conservation* 14, 45–62.

KALKMAN, C. (1965). The old world species of *Prunus* subgenus *Laurocerasus* including those formerly referred to *Pygeum*. *Blumea* 13, 1–115.

KALKMAN, C. (1988). The phylogeny of the Rosaceae. *Botanical Journal of the Linnean Society* 98, 37–59.

KERFOOT, O. (1961). *Juniperus procera* Endl. (The

African Pencil Cedar) in Africa and Arabia I. Taxonomic affinities and geographical distribution. *East African Agricultural and Forestry Journal* **26**, 170–7.

KERFOOT, O. (1964). The distribution and ecology of *Juniperus procera* Endl. in East Central Africa, and its relationship to the genus *Widdringtonia* Endl. *Kirkia* **4**, 75–86.

KERFOOT, O. (1966). Distribution of the Coniferae: the Cupressaceae of Africa. *Nature* **212**, 961.

KERFOOT, O. (1968). Mist precipitation on vegetation. *Forestry Abstracts* **29**, 8–20.

KERFOOT, O. (1975). Origin and speciation of the Cupressaceae in sub-Saharan Africa. *Boissiera* **24**, 151–72.

KILLICK, D. J. B. (1978). The Afro-alpine region. In *Biogeography and Ecology of Southern Africa*, ed. M. J. A. Werger, pp. 515–60. The Hague: Junk.

LEEUWENBERG, A. J. M. (1979). The Loganiaceae of Africa XVIII. *Buddleja* L. II. Revision of the African and Asiatic species. *Mededeelingen van de Landbouwhoogeschool te Wageningen* **79**(6), 1–163.

LEROY, J. F. (1978). Composition, origin and affinities of the Madagascan vascular flora. *Annals of the Missouri Botanical Garden* **65**, 535–89.

LOVETT, J. C. (1985). Moist forests of Tanzania. *Swara* **8**(5), 8–9.

LOVETT, J. C. (1988). Endemism and affinities of the Tanzanian montane forest flora. In *Modern Systematic Studies in African Botany*, ed. P. Goldblatt and P. P. Lowry, pp. 591–8. St Louis: Missouri Botanical Garden.

LOVETT, J. C. (1990a). Classification and status of the Tanzanian forests. *Mitteilungen aus dem Institut für Allgemeine Botanik in Hamburg* **23A**, 287–300.

LOVETT, J. C. (1990b). Altitudinal variation in large tree community associations on the West Usambara mountains. In *Research for Conservation of Tanzania Catchment Forests*, ed. I. Hedberg and E. Persson, pp. 48–53. Uppsala: Uppsala Universitet.

LOVETT, J. C. (1992). An ordination of the large tree associations in the moist forests of the West Usambara mountains, Tanzania. *Proceedings of the IUFRO meeting, Arusha, 3–7 August 1991*. Morogoro: Sokoine University (in press).

LOVETT, J. C., BRIDSON, D. M. & THOMAS, D. W. (1988). A preliminary list of the moist forest angiosperm flora of the Mwanihana forest reserve, Tanzania. *Annals of the Missouri Botanical Garden* **75**, 874–88.

LOVETT, J. C. & CONGDON, T. C. E. (1989). Notes on the Ihangana forest and Luhega forest near

Uhafiwa, Uzungwa mountains, Tanzania. *East Africa Natural History Society Bulletin* **19**, 30–1.

LOVETT, J. C. & CONGDON, T. C. E. (1992). Notes on *Syzygium* in Mufindi, Tanzania. *Annals of the Missouri Botanical Garden* (in press).

LOVETT, J. C. & GEREAU, R. E. (1990). Notes on the floral morphology and ecology of *Margaritaria discoidea* (Euphorbiaceae) at Mufindi, Tanzania. *Annals of the Missouri Botanical Garden* **77**, 217–18.

LOVETT, J. C. & MINJA, T. R. A. (1990). Notes on a visit to Ukwiva forest, Tanzania. *East Africa Natural History Society Bulletin* **20**, 4–5.

MABBERLEY, D. J. (1987). *The Plant-Book: A Portable Dictionary of the Higher Plants*. Cambridge: Cambridge University Press.

MOLL, E. J. & WHITE, F. (1978). The Indian Ocean coastal belt. In *Biogeography and Ecology of Southern Africa*, ed. M. J. A. Werger, pp. 561–98. The Hague: Junk.

MONOD, T. (1957). *Les grandes divisions Chorologiques de l'Afrique*. Publication no. 24. London: CCTA.

MOREAU, R. E. (1935). A synecological study of Usambara, Tanganyika Territory, with particular reference to birds. *Journal of Ecology* **23**, 1–43.

NEILL, D. A. (1988). Experimental studies on species relationships in *Erythrina* (Leguminosae: Papilionoideae). *Annals of the Missouri Botanical Garden* **75**, 886–969.

PENNINGTON, T. D. & STYLES, B. T. (1975). A generic monograph of the Meliaceae. *Blumea* **22**, 419–540.

PHILLIPSON, P. B. & RUSSELL, S. (1988). Phytogeography of the Alexandria Forest (southeastern Cape Province). In *Modern Systematic Studies in African Botany*, ed. P. Goldblatt and P. P. Lowry, pp. 661–70. St Louis: Missouri Botanical Garden.

PITT-SCHENKEL, C. J. W. (1938). Some important communities of warm temperate rain forest at Magamba, West Usambara, Tanganyika Territory. *Journal of Ecology* **26**, 50–81.

POCS, T. (1974). Bioclimatic studies in the Uluguru mountains (Tanzania, East Africa) I. *Acta Botanica Academiae Scientiarum Hungaricae* **20**, 115–35.

POCS, T. (1975). Affinities between the bryoflora of East Africa and Madagascar. *Boissiera* **24**, 125–8.

POCS, T. (1976a). The role of the epiphytic vegetation in the water balance and humus production of the rain forests of the Uluguru mountains, East Africa. *Boissiera* **24**, 499–503.

POCS, T. (1976b). Bioclimatic studies in the Uluguru mountains (Tanzania, East Africa) II. Correlations between orography, climate and vegetation. *Acta Botanica Academiae Scientiarum Hungaricae* **22**, 163–83.

POCS, T. (1976c). Vegetation mapping in the Uluguru mountains (Tanzania, East Africa). *Boissiera* **24**, 477–98.

POCS, T. (1985). East African bryophytes, VIII. The Hepaticae of the Usambara rain forest project expedition, 1982. *Acta Botanica Hungarica* **31** (1–4), 113–33.

POCS, T., TEMU, R. P. C. & MINJA, T. R. A. (1990). Survey of the natural vegetation and flora of the Nguru mountains. In *Research for Conservation of Tanzanian Catchment Forests*, ed. I. Hedberg and E. Persson, pp. 135–46. Uppsala: Uppsala Universitet.

POLHILL, R. M. (1989). Speciation patterns in African Loranthaceae. In *Tropical Forests: Botanical Dynamics, Speciation and Diversity*, ed. L. B. Holm-Nielsen, J. C. Nielsen and H. Balslev, pp. 221–36. London: Academic Press.

POOLE, M. M. (1981). Pollen diversity in *Zimmermannia* (Euphorbiaceae). *Kew Bulletin* **36**(1), 129–38.

PRANCE, G. T. (1987). Biogeography of Neotropical plants. In *Biogeography and Quaternary History in Tropical America*, ed. T. C. Whitmore and G. T. Prance, pp. 46–65. Oxford: Oxford University Press.

PRELL, W. L., HUTSON, W. H., WILLIAMS, D. F., BÉ, A. W. H., GEITZENHAUER, K. & MOLFINO, B. (1980). Surface circulation of the Indian Ocean during the last glacial maximum, approximately 18,000 yr BP. *Quaternary Research* **14**, 309–36.

RAVEN, P. H. & AXELROD, D. I. (1974). Angiosperm biogeography and past continental movements. *Annals of the Missouri Botanical Garden* **61**, 539–673.

RODGERS, W. A. & HOMEWOOD, K. M. (1982). Species richness and endemism in the Usambara mountain forests, Tanzania. *Biological Journal of the Linnean Society* **18**, 197–242.

SIESSER, W. G. (1980). Late Miocene origin of the Benguela upswelling system off northern Namibia. *Science* **208**, 283–5.

SMITH, A. R. (1987). Euphorbiaceae (Part 1). In *Flora of Tropical East Africa*. Rotterdam: Balkema.

THORNE, R. F. (1973). Floristic relationships between tropical Africa and tropical America. In *Tropical Forest Ecosystems in Africa and South America: A Comparative Review*, ed. B. J. Meggers, A. S. Ayensu and W. D. Duckworth, pp. 27–47. Washington, DC: Smithsonian Institution Press.

TROUPIN, G. (1966). *Etude phytocenologique du Parc National de l'Akagera et du Rwanda Oriental. Recherche d'une Methode d'analyse appropriée à la Vegetation d'Afrique Intertropicale*. Republique Rwandaise: Institut National de Recherche Scientifique, Butare.

VAN CAMPO, E., DUPLESSY, J. C., PRELL, W. L., BARRATT, N. & SABATIER, R. (1990). Comparison of terrestrial and marine temperature estimates for the past 135 kyr off southeast Africa: a test for the GCM simulations of palaeoclimate. *Nature* **348**, 209–12.

VERDCOURT, B. (1958). Melianthaceae. In *Flora of Tropical East Africa*. London: Crown Agents for Overseas Governments and Administrations.

VERDCOURT, B. (1976). Rubiaceae. In *Flora of Tropical East Africa*. London: Crown Agents for Overseas Governments and Administrations.

VERDCOURT, B. (1987). A new species of *Mitriostigma* (Rubiaceae – Gardenieae) from Tanzania. *Kew Bulletin* **42**(1), 245–50.

VOLLESEN, K. (1982). A new species of *Seychellaria* (Triuridaceae) from Tanzania. *Kew Bulletin* **36**(4), 733–6.

WEBSTER, G. L. (1965). A revision of the genus *Meineckia* (Euphorbiaceae). *Acta Botanica Neerlandica* **14**, 323–65.

WERGER, M. J. A. (1978). The Karoo–Namib region. In *Biogeography and Ecology of Southern Africa*, ed. M. J. A. Werger, pp. 231–99. The Hague: Junk.

WHITE, F. (1970). Floristics and plant geography. In *The Evergreen Forests of Malawi*, ed. J. D. Chapman and F. White, pp. 38–77. Oxford: Commonwealth Forestry Institute.

WHITE, F. (1978a). The Afromontane region. In *Biogeography and Ecology of Southern Africa*, ed. M. J. A. Werger, pp. 463–513. The Hague: Junk.

WHITE, F. (1978b). Myrtaceae. In *Flora Zambesiaca*, Vol. 4, ed. E. Launert, pp. 183–212. London: Flora Zambesiaca Managing Committee.

WHITE, F. (1981). The history of the Afromontane archipelago and the scientific need for its conservation. *African Journal of Ecology* **19**, 33–54.

WHITE, F. (1983). *The Vegetation of Africa*. Paris: United Nations Education Scientific and Cultural Organization.

WHITMORE, T. C. (1984). *Tropical Rain Forests of the Far East*. Oxford: Clarendon Press.

WICKENS, G. E. (1975). Melastomataceae. In *Flora of Tropical East Africa*. London: Crown Agents for Overseas Governments and Administrations.

5 East African coastal forest botany

W. D. HAWTHORNE

Abstract

For more than a century collections and observations have been made in the Kenyan and Tanzanian coastal forests.

Coastal forest is defined in terms of its geographical position and geomorphological association. Environmental and human influences are discussed.

Major patterns in the flora are discussed in terms of five ecogeographical elements, relating to distribution of species within and adjacent to coastal forests. The distribution of species outside the coastal region is summarised as well.

Forests are described informally, emphasising local variation and complex patterns. The range of association within both Moist and Dry forests locally is compounded by Northern and Southern elements, defined by species of restricted distribution.

Origins of the observed patterns are discussed. Apart from environmental and human influences, 'Gleasonian' factors are likely to have been important. Priorities for conservation measures are suggested.

Historical background

Botanical collections from East Africa's coastal regions started with missionary activity in the mid-19th century. The first mission station on mainland East Africa was established in 1855 close to Mombasa (Krapf, 1860, 1882), and the Rev. Thomas Wakefield soon sent plant specimens to England from this area (see Brewin, 1879; Wakefield, 1904). A few years later, Rev. Charles New provided a vivid, first account (New, 1873) of the natural history of the coastal vegetation (*nyika*) and of the relationship between the forests and the Mijikenda (the 'Wanyika') who lived in and around them.

Plant collections were made in a few coastal areas of Tanzania in the second half of the last century, for example by Holst in the Pugu Hills, Stuhlmann in the Pangani area and Kirk in 1884 in the Tanga area (see Gillett, 1961b). Other less prolific collectors included Baron Hofmarschal Saint Paul who discovered the first African violets (*Saintpaulia ionantha*) amongst the Tanga limestone outcrops (Burtt, 1948). More recently, collectors such as Faulkner in the Pangani area have filled gaps in our knowledge (Polhill, 1980, in Hubbard *et al.*, 1952–83), but collections from many Tanzanian coastal forests remain far from complete. Many new species have been discovered in the last 20 years, several yet unnamed. Almost certainly others lie undiscovered.

Accounts of vegetation by New and others (Farler, 1875; Hildebrandt, 1879; Hobley, 1895, 1905) give some idea of the forest in the last century. A more thorough survey of the coastal forests was made for Kenya by Dale (1939) and more recently by Moomaw (1960), but no equivalent treatment of the Tanzanian coastal forests exists. For parts of Tanzania, Engler's (1910) account is the only one available. More general summaries of the coastal forests have been made by Phillips (1931), Rea (1935), Burtt Davy (1935, 1938), Gillman (1949), Greenway (1973), Hamilton & Faden (1973) and White

(1983). Polhill (1968) and Lucas (1968) provide floristic data for several coastal forests. Specific forests or vegetation types have been described by Milne (1947), Burtt (1942), Welch (1960), Birch (1963), Glover (1969, 1970), Johannson (1978), Andrews, Grove & Horne (1975), Howell (1981), Rodgers *et al.* (1983, 1984) and Hall *et al.* (1984). More recently an attempt has been made by Hawthorne (1984) to summarise this information and to relate the Tanzanian coastal forests to those of Kenya.

What is a coastal forest?

The coastal forests are part of White's (1983) Zanzibar–Inhambane regional mosaic. There is a very broad spectrum of plant communities within this mosaic, between mangrove and other vegetation close to the sea, and higher altitude forest or non-forest vegetation inland. Afromontane forest and various Somali–Masai or Zambezian vegetation types merge and interdigitate with the coastal mosaic. Any definition of 'coastal forest' is arbitrary regarding delimitation of both area, i.e. the boundaries of the 'coast', and physiognomy, i.e. the definition of forest. As this chapter should make clear, coastal forest has no simple and watertight definition regarding floristic content.

For the 'coast' I have used a definition based on geomorphology. The coast is the land over the sedimentary (and intrusive volcanic) rocks of the coastal plains and plateau, to the east of the exposed basement complex land. All forest in the coast is coastal forest. Certain other lowland forests, like the forest reserves at Kimboza (Rodgers *et al.*, 1983) or Mwanihana (Lovett, Bridson & Thomas, 1988) and in the Usambara foothills (Rodgers & Homewood, 1982), resemble certain coastal forests in species composition, yet are not included because they occur on the basement complex lands inland. Forests running inland along large rivers, for example the Tana River forests (Andrews *et al.*, 1975; Hughes, 1990), traverse any arbitrary borderline and again serve to remind us that coastal forest communities merge gradually into those of surrounding areas.

The definition of forest adopted here is that of a continuous stand of trees 10 m or more tall

(White, 1983). However, just as minor deviations from this ideal in the form of natural treefall gaps and streams are tolerable in any definition of forest, so it is useful to include as part of our definition of coastal forests other vegetation types intimately associated with the forest proper. The coastal forests enclose and merge with coastal thickets and scrub forest, bushland and woodland to such an extent that to disentangle too assiduously the true forest from the rest is to risk missing an essential grainy heterogeneity in the coastal forests, and certainly excludes many unusual and rare species. Species typical of certain non-forested components of the mosaic will, however, be excluded from the sample of coastal forest species discussed below.

Many coastal forests are still not well known. Forests of the offshore islands (Pemba, Mafia, Zanzibar) do not feature in the following discussion. However, from Robins (1976), Greenway *et al.* (1988) and Beentje (1990) it can be deduced that although Mafia's forest corresponds closely to that found, for instance, at Kisiju and in part of the Pugu Hills on the adjacent mainland, parts of the forest communities of Zanzibar and Pemba include species not recorded on the mainland, probably because of greater rainfall on Zanzibar. Similarly, apart from information in Vollesen (1980) concerning the Selous, and Moll & White's (1978) description of forests further south in the Indian Ocean coastal belt, data from forests south of the Rufiji River are very incomplete. This chapter should therefore be seen as an interim summary, pending the more complete botanical exploration of the coast.

Figures 5.1 and 5.2 give a summary of coastal forests discussed here, with some neighbouring non-coastal forests in brackets. Many small coastal forests are not shown.

Coastal forest environment

Coastal regions of Kenya and Tanzania consist of series of plateaux studded with hills, swamps and rocky outcrops, and traversed by many rivers and streams. Throughout, there are farms and fallow grassland with or without scattered trees, thickets, coconut, mango and cashew groves, and forests in

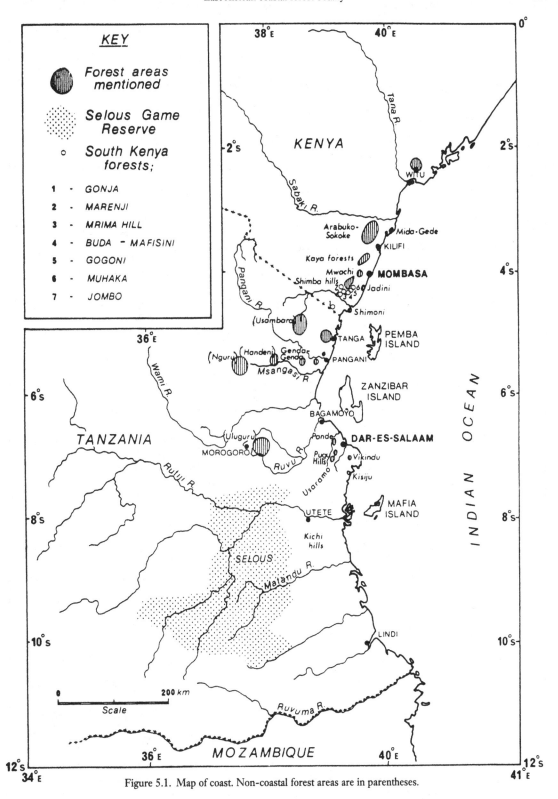

Figure 5.1. Map of coast. Non-coastal forest areas are in parentheses.

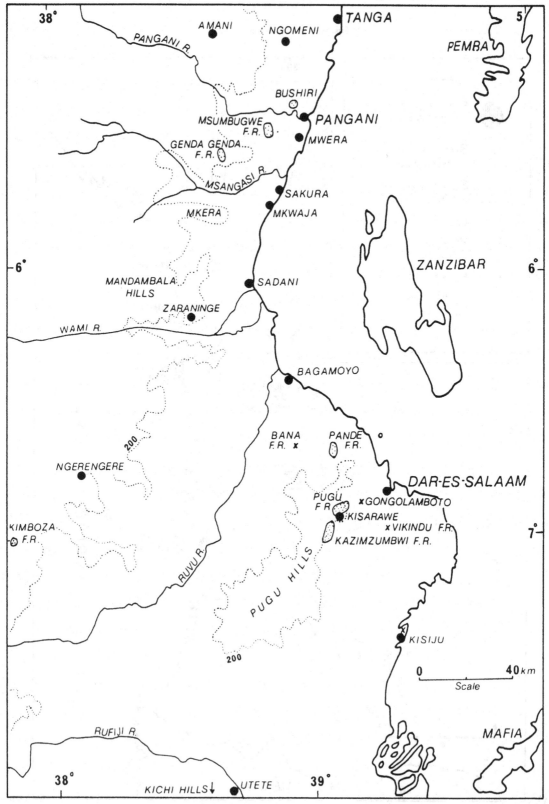

Figure 5.2. Map of Southern coast, showing location of various coastal forests and Kimboza Forest Reserve (non-coastal).

various stages of development and degradation.

Variation in the forest is correlated with heterogeneity in the physical environment. Four idealised transects from west to east are given in Figure 5.3, showing the major geomorphological features associated with coastal forests (simplified from Gregory, 1921; Oates, 1933; Teale, 1936; Stockley, 1948; King, 1951; Miller, 1953; Caswell, 1953, 1956; Thompson, 1956; Spence, 1957; Halligan, 1958; Dighton-Thomas, 1963; McKinlay, 1963; Moore, 1963; Cooke, 1974; Miyata & Saka, 1979). Excepting some intrusive, volcanic peaks such as Mrima Hill in the coastal range of Kenya, the coastal forests occur on sedimentary rocks built up since the drift of Madagascar away from Africa (Rust, 1973). Tilting and erosion have led to the exposure of rocks dating from the Karroo series in the Kenyan coastal range, through Jurassic limestones and Cretaceous sandstones, to Quaternary deposits such as the sands and raised coral reefs of the coastal plain.

Rainfall varies in monthly distribution and total amount throughout the coastal forests (Walter, 1952; Glover, Robinson & Henderson, 1954; Griffiths, 1958, 1972; Walter & Lieth, 1967). From the south to the north there is a general increase in tendency towards two rainy and two dry seasons, with a single dry season being marked south of the Rufiji River. Reliability of annual rainfall declines towards the north of the Kenyan coast, but with anomalous high rainfall (1085 mm) around the Witu forests near the Tana delta (Dale, 1939; Moomaw, 1960). Average annual rainfall is in some areas as low as about 950 mm (near Kilifi, for instance). Although parts of the offshore islands receive more than 1500 mm per year, no mainland coastal forest receives more than an average of 1400 mm (1320 mm near Tanga).

Hills attract significant orographic precipitation and support most of the moister patches of coastal forest. The landward side of some hills (e.g. Genda-Genda) is grassland whereas the seaward side supports forest. In some areas, like the Pugu Hills with an annual rainfall of 1236 mm (at Kisarawe: Howell, 1981), the Moister forest types are found in the valley bottoms. Elsewhere, for example the Shimba Hills (with 1090 mm per year) and the *kaya* forests, Moist forest is found on higher land as well, with much of the adjacent low-lying land supporting Dry forest.

There is a suggestion (map in Anon., 1947) of low rainfall around Bagamyoyo (and the Msangasi River) in the shadow of Zanzibar, which could explain the apparently abrupt change in the flora across this area, described below.

Drainage, geomorphology and precipitation interact in their influence on forest type, and although soil type is often dependent on this complex of relationships (Milne, 1947), soil sometimes exerts a more independent influence. Calton, Tidbury & Walker (1955), for instance, have suggested that past climates continue to exert influence on the present coastal vegetation by having helped create deep, heavily leached red soils. These rather infertile, low phosphate soils have persisted and now support coastal forests distinct from others, under similar conditions of climate and geomorphology but with younger soils. Similarly, alluvial soils, transported from different areas, have a profound influence on tree distribution according to their nature and disposition, as outlined for the Tana flood plain by Andrews *et al.* (1975).

The influence of rainfall alone on the vegetation is apparent from differences between patches of forest in areas of similar geology but different rainfall. Coral rag is a common landscape unit on the coastal plain, very close to the sea. The forests growing over it are very varied, with forests in areas of greater rainfall being of a Moister type (Moomaw, 1960; Birch, 1963; Hall *et al.*, 1984).

This web of physical factors has a profound influence on the species composition of the coastal forests, but cannot account for all floristic variation. Part of the reason for this is the influence of other environmental influences, particularly the influence of humans.

Human influence

People have undoubtedly been influencing the coastal ecology for millennia. Trade in many forest products, like gum copal (from *Hymenaea verrucosa*) started even before the arrival of the first

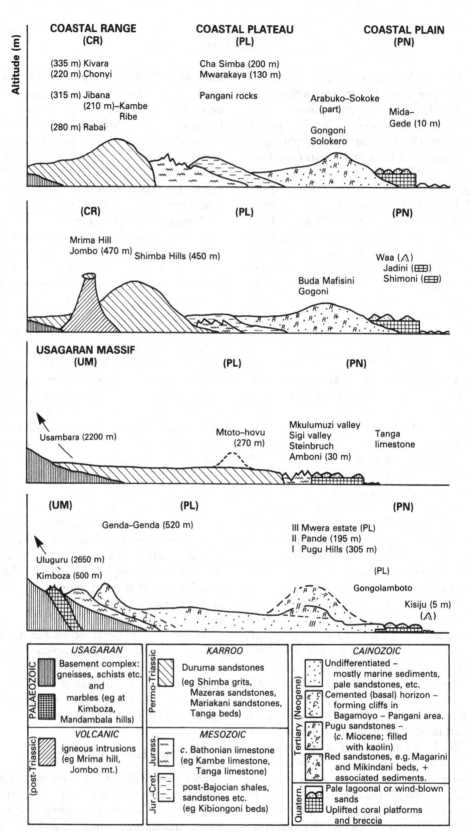

Figure 5.3. Idealised east–west sections showing basic pattern of coastal geology. The horizontal axis is not to scale.

Arab town-builders of *c.* AD 900, and involved the ancient Egyptian and Phoenician economies. The Swahili and Mijikenda were still actively trading in a variety of forest products in the 19th century (see Prins, 1952, 1972; Cashmere, 1961; Parkin, 1972; Salim, 1973; Spear, 1978). Gede forest, which stands over the 16th century ruins of an Arab sea-front town, is a striking symbol of the changing fortunes of coastal populations and of coastal forests (Gerhardt & Steiner, 1986). Hartnoll & Fuggles-Couchman (1937) commented on the influence of shifting cultivation on vegetation in the Pugu Hills, and Hunt *et al.* (1981) have summarised the wide-ranging influences of the Mijikenda on the *kaya* forests. Indeed, there can be very few areas of coastal forest that have not been influenced by human activity at some time.

Some coastal forest plants, like the baobab (*Adansonia digitata*) and *Synadenium pereskiifolium*, are planted in sacred places and on graves within forests and survive into closed-canopy forest (see also Vabmah & Vaid, 1978). Because of the long history of human influence, the origins of some plants in some forests are open to speculation. It is probable that *Mkilua fragrans* has had its natural range extended by humans, for it was previously traded by the coastal tribes for its strong perfume, like lemon-scented soap. The flamboyant shrub *Hibiscus schizopetalus* may be restricted naturally to the rocky streams of forests near the ancient trade centre of Mombasa and possibly Lindi, in which case it has colonised the tropics from its coastal forest base with human help (see Kirk & Oliver, 1876; Wild (in Brenan *et al.* 1960–83), 1965; Waalkes, 1966; Kachecheba, 1972). However, it is more likely that *H. schizopetalus* is a long-standing garden escape into, not out of, the coastal forests, because it has no less tenuous a foothold there than elsewhere. *Thunbergia kirkii* seems a better example of a coastal forest shrub species taken elsewhere as an ornamental. Most readers will be familiar with the most widely dispersed plant of the coastal forests – the African violet (*Saintpaulia ionantha*).

Many patches of forest north of the Pangani River enclosed *kaya* or *kaya*-like clearings once. These include the Shimba Hills and areas within the Usambara foothills, but especially nowadays in the Rabai–Chonyi hills (see Spear, 1978; Hunt *et al.*, 1981; Robertson, 1987). *Kaya* is the Mijikenda word for their traditional home settlement, a distinctive arrangement of grass huts in a stockaded forest clearing with two opposing gateways. *Kaya* forests are the home of ancestral spirits and some are still used as burial groves. In 1981 only two of the Rabai–Chonyi *kaya* forests (Kambe and Rabai) were inhabited, although others are still managed respectfully by the Mijikenda and they remain the focal points of clans within the Mijikenda. Without their utilisation or sanctity it is likely that few such forests would have survived. Not all sacred forests are *kaya* forests, however. The Mijikenda are not known to have spread south of the Pangani River in Tanzania where some forests, like Genda-Genda South and Gongolamboto burial grove, have local sanctity.

In recent times some forests, especially the larger ones like the Arabuko–Sokoke, Msumbugwe and Pugu Hills forests, have been gazetted as Forest Reserves and systematically exploited for timber, charcoal or fuelwood. Although management has not placed a great priority on maintenance of the diversity of wildlife, there is still at least some forest standing in these areas. On the other hand, the Shimba Hills reserve and Gede Forest are nowadays managed partly to encourage wildlife and tourism.

Increasing demand for land and wood and growing disrespect for traditional values are paralleled by increasing erosion of the forest edge, even in sacred forests. Aerial photographs taken over the last 30 years invariably show that forests have declined in area (Hunt *et al.*, 1981). Moomaw (1960) estimated that less than 10% of potentially suitable areas were forested with the *Sterculia–Chlorophora*[1] forest type. Large areas of forest must have been burnt and lost throughout history to clear land for hunting and agriculture.

---

[1] *Chlorophora excelsa* is now *Milicia excelsa.* Moomaw's forest type will, however, be referred to as *Sterculia–Chlorophora* forest here to distinguish it from any more specific *Sterculia–Milicia* association.

Burning of coastal vegetation has been an annual event at least since the last century (New, 1873). Humans are the main causes of fire, formerly for hunting but nowadays for farm clearance. Where forest is not destroyed entirely by fire it is often influenced by it. In Pande and Msumbugwe Forest Reserves, for instance, regular fires have produced a broad, highly convoluted border zone of vegetation with both forest and fire-tolerant species in close association. Forest-like and thicket vegetation occur in small clumps in the surrounding grassland, often on termite hills. The coastal plateau in these areas is dominated by seasonally waterlogged black cotton soils. It is likely that termites, drainage, fire and the prevailing winds interact to produce the peacock feather patterns in these 'forest' and thicket clumps (Hesse, 1955; Glover, Trump & Wateridge, 1964).

Human activity of one type or another has therefore been a pronounced and long-standing influence on the forests. Because of the small size of most forests and consequent high perimeter length relative to total area, disturbance, even if restricted to areas near the forest edge, influences strongly their species composition. Many remaining forest patches are, in effect, all edge. Internal and edge disturbance makes the forest richer in fire-tolerant, pioneer and other sun-loving species many of which are more typical of the vegetation surrounding the forest. This creates difficulty in defining a strictly forest flora: the coastal forests may appear as islands of vegetation but they are not isolated completely from their matrix.

Floristic patterns

Patterns in the coastal forest flora were examined with a sample of species gleaned from field visits, herbaria and the literature (Hawthorne, 1984). From a complete and indiscriminate list of 841 species, half were excluded as they are non-forest species, restricted to heavily disturbed forest or forest edge. The excluded group is nevertheless a common component of many coastal forests for reasons outlined above. Species of riverine or swamp forest, or species found growing over rocks in or amongst forest, were kept in the sample, but specialised mangrove species and weedy species sometimes found on rocks were excluded. This sample has been updated (and the categories simplified): the 483 included taxa and a representative sample of 73 excluded species are listed in Appendix 5.1.

The included species have been classified according to their distribution within the coastal forests and secondarily from records of the ecology of the species elsewhere in the Zanzibar–Inhambane regional mosaic into five major ecogeographical elements. Some of these elements are subdivided into more precise categories. Additionally, the global distribution of each species is summarised in crude chorological categories. The species in these ecogeographical and chorological categories are listed in Appendix 5.1 and summarised in Table 5.1. Fifty-two infraspecific taxa (some of which would have specific rank in the hands of different taxonomists) and a few unnamed, but distinctive plants have been included, making a total sample of 483 plant taxa.

Chorological categories

Chorological categories have been determined from floras and monographs. Species, or infraspecific taxa, are assigned to whichever of the following categories fits the distribution best.

1. **Coastal**: endemic to the Indian Ocean coastal belt (see Moll & White, 1978), or extending (e.g. up rivers) only a short distance from it. As well as the Zanzibar–Inhambane regional mosaic, this zone covers the Tongaland–Pondoland Regional Mosaic further south.
2. **Eastern**: extending, for instance, along rivers to the Lake Victoria Regional Mosaic; elsewhere in eastern central Africa, e.g. Malawi (see Dowsett-Lemaire, 1990), often in or near the Rift Valley; or species in similar, scattered localities not qualifying as 4–6.
3. **Oceanic**: as 1 or 2 but found also on Indian Ocean islands; including Madagascar, but excluding offshore islands like Zanzibar (see Vaughan & Wiehe, 1937; Renvoize, 1976, 1979).

Table 5.1. *Summary of ecogeographical (rows) and chorological (columns) categories. In each category, No. spp. represents numbers at species level and No. vars., if any, are numbers of taxa belonging to that category only at the level of subspecies or variety. All percentages are of row totals, and combine species and infraspecific taxa. Ecogeographic categories with the greatest proportion of coastal endemics are listed higher in the table. Chorological categories are listed with the best represented categories on the left. Categories are explained in the text*

Category	Coastal			Afrotropic			Eastern			Zambezian			Oceanic			Tropical			?			Total	
	No. spp.	No. var.	%	No. spp.	No. var.	%	No. spp.	No. var.	%	No. spp.	No. var.	%	No. spp.	No. var.	%	No. spp.	No. var.	%	No. spp.	No. var.	%	No. spp.	No. var.
Northern (N)	33	15	96	1		2	0			0			1			0			0			37	15
Southern (S)	36	7	91	0			3		6	1			0			0			0			40	7
Dry forest (D)	46	11	50	22		19	11		10	9		8	8		7	3			3			102	11
Moist-high (MH)	15	1	50	11		33	5		15	0			0			0			0			31	1
Dry-maritime-riverine (LD)	30	4	43	12		15	11	1	15	6		8	5	1	8	5		6	4			73	6
Moist-maritime-riverine (LM)	13		31	12		29	6	3	21	2		5	1		2	2		5	3			39	3
Rocks/rivers (LR)	11	2	28	14		30	4	1	11	4		8	4		8	3		6	4			44	3
Moist forest (M)	17	2	27	29	2	44	7	1	11	2		3	2		3	7		10	2			65	5
Total	203	42	51	100	3	21	47	6	11	24	0	5	21	1	5	20	0	4	16	0	3	431	52

4. **Zambezian**: widespread in Zambezian region, but not in Sudanian, Somali–Masai nor (widespread in) Guineo–Congolian region.

5. **Guineo–Congolian**: common in at least Congolian region, extending sometimes into part of adjacent regions (see White, 1979). Including Guineo–Congolian species extending beyond Africa to Madagascar, but no further. (In the summary in Table 5.1, category 5 has been joined with 6, under 'Afrotropical'.)

6. **Wide African**: widespread, but not 1–5. Including Afromontane species and savanna species from north and south of the Equator, but excluding species well established in the Guineo–Congolian region. (Joined with 5 in Table 5.1.)

7. **Tropical**: species found beyond Indian Ocean islands (even if only to India).

8. **Unknown**: especially for genera needing revision, although many of these have been excluded from the sample.

There are several borderline cases, for instance between 1 and 2. The aim of these categories is to highlight only the strongest of trends in the sample, and the main trends are not altered significantly by redesignation of borderline cases.

Geographical elements

Although most species are widespread in the coastal forests within the broad environmental constraints outlined above, some species are limited to a small range of neighbouring coastal forests even though apparently suitable habitats occur outside their range. Ninety-nine taxa of restricted distribution within the coast have been put in the Northern (52) and Southern (47) geographical elements. Northern and Southern elements occur in the coast only north and south of the Msangasi River. Most species (91/99) in the geographical elements are coastal endemics. Exceptions include the Northern species *Dichapetalum fructuosum* and the Southern species *Millettia eetveldeana* and *Diospyros verrucosa*. Several Southern species, like *Drypetes arguta*, have a fairly wide range extending southwards.

Pachystela msolo could have qualified as a Northern species but it seems so widespread elsewhere, and so rare in the coast, that this status probably reflects insufficient sampling. For similar reasons *Cnestis corniculata* (previously *C. confertiflora*) does not quite qualify as a Southern species, in spite of not having been recorded in the Northern forests.

Membership of a geographical element takes precedence over membership of an ecological element. The three ecological elements are the Dry forest, Moist forest and Maritime–Riverine elements. Species in the ecological elements occur north and south of the Msangasi River.

Ecological elements

The definition of ecological elements goes hand in hand with definition of forest types. Species in the Moist forest element (labelled M in Appendix 5.1) are restricted to areas where the forest, if undisturbed, is of a Moist type. More specialised Moist forest species that occur in higher altitude or higher rainfall areas are distinguished in a separate category (MH).

Species of the Dry forest element (D) are characteristic of a wide range of Dry forest types but some are also common in Moist forest, or in thickets or woodland. This element is therefore well represented in all coastal forests.

Maritime–Riverine species favour, within the coast at least, riversides, rocky, saline or base-rich soils, termite hills or other edaphically more extreme sites. Most Maritime–Riverine species favour various combinations of such habitats. Many categories of the Maritime–Riverine element could be defined (Hawthorne, 1984), but here they are simplified to just three:

1. Maritime–Riverine species also characteristic of various types of Moist forest (LM in Appendix 5.1);
2. Maritime–Riverine species that occur often in Dry forest (LD);
3. the remainder, which are more specialised and found mostly on stream banks or in very rocky areas (LR).

It should be emphasised that, within the coastal forests, species in the first two Maritime–Riverine categories are not confined to swamps, river banks and similarly extreme habitats. Their general preference for such habitats throughout the coast and beyond, however, points to some niche specialisation that distinguishes them from species in the other ecological elements. Many species in the Maritime–Riverine element, e.g. *Parkia filicoidea*, are widespread outside the Zanzibar–Inhambane regional mosaic along rivers, but often in a wide range of other vegetation types (see Hopkins & White, 1984).

Figure 5.4 shows the representation of different elements in some coastal forest samples. These 'ecogeographic spectra' indicate the relative importance of the various ecogeographic categories within samples of coastal forest. Samples were plotless, and defined by position in the landscape. For instance, samples might be of all species recorded in forest along a particular river or stream, or all species on a certain north-facing slope. Most samples include 40 or more species. Within these samples, all species were scored 1–3 depending on their abundance (most scored 1). Each species' score is expressed as a percentage of the total for all species in the sample, then the scores are added for all species in each ecogeographic category.

The variation in these spectra over many samples of coastal forest expresses in a simplified and meaningful way the same patterns that are revealed by multivariate analyses, for instance Detrended Correspondence Analysis (Hawthorne, 1984). Analysis of a sample of forests (unfortunately without comparable samples from the south Kenyan coast) showed that variation along a north–south axis is most significant, with subsidiary variation according to position in landscape.

Further details on the various ecogeographical elements will be given after a summary of forest types.

Types of forest

In the classifications of Kenyan coastal forests by Dale (1939) and Moomaw (1960) large areas of forest are defined under one forest type, despite great variation over short distances. Informal,

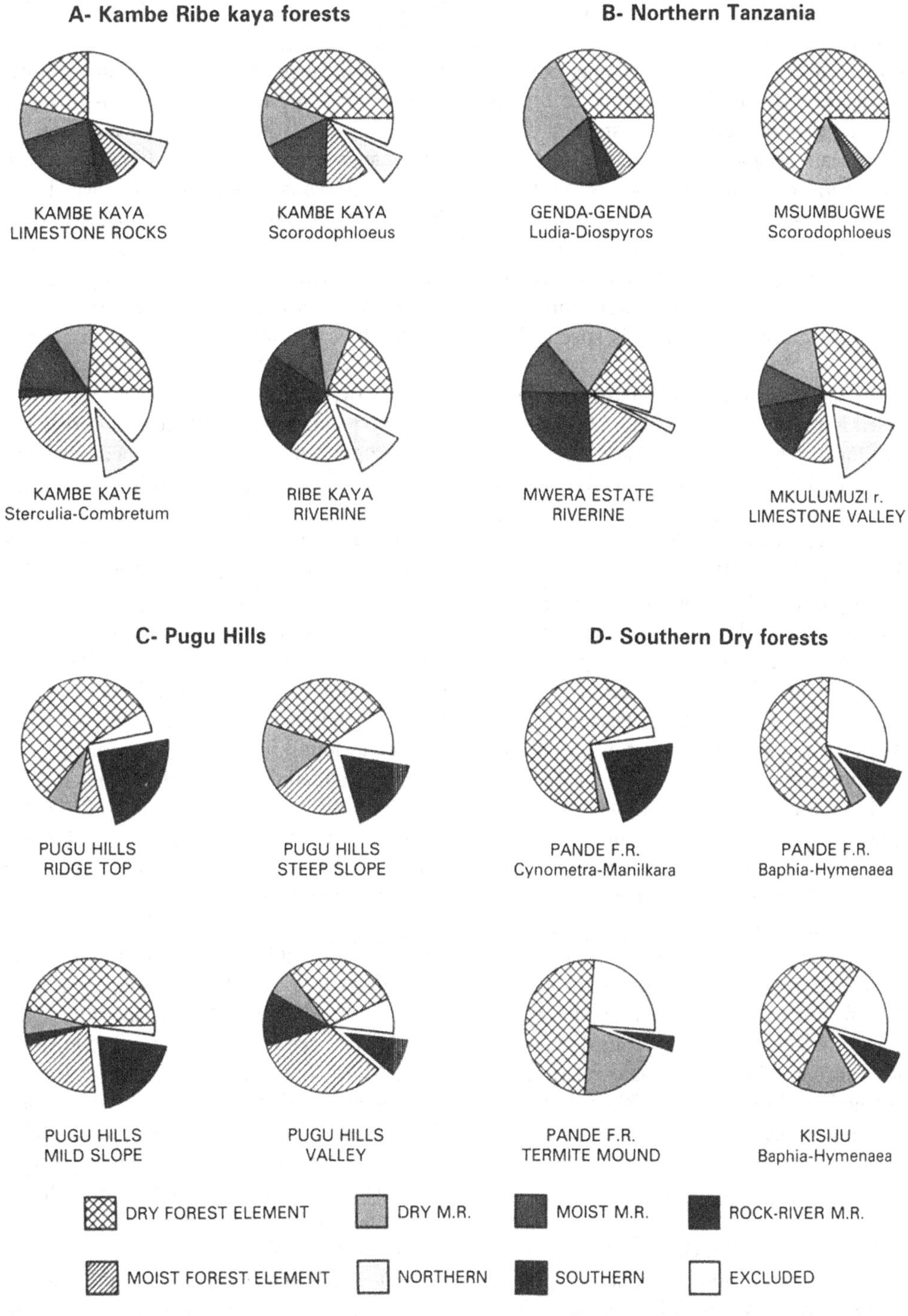

A- Kambe Ribe kaya forests

KAMBE KAYA
LIMESTONE ROCKS

KAMBE KAYA
Scorodophloeus

KAMBE KAYE
Sterculia-Combretum

RIBE KAYA
RIVERINE

B- Northern Tanzania

GENDA-GENDA
Ludia-Diospyros

MSUMBUGWE
Scorodophloeus

MWERA ESTATE
RIVERINE

MKULUMUZI r.
LIMESTONE VALLEY

C- Pugu Hills

PUGU HILLS
RIDGE TOP

PUGU HILLS
STEEP SLOPE

PUGU HILLS
MILD SLOPE

PUGU HILLS
VALLEY

D- Southern Dry forests

PANDE F.R.
Cynometra-Manilkara

PANDE F.R.
Baphia-Hymenaea

PANDE F.R.
TERMITE MOUND

KISIJU
Baphia-Hymenaea

 DRY FOREST ELEMENT   DRY M.R.   MOIST M.R.   ROCK-RIVER M.R.

 MOIST FOREST ELEMENT   NORTHERN   SOUTHERN   EXCLUDED

Figure 5.4. 'Ecogeographic spectra' showing the representation of ecogeographical categories in 16 samples of coastal forest. Representation is calculated as the sum of species ($\times 2$
or 3 if common) in each category, expressed as a percentage of the total scores. Geographical elements are cut. F.R., Forest Reserve; M.R., Maritime–Riverine element.

nodal nomenclature (Poore, 1962) allows for continual variation in, and incomplete knowledge of, vegetation. Convenient noda in a spectrum of communities are defined by reference to one or more of the commonest canopy species. This terminology can be applied to patches of forest from a few hundred square metres to several tens of hectares and is ideal for describing the coastal forests. The floras of coastal communities generally correlate with their dominant species, so the noda provide a flexible and useful shorthand, especially if geographical position is specified as well.

Variation in physiognomy of coastal forest is emphasised here by two contrasting broad designations of forest type – Dry forest and Moist forest – to parallel the definition of ecological elements outlined above. These generic terms help impose some order on the otherwise rather anarchic nodal names. Riverine forest can also often be described as Moist or Dry, depending on the dominant trees. However, especially where the gradation in the vegetation is steep or complex, it is often best left unspecified as 'riverine forest'. Although physiognomic characters are implied in the designation of Moist or Dry forest, species composition alone is a more reliable and practical basis for any formal summary of forest types. However, as the forest becomes Moister there is less of a tendency towards local dominance and choice of a reference species becomes more arbitrary (*Sterculia appendiculata* is convenient even in very mixed communities because of its conspicuous, smooth yellow bole). Therefore, for the Moistest coastal forests, 'mixed Moist forest' or simply Lowland Rain Forest are convenient terms. The approximate relationship of various noda to Moomaw's (1960) and White's (1983) terminologies is shown in Table 5.2.

Moist forest

Moister forest types have a canopy of 20–35 m in less disturbed patches and generally larger and less sclerophyllous leaves than Dry forests. Moist forests occur on scattered coastal hills or in other areas where precipitation is high, or where groundwater and nutrients are abundant.

Remaining patches are found in the valley bottoms or less steep slopes of the Pugu Hills, along the coastal range in Kenya, at Witu in northern Kenya and along the valley bottoms of the limestone forests of the Mkulumuzi and Sigi rivers. Moomaw noted that *Sterculia–Chlorophora* (Moist) forest requires an annual rainfall of over 1000 mm (40 inches) precipitation, but it may also occur in drier areas with compensating factors, such as a high water table.

Transects through Moist coastal forest in the Pugu Hills are shown in Figures 5.5 and 5.6. Figure 5.7 shows the variation over a short distance between Moist and Drier forest.

Most patches of Moist coastal forest have a high proportion of canopy tree species which lose their leaves simultaneously during the dry season (most emergents, but fewer lower canopy species are deciduous). Epiphytes are abundant only by rivers, and perhaps locally on some of the higher hills. The range of physiognomy of Moist forest is comparable to the drier extremes of White's Guineo–Congolian rain forest. The wetter extremes of Guineo–Congolian rain forest are not even remotely matched by any coastal forest.

Species found in a broad range of the Moister forest types include the tall, deciduous trees *Antiaris toxicaria*, *Milicia excelsa* and *Ricinodendron heudelotii*, and the grass *Olyra latifolia*. These constitute a catholic category within the Moist forest element, comprising 70 species in the current sample (14%). Almost half of these Moist forest species (see Table 5.1) are common in forest throughout tropical Africa, especially in secondary forest and forest defined in other contexts (e.g. Ghana: see Hall & Swaine, 1981) as 'Dry'.

All Moist forests are rich in Moist forest element species. The catholic Dry forest element also makes an important contribution to Moist coastal forests. However, Moist forests vary considerably in the representation of the Maritime–Riverine element (see Figure 5.4).

Moomaw and Dale distinguish between *Sterculia–Chlorophora* rain forest and *Combretum schumannii–Cassipourea euryoides* Dry forest. However, Moist forest intermediate between these extremes is common along the coast, particularly over rocky, shallow (rendzina) soils in

Table 5.2. *The spectrum of coastal forest types. Empty boxes reflect areas where the correspondence between columns is in doubt*

Moomaw (1960)		White (1983)	Corresponding 'noda'	Comments/examples
Lowland evergreen dry forest	*Manilkara–Diospyros*	Z–I scrub forest	*Manilkara–Diospyros*	N. Kenya coast: Boni
			Ludia–Diospyros greenwayi	Genda-Genda South FR: parts of slope
	Cynometra–Manilkara	Z–I undifferentiated forest	Northern or Southern *Cynometra–Manilkara* *Scorodophloeus* *Julbernardia*	Parts of Arabuko–Sokoke, Msumbugwe (Northern); Pande, Pugu Hills (Southern). Local dominance of certain other spp. can be included or treated separately
Lowland woodland	*Brachystegia–Afzelia*	Z–I transition woodland	*Brachystegia–Afzelia*	Mida-Gede; parts of Arabuko–Sokoke
			Brachystegia–Baphia kirkii	(Seral) Woodland at Gongolamboto: remnant of Mogo FR
			Baphia kirkii Hymenaea	Dry forest at Kisiju and Pande
			Dialium holtzii–(Baphia)	Moist forest Pugu Hills (+ Kambe *kaya*); mild slopes
			Antiaris–Milicia	Moist forest Pugu Hills valley basin
Lowland dry forest on coral rag	*Combretum–Cassipourea*	Z–I undifferentiated forest	*Combretum–Cassipourea*	Kenyan coral rag (e.g. Shimoni)
			Combretum (Sterculia)	Spectrum from Drier types (Genda-Genda; parts Kaya Jibana) . . . to Coastal rain forest (e.g. parts of Mrima Hill)
			Combretum–Sterculia	
Lowland rain forest	*Sterculia Chlorophora*		Mixed (*Sterculia–Milicia,* etc.)	
		Z–I lowland rain forest	(Moist–High spp.)	

moister areas. The best developed areas of *Combretum–Sterculia* forest are associated with the limestones, both Jurassic limestones and recent coral, of Kenya and northern Tanzania. Drier extremes become increasingly rich in *Combretum schumannii*, with almost pure stands in parts of the *kaya* forests and on Genda-Genda Hill, on well-drained soils. *Cassipourea euryoides* is not common

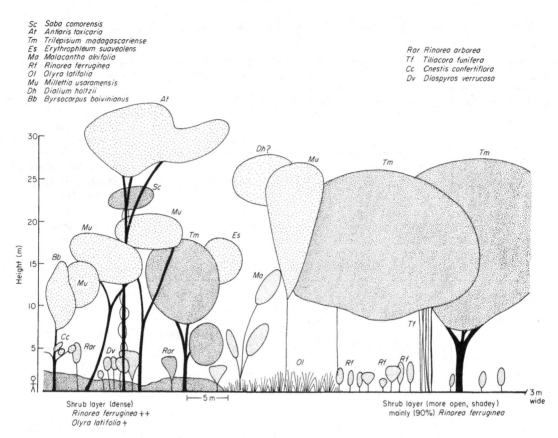

Sc Saba comorensis
At Antiaris toxicaria
Tm Trilepisium madagascariense
Es Erythrophleum suaveolens
Ma Malacantha alnifolia
Rf Rinorea ferruginea
Ol Olyra latifolia
Mu Millettia usaramensis
Dh Dialium holtzii
Bb Byrsocarpus boivinianus

Rar Rinorea arborea
Tf   Tiliacora funifera
Cc   Cnestis confertiflora
Dv   Diospyros verrucosa

Shrub layer (dense)
Rinorea ferruginea ++
Olyra latifolia +

Shrub layer (more open, shadey)
mainly (90%) *Rinorea ferruginea*

Figure 5.5. Transect through Moist *Antiaris–Milicia* forest in the Pugu Hills Forest Reserve. + = common, + + = very common in understorey.

in many areas where the other recorded associates are, and historical factors may be responsible for the local abundance over the coral rag north of Mombasa.

Combretum–Sterculia forest is not well developed in the Tanzanian coastal forests south of the Msangasi River (and north of the Rufiji). This may be attributable to the relative scarcity of certain sedimentary rocks, like raised coral and limestones, or to a less even distribution of rain throughout the year. Wherever undisturbed vegetation does occur on coral, for example on Mbudya Island (Hall *et al.*, 1984) and on Mafia (Greenway *et al.*, 1988) it is generally scrub forest or thicket.

Many species characteristic of the *Combretum–Sterculia* forests in Kenya, including trees like *Lecaniodiscus fraxinifolius* subsp. *vaughanii*, are grouped in the Moist category of the Maritime–

Riverine element. Although these species are inconspicuous in the Moist forest of the Pugu Hills, they are abundant in Kimboza Forest, over basement complex rocks to the west of the Pugu Hills (Rodgers *et al.*, 1983). The Dry Maritime–Riverine element (e.g. *Uvariodendron kirkii*, *Cordia goetzei* and the *Combretum schumannii* already mentioned) is often well represented in Moist forests, but this category is increasingly important in drier conditions, and is especially important on drier parts of the coral rag (see Birch, 1963).

Moist coastal forests along rivers and in swampy or rocky areas include many Maritime–Riverine species typical of Moist limestone forests (e.g. *Trichilia emetica* and *Lecaniodiscus fraxinifolius*), alongside the more specialised riverine species (e.g. *Barringtonia racemosa* and *Mascarenhasia arborescens*).

Kimboza, and some of the other Moister

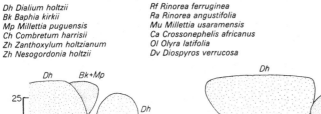

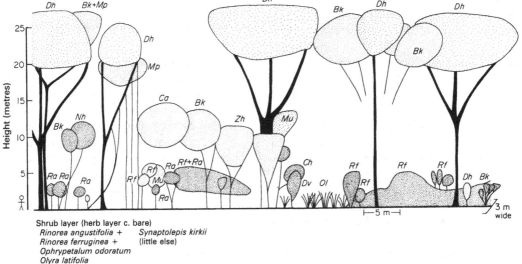

Figure 5.6. Transect through Moist *Dialium–Baphia* forest in the Pugu Hills Forest Reserve. *Diospyros* sp. aff. *verrucosa* is now *D. capricornuta*.

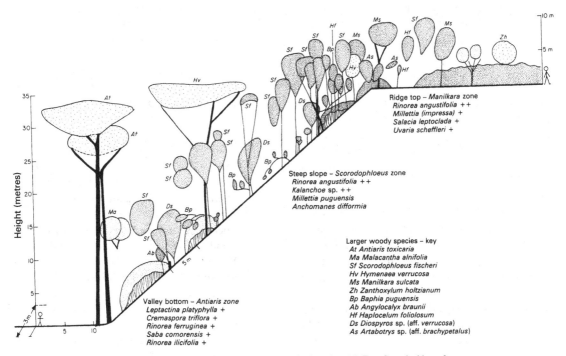

Figure 5.7. Transect from *Manilkara* ridge top thicket, through Dry *Scorodophloeus* forest to *Antiaris–Milicia* forest in the valley bottom. Pugu Hills Forest Reserve.

coastal forests are distinguished also by the Moist–High (MH in Appendix 5.1) forest species within the Moist forest element. Moist–High species such as *Uvariodendron gorgonis* are restricted to the few higher altitude, or higher precipitation coastal forests like Mrima Hill and the Shimba Hills. However, most are widespread at similar or higher altitudes in the forests of the basement complex mountains like the Usambaras and Ulugurus to the west, a fact noted by Verdcourt (1980). These Moister coastal forests, therefore, are nearly a microcosm of the coastal forest flora, with representatives of a wide range of ecological categories separated incompletely by habitat specialisation. Inevitably, though, these mixed, Moist forests are deficient in certain geographical elements.

Although the less specialised Moist forest species tend to be widespread in Africa, about half of the Moist–High species in the coastal forests are coastal endemics, or almost so (Table 5.1). Maritime–Riverine species often have an oceanic distribution, but this is less marked than might be anticipated. The tendency for such species to be found around the Rift Valley, in Malawi and elsewhere in 'Eastern' Africa is more marked.

Dry forest

Whereas Moist forests are of sporadic occurrence along the coast, it is likely that Dry forest types previously occurred in extensive blocks. Undisturbed stands of Dry forest generally have a dense canopy usually less than 20 m tall. Moomaw (1960) notes an association between low phosphate and other nutrient levels in the Dry evergreen forest of the Arabuko–Sokoke Forest Reserve (a factor associated with dry forests elsewhere: e.g. Beadle, 1962, 1966). Water supply is also relevant and *Cynometra webberi*, for instance, a typical Dry forest tree, is favoured by soils of impeded drainage, with an unpredictable water supply in the Tana delta (Andrews *et al.*, 1975). Local dominance of one or two canopy species is common in undisturbed stands of Dry forest (see Figures 5.8 and 5.9).

The 32 Dry forest element canopy trees are varied in ecology, and can be divided into four

types, described below. Members of the first two types are often locally dominant, and often characterise types of forest. Members of the second two types are much less gregarious, but are sometimes the most important trees in secondary forest.

1. Evergreen trees of the Caesalpiniaceae, especially *Scorodophloeus fischeri* and *Cynometra webberi* are typical dominants of some types of Dry forest, often with *Manilkara sulcatas*. These evergreen species are often broadly associated with each other, and are ecologically similar, so it is convenient to extend Moomaw's *Cynometra–Manilkara* forest (below) to include patches dominated by *Scorodophloeus fischeri*. *Julbernardia magnistipulata* and *Hymenaea verrucosa* are also in some areas locally dominant, but elsewhere occur in savanna. These caesalpinioid dominants are often to be found with many of their saplings tolerating the parental shade. *Paramacrolobium coeruleum* seems similar to these species but is rarer, and seems restricted to clumps in savanna near Moist forest.

2. *Brachylaena huillensis* and *Manilkara discolor* are more restricted to Dry forest than type 1. *Brachylaena* has been removed from many areas for timber, but both types can otherwise be locally very abundant.

3. Where clearings are made, deciduous species like *Erythrina sacleuxii* and *Sterculia schliebenii* make more rapid growth and often thrive as emergents above the evergreen canopy.

4. Species more typical of savanna or open woodland, like the baobab *Adansonia* (see Wickens, 1982) and *Margaritaria discoideus*, are nevertheless often encountered as healthy trees in coastal forests, especially drier types. These constitute about half of the Dry forest element trees (DX in Appendix 5.1). Whereas type 4 species regenerate only in medium to large gaps, or savanna, type 3 species are typical

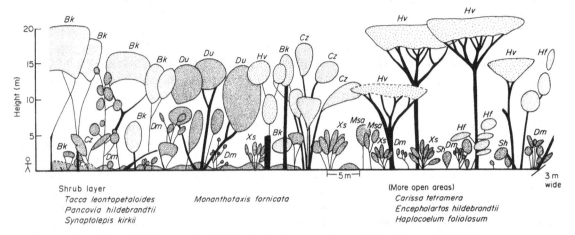

Bk Baphia kirkii
*Cz Craibia zimmermannii**
Hv Hymenaea verrucosa
Du Dracaena usambarensis
Hf Holarrhena pubescens
Dm Diospyros mafiensis
*Xs Xylopia sp.**
Msa Manilkara sansibarensis
Sh Strychnos henningsii

Shrub layer
   Tacca leontopetaloides   *Monanthotaxis fornicata*
   Pancovia hildebrandtii
   Synaptolepis kirkii

(More open areas)
   Carissa tetramera
   Encephalartos hildebrandtii
   Haplocoelum foliolosum

Figure 5.8. Transect through Dry *Baphia–Hymenaea* forest at Kisiju. *Xylopia* sp.* is *Xylopia* sp. B in FTEA. *Craibia zimmermannii** is possibly an unnamed subspecies of this species.

Jm Julbernardia magnistipulata
Ts Tessmannia martiniana
Hv Hymenaea verrucosa
Ti Tamarindus indica
*Xs Xylopia sp.**
Md Manilkara discolor
*Cz Craibia zimmermannii**
Mm Monodora minor
Hz Hunteria zeylanica
Da Drypetes arguta
Hfo Haplocoelum foliolosum

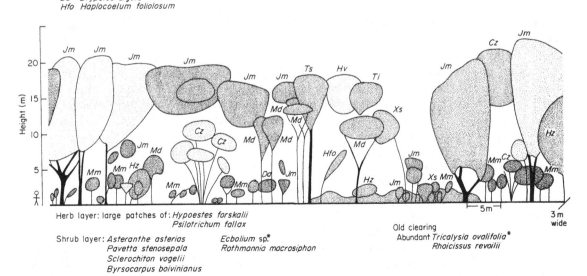

Herb layer: large patches of: *Hypoestes forskalii*
                     Psilotrichum fallax

Shrub layer: *Asteranthe asterias*   *Ecbolium sp.**
           Pavetta stenosepala   *Rothmannia macrosiphon*
           Sclerochiton vogelii
           Byrsocarpus boivinianus

Old clearing
   Abundant *Tricalysia ovalifolia**
         Rhoicissus revoilii

Figure 5.9. Transect through Dry evergreen forest dominated by *Julbernardia magnistipulata* in Pande Forest Reserve.

pioneers of small to medium gaps in forest
(and are probably less fire-tolerant),
although this distinction is not always easy
in practice.

Moomaw distinguishes three types of Dry coastal
forest in Kenya, including the *Combretum–Cassi-pourea* forest on coral rag mentioned under
Moist forest. The most widespread type, if inter-
preted broadly (to include *Scorodophloeus* forest,
often found on slopes) is the *Cynometra–Manilkara* Dry evergreen forest, although areas
dominated by *Cynometra webberi* and/or *Manilkara
sulcata* themselves are much less extensive.
Cynometra–Manilkara forest is usually associated
with hills or gentle undulations typical of the
coastal plateau, usually with reddish soils. Parts of
the Arabuko–Sokoke Forest are of this type, as
are parts of Msumbugwe, Pande and Pugu forests
in Tanzania. These forests are normally dominated by the Dry forest element, with a lesser contribution from the Maritime–Riverine element
than the other Dry forest types. *Cynometra–Manilkara* forests around the Pugu Hills and
Pande include several Southern species, possibly
including the locally dominant *Cynometra* cf. *web-beri*. Type 1 and type 2 trees other than *Cynometra*
and *Manilkara sulcata* are often found in small
patches associated with this forest type.

Dry forest trees of types 1 and 3 are often
found in Moist forests. Whereas type 3 trees are
pioneers in Moist forest, type 1 trees are
sometimes locally dominant in Moist forest on
steep slopes or rocky soils, or are common as
lower storey trees. In the former case, the pockets
of apparently Dry forest often have a flora rich in
members of the Moist forest element.

As is true in Moist forests, Dry forests are often
distinctively 'flavoured' by members of the
Maritime–Riverine element. A spectrum of communities has already been mentioned, from Moist
Combretum–Sterculia forest to Dry *Combretum–Cassipourea* forest on coral rag (to coral rag thicket
and other communities close to the sea). The
drier extremes of this progression are dominated
by Dry Maritime–Riverine species.

Moomaw's *Manilkara–Diospyros* evergreen forest, called scrub forest by White (1983), is not
widespread. It is associated with regions of lower
rainfall, particularly on the northern Kenyan
coast, in areas of impeded drainage or high cal-
cium–magnesium concentrations (Moomaw,
1960). In Tanzania *Ludia–Diospyros greenwayi* forest on Genda-Genda Hill shows some affinities
with Moomaw's *Manilkara–Diospyros* forest,
especially the low sclerophyllous-leaved canopy
and commonness of species of the Dry Maritime–
Riverine category, like *Macphersonia hildebrandtii*.
These forests are probably associated with similar
soils, perhaps with better drainage or water availability in the *Ludia* type.

Moomaw also describes *Brachystegia–Afzelia*
woodland that is in some areas successional but
elsewhere, particularly on infertile sands or
shales, is more stable (White, 1983). In Tanzania,
around the 'sachsenwald' (on the coastal plateau
between Dar es Salaam and the Pugu Hills) of
Engler (1910), various forested phases related to
this woodland can be recognised. For example a
Baphia kirkii–Hymenaea forest at Kisiju (and on
Mafia Island) and Pande, and a *Dialium–Baphia
kirkii* forest on milder slopes within the Pugu
Hills. The Pugu Hills *Dialium–Baphia* community
can be defined as Moist forest because of a significant presence of the Moist forest element, but
this is less conspicuous in the *Baphia–Hymenaea*
forest, which is a type of Dry forest. Many species
in these forests, including *Dialium holtzii*, *Dracaena usambarensis*, *Croton jatrophoides* and *Baphia
kirkii* have a Maritime–Riverine ecology, although
Hymenaea is a more catholic Dry forest species.
Ellipanthus hemandradenoides and certain other
uncommon species are recorded from coastal
vegetation over white sands even in the north and
perhaps have an equivalent Maritime–Riverine
ecology. However, several such species have been
excluded from the sample either because they are
(at least nowadays) non-forest species or because
too little is known. The *Dialium–Baphia* and
Baphia–Hymenaea forests include many Southern
species, and for this reason at least are not
matched by any Kenyan forests.

Eighty-two small tree, shrub, herb and liane
species are in the Dry forest element. These
include *Asteranthe asterias*, the cycad *Encephalartos
hildebrandtii* and the liane *Strychnos mada-*

gascariense. These non-tree species are often found in thickets where the canopy trees mentioned above are inconspicuous or absent. However, savanna shrubs (analogous to type 4 trees, above) are not included if, as is usually true, they die out very rapidly in closed forest. Several Dry forest species including *Asteranthe asterias*, *Clerodendrum capitatum* and *Strychnos henningsii* are quite variable in appearance, sometimes in a way apparently correlated with habitat but often without clearly defined subspecies or varieties. Some Dry element species, such as *Tricalysia ovalifolia*, have a more discontinuous variation pattern. Six infraspecific variants of Dry forest species are included the geographical elements.

The Dry coastal forests are more difficult to compare with forests outside the Zanzibar–Inhambane regional mosaic than is the case with the Moist forests. In the Guineo–Congolian region and its transition zones small patches of structurally similar forest of limited extent are known. *Cynometra megalophylla–Manilkara obovata* Dry evergreen coastal forest is found in the Sudanian transition zone of the coastal plains of Ghana (Hall & Swaine, 1981). Close to the Volta lake, *Talbotiella gentii*-dominated forest (see Swaine & Hall, 1981) is similar in physiognomy to caesalpinioid-dominated East African coastal forest. Indeed *T. gentii* is strikingly similar in its vegetative appearance to *Cynometra webberi*. It is possible that other isolated pockets of Sudanian dry forest may be similar to coastal forest as well, but overlap in species composition is negligible. Most Zambezian Dry evergreen forest has been destroyed by cultivation and fire (White, 1983) and it is not clear how similar in structure these would be to the coastal forests.

Closer to the coast, *Brachylaena huillensis*, *Afzelia quanzensis*, *Scorodophloeus fischeri* and *Manilkara sulcata* are found in Somali–Masai scrub forest (White, 1983), and some species of the Dry Maritime–Riverine category occur there along rivers (*Tamarindus indica*, *Acacia robusta*). Half of the Dry forest element are coastal endemics, and a further 10% are Eastern species. No other ecological category of forest species has such a high proportion of coastal endemics except the 'Moist–High' category. The percentage of endemic Dry forest species is even greater if one excludes trees of more disturbed forest (especially type 4, above), most of which are found in the Zambezian region.

Dry Maritime–Riverine species also show a greater tendency to be coastal endemics than Moist Maritime–Riverine species, although the former more often have an oceanic distribution.

Many species that occur in a variety of coastal communities are very limited in their association elsewhere on the continent, and vice versa. For example, *Asteranthe asterias* and *Scorodophloeus fischeri* are common in most coastal forests, yet *Asteranthe* is a genus and *S. fischeri* is a species endemic to the Zanzibar–Inhambane regional mosaic. Yet the trees *Celtis mildbraedii* and *Paramacrolobium coeruleum* are rare in the coastal forests but widespread on a continental scale. This may arise from several factors, including the 'Jack-of-all-trades or master of one' dilemma and the fact that widespread species are often adapted only to a narrow band of the conditions available locally. Gleasonian influences, which are mentioned below, have a separate and important influence on this broad scale and more locally.

Geographical elements in the coastal forest flora

Geographical elements are defined entirely within the context of the coast. The Msangasi River is a convenient, rather arbitrary boundary, chosen to separate as many species of restricted distribution as possible. Species found only north or only south of this river constitute the Northern and Southern elements, respectively.

Few rare species or species of restricted distribution straddle this river. Most of these are Maritime–Riverine species of specialised ecology, thriving alongside *Vitellariopsis kirkii* in the riverine thicket or forest typical of the many watercourses draining the coastal plateau into the Pangani and adjacent rivers. Examples include: *Stuhlmannia moavi*, an evergreen tree, representing a genus endemic to the coastal area; *Polysphaeria braunii* (Verdcourt, 1980) and *Byttneria fruticosa*, which have close relatives elsewhere in the coastal forests; and *Anisocycla blepharosepala*

subsp. *tanzaniensis*, a riverine forest liane in the
Selous Game Reserve and along the Tana and
Pangani rivers (subsp. *blepharosepala*, however,
occurs in the Zambezian region). *Baphia kirkii* and
Schefflerodendron usambarense occur on both sides
of the river but, surprisingly, are not known in
Kenya.

Apart from these few specialised cases, all species with a restricted distribution, comprising
20% of the sample, can be assigned to the Northern or Southern element. Most samples of, say,
40 or more forest species either north or south of
the Msangasi River include representatives of
their respective element (see Figure 5.4),
although Northern species are sparsely distributed in Dry forest. Some Northern or
Southern species have a very distinctive form or
behaviour, giving a distinctive atmosphere to the
whole vegetation. These include *Millettia puguensis*, a ribbon-stemmed liane, and *Sapium* sp. nov.
(FTEA: see Hubbard *et al.*, 1952–83), a sprawl-
ing, half-climbing treelet in Southern forests; and
the spiny, succulent *Euphorbia wakefieldii* and
beautifully scented *Mkilua fragrans* in the Northern forests.

The Northern element

Most (57%) of the Northern element are often
associated with the Maritime–Riverine element
and have an equivalent ecology within their range
(NL in Appendix 5.1). Some are species associated strongly with rocks or rocky soils (e.g.
Euphorbia wakefieldii), or rivers (e.g. *Oxystigma
msoo*: Rodgers, 1983). As such, they are well
represented in the *Combretum–Sterculia* forests of
the limestone areas, associations that are not
represented in the Southern forests. This suggests an environmental basis for the north–south
differences.

Not all Northern species, however, are of this
broad ecological type. Twenty-one per cent (NM
in Appendix 5.1), including *Mkilua fragrans* and
Cola porphyrantha, are more associated with members of the Moist forest element, though *Mkilua*
may well have had its range extended (e.g. from
the Usambaras) by human influence. Most of
these are restricted to the most mixed, Moist

coastal forests, especially those on the coastal
range south of Mombasa. Bridson (1980) segre-
gates the population of *Oxyanthus pyriformis* there
in subsp. *longitubus*, distinct from populations to
the south. *Drypetes usambarica* var. *mrimii* is
known only from Mrima Hill and the *kaya* forests,
and has a facies very different from other coastal
forest varieties of this species. *Diospyros shimbaen-
sis*, *Dichapetalum fructuosum* and *Ancistrocladus
robertsionorum* (Leonard, 1984) are well-defined
taxa restricted to forests in this area, but their
ecology is less clear.

Sixteen per cent of the Northern element
(ND), including *Dialium orientale* and *Croton
talaeporos* (Radcliffe-Smith, 1972) can be related
to members of the Dry forest element. These two
tree species are, however, barely forest species,
equivalent at best to the type 3 or 4 Dry forest
trees mentioned above. Although most Northern
species recorded sometimes in Dry forest, like
Aristogeitonia monophylla (see Shaw, 1971), prefer
edaphically unusual sites, *Tricalysia ovalifolia* var.
taylorii seems a better example of a Northern Dry
forest species.

Some Northern species have close relatives
elsewhere in the Zanzibar–Inhambane regional
mosaic, but these are mostly inland from the
Southern forests. For example, *Saintpaulia rupicola* (Burtt, 1948, 1958, 1964) and *Chlamydacan-
thus lindavianus* are both apparently restricted to
Moist or Riverine limestone forests, yet the close
relatives of both occur to the west or south of the
Southern coastal forests (*Saintpaulia* spp. and *C.
dichrostachyus*). *Pavetta sphaerobotrys* has different
subspecies in two of the Northern rivers (Bridson,
1978). Other Northern species with obvious close
relatives include the apomictic *Dorstenia tayloriana*
var. *tayloriana* (M. Hijman, personal communication); *Euphorbia wakefieldii* from the *kaya* forests, a
sister species of *E. quadrialata* from more
southerly coastal forests (S. Holmes, personal
communication); and *Mitriostigma greenwayi*
(Bridson, 1980; FTEA: see Hubbard *et al.*, 1952–
83). Some may have only recently split from their
closest relatives (*Euphorbia wakefieldii*). Several of
these have small populations gregarious in rocky
places and are capable of vegetative reproduction
(*Dorstenia*, *Saintpaulia*, *Euphorbia*, *Byttneria*).

Most Northern species, on the other hand, are more isolated, geographically or taxonomically, from their closest relatives. These include *Mkilua fragrans* (the only species in the genus); *Savia fadenii* (although other sections of this genus occur on Indian Ocean islands); *Aristogeitonia monophylla*, which is related to a riverine species in Angola (Shaw, 1945, 1971); *Cola octoloboides* (Brenan, 1978); and *Caesalpinia dalei* (Brenan, 1963). Such distant relationships suggest evolution of the Northern element over a long period, although quirky long-distance dispersal followed by rapid divergence caused by the founder effect has possibly contributed. The evolution of some of the Northern Dry forest–savanna species is no doubt best understood in terms of a history of isolation of vegetation other than coastal forest, possibly based in what is now the Somali–Masai region.

The Southern element

By contrast with the Northern element, many species in the Southern element occur in Dry forest, thicket or woodland, sometimes in associations similar to some in the North. The only exceptions come from the Moist (or riverine) forest in or near the Pugu Hills, but few of even these species are confined to Moist forest.

The Pugu Hills forest is mainly on steep slopes and ridge tops, with Moist forest in the valleys and milder slopes (see Fig. 5.6). *Dialium holtzii–Baphia kirkii* forest on milder slopes, *Scorodophloeus* forest on steep slopes and forest or thicket on some ridge tops are rich in species endemic or nearly endemic to the Pugu Hills (21% of the Southern element: SP in Appendix 5.1). These include the small trees *Baphia puguensis* and *Diospyros capricornuta*, the liane *Millettia puguensis*, and the forest grass *Humbertochloa greenwayi*. An unnamed *Sapium* sp. and *Tricalysia bridsoniana* var. *pandensis* are known only from Pande forest, a rather different Dry forest nearby.

More widespread, Southern, Dry forest species are common in the Southern forests. These include 'Usaramo endemics' (26% of Southern element: SU), unknown south of the Rufiji River and Kichi Hills, for example *Combretum harrisii*,

Tricalysia allocalyx (Robbrecht, 1982) and *Xylopia* sp. B. Most widespread Southern species are at the northern limit of a range stretching from far south of the Rufiji River (47%: ST). Some species, like *Croton steenkampianus*, reach as far south as Natal. Like members of the Dry forest element many of these are typical of thicket or other non-forest vegetation. Although they often occur in Dry forest as relics from earlier successional stages, they are more specialised than forest pioneers like *Trema*. It is likely that many of these widespread Southern species previously flourished in a woodland, forest and thicket mosaic (Engler's 'Sachsenwald') of a type never well developed around the Northern forests, on or near gently undulating, sometimes seasonally waterlogged coastal sediments. The edges and recently closed gaps of the forested parts of this mosaic are today the best known habitat of such species, but provide an uncertain refuge for them.

A few Southern species have relatives that are more widespread in the Zanzibar–Inhambane region, although only a few have close relatives in the Northern forests (*Cynometra* sp. cf. *webberi* and *Tricalysia ovalifolia* (varieties)). Southern species, with relatives in basement complex forests and not typical of Northern coastal forests, include *Lasiodiscus holtzii* (related to *L. usambarensis*) and *Garcinia acutifolia* (closely related to the Kimboza endemic *G. bifasciculata*). Some pairs of Southern taxa are closely related; for example, *Sapium triloculare*, a Pugu Hills endemic, is at least closely related to *S. armatum*; and *Pseudoprosopis euryphylla* is a widespread Southern species with the Pugu Hills population differentiated under subsp. *puguensis* (Brenan, 1984).

Several species have close relatives in the Guineo–Congolian region, including *Diospyros mafiensis* (White & Caveney, 1980), *D. verrucosa* (White, 1955); *Millettia impressa* subsp. *goetzeana* and *M. puguensis* (Gillett, 1961a); *Platysepalum inopinatum* (Gillett, 1960); *Baphia puguensis* (Brummitt, 1968; Soladoye, 1985); and *Combretum harrisii* (Wickens, 1976). The genera *Tessmannia* and *Guibourtia*, found in some Southern forests, especially the Kichi Hills (K. Vollesen, personal communication) and elsewhere in continental Africa, are absent from the Northern coast. It is hard to envisage many of the legumes being dispersed other than across land, even other than within forest. The Southern element, therefore, may include remnant species or lineages from a dry forest or thicket belt spreading to what is now the Guineo–Congolian or Zambezian regions, but which has been almost destroyed by fire (as discussed by Hamilton, 1976). Such species may have found a safer refuge under the more seasonally dry conditions of the Southern coast than further north.

Although affinities of the Southern element seem strongest with more central regions of continental Africa, the only close relative of the Southern grass *Humbertochloa greenwayi* occurs on Madagascar in steep limestone ravines by the sea (Camus, 1961). The affinities of the Southern element are as diverse as those of the Northern element, with perhaps a greater bias away from lineages that thrive on the coastline of continental Africa. This may be because of a less extensive or less Moist limestone/coral rag forest along the Southern coast; it may also be because of inadequate sampling of Southern riverine vegetation, as suggested by the recent discovery of *Foetidia africana* (an Oceanic genus) in the area.

Discussion

The wide range of floristic association in the coastal forests invites consideration of the factors causing such diversity. It does not seem likely that the coastal forests have arisen simply through fragmentation of a previously more extensive, more homogeneous community, even though within certain areas this surely applies. Rather, even before human intervention there must have been a patchwork of communities, much of which was even then not forest. The existence of many non-forest species (e.g. *Rothmannia ravae* and *Phillipia mafiensis*) which are endemic to the mosaic supports this view.

The two main causes of variation in any community are habitat availability and 'Gleasonian' or logistic/historical factors. Other factors, like disturbance, play a subordinate role.

1. Habitat availability. The pattern in the physical environment determines the potential range for species and thereby creates the fundamental trends in variation of the forests. The coastal landscape is complex and patchy, and so are the coastal forests. Many species are restricted to, or strongly favour, certain habitats; for instance, the Northern endemics *Saintpaulia rupicola* and *Euphorbia wakefieldii* grow only on Jurassic limestone outcrops. Hard sedimentary rocks supporting forest are more exposed towards the north, whereas in the south, Dry forest of a variety of types is better developed. The variation in rainfall patterns from south to north probably has some influence on the south–north floristic trends. The possibility that Zanzibar steals the rain destined for the coast and makes a barrier to the free migration of some species has been suggested above.

The Kimboza Forest on metamorphosed limestone under a high rainfall west of the Southern coastal forests is similar to some of the Northern coastal forests. However, absence of the Northern element from Kimboza Forest, despite the occurrence in Kimboza of close relatives of the Northern element, suggests that environmental determinism alone cannot explain all the observed patterns. If, as seems likely, isolation has contributed to differences between Kimboza and the Northern coastal forests, then one can expect that the isolation between the Northern and Southern coastal forests, which are sometimes more disparate environmentally, will also have contributed to their differences. Similarly, Dry forest in the North lacks Southern species. Such considerations lead one to the conclusion that environmental factors do not explain all the patterns in the coastal forests.

2. History and logistics. Isolation and limitations on dispersal in a changeable and patchy environment will have contributed to the development of different taxa and different communities in the Northern and Southern coastal forests, as well as on more local levels.

Taxa such as *Saintpaulia rupicola* and *Oxyanthus pyriformis* subsp. *longitubus* have apparently diverged within blocks of coastal forest from close relatives elsewhere. If two communities are apart they also tend to diverge, through differences in patterns of local extinction and immigration (Gleason, 1926; Macarthur & Wilson, 1967). Species are not always found in all apparently suitable coastal forests, especially where such suitable habitats are isolated from others of the same type. Conversely, some species of limited geographic range occur in a wide range of habitats locally. The Northern species *Aristogeitonia monophylla* and *Mitriostigma greenwayi*, for instance, occur in a broad spread of coastal forest and thicket habitats around Mombasa, but not in the Southern forests. This is no doubt because they flourish in certain (limestone) areas in the North, and from such bridgeheads make incursions into temporary habitats (e.g. forest gaps and stream banks over sandstones) in the surrounding landscape. Similarly, *Baphia puguensis* which favours steep sided valleys in the Pugu Hills occurs from Moist valley forest to dry ridge top thicket there, yet in spite of this local versatility is unknown elsewhere.

3. Disturbance and competition. The influences discussed in previous paragraphs shape the distinctive species pool in each area: disturbance, competition and, on the level of single species, more specific interactions influence on how the members of this species pool associate on a finer-grained level. The edge of environmental determinism is sharpened by competition, and especially by diffuse competition with the whole of the surrounding forest. Conversely, the relationship between landscape and species distribution is generally made less close by disturbances, especially canopy gaps. Gaps often allow the invasion of species less well suited to the site if such species happen to arrive, for instance from a neighbouring patch in the mosaic, before a better adapted species. The invasion of coastal forest gaps by savanna trees like *Brachystegia spiciformis*, or of Pugu Hills valley bottom forest by *Baphia puguensis*, are manifestations of this first-come, first-served influence. The relative contribution of this influence, i.e. the influence of the local mosaic on one patch within it, is particularly important in the coastal forests because of the

environmental heterogeneity, and because of regular human and other disturbances.

These three factors, and particularly the first two can account in principle for the complexity observed in the coastal forests, at all levels. The north–south division of geographical elements expresses part of this diversity only at the broadest level, and even so only rather crudely. The Gleason individualistic model of the community is most apt and evident in the coastal forests. It is for this reason that coastal forest as a whole has to be defined not by floristic content but by geographical position.

There is little doubt that the Tanzanian mountain arc has been a refuge for coastal forest species during inclement times. Species surviving climatic change in shifting belts of vegetation in the mountains, including species that evolved there in isolation, will have repeatedly recolonised the coast when favourable conditions returned. It has been seen, however, that plant species have almost certainly evolved within the coastal mosaic itself. Coastal forests are today the only refuge for a wide variety of plants, especially Dry forest, riverine or rock-loving ones. There must have been times, possibly during drier climates, when coastal forests have enriched the mountain communities. The coastal vegetation is a buffer zone between rain forest and maritime communities, and it seems likely that at least certain Oceanic or coastal lineages like *Gigasiphon*, *Byttneria* and *Mitriostigma* reached the mountain forests via a coastal forest form. At least the first of these is capable of long distance dispersal across the sea (Verdcourt, 1981).

Some Northern species may have previously occurred in the South, and vice versa. Further exploration will probably show that some species are distributed more widely than thought now. However, the complex patterns and the north–south differences in the coastal forest flora cannot be attributed entirely to inadequate sampling or random extinctions. Species, even genera, have been weaving themselves into and out of this tapestry of communities for many millennia. There is little chance that any further development will be one of enrichment, although one might hope to ensure that degradation is kept to a minimum.

Conservation

Because of the limited area and patchy distribution of the coastal forests, and because of the striking individuality of many of them, all must be given high priority for conservation on an international level. Most coastal forests include rare species (most species of high conservation priority in the sample are marked with two or three more exclamation marks in Appendix 5.1). Some coastal forests are particularly rich in rare species. Most of the remaining forests of these types are seriously threatened by human activities.

1. The Southern Dry forests, especially Pugu Hills and Pande, include several endangered species threatened by charcoal making and other destructive activities.
2. The small remaining patches of limestone forest in the Northern coastal forests are seriously threatened by farming.
3. The higher peaks in southern Kenya, e.g. Mrima Hill, also include several very rare or endemic species.

The pattern described here covers the forests of only a small part of a chain of the East African coasts. There is much to be discovered taxonomically, phytogeographically and ecologically by more detailed examination of these forest communities, and by exploration of the lesser known vegetation south of the Rufiji River. Even if that exploration were complete, however, the Tanzanian coastal vegetation south of the Rufiji River would almost certainly warrant a new chapter in the story of the East African forests.

Acknowledgements
I thank Dr D. Mabberley for supervising the thesis on which this work is based, Professor Whatley for providing facilities in Oxford, the Science and Engineering Research Council for financial support and Jon Lovett for editorial comments. Thanks also to the staff at Kew Gardens, particularly Drs R. Polhill and D. Bridson, who have provided indispensible help in the herbarium.

References

ANDREWS, P., GROVE, C. P. & HORNE, J. F. M. (1975). Ecology of the lower Tana floodplain (Kenya). *Journal of the East African Natural History Society and National Museum* **151**, 1–30.

ANON. (1947). *Atlas of Tangyanika Territory*. London: Government Printer.

BEADLE, N. C. W. (1962). Soil phosphate and the delimitation of plant communities in eastern Australia. *Ecology* **43**, 281–8.

BEADLE, N. C. W. (1966). Soil phosphate and its role in molding segments of the Australian flora and vegetation, with special reference to xeromorphy and sclerophylly. *Ecology* **47**, 992–1007.

BEENTJE, H. J. (1990). A reconnaissance survey of Zanzibar forests and coastal thicket. Report for Department of Environment, Commission of Lands and Environment, under the Zanzibar Integrated Land Use project of FINNIDA.

BIRCH, W. R. (1963). Observations on the littoral and coral vegetation of the Kenya coast. *Journal of Ecology* **51**, 603–15.

BRENAN, J. P. M. (1963). Notes on African Caesalpinioideae. *Kew Bulletin* **17**, 197–226.

BRENAN, J. P. M. (1978). A new species of *Cola* (Sterculiaceae) from East Africa. *Kew Bulletin* **33**, 283–6.

BRENAN, J. P. M. (1984). A new record and a new taxon in the genus *Pseudoprosopis* (Leguminosae) from Africa. *Kew Bulletin* **39**, 657–8.

BRENAN, J. P. M., EXCELL, A. W., FERNANDES, A., LAUNERT, E. & WILD, H. (ed.) (1960–1983). *Flora Zambesiaca* London.

BREWIN, REV. R. (1879). *The Memoirs of Mrs. Rebecca Wakefield*, 2nd edn. London.

BRIDSON, D. M. (1978). Studies in *Pavetta* (Rubiaceae subfam. Cinchonioideae) for part 2 of FTEA – Rubiaceae. *Kew Bulletin* **32**, 609–52.

BRIDSON, D. M. (1980). Studies in *Oxyanthus* and *Mitriostigma* (Rubiaceae subfam. Cinchonioideae) for part 2 of FTEA – Rubiaceae. *Kew Bulletin* **34**, 113–30.

BRUMMITT, R. K. (1968). The genus *Baphia* (Leguminosae) in east and northeast tropical Africa. *Kew Bulletin* **22**, 513–36.

BURTT, B. D. (1942). Some East African vegetation communities. *Journal of Ecology* **30**, 65–146.

BURTT, B. L. (1948). Species of *Saintpaulia*. *East African Agricultural Journal* (1948), 183.

BURTT, B. L. (1958). Studies in Gesneriaceae of the Old World. XV. The genus *Saintpaulia*. *Notes from the Royal Botanic Garden, Edinburgh* XXII, 547–68.

BURTT, B. L. (1964). Studies in Gesneriaceae of the Old World. XXV. Additional notes on *Saintpaulia*. *Notes from the Royal Botanic Garden, Edinburgh* XXV, 191–5.

BURTT DAVY, J. (1935). A sketch of the forest vegetation and flora of tropical East Africa. *Empire Forestry Journal* 14, 191–204.

BURTT DAVY, J. (1938). The classification of tropical woody vegetation types. *Imperial Forestry Institute Paper* No. 13.

CALTON, W. E., TIDBURY, G. & WALKER, G. (1955). A study of the more important soils of Zanzibar protectorate. *East African Agricultural Journal* July (1955), 53.

CAMUS, A. (1961). Sur quelques graminées de Madagascar. *Bulletin de la Société Botanique de France* 108, 158–63.

CASHMERE, M. A. (1961). A note on the chronology of the Wanyika of the Kenya coast. *Tanganyika Notes and Records* 57, 153.

CASWELL, P. V. (1953). Geology of the Mombasa–Kwale area. *Geological Survey of Kenya* 24.

CASWELL, P. V. (1956). Geology of the Kilifi–Mazeras area. *Geological Survey of Kenya* 34.

COOKE, M. J. (1974). Cave systems of the Tanga limestone. *Tanganyika Notes and Records* 67, 1–14.

DALE, I. R. (1939). The woody vegetation of the coast province of Kenya. *Imperial Forestry Institute Paper* No. 18, 1–38.

DIGHTON-THOMAS, H. (1963). *Corals and the Correlation of the Tanga Limestone*. London.

DOWSETT-LEMAIRE, F. (1990). The flora and phytogeography of the evergreen forests of Malawi II: Lowland forests. *Bulletin du Jardin Botanique National de Belgique* 60, 9–71.

ENGLER, A. (1910). Die Pflanzenwelt Afrikas, inbesondere seiner tropischen Gebiete, Bd 1. In *Die Vegetation der Erde*, vol. 9, ed. A. Engler and O. Drude. Leipzig.

FARLER, J. P. (1875). The Usambara country in East Africa. *Proceedings of the Royal Geographical Society*, New Series 1, 81–97.

GERHARDT, K. & STEINER, M. (1986). *An Inventory of a Coastal Forest in Kenya at Gedi National Monument*. Uppsala: Swedish University of Agricultural Sciences IRDC.

GILLETT, J. B. (1960). A key to the species of *Platysepalum* with notes. *Kew Bulletin* 14, 464–7.

GILLETT, J. B. (1961a). Notes on *Millettia* Wight and Arn. in East Africa. *Kew Bulletin* 15, 19–40.

GILLETT, J. B. (1961b). The botanical exploration of the area of the *Flora of Tropical East Africa*. In *Comptes rendues de la IV Reunion plenaire de l'AETFAT*, ed. A. Fernandes, pp. 205–29. Lisbon: Association pour l'Etude Taxonomique de la Flore d'Afrique Tropicale.

GLEASON, H. A. (1926). The individualistic concept of the plant association. *Bulletin of the Torrey Botanical Club* 53, 7–26.

GLOVER, J., ROBINSON, P. & HENDERSON, J. P. (1954). Provisional maps of the reliability of annual rainfall in East Africa. *Quarterly Review of the Meteorological Society* 80, 607.

GLOVER, P. E., TRUMP, E. C. & WATERIDGE, L. E. D. (1964). Termitaria and vegetation patterns on the Loita plains of Kenya. *Journal of Ecology* 52, 367–77.

GLOVER, P. E. (1969). *Report on an Ecological Survey of the Proposed Shimba Hills National Reserve*. Nairobi: Kenya National Parks.

GLOVER, P. E. (1970). *List of Plants Numbered in the Shimba Hills Reserve*. Nairobi: Kenya National Parks.

GREENWAY, P. J. (1973). A classification of the vegetation of East Africa. *Kirkia* 9, 1–69.

GREENWAY, P. J. WITH RODGERS, W. A., WINGFIELD, R. J. & MWASUMBI, L. B. (1988). The vegetation of Mafia Island, Tanzania. *Kirkia* 13, 197–238.

GREGORY, J. W. (1921). *The Rift Valley and Geology of East Africa*. London.

GRIFFITHS, J. F. (1958). Climatic zones of East Africa. *East African Agricultural Journal* 23, 179.

GRIFFITHS, J. F. (1972). Climates of Africa. In *World Survey of Climatology*, vol. 10, ed. H. E. Landsberg, Amsterdam: Elsevier.

HALL, J. B. & SWAINE, M. D. (1981). *Distribution and Ecology of Vascular Plants in a Tropical Rainforest: Forest Vegetation in Ghana*. The Hague: Junk.

HALL, J. B., RODGERS, W. A., MWASUMBI, L. B. & SWAI, D. K. (1984). *The Vegetation of the Tanzanian Coral Rag*. Unpublished MS.

HALLIGAN, R. (1958). Geology of the Tanga area. *Records of the Geological Survey of Tanganyika* 6, 17–29.

HAMILTON, A. C. (1976). The significance of patterns of distribution shown by forest plants and animals in tropical Africa for the reconstruction of

Upper Pleistocene palaeoenvironments: a review. *Palaeoecology of Africa* 9, 63–97.

HAMILTON, A. & FADEN, R. B. (1973). The history of the vegetation. In *East African Vegetation*, ed. E. M. Lind and M. E. S. Morrison, pp. 188–209. London: Longman.

HARTNOLL A. V. & FUGGLES-COUCHMAN, N. R. (1937). The Mashokoro cultivations of the coast. *Tanganyika Notes and Records*, April 1937, p. 34.

HAWTHORNE, W. D. (1984). Ecological and biogeographical patterns in the coastal forests of East Africa. DPhil thesis, University of Oxford.

HESSE, P. R. (1955). A chemical and physical study of the soils of termite mounds in East Africa. *Journal of Ecology* 43, 449–61.

HILDEBRANDT, J. M. (1879). Von Mombasa nach Kitui. *Zeitschrift für allgemeinde Erdkunde (Berlin)* XIV, 214 and 314.

HOBLEY, C. W. (1895). Upon a visit to Tsavo and the Taita highlands. *Geographical Journal* V, 545.

HOBLEY, C. W. (1905). Mr. C. W. Hobley's recent journey in East Africa. *Geographical Journal* XXV, 292.

HOPKINS, H. C. & WHITE, F. (1984). The ecology and chorology of *Parkia* in Africa. *Bulletin du Jardin Botanique National de Belgique* 54, 235.

HOWELL, K. M. (1981). Pugu forest reserve: biological values and development. *African Journal of Ecology* 10, 73–81.

HUBBARD, C. E., MILNE-REDHEAD, E. POLHILL, R. M. & TURRILL, W. B. (ed.) (1952–1983). *Flora of Tropical East Africa*. London: HMSO.

HUGHES, F. M. R. (1990). The influence of flooding regimes on forest distribution and composition along the Tana River, Kenya. *Journal of Applied Ecology* 27, 475–91.

HUNT, K. J., HAWTHORNE, W. D., RUSSELL, A. & JONES, A. (1981). Kaya: an ethnobotanical perspective. Unpublished report of Oxford Ethnobotanical expedition to Kenya, January–June 1981.

JOHANSSON, D. R. (1978). *Saintpaulias* in their natural environment with notes on their present status in Tanzania and Kenya. *Biological Conservation* 14, 45–62.

KACHECHEBA, J. L. (1972). The cytotaxonomy of some species of *Hibiscus*. *Kew Bulletin* 27, 425–33.

KING, A. J. (1951). Notes on the Jurassic rocks of part of the Morogoro and Bagamoyo districts. *Records of the Geological Survey of Tanganyika* 1, 15.

KIRK, J. & OLIVER, (1876). *Hibiscus schizopetalus*. *Journal of the Linnean Society, Botany* 15, 478–81.

KRAPF, J. L. (1860). *Travels, Researches and Missionary Labours During an Eighteen Years' Residence in East Africa*, ed. R. C. Bridges London.

KRAPF, J. L. (1882). *The missionary career of Dr. Krapf*. Reprint, Church Missionary Intelligencer. London.

LEONARD, J. (1984). *Ancistrocladus robertsoniorum* J. Leonard (Ancistrocladaceae), espèce nouvelle du Kenya. *Bulletin du Jardin Botanique National de Belgique* 54, 465–71.

LOVETT, J. C., BRIDSON, D. M. & THOMAS, W. (1988). A preliminary list of the moist forest angiosperm flora of Mwanihana Forest Reserve, Tanzania. *Annals of the Missouri Botanical Garden* 75, 874–85.

LUCAS, G. L. (1968). Kenya. In *Conservation of Vegetation in Africa South of the Sahara*, ed. I. Hedberg and O. Hedberg. *Acta Phytogeographica Suecica* 54, 152–63.

MACARTHUR, R. H. & WILSON, E. O. (1967). *The Theory of Island Biogeography*. Princeton, NJ: Princeton University Press.

MCKINLAY, A. C. M. (1963). The Karroo rocks of the north east corner of Tanganyika. *Records of the Geological Survey of Tanganyika* 8, 41–5.

MILLER, J. M. (1953). Geology of the Mariakani–Mackinnon road area. *Geological Survey of Kenya Report* No. 20.

MILNE, G. (1947). A soil reconnaissance through parts of Tanganyika territory, Dec. 1935–Feb. 1936. *Journal of Ecology* 35, 192–265.

MIYATA T. & SAKA, Y. (1979). Deformed cross-lamination in the Karroo system, near Mombasa, Kenya. *4th Preliminary report, African studies, Nagoya University*, pp. 63–71.

MOLL, E. J. & WHITE, F. (1978). The Indian Ocean coastal belt. In *Biogeography and Ecology of Southern Africa*, ed. M. J. A. Werger, pp. 561–98. The Hague: Junk.

MOOMAW, J. C. (1960). *A study of the Plant Ecology of the Coast Region of Kenya, East Africa*. Nairobi.

MOORE, W. R. (1963). Geology of quarter degree sheet 168, Bagamoyo and a compilation map of Tanganyika coastal sediments north of Dar es Salaam. *Records of the Geological Survey of Tanganyika* 10, 1–6.

NEW, REV. C. (1873). *Life, Wanderings and Labours in Eastern Africa*. London.

OATES, F. (1933). The limestone deposits of Tanganyika territory. *Geological Survey of Tanganyika Bulletin* No. 4, pp. 1–120. Dar es Salaam.

PARKIN, D. J. (1972). *Palms, Wines and Witnesses.* London.

PHILLIPS, J. F. V. (1931). A sketch of the floral regions of Tanganyika territory. *Transactions of the Royal Society of South Africa* 19, 363–72.

POLHILL, R. M. (1968). Tanzania. In *Conservation of Vegetation in Africa South of the Sahara*, ed. I. Hedberg and O. Hedberg, pp. 166–78. Acta Phytogeographica Suecica, Vol. 54.

POLHILL, R. M. (1980). Helen Faulkner, 1888–1979. *Kew Bulletin* 34, 619.

POORE, M. E. D. (1962). The method of successive approximation in descriptive ecology. *Advances in Ecological Research* 1, 35–68.

PRINS, A. J. H. (1952). *The Coastal Tribes of the Northeastern Bantu.* London: International African Institute.

PRINS, A. H. J. (1972). The Shungwaya problem. *Anthropos* 67, 9–36.

QUENNEL, A. M. (1956). Summary of the geology of Tanganyika. *Tanganyika Geological Survey Memoir* 1.

RADCLIFFE-SMITH, A. (1972). New and noteworthy Euphorbiaceae from tropical Africa I. *Kew Bulletin* 27, 505–9.

REA, H. J. A. (1935). The forest types of vegetation in Tanganyika territory. *Empire Forestry Journal* 14, 202.

RENVOIZE, S. A. (1976). A floristic analysis of the western Indian Ocean coral islands. *Kew Bulletin* 30, 133–52.

RENVOIZE, S. A. (1979). The origin of the Indian Ocean island floras. In *Plants and Islands*, ed. D. Bramwell, pp. 107–29. London: Academic Press.

ROBBRECHT, E. (1982). The African genus *Tricalysia* A. Rich. (Rubiaceae – Coffeeae) (2) *Ephedranthera*, a new section of subgenus *Tricalysia*. *Bulletin du Jardin Botanique National de Belgique* 52, 311–39.

ROBERTSON, S. A. (1987). Preliminary floristic survey of kaya forests of coastal Kenya. Report to the Director, National Museums of Kenya.

ROBINS, R. J. (1976). The composition of the Josani forest, Zanzibar. *Journal of the Linnean Society, Botany* 72, 223–34.

RODGERS, W. A. (1983). A note on the distribution and conservation of *Oxystigma msoo* Harms (Caesalpiniaceae). *Bulletin du Jardin Botanique National de Belgique* 53, 161–4.

RODGERS, W. A. & HOMEWOOD, K. M. (1982). Species richness and endemism in the Usambara mountain forests, Tanzania. *Biological Journal of the Linnean Society* 18, 197–242.

RODGERS, W. A., HALL, J. B., MWASUMBI, L. B., GRIFFITHS, C. J. & VOLLESEN, K. (1983). The conservation values and status of Kimboza forest reserve, Tanzania. Unpublished report of the Forest conservation working group, University of Dar es Salaam.

RODGERS, W. A., HALL, J. B. & MWASUMBI, L. (1984). The floristics of three coastal forests near Dar es Salaam. MS, University of Dar es Salaam.

RUST, I. C. (1973). Tectonic and sedimentary framework of the Gondwana basin in southern Africa. In *Gondwana Geology*, ed. K. S. W. Campbell. Australian National University Press.

SALIM, A. I. (1973). *The Swahili-speaking peoples of Kenya's coast.* Nairobi.

SHAW, H. K. AIRY (1945). The vegetation of Angola. *Journal of Ecology* 35, 23–48.

SHAW, H. K. AIRY (1971). A second species of the genus *Aristogeitonia* Prain (Euphorbiaceae) from East Africa. *Kew Bulletin* 26, 495.

SOLADOYE, M. O. (1985). A revision of *Baphia* (Leguminosae – Papilionoideae). *Kew Bulletin* 40, 291–386.

SPEAR, T. T. (1978). *The Kaya Complex.* Nairobi: Kenya Literature Bureau.

SPENCE, S. (1957). The Geology of the Eastern Province of Tanganyika. *Geological Survey of Tanganyika Bulletin* No. 28.

STOCKLEY, J. (1948). The geology and mineral resources of Tanganyika territory *Geological Survey of Tanganyika Bulletin* No. 20, pp. 1–70.

SWAINE, M. D. & HALL, J. B. (1981). The monospecific tropical forest of the Ghanaian endemic tree *Talbotiella gentii*. In *The Biological Aspects of Rare Plant Conservation*, ed. H. Synge, p. 355 London.

TEALE, E. O. (1936). A Provisional geological map of Tanganyika territory with explanatory notes. *Geological Division, Tanganyika Bulletin* No. 6.

THOMPSON, A. O. (1956). Geology of the Malindi area. *Geological Survey of Kenya* No. 36.

VABMAH, J. C. & VAID, K. M. (1978). Baobab – the historic African tree at Allahabad. *Indian Forester* 104, 464.

VAUGHAN, R. E. & WIEHE, P. O. (1937). Studies on the vegetation of Mauritius. I. A preliminary survey of the plant communities. *Journal of Ecology* 25, 289–343 (continued in *J. Ecol.* 27, 263).

VERDCOURT, B. (1980). A conspectus of *Polysphaeria* (Rubiaceae). *Kew Bulletin* 35, 97–130.

VERDCOURT, B. (1981). *Gigasiphon humblotianum* (Leguminosae – Caesalpiniodeae – Bauhiniae) as a drift seed. *Kew Bulletin* **36**, 659.

VOLLESEN, K. (1980). Annotated check-list of the vascular plants of the Selous game reserve, Tanzania. *Opera Botanica* **59**, 1–117.

WAALKES, J. VAN BORSUM (1966). Malesian Malvaceae revised. *Blumea* **14**, 1–213.

WAKEFIELD, E. S. (1904). *Thomas Wakefield, Missionary and Geographical Pioneer in East Equatorial Africa*, 2nd edn. London.

WALTER, M. W. (1952). A new presentation of the seasonal rainfall of East Africa. *East African Agricultural Journal* **18**, 11.

WALTER, H. & LIETH, H. (1967). *Klimadiagramm–Weltatlas*. Jena: Fischer.

WELCH, J. R. (1960). Observations on deciduous woodland in the Eastern province of Tanganyika. *Journal of Ecology* **48**, 557–73.

WHITE, F. (1955). Distribution of the African species of *Diospyros*. *Webbia* **11**, 525.

WHITE, F. (1979). The Guineo–Congolian region and its relationship to other phytochoria. *Bulletin du Jardin Botanique National de Belgique* **49**, 11–55.

WHITE, F. (1983). *The Vegetation of Africa*. Paris: UNESCO.

WHITE, F. & CAVENEY, A. N. (1980). Notes on the Ebenaceae VII. Some new names and taxa in Africa. *Bulletin du Jardin Botanique National de Belgique* **50**, 393–9.

WICKENS, G. E. (1976). A new species of *Combretum* in Tanzania. *Kew Bulletin* **31**, 152.

WICKENS, G. E. (1982). The Baobab – Africa's upside-down tree. *Kew Bulletin* **37**, 173–209.

Appendix 5.1: A sample of taxa from the coastal forests

NOTES:

Species names which are excluded from statistics because infraspecific taxa are specified on subsequent lines are followed by curly brackets, enclosing Chorological category at species level. Subspecies or varietal names on subsequent lines (included in statistics) are followed by '⟨'.

Codes in right columns except for those enclosing species are in the order:
| Habit Ecogeographic category | Chorological category |
The first and last | are replaced by ⟨ and ⟩ for infraspecific taxa.

Habit codes are:
C=large climber; H=large herb; S=large shrub; T=canopy tree; c=small climber, etc.

Ecogeographic codes, explained in text, are:
D=Dry forest; LD=Dry (forest) Maritime–Riverine; LM=Moist Maritime–Riverine; LR=Rocks & rivers; M=Moist forest; MH=Moist–High (rain forest); N=Northern; S=Southern; ND=Northern, Dry forest, etc. SP=Pugu Hills endemic; SQ=Pande endemic; ST=Southern-wide; SU=Southern, north of Rufiji and Kichi hills.
X=excluded; X?=Excluded for lack of information; XD=Excluded, nearly Dry forest, etc.

Chorological codes are:
!=Coastal endemic; !!=restricted within coast; !!!=very restricted within coast.
¦ =Eastern, ¦z=Eastern (reaching Zambezian); ¦n=Eastern (north); ¦v=Eastern (around Lake Victoria); ¦s=Eastern (southwards).
?=unknown. ·
A=widespread African; GC=Guineo–Congolian; GC+=Guineo–Congolian and beyond.
O=Oceanic; OM=Oceanic (Madagascar).
Z=Zambesian.

ACANTHACEAE

Barleria sp. aff. *amaniensis* = (Faulkner 672)	c	?N	!!
Chlamydacanthus lindavianus H.Winkl.	H	NL	!!
Crossandra pungens Lindau	h	X?	?
Ecbolium amplexicaule S.Moore	H	XD	!
Elytraria lyrata Vahl	h	D	GC
Elytraria minor Dokosi	h	D	!
Hypoestes forskalei (Vahl) Roem. & Schult.	H	D	A
Justicia aff. *schimperianum* Hoechst.	S	?N	!
Justicia engleriana Lindau	s	M	!
Justicia fittonioides S.Moore	h	LR	!
Justicia pseudorungia Lindau	H	LM	¦
Lankesteria alba Lindau	h	D	?
Megalochlamys tanaensis Vollesen	H	NL	!!!
Pseuderanthemum hildebrandtii Lindau	S	LD	!
Pseuderanthemum tunicatum (Afzel.) Milne–Redhead	S	M	GC
Sclerochiton boivinii C.B.Cl.	t	LM	!
Sclerochiton vogelii (Nees) T.Anders.			{GC}
Sclerochiton vogelii subsp. holstii (Lindau) Napper	⟨ S	D	! ⟩
Thunbergia kirkii Hook f.	H	LR	!
Whitfieldia elongata (Beauv.) C.B.Cl.	S	M	GC+
Whitfieldia stuhlmannii (Lindau) C.B.Cl.	S	D	¦z

AGAVACEAE
Dracaena aletriformis Engl.   t   M   ?
Dracaena usambarensis Engl.   T   LD   GC
Sanseviera arborescens Cornn.   H   LR   ?
Sanseviera conspicua N.E.Br.   H   LR   ?

AMARANTHACEAE
Psilotrichum fallax C.C.Townsend   H   D   !
Pupalia lappacea (L.) A.Juss.   h   X   ?
Scadoxus multiflorus (Martyn) Raf.   m   XW   A

ANACARDIACEAE
Lannea welwitschii (Hiern) Engl.   T   LM   GC
Sorindeia madagascariensis DC.   T   LD   ¦z

ANCISTROCLADACEAE
Ancistrocladus robertsoniorum J.Leonard | T   NM |!!! |

ANNONACEAE
Artabotrys modestus Diels.   {Z}
  Artabotrys modestus subsp. macranthus Verdc.   ⟨ C   D   !   ⟩
Artabotrys monteiroae Oliv.   C   MH   Z?
Artabotrys sp. aff. *brachypetalus* Benth.   C   SU   !!
Asteranthe asterias (S.Moore) Engl. & Diels   s   D   !
Lettowianthus stellatus Diels   t   MH   !
Mkilua fragrans Verdc.   t   NM   !!
Monanthotaxis buchananii (Engl.) Verdc.   s   D   ¦z
Monanthotaxis fornicata (Baill.) Verdc.   s   D   !
Monanthotaxis trichocarpa (Engl. & Diels) Verdc.   c   D   !
Monodora grandidieri Baill.   t   LD   !
Monodora minor Engl. & Diels   t   ST   !!
Ophrypetalum odoratum Diels   s   LD   !
Polyalthia stuhlmannii (Engl.) Verdc.   t   D   !
Sphaerocoryne gracilis (Engl. & Diels) Verdc.   S   X   A?
Toussaintia orientalis Verdc.   t   XS   !!
Uvaria acuminata Oliv.   C   D   O
Uvaria faulknerae Verdc.   C   LD   !!
Uvaria sp. aff. *decidua* Diels.   C   SU   !!
Uvaria welwitschii (Hiern.) Engl. & Diels   C   LR   ¦v
Uvariodendron gorgonis Verdc.   t   MH   !
Uvariodendron kirkii Verdc.   t   LD   !
  Uvariodendron kirkii var. (Unnamed plant from limestone/coral area)   s   XN   !!
Xylopia arenaria Engl.   t   D   !
Xylopia parviflora (A.Rich.) Benth.   T   D   A
Xylopia sp. B. FTEA   t   SU   !!

APOCYNACEAE
Alafia microstylis K.Schum.   C   LD   ?
Dictyophleba lucida (K.Schum.) Pierre   C   M   GC
Holarrhena pubescens (Buch.-Ham.) Wall.   t   D   Z
Hunteria zeylanica Gard. ex Thw.   t   D   O
Landolphia kirkii Dyer ex Hook f.   C   D   ¦z
Mascarenhasia arborescens A.DC.   t   LR   O
Pleiocarpa pycnantha (K. Schum) Stapf   t   MH   GC

Rauvolfia mombasiana Stapf	t	M	!
Saba comorensis (Bojer) Pichon	C	M	GC+
Schizozygia coffaeoides (Bojer) Baillon	s	M	¦ z
Stephanostemma stenocarpum K. Schum.	s	SU	!!!
Strophanthus petersianus Klotzsch	C	D	¦
Tabernaemontana pachysiphon K.Schum.	t	MH	GC
Zygodia melanocephala (K.Schum.) Stapf	c	M	?

ARACEAE

Anchomanes abbreviatus Engl.	M	D	!
Callopsis volkensii Engl.	m	M	!?
Culcasia orientalis S.Mayo	R	LR	!
Gonatopus boivinii (Dcne.) Engl.	M	LR	Z
Gonatopus petiolatus (Peter) Bogner	M	ST	!!
Stylochiton salaamicus N.E.Br.	m	LD	!
Zamioculcas zamiifolia (Lodd) Engl.	M	LR	Z

ARALIACEAE

Cussonia zimmermannii Harms	T	D	!

BALSAMINACEAE

Impatiens walleriana Hook f.	h	LM	¦ z

BIGNONIACEAE

Fernandoa magnifica Seem.	T	D	¦ z
Kigelia africana (Lam.) Benth.	T	LD	A

BOMBACACEAE

Adansonia digitata L.	T	DX	T
Rhodognaphalon schumannianum A.Robyns	T	D	¦ z

BORAGINACEAE

Cordia goetzei Guerke	t	LD	?
Ehretia bakeri Britten	t	D	?

BREXIACEAE

Brexia madagascariensis (Lam.) Ker–Gawl.	t	LR	OM

BURSERACEAE

Commiphora eminii Engl.			{?}
Commiphora eminii subsp. *boiviniana* (Engl.) Gillett ined	⟨ T	XL	? ⟩
Commiphora eminii subsp. *zimmermannii* (Engl.) Gillett ined	⟨ T	LM	? ⟩
Commiphora pteleifolia Engl.	C	LD	Z

BUXACEAE

Notobuxus obtusifolius Mildbr.	s	LM	!

CANELLACEAE

Warburgia stuhlmannii Engl.	t	LD	!

CAPPARACEAE

Capparis erythrocarpos Isert	c	M	GC
Cladostemon kirkii (Oliv.) Pax & Gilg	t	LD	Z
Thylachium thomasii Gilg	S	ND	!

CELASTRACEAE

Maytenus undata (Thunb.) Blakelock	t	D	A
Mystroxylon aethiopicum (Thunb.) Loess	S	LD	A

Salacia erecta   (G.Don) Walp. | s | M | GC+ |
Salacia leptoclada   Tul. | s | D | A |
Salacia madagascariensis   (Lam.) DC. | C | D | OM |

COMBRETACEAE
Combretum butyrosum   (Bertol.f.) Tul. | C | LD | ! |
Combretum chionanthoides   Engl. & Diels | C | LD | ! |
Combretum harrisii   Wickens | t | SU | !! |
Combretum illairii   Engl. | C | D | ! |
Combretum padoides   Engl. & Diels | c | LD | Z |
Combretum pentagonum   Laws. | c | X | ¦z |
Combretum pisoniiflorum   (Klotsch) Engl. | c | LD | ¦z |
Combretum schumannii   Engl. | T | LD | ¦z |
Combretum tenuipetiolatum   Wickens | t | N? | !! |
Combretum xanthothyrsum   Engl. & Diels | C | D | ! |
Pteleopsis myrtifolia   (Laws) Engl. & Diels | T | DX | Z |
Pteleopsis tetraptera   Wickens | T | NL | !! |
Terminalia boivinii   Tul. | T | LD | OM |
Terminalia sambesiaca   Engl. & Diels | T | LM | ¦z |

COMMELINACEAE
Commelina diffusa   Burm f. | m | LD | T |

COMPOSITAE
Blepharispermum zanguebaricum   Oliv. & Hiern | C | D | ! |
Brachylaena huillensis   O.Hoffm. | T | D | ¦ |
Gynura colorata   F.G.Davies | H | NL | !! |

CONNARACEAE
Byrsocarpus boivinianus   (Baill.) Schellenb. | t | D | ! |
Cnestis corniculata   Lam. [*C. confertiflora* in FTEA] | C | D | GC |
Cnestis polyphylla   Lam. | C | NL | !!! |
Connarus longistipitatus   Gilg | C | MH | ¦g |
Ellipanthus hemandradenioides   Brenan | t | XL | ! |
Santaloides splendidum (Gilg) Schellenb. ex Engl. | C | MH | GC |

CRASSULACEAE
Kalanchoe crenata   (Andrews) Haw. | H | D | A |
Kalanchoe lateritia   Engl. | | | {Z} |
   Kalanchoe lateritia var. *prostrata*   Raadts | h | NM | !! |
Kalanchoe obtusa   Engl. | h | LM | ! |

CUCURBITACEAE
Coccinia sp. B. [FTEA] | c | X? | ! |
Diplocyclos tenuis   (Klotsch) C.Jeffrey | c | XL | ! |
Momordica anigosantha   Hook f. | c | M | ! |
Momordica charantia   L. | c | X | T |
Momordica leiocarpa   Gilg | c | ?N | ! |
Momordica peteri   A. Zimm. | c | M | ! |
Momordica trifoliolata   Hook f. | c | XD | A |
Zehneria scabra   (Linn.f.) Sond. | c | X | T |

CYCADACEAE
Encephalartos hildebrandtii   A.Br. & Bouché | S | D | ! |

DICHAPETALACEAE
Dichapetalum arenarium Bret.	C	D	!
Dichapetalum fructuosum Hiern.	C	NM	GC
Dichapetalum madagascariense Poir. [An excessively broad species]	C	M	GC
Dichapetalum mossambicense (Klotsch) Engl.	c	D	!
Dichapetalum ruhlandii Engl.	c	MH	¦
Dichapetalum stuhlmannii Engl.	s	ST	¦z
Dichapetalum zenkeri Engl.	S	LM	GC
Tapura fischeri Engl.	t	D	GC

DILLENIACEAE
Tetracera boiviniana Baill.	c	XL	!
Tetracera litoralis Gilg	C	LD	!

DIOSCOREACEAE
Dioscorea sansibarensis Pax	c	M	GC

EBENACEAE
Diospyros abyssinica (Hiern) F.White [sub.sp. *chapmaniorum* in Malawi]	T	LD	A
Diospyros amaniensis Gurke	T	MH	!
Diospyros capricornuta F.White	T	ST	!!
Diospyros consolatae Chiov.	t	LD	!
Diospyros greenwayi F. White	T	LM	¦z
Diospyros kabuyeana F.White	t	M	!
Diospyros mafiensis F.White	t	ST	!!
Diospyros natalensis (Harv.) Brenan	t	LD	¦z
Diospyros shimbaensis F.White	T	NM	!!!
Diospyros squarrosa Klotzsch	T	LD	Z
Diospyros verrucosa Hiern	s	ST	¦z

ERYTHROXYLACEAE
Erythroxylum emarginatum Thonn.	t	LD	A
Nectaropetalum kaessneri Engl. [var. *parvifolia* Engl. in Somalia & NE Kenya]	t	D	!n

EUPHORBIACEAE
Acalypha engleri Pax	t	M	!
Acalypha fruticosa Forsk.	s	LD	T
Acalypha gillmanii A.R.–SM.	S	ST	!!
Acalypha neptunica Muell. Arg.	S	D	GC
Alchornea laxiflora (Benth.) Pax & K.Hoffm.	s	D	A
Antidesma venosum Tul.	S	LD	A
Antidesma vogelianum Muell. Arg.	t	LR	GC+
Aristogeitonia monophylla Airy Shaw	S	NL	!!
Bridelia atroviridis Muell. Arg.	t	MH	GC
Cavacoa aurea (Cavaco) J.Leon	t	X?	!
Croton jatrophoides Pax	T	LD	!!
Croton macrostachyus Del.	T	MH	Af
Croton pseudopulchellus Pax	s	D	A
Croton steenkampianus Gerstner	T	ST	!s
Croton sylvaticus Hochst.	T	LR	GC+
Croton talaeporos A.Radcliffe-Smith	T	ND	!!
Drypetes arguta (Muell. Arg.) Hutch.	t	ST	!s
Drypetes natalensis (Harv.) Hutch.			{A}
Drypetes natalensis var. *leiogyna* Brenan	⟨ t	N?	¦ ⟩
Drypetes natalensis var. *natalensis*	⟨ t	M	A ⟩

Taxon			
Drypetes parvifolia (Muell. Arg.) Pax & K.Hoffm.	S	LM	GC
Drypetes reticulata Pax	T	LD	!s
Drypetes usambarica (Pax) Hutch.			{!}
Drypetes usambarica var. *mrimii* A.Radcliffe-Smith	⟨ t	NM	!! ⟩
Drypetes usambarica var. *usambarica*	⟨ t	MH	!! ⟩
Erythrococca berberidea Prain	S	ST	!s
Erythrococca sp.C. FTEA	S	N?	!
Erythrococca usambarica Prain	S	LM	!
Euphorbia nyikae Pax	t	D	!
Euphorbia quadrialata Pax	t	LR	!!
Euphorbia wakefieldii N.E.Br.	t	NL	!!!
Excoecaria madagascariensis (Baill.) Muell. Arg.	s	LR	O
Hymenocardia ulmoides Oliv.	t	XD	A
Macaranga capensis (Baill.) Sim	T	MH	A
Mallotus oppositifolius (Geisler) Muell. Arg.	t	LD	GC+
Meineckia fruticans (Pax) Webster			{!}
Meineckia fruticans var. *engleri* (Pax) Webster	⟨ S	NL	!! ⟩
Meineckia fruticans var. *fruticans*	⟨ S	LM	! ⟩
Micrococca scariosa Prain	S	LM	!!
Mildbraedia carpinifolia (Pax) Hutch.			{z}
Mildbraedia carpinifolia var. *carpinifolia*	⟨ S	D	! ⟩
Mildbraedia sp.A. FTEA	S	N?	!!!
Neoholstia tenuifolia (Pax) Rauschert			{¦z}
Neoholstia tenuifolia var. *glabrata* (Prain) A.R.-Sm.	⟨ ¦s	LR	⟩
Oldfieldia somalensis (Chiov.) Milne-Redhead	T	XD	!
Pycnocoma littoralis Pax	t	LD	!
Ricinodendron heudelotii (Baill.) Pierre ex Pax			{A}
R. heudelotii subsp. *africanum* (Muell. Arg.) J.Leon.	⟨ T	M	GC+ ⟩
Sapium armatum Pax & K.Hoffm.	t	ST	!
Sapium sp.nov. FTEA	C	SQ	!!!
Sapium triloculare Pax & K.Hoffm.	t	ST	!!
Savia fadenii A. Radcliffe-Smith	t	NL	!!!
Spirostachys africana Sond.	T	LD	¦n
Suregada zanzibariensis Baillon	s	D	O
Synadenium pereskiifolium (Baill.) Guill	t	LD	!
Thecacoris usambarensis Verdc.	S	NL	!!

FLACOURTIACEAE

Bivinia jalbertii Tul.	S	LD	OM
Casearia gladiiformis Mast.	t	LD	¦z
Dovyalis hispidula Wild.	S	XL	¦z
Dovyalis sp.A. FTEA	t	N?	!!
Grandidiera boivinii Jaub.	S	D	!
Homalium abdessammadii Aschers. & Schweinf.	t	LR	A
Homalium longistylum Mast.	t	LR	GC+
Ludia mauritiana Gmelin.	T	LD	O
Rawsonia lucida Harv. & Sond.	t	MH	Af
Scolopia rhamnophylla Gilg	t	MH	Af
Xylotheca tettensis (Klotsch) Gilg	C	X	¦z

GESNERIACEAE

Saintpaulia diplotricha Burtt	h	NL	!!
Saintpaulia intermedia Burtt.	h	NL	!!

Saintpaulia ionantha H. Wendl.	h	LR	!
Saintpaulia rupicola Burtt	h	NL	!!!

GRAMINEAE

Humbertochloa greenwayi C.E.Hubbard	M	SP	!!!
Leptochloa uniflora A.Rich.	m	M	T
Megastachya mucronata (Poir.) P.Beauv.	m	D	A
Olyra latifolia L.	M	M	T
Oplismenus compositus (L.) P.Beauv.	m	D	T
Panicum brevifolium L.	m	M	T
Panicum laticomum Nees	m	M	¦z
Panicum pleianthum Peter	m	M	!
Panicum trichocladum K.Schum.	m	D	T
Setaria megaphylla (Steud.) Th. Dur. & Schinz	m	LR	T

GUTTIFERAE

Garcinia acutifolia N.Robson	t	ST	!!
Garcinia livingstonei T.Anders.	t	LR	A
Harungana madagascariensis Poir.	t	X	GC

HERNANDIACEAE

Gyrocarpus americanus subsp. *americanus* Jacqu.	⟨ T	LM	T ⟩

ICACINACEAE

Apodytes dimidiata E.Meyer ex Arn	t	XL	A
Iodes usambarensis Sleumer	c	NL	!!

LECYTHIDACEAE

Barringtonia racemosa Blume	t	LR	O
Foetidia africana Verdc.	T	SL	!!!

LEGUMINOSAE subfamily Caesalpinioideae

Afzelia quanzensis Welw.	T	DX	Z
Brachystegia spiciformis Benth.	T	DX	Z
Caesalpinia dalei Brenan & J.B.Gillet	t	NL	!!!
Caesalpinia volkensii Harms	C	LR	!
Cordyla africana Lour.	T	LM	Z
Cynometra lukei Beentje	T	NL	!!!
Cynometra suaheliensis (Taubert) Baker f.	T	LD	!
Cynometra webberi Baker f.	T	D	!
*Cynometra webberi** subsp. nov.?	T	SU	!!!
Dialium holtzii Harms	T	LD	!
Dialium orientale Baker f.	T	ND	!!
Erythrophleum suaveolens (Guill. & Perr.) Brenan	T	M	GC
Gigasiphon macrosiphon (Harms) Brenan	t	MH	!!
Guibortia schliebenii (Harms) J. Leonard	T	ST	!!
Hymenaea verrucosa Gaertner	T	D	O
Julbernardia magnistipulata (Harms) Troupin	T	D	!
Oxystigma msoo Harms	T	NL	!!
Paramacrolobium coeruleum (Taubert) Leonard	T	M?	GC
Scorodophloeus fischeri (Taubert) J.Leonard	T	D	!
Stuhlmannia moavi Taubert	T	LD	!!!
Tamarindus indica L.	T	LD	T
Tessmannia martiniana Harms.	T	ST	!!
Zenkerella egregia J. Leonard	T	MH	!!

LEGUMINOSAE subfamily Mimosoideae

Acacia adenocalyx Brenan & Exell	t	X	!
Acacia robusta Burchell			{A}
⟨ A. robusta subsp. *usambarensis* (Taubert) Brenan	T	LD	¦ n ⟩
Acacia tephrodermis Brenan	C	SU	!!
Albizia adianthifolia (Schumach.) W.F.Wright	T	M	A
Albizia glaberrima (Schumach. & Thonn.) Benth. [var. *glabrescens* (Oliv.) Brenan]	T	LD	A
Albizia gummifera (J.F. Gmel.) C.A.SM.	T	M	A
Albizia petersiana (Bolle) Oliv.	T	LD	¦ z
Entada pursaetha DC.	C	M	T
Newtonia paucijuga (Harms) Brenan	T	M	!
Parkia filicoidea [Welw. ex] Oliv.	T	LR	GC
Pseudoprosopis euryphylla Harms			{!}
⟨ *Pseudoprosopis euryphylla* subsp. *puguensis* Brenan	C	SP	!!! ⟩
Tetrapleura tetraptera (Schumach. & Thonn.) Taub.	T	X?	GC

LEGUMINOSAE subfamily Papilionoideae

Angylocalyx braunii Harms	t	M	!
Baphia Kirkii Baker			{!}
⟨ *Baphia kirkii* subsp. *kirkii* [subsp. *ovata* occurs further south]	T	LD	!! ⟩
Baphia puguensis Brummitt	t	SP	!!!
Craibia brevicaudata (Vatke) Dunn			{A}
⟨ *Craibia brevicaudata* subsp. *brevicaudata* (Vatke) Dunn	T	LM	¦ n ⟩
Craibia zimmermannii (var.) (Harms) Dunn	T	ST	¦ ?
Erythrina sacleuxii Hua	T	D	!
Millettia eetveldeana (Micheli) Hauman	T	ST	Z
Millettia impressa Harms			{GC}
⟨ *Millettia impressa* subsp. *goetzeana* (Harms) J.B.Gillet	C	ST	!! ⟩
Millettia puguensis J.B.Gillett	C	SP	!!!
Millettia usaramensis Taubert [subsp. *australis* in Malawi, Mozambique]			{ z }
⟨ *Millettia usaramensis* subsp. *usaramensis* Taubert	t	D	! ⟩
Mucuna gigantea (Willd.) DC.	C	LR	T
Platysepalum inopinatum Harms	C	ST	!!
Rhynchosia holtzii Harms	C	SP	!!!
Schefflerodendron usambarense Harms [Only recorded in Gabon in G–C region]	T	D	GC

LILIACEAE

Asparagus falcatus L.	c	D	?
Chlorophytum filipendulum Bak.	m	LM	?

LINACEAE

Hugonia castaneifolia Engl.	C	D	!

LOGANIACEAE

Mostuea brunonis Didr.	S	D	GC
Mostuea microphylla Gilg	S	D	¦ v
Strychnos henningsii Gilg	t	D	¦ z
Strychnos mitis S.Moore	T	M	¦ z
Strychnos panganensis Gilg	C	D	!
Strychnos scheffleri Gilg ex Bak.	C	M	GC
Strychnos usambarensis Gilg	t	LD	GC

MALPIGHIACEAE

Acridocarpus alopecurus Sprague	C	X?	¦ g

Acridocarpus chloropterus Oliv.	C	LR	¦z
Acridocarpus zanzibaricus A.Juss.	C	XL	!
Triaspis mozambica A.Juss	c	XD	!
Tristellateia africana S.Moore	c	XL	!

MALVACEAE

Gossypioides kirkii (Mast.) J.B.Hutch	S	X	!
Hibiscus faulknerae Vollesen	s	LD	!
Hibiscus schizopetalus Hook f.	t	LR	?!
Thespesia danis Oliv.	t	LD	!

MELASTOMATACEAE

Memecylon amaniense A. & R. Fernandes	t	M	!
Memecylon fragrans A. & R. Fernandes	t	ND	!!
Memecylon mouririifolium Brenan	t	LM	!!
Memecylon sansibaricum Taub.	t	D	Z
Memecylon sp.aff. *flavovirens* Bak.	t	D	!!
Memecylon verruculosum Brenan	t	MH	!

MELIACEAE

Lovoa swynnertonii Baker f.	T	MH	¦
Pseudobersama mossambicensis (Sim) Verdc.	T	D	!
Trichilia emetica Vahl	T	LD	A
Turraea mombassana Hiern ex DC.	s	D	¦z
Turraea wakefieldii Oliv.	T	DX	!

MENISPERMACEAE

Anisocycla blepharosepala Diels.			{ z}
Anisocycla blepharosepala subsp. *tanzaniensis* Vollesen	⟨ C	LD	!
Dioscoreophyllum volkensii Engl.	C	LR	GC
Jateorhiza palmata (Lam.) Miers	C	D	GC
Tiliacora funifera (Miers) Oliv.	C	D	GC
Tinospora caffra (Miers) Troupin	c	D	|
Tinospora oblongifolia (Engl.) Troupin	c	D	!
Triclisia sacleuxii (Pierre) Diels	c	XL	GC?

MONTINIACEAE

Grevea madagascariensis Baill.			{O}
Grevea madagascariensis subsp. *keniensis* Verdc.	⟨ S	NL	!! ⟩

MORACEAE

Antiaris toxicaria Leschenault	T	M	GC
Dorstenia alta Engl.	S	LM	¦g
Dorstenia goetzei Engl.	H	LM	!
Dorstenia hildebrandtii Engl.	h	LR	¦v
Dorstenia kameruniana Engl.	S	M	GC
Dorstenia tayloriana Rendle			{!}
Dorstenia tayloriana var. *laikipiensis* (Rendle) Hijman	⟨ S	LM	! ⟩
Dorstenia tayloriana var. *tayloriana*	⟨ H	NL	!! ⟩
Dorstenia warneckei Engl.	h	MH	!
Ficus bubu Warb.	T	LR	GC
Ficus bussei Mildbr. & Burret	F	D	Z
Ficus craterostoma Mildbr. & Burret	F	LM	A
Ficus exasperata Vahl	T	M	T
Ficus faulkneriana C.C.Berg	F	XN	!!

Ficus ingens (Miq.) Miq.	F	D	A
Ficus kirkii Hutch.	T	LR	GC
Ficus lingua De Wild. & T.Durand			{A}
⟨ *Ficus lingua* subsp. *depauperata* (Sim.) C.C.Berg	F	LM	¦z ⟩
Ficus lutea Vahl	F	M	A
Ficus natalensis Hochst.	F	M	GC
Ficus ottonifolia (Miq.) Miq.			{A}
⟨ *Ficus ottonifolia* subsp. *ulugurensis* (Mildbr. & Burret) C.C.Berg	F	LM	! ⟩
Ficus sansibarica Warb.	F	LM	Z
Ficus sur Forssk.	T	M	A
Ficus sycomorus L.	T	LR	A
Ficus tremula Warb. [inc. subsp. *kimuezensis* (Warb.) C.C.Berg]			{GC}
⟨ *Ficus tremula* subsp. *tremula* Warb.	F	D	!s ⟩
Ficus wakefieldii Hutch.	T	DX	Z
Maclura africana (Bureau) Corner	t	XL	Z
Milicia excelsa (Welw.) C.C.Berg	T	M	GC
Sloetiopsis usambarensis Engl.	S	LR	GC
Trilepisium madagascariensis DC.	T	M	GC

MYRTACEAE

Eugenia capensis (Ecklon & Zeyher) Sond.			{ }
⟨ *E. capensis* subsp. *aschersoniana* (F.Hoffm.) F.White	t	D	! ⟩
Syzygium cordatum Hochst. ex Krause	T	XW	¦¦

OCHNACEAE

Ochna thomasiana Engl.	S	LD	!

ORCHIDACEAE

Aerangis hololglottis (Schtr.) Schltr. [Also in Sri Lanka]	E	M	O?
Aerangis kirkii (Reichb. f) Schlctr.	E	D	!
Angraecum dives Rolfe [Also on Socotra]	E	LD	!
Microcoelia exilis Lindl.	E	M	Z
Microcoelia smithii (Rolfe) Summerh.	E	D	!
Oeceoclades saundersiana (Reichb.f.) Geray & Taylor	E	M	GC
Solenangis aphylla (Thou.) Summerh.	E	X	ZO
Solenangis wakefieldii (Rolfe) Cribb & J.Stewart	E	X	!

OXALIDACEAE

Biophytum helenae Buscal. & Muschl.	h	LR	Z

PANDANACEAE

Pandanus rabaiensis Rendl.	t	LR	!

PASSIFLORACEAE

Adenia gummifera (Harv.) Harms	C	LD	¦z
Adenia rumicifolia Engl.	c	M	GC
Paropsia braunii Gilg	t	ST	!!
Schlecterina mitostemmatoides Harms	c	D	!

POLYGALACEAE

Carpolobia goetzei Guerke	S	D	¦z

RHAMNACEAE

Lasiodiscus holtzii Engl.	s	SP	!!!
Lasiodiscus mildbraedii Engl.			{GC}
⟨ *Lasiodiscus mildbraedii* subsp. *ferrugineus* (Verdc.) Faden	S	LR	! ⟩
Ziziphus mucronata Willd.	T	XD	A

RHIZOPHORACEAE
Cassipourea euryoides   Alston | T LD | ! |

RUBIACEAE

Breonadia salicina (Vahl) Hepper & Wood	t	LR	A
Calycosiphonia spathicalyx (K.Schum.) Bremek.	S	MH	GC+
Canthium bibracteatum (Baker) Hiern	t	LD	¦ z
Canthium micans Bullock	C	ST	!!
Canthium mombazense Baill.	s	D	!
Catunaregam spinosa (Thunb.) Tirvengadum	S	X?	T
Chassalia umbraticola Vatke	S	X	!
Chazaliella abrupta (Hiern) Petit & Verdc.	s	LM	GC
Cladoceras subcapitata (K.Schum. & K.Krause) Bremek.	c	XD	!
Coffea pseudozanguebariae Bridson	s	D	!
Coffea sessiliflora Bridson			{!}
Coffea sessiliflora subsp. *mwasumbii* Bridson	⟨ s	SP	!!! ⟩
Coffea sessiliflora subsp. *sessiliflora*	⟨ s	NL	!!! ⟩
Cremaspora triflora (Thonn.) K.Schum.			{GC}
Cremaspora triflora subsp. *confluens* (K.Schum) Verdc.	⟨ t	M	!
Didymosalpinx norae (Swynnerton) Keay	t	M	¦ z
Feretia apodanthera Del.			{A}
Feretia apodanthera subsp. *keniensis* Bridson	⟨ s	XN	!! ⟩
Gardenia posoqueroides S.Moore	S	LM	¦ z
Gardenia transvenulosa Verdc.	s	D	!
Geophila obvallata (Schumach.) F.Didr.			{GC}
Geophila obvallata subsp. *iodes* (K.Schum.) Verdc.	⟨ h	D	! ⟩
Heinsia crinita (Afzel.) G. Taylor	S	XD	GC
Heinsia zanzibarica (Bojer) Verdc.	s	X	!
Ixora narcissodora K.Schum.	t	LR	¦ z
Keetia venosa (Oliver) Bridson	S	X	?
Keetia zanzibarica (Klotsch) Bridson	t	X	Z
Kraussia speciosa Bullock	S	MH	!
Leptactina platyphylla (Hiern) Wernham	t	M	GC
Leptactina sp.A FTEA	s	SU	!!!
Mitriostigma greenwayi Bridson	S	NL	!!!
Oxyanthus pyriformis (Hoscht.) Skeels			{ z}
Oxyanthus pyriformis subsp. *longitubus* Bridson	⟨ s	NM	!!! ⟩
Oxyanthus sp.A FTEA	s	SU	!!!
Oxyanthus zanguebaricus (Hiern) Bridson	s	D	!
Pavetta crebrifolia Hiern			{!}
Pavetta crebrifolia var. *crebrifolia*	⟨ s	D	! ⟩
Pavetta crebrifolia var. *pubescens* Bridson	⟨ s	ND	!! ⟩
Pavetta linearifolia Bremek.	S	XL	!
Pavetta macrosepala Hiern			{!}
Pavetta macrosepala var. *puberula* K.Schum.	⟨ s	XS	!! ⟩
Pavetta sansibarica K.Schum.			{!}
Pavetta sansibarica subsp. *trichosphaera* (Bremek.) Bridson	⟨ s	NM	!! ⟩
Pavetta sphaerobotrys K.Schum.			{!}
Pavetta sphaerobotrys subsp. *lanceisepala* (Bremek.) Bridson	⟨ S	NL	!!! ⟩
Pavetta sphaerobotrys subsp. *tanaica* (Bremek.) Bridson	⟨ S	NL	!!! ⟩
Pavetta stenosepala K.Schum			{!}
Pavetta stenosepala subsp. *kisarawensis* (Bremek.) Bridson	⟨ S	S?	!! ⟩

Pavetta stenosepala subsp. *stenosepala* K.Schum.	⟨	s	D	!		⟩
Pavetta tarennoides S.Moore		s	NM	!!		
Polysphaeria braunii K.Krause		t	LD	!		
Polysphaeria multiflora Hiern				{O}		
Polysphaeria multiflora subsp. *multiflora*	⟨	t	LD	O		⟩
Polysphaeria multiflora subsp. *pubescens* (Verdc.) Bridson	⟨	T	LR	¦ z		⟩
Polysphaeria parvifolia Hiern		S	X	¦ n		
Porterandia penduliflora (K.Schum.) Keay		S	X?	!		
Psychotria amboniana K.Schum.		S	X	!		
Psychotria holtzii (K.Schum.) Petit		S	LD	!		
Psychotria lanceolata Hiern		S	LD	Z		
Psychotria lauracea (K.Schum.) Petit		S	M	¦ g		
Psychotria leucopoda Petit		S	X	!		
Psychotria punctata Vatke (Several varieties)		s	D	O		
Psychotria riparia (K.Schum. & K.Kraus.) Petit [2 vars. in coast]		t	LD	Z		
Psychotria schliebenii Petit		S	LR	!		
Psydrax sp.A (Hawthorne 273)		s	N?	!!!		
Rothmannia macrosiphon (Engl.) Bridson		S	D	!		
Rothmannia manganjae (Hiern) Keay		T	M	!		
Rothmannia ravae (Chiov.) Bridson		S	XD	!		
Tarenna drumondii Bridson		t	D	!		
Tarenna littoralis (Hiern) Bridson		t	XL	¦ z		
Tarenna nigrescens (Hook f.) Hiern		S	D	O		
Tarenna supra-axillaris (Hemsley) Bremek.				{!}		
Tarenna supra-axillaris subsp. *supra-axillaris*	⟨	t	LD	!s		⟩
Tricalysia allocalyx Robbrecht		t	SU	!!		
Tricalysia bridsoniana Robbrecht				{!}		
Tricalysia bridsoniana var. *bridsoniana* Robbrecht	⟨	S	ND	!!		⟩
Tricalysia bridsoniana var. *pandensis* Robbrecht	⟨	S	SQ	!!!		⟩
Tricalysia elegans Robbrecht		t	NL	!!!		
Tricalysia microphylla Hiern		t	M	!		
Tricalysia ovalifolia Hiern				{O}		
Tricalysia ovalifolia var.A. FTEA	⟨	s	SU	!!		⟩
Tricalysia ovalifolia var. *glabrata* (Oliv.) Bremek.	⟨	S	D	!		⟩
Tricalysia ovalifolia var. *ovalifolia*	⟨	S	XL	O		⟩
Tricalysia ovalifolia var. *taylorii* (S.Moore) Bremek.	⟨	S	ND	!!		⟩
Tricalysia ruandensis Bremek.		S	MH	¦ z		
Uncaria africana G.Don		c	X	GC		

RUTACEAE

Diphasia morogorensis Kokwaro [2 vars. of this Southern species]		S	SU	!!!		
Diphasia sp.A FTEA		s	NL	!!!		
Teclea amaniensis Engl.		t	MH	!		
Teclea simplicifolia (Eng.) Verdoorn			LD	Af		
Teclea trichocarpa (Engl.) Verdoorn		S	D	Af		
Toddaliopsis sansibarensis (Engl.) Engl.		S	XD	!		
Vepris eugeniifolia (Engl.) Verdoorn		S	D	!s		
Vepris lanceolata (Lam.) G.Don		S	D	O		
Zanthoxylum chalybeum Engl.		t	X	A		
Zanthoxylum holtzianum (Engl.) Waterman				{!}		
Zanthoxylum holtzianum subsp. *holtzianum*	⟨	t	D	!		⟩
Zanthoxylum holtzianum subsp. *tenuipedicellatum* Kokwaro	⟨	t	XS	!!!		⟩
Zanthoxylum paracanthum (Mildbr.) Kokwaro		C	MH	!		

SALVADORACEAE

Dobera loranthifolia   (Warb.) Warb. ex Harms | T LD | ¦ n |

SAPINDACEAE

Allophylus pervillei Blume	S M	!
Aporrhiza nitida Gilg.	T LR	Z
Aporrhiza paniculata Radlk.	T LR	!
Blighia unijugata Bak.	t M	GC+
Chytranthus obliquinervis Radlk.	t LR	!
Crossonephelis africanus (Radlk.) Leenh.	T M	A
Haplocoelum foliolosum (Hiern) Bullock [name possibly too broadly applied]	t D	Z
Haplocoelum inopleum Radlk.	t D	!
Haplocoelum trigonocarpum Radlk.	t LD	?
Lecaniodiscus fraxinifolius Baker		{ v}
Lecaniodiscus fraxinifolius subsp. *scassellatii* Friis	⟨ T ND	!! ⟩
Lecaniodiscus fraxinifolius subsp. *vaughanii* (Dunkley) Friis	⟨ T LM	¦ v ⟩
Lepisanthes senegalensis Radlk.	T LD	A
Macphersonia hildebrandtii O.Hoffm.	T LD	O
Majidea zanguebarica Kirk ex Oliv.	T LM	!
Pancovia golungensis (Hiern) Exell & Mendonza	T LM	A
Paullinia pinnata L.	c D	T
Stadmania oppositifolia Poir	t LD	¦ z

SAPOTACEAE

Afrosersalisia kassneri (Engl.) J.H.Hemsley	S M	!
Aningeria pseudoracemosa J.H.Hemsley	T LR	!
Bequaertiodendron magalismontanum (Sond.) Heine & J.Hemsley (*zeyherella*)	t LR	A
Chrysophyllum viridifolium Wood & Franks	t MH	Af
Inhambanella henriquesii (Engl. & Warb.) Dubard	T MH	¦ z
Malacantha alnifolia (Baker) Pierre	T M	GC
Manilkara discolor (Sond.) J.H.Hemsley	T D	Af
Manilkara sansibarensis (Engl.) Dubard	D LD	!
Manilkara sulcata (Engl.) Dubard	T D	!
Mimusops aedificatoria Mildbr.	T MH	!
Mimusops fruticosa Bojer ex DC.	T LD	O
Pachystela brevipes (Bak.) Engl.	T LM	GC
Pachystela msolo (Engl.) Engl.	T LR	GC
Pachystela subverticillata E.A.Bruce	t NL	!!
Sideroxylon inerme L.		{O}
Sideroxylon inerme subsp. *diospyroides* (Baker) Hemsley	⟨ T LD	! ⟩
Vitellariopsis kirkii (Baker) Dubard	t LD	!

STERCULIACEAE

Byttneria fruticosa K.Schum.	t LD	!!
Byttneria sp.aff. *fruticosa*	t NL	!!!
Cola clavata Mast.	T D	!
Cola microcarpa Brenan	t M	!
Cola octoloboides Brenan	t NL	!!!
Cola porphyrantha Brenan	T NM	!!
Cola sp.* [Pugu Hills]	S X?	?
Cola uloloma Brenan	T M	!
Leptonychia usambarensis K.Schum.	T MH	A
Nesogordonia holtzii (Engl.) Caperon [inc. *N. parvifolia*]	T D	!
Pterygota sp.	T LR	?

Sterculia appendiculata K.Schum.	T	LM	!
Sterculia schliebenii Mildbr.	T	D	!

TACCACEAE
Tacca leontopetaloides (L.) O.Ktze.	M	X	T

THYMELEACEAE
Synaptolepis kirkii Oliv.	s	D	!

TILIACEAE
Carpodiptera africana Mast.	T	D	!
Grewia calymmatosepala K.Schum.	S	MH	!
Grewia conocarpa K.Schum.	t	ST	!!
Grewia goetzeana K.Schum.	T	LD	!
Grewia holstii Burret	C	D	Z
Grewia vaughaniae Exell	S	LD	!

ULMACEAE
Celtis mildbraedii Engl.	T	LM	GC
Celtis wightii Planch.	T	LM	GC
Trema orientalis (L.) Bl.	t	X	GC

URTICACEAE
Laportea lanceolata (Engl.) Chew	H	XD	!
Pilea holstii Engl.	h	M	¦ v
Pouzolzia fadenii Friis & Jellis	H	XN	!!
Urera sansibarica Engl.	C	LM	!
Urera trinervis (Hochst.) Friis & Immelma	R	LM	A

VERBENACEAE
Clerodendrum capitatum (Willd.) Schum. & Thonn.	s	D	GC
Holmskioldia sp.nov.	T	NL	!!?
Premna hildebrandtii Guerke	t	LD	?
Vitex zanzibarensis Vatke	T	ST	!!

VIOLACEAE
Hybanthus enneaspermus (L.) F.Muell.	t	X	A
Rinorea angustifolia (Thouars) Baill.			{GC}
Rinorea angustifolia subsp. *ardisiiflora* (Oliv.) Grey-Wilson	⟨ t	M	¦ z ⟩
Rinorea arborea (Thouars) Baill.	t	M	OM
Rinorea elliptica (Oliv.) O.Kuntze [incl. subsp. *comorensis* (Oliv.) Grey-Wilson]	t	LM	O
Rinorea ferruginea Engl.	S	M	¦ z
Rinorea ilicifolia (Welw. & Oliv.) Kuntze	S	LM	GC
Rinorea squamosa (Tul.) Baill.			{OM}
Rinorea squamosa subsp. *kassneri* (Engl.) Grey-Wilson	⟨ S	M	! ⟩
Rinorea welwitschii (Oliv.) Kuntze			{GC}
Rinorea welwitschii subsp. *tanzanica* Grey-Wilson	⟨ t	ST	!! ⟩

VITACEAE
Cissus quadrangularis L.	c	x	?
Cyphostemma hildebrandtii (Gilg) Wild & R.B.Drumm.	c	X	Z
Rhoicissus revoilii Planch.	C	D	A

ZINGIBERACEAE
Aframomum alboviolaceum (Ridley) K.Schum.	H	X	A
Aframomum amaniense Loes.	H	MH	!

Aframomum angustifolium (Sonnerat.) K.Schum.	H	LM	A
Aframomum orientale J.M.Lock	M	MH	!
Clerodendrum incisum Klotsch	S	X	?
Costus afer Ker-Gawl	H	X	GC
Costus sarmentosus Bojer	H	MH	!
Siphonochilus brachystemon (K.Schum.) B.L.Burtt	h	X?	¦
Siphonochilus kirkii (Hook f.) B.L.Burtt	h	X	¦

Pteridophytes (misc. families)

Actiniopteris radiata (Sw.) Link	h	LD	T
Adiantum incisum Forssk.	h	LM	T
Asplenium buettneri Hieron.	h	M	A
Asplenium emarginatum Beauv.	h	LM	?
Asplenium nidus L.	h	LR	T
Lindsaea ensifolia Swartz	h	M	A
Microgramma owariensis (Desv.) Alston	R	M	T
Microsorium scolopendrium (Burm. f) Copel.	F	M	T
Pellaea doniana Hool	h	M	A
Pellaea viridis (Forssk.) Prantl	h	LD	T

PART III

Forest fauna of eastern Africa

6 Biogeography of East African montane forest millipedes

RICHARD L. HOFFMAN

Abstract

Knowledge of the montane diplopod fauna of East Africa dates back only to the collections made by Sjoestedt on Mount Kilimanjaro in 1905–6 and described by Attems in 1909. Little was added to that beginning until the onset of explorations since about 1964 by personnel from the universities of Dar es Salaam and Copenhagen. Since the great majority of species discovered in the Tanzanian Eastern Arc mountains are undescribed endemics, it is not possible to present a comprehensive chorographic analysis. Only the species of Oxydesmidae have been worked out (Hoffman, 1990) for the entire region, and of the various geographic units only the fauna of the East Usambaras is at all well known. Nonetheless it is possible to indicate some generalities of interest: (i) with few exceptions most of the genera occurring in these mountains are endemic, so that lines of affinity with other regions must be sought at the level of tribe or higher; (ii) in most cases such genera appear to be the result of local derivation from formerly widespread ancestral stocks; (iii) the postglacial condensation of montane forest to higher mountains surrounded by seasonally arid savanna or scrub forest has resulted in profuse local speciation on individual ranges or close clusters, involving, so far as can be deduced, both sympatric and allopatric isolating mechanisms. These constellations of local species may thus be classified as neoendemics in the East African fauna.

In the few groups of Diplopoda so far studied in any detail, the taxa of the Tanzanian montane forests are biogeographically related to the fauna of the Congo and West African rain forests and thus relictual from previous transcontinental sylvan habitats.

Introduction

Species of the class Diplopoda, and particularly those of the polydesmoid family Oxydesmidae, exhibit an astonishing display of endemism and local insular speciation in the Tanzanian mountain ranges considered in this volume. Perhaps no other group of organisms has responded to the effects of fragmentation and isolation to the extent revealed by recent studies.

Regrettably, as will become evident, knowledge of these phenomena has only now reached a threshold level, and for the most part has accumulated only during the past two decades. Owing to the paucity of interested specialists, the collection of fresh material has far outpaced the rate of taxonomic analysis. Present information permits only a tantalising glimpse of an iceberg tip, a shadowy perception of what awaits revelation as both field and laboratory studies proceed.

Although the first millipedes endemic to the montane forests of East Africa were collected as long ago as the 1890s, the first real contribution of our knowledge of that fauna was made in 1909, when Graf Carl Attems published the results of his study of material collected by Ynge Sjoestedt on Mount Kilimanjaro, totalling 19 species. In the following decades a few other species were described from the East Usambaras, the Ulugurus, and the Uzungwas, but no real impetus was gained until around 1964, when intensive field studies were inaugurated by zoologists from the universities of Dar es Salaam and Copenhagen. Since then collecting activities have been extended, especially by the former group, to many areas and biotic regions in Tanzania, giving for the first time a base-line concept of the East African diplopod fauna from which to extract and contrast the true montane fauna.

At the present, approximately 212 nominal species are known from Tanzania and about twice that number are still undescribed: extrapolation allows an estimate of at least 1000 endemic Tanzanian millipede species, the majority of them probably being montane species. During the early period of alpha taxonomy in this group and region, roughly 1864–1964, the majority of species were described (and have remained known) from single specimens with faulty or inadequate locality data. Only through the recent efforts of K. M. Howell and his colleagues and students has it been possible to perceive the remarkable biogeographic–evolutionary potential of this very rich and diversified fauna. Specialists working in better-known groups of both animals and plants are especially reminded that so far only one modern revision of an African family of Diplopoda has been published, and that, perforce, the few recent generic revisions have been of limited biogeographic significance.

About Diplopoda

Millipedes, in general, comprise one of the least known of the larger classes of arthropods – the present total of nearly 10 000 species is organised into no fewer than 115 families and 15 orders – despite which our knowledge of the group remains at a level approximated by entomology 150 years ago. Millipedes are tracheate arthropods and virtually all of them exist as detritivores in the moist litter and upper soil horizons of deciduous forests around the globe. Their fossil record, although fragmentary, extends back to the Lower Devonian and some groups appear to have changed but little down to the Recent. Such long lineages and evolutionary stability, plus their somewhat parochial preferences of biotope and

lifestyle, render millipedes useful for studies of evolution and dispersal, the more so since all are intolerant of exposure to seawater and most to the smallest desiccation. Alternatively, they share with other organisms the ability to exploit adaptive opportunities by both sympatric and allopatric mechanisms, and impressive bursts of speciation can occur within the same higher taxa that include stabilised and even relictual species.

In other respects millipedes exhibit a wide range of anatomical and biological singularities. They are the only arthropods in which the original body somites have become fused in pairs, forming diplosomites provided with two pairs of legs. The majority of species are equipped with serially segmental defence glands, producing allomones capable of discouraging predators as well as inhibiting fungal growth. Each order of millipedes produces highly characteristic allomones, and the spectrum includes such ingredients as hydrogen cyanide, hydrochloric acid, quinones and phenols. Only the merest beginning has been made so far in exploring the variety, uses and manufacture of these bizarre compounds: some may yet be shown to possess medical properties. Many of the larger species discharge their allomone readily when picked up, sometimes (as in *Dendrostreptus macracanthus* Attems) in the form of a fine spray. Aside from this fairly passive means of defence, millipedes are totally non-aggressive creatures: they do not bite, sting, transmit diseases, nor – except in unusual cases – eat the crops.

The procedures of reproduction are fairly standardised throughout the class. All species practise internal fertilisation and the young hatch from shelled eggs, usually in a larval stage with three pairs of legs. Subsequent growth occurs through a number of stadia and periods of diapause, during which additional segments are produced in a posterior germination area (teloblasty); legs are then provided for the new segments of the previous moult. Some species pass through six or seven immature stages and become sexually mature at the last moult, when one or two pairs of legs are dramatically transformed (in the males) into copulatory organs of sometimes surpassing complexity. The outer ends of the oviducts are correspondingly equipped with sclerotised female counterparts for sperm storage. In other groups, transformation of legs into genitalia begins early in life, and proceeds gradually throughout the young stadia. Adult life may last only a few months, or in some groups for several years with a continued pattern of moulting into larger body size but without increasing the number of segments.

By far the great majority of known species belong to the subclass Helminthomorpha, in which one or both pairs of legs of the 7th segment are modified in males into sperm transfer structures called gonopods. Before mating, the male curves his body anteriorly so that the openings of the vasa deferentia, located on or near the base of the 2nd pair of legs, can place a droplet of semen (or a spermatophore) on the gonopod, from which it will later be transferred to the female genitalia (also located just behind the 2nd pair of legs) during the copulatory phase. Although the study of millipede sexual biology is still only at an early stage, it is now known that a number of species practise fairly complex courtship rituals, some involving stridulation by the male, or the provision of an edible secretion with which a female is lured into a stance that will facilitate further advance by the host animal. Of course, in a large number of species, the male effects copulation simply by the exercise of brute force, although even here complicity of the female may often be noted.

It is well known that amongst many animal groups the reproductive systems may remain the most conservative element of internal anatomy, so that any modifications reflect evolutionary events of major importance; yet, on the other hand, those parts involved with sperm transfer may represent distinctions at the species level with utter fidelity. This is especially well illustrated in arthropods: the pedipalps of male spiders, the first pleopods of larger crustaceans, and the terminalia of insects, each of which are indispensable to the systematist working with such creatures. To this spectrum may be added the gonopods of most millipedes, some of which have developed patterns of such complexity as to challenge our comprehension of their possible mode of function. So reliable are the genitalia as indices of the validity of species, that they have understandably provided the raw

material of modern classifications and, to a regrettable extent, have been abused by overworked taxonomists of the past whose descriptions of new species often consisted of a sentence or two about size and coloration, with the inevitable 'Gonopods as figured'. Far too often the figure was totally inadequate, and in the early days of diplopodology, before the value of gonopods was appreciated, they were generally not illustrated (or even mentioned) at all. It can be appreciated that when (about a century ago) species, genera and families began to be based on gonopod structure, many taxa published without any knowledge of their genitalia came to represent major problems especially as regards matters of priority in nomenclature. A large part of modern diplopod taxonomy still consists of redescribing the type material (when it exists at all) of the earlier species which were the basonyms of genera and families.

An additional difficulty incurred by obsessive reliance on gonopods in distinguishing species was the tendency to neglect other character systems that often reflected specific differences more strikingly than genital features would do. Realisation of this interesting state of affairs resulted in part from the study of Tanzanian material, amongst which several species of a particular genus from a single locality were found to differ from each other in size, coloration and details of the body surface in very clear-cut ways, whilst the gonopods showed little or no differentiation. Further studies have suggested that this situation represents an early stage in the process of sympatric speciation, perhaps initiated by adaptation to microniches, in which selective pressures have affected overall body form more rapidly than the genitalia have had the ability (or necessity) to 'catch up'. This phenomenon is especially evident in the montane fauna of oxydesmid millipedes and will be mentioned especially in connection with that family, but it is noted also amongst genera of the Spirostreptidae, for instance.

In general, however, most millipede taxonomy remains heavily based on characters of the male genitalia, and for this reason the collector is urged to continue his efforts afield to the point of ensuring that at least a single male is preserved if at all possible! Living specimens can be gently unrolled sufficiently that the 7th segment is visible from below: in polydesmoid species the two gonopods will be readily visible as large sclerotised objects. In the 'juliform' groups the genitalia are contained inside the 7th segment but the apices are usually visible and the 6th and 7th segments are somewhat enlarged over the others.

Anyone collecting in tropical parts of the world, especially if somewhat distant from settlements and plantations, can be certain that even the most casual collections of millipedes will contain a large proportion of new or rare species, and new genera are still commonplace. Probably only one-eighth of the existing millipede fauna has so far been described, and most of the tropical species are still known only from the original locality.

Survey of selected montane forest diplopod taxa

The following account of millipede distribution in the montane forests of East Africa is of necessity based largely upon the conditions that prevail in Tanzania, the only country in that region whose diplopod fauna is at all well known. Reference to other African states is generally made only in the most general terms, for reasons which will be only too evident.

Presently five orders and 11 families of Diplopoda have been found in Tanzania. Of the latter, current knowledge of the classification and distribution permits reference to only about six, the other families being either poorly collected or in need of complete revision (or both); this limitation applies especially to those families that include small to tiny species. Of the larger and relatively better-known taxa, only the Oxydesmidae has been studied in detail (Hoffman, 1990) and will therefore claim the lion's share of attention in the following account. Other work (R.L.H., unpublished data) suggests that the detailed distributional patterns in other families are of at least equal interest and likely to yield substantiating data as soon as their systematic status has been upgraded.

For the present, it must suffice to present a brief indication of the condition of these five abundant and widespread Tanzanian families in the light of their occurrence in the arc of block-fault mountains.

Family Spirostreptidae. One of the largest of all diplopod families, the Spirostreptidae is amphiatlantic with about equal diversification in the Neotropical and Ethiopian regions. The number of African genera is not yet established, but 60 is a minimal estimate. Spirostreptids tend to favour wooded habitats, and only a few of the larger species have adapted to semi-arid savanna or scrub forest habitats. So far, 21 genera have been found in Tanzania, several of which appear to be partly or entirely confined to montane forest. Unfortunately, three of these taxa remain to be defined and published.

Haplogonopus (Verhoeff, 1941) is represented by a single species, *H. inflatannulus* Verhoeff, in the Uluguru Mountains. This large animal is characterised by the distinct elevation of the metazona and simplified gonopod telopodites, and has, so far as is now known, no close relatives.

Pseudotibiozus (Demange, 1978) is known from *P. cerasopus* (Attems) in the East Usambaras, *P. anaulax* (Attems) from the Wami River valley east of the Ngurus, and *P. sulcatulus* (Pocock) from central eastern Kenya. There are also undescribed species in the West Usambaras, Ulugurus, Uzungwas, and northern Mbarikas in the montane arc, and from Pugu Forest and Zanzibar Island in coastal rain forest. The gonopods of all these species are quite similar but marked differences occur in size, colour, body texture, collum shape, and other peripheral characters. A fairly young evolutionary status is assumed for this genus, which has not yet been placed with reference to close relatives. Undoubtedly many more undescribed species will be discovered.

'Undescribed genus I' includes only *Spirostreptus montanus* Attems, from Kilimanjaro. 'Undescribed genus II' will be based on *S. strongylotropis* Attems, which is known from the East Usambaras and Ngurus, as well as Pugu Forest on the coast. 'Undescribed genus III' will encompass *Epistreptus austerus* Attems and *E. hamatus* Demange from the East Usambaras, plus a number of undescribed species from Zanzibar, the Uzungwas (at Mwanihana and Mufindi), and the Ruaha lowlands near Sanje. Although montane conditions are not obligatory for the existence of these species they are all associated with rain forest, and at least the majority appear to be restricted to elevations above 1000 m a.s.l. As in the case of *Pseudotibiozus*, gonopod structure is relatively uniform whilst substantial differences are manifest in non-sexual character systems. This genus is closely related to *Charactopygus* (DeSaussure & Zehntner, 1902), which is known from several insular species in the Indian Ocean; some specialists might combine them. Otherwise, it is endemic to Tanzania with no close relatives elsewhere on the continent.

The numerous genera with mostly lowland Tanzanian species have geographic affinities with the south (e.g. *Doratogonus*, *Plagiotaphrus*) or west (*Triaenostreptus*); it is not yet possible to postulate either phylogenetic or chorographic antecedents for any of them.

Family Harpagophoridae. Generally considered to be the sister group for the preceding taxon, this family is dominantly represented in Southeast Asia and the East Indies; only a few genera occur in East and South Africa. Clear-cut distinctions between the two families have not been established, and possibly several 'spirostreptid' genera will later be transferred to the Harpagophoridae.

Three genera occur in Tanzania. *Obelostreptus* (Attems, 1909) is represented by *O. proximospinosus* Krabbe in the East Usambaras; the type and only other known species (*O. acifer* Attems) was found in the central highlands of Ethiopia. The Usambaras harbour an endemic genus, *Apoctenophora* Hoffman & Howell (1980), with one species, *A. enghoffi* Demange in the eastern segment at Amani, and a second, *A. trachypyga* Hoffman & Howell in the western at Mazumbai. Both of these genera seem taxonomically quite disjunct with no known closely related groups. The third genus is *Zinophora* (Chamberlin, 1927), which occupies a wide range in South Africa and has species adapted to rain forest conditions, although most occur in semi-arid habitats. One undescribed species has been found

in the southern interior at Matemanga (Tunduru District), presumably in savanna habitat. *Z. knipperi* (Kraus) was described from Mufindi in the southern Uzungwas; it or a closely allied sibling species has been found also at Sao Hill and Sumbawanga. This (or these) species is very closely related to *Z. distincta* (Carl) of southern Katanga and probably does not represent a montane forest obligate. Nor, apparently, are any of the dozen or so other members of the genus.

Other kinds of harpagophorids should occur in Tanzania, including most of the Eastern Arc mountains. A condensed or fragmented distributional area is implied by the isolated localities of the rather disjunct taxa.

Family Odontopygidae. This third component of the spirostreptomorph millipedes is strictly endemic to Africa, ranging from Sierra Leone to Ethiopia and south to Cape Peninsula; the greatest generic diversification appears to occur in central Africa (Zaire–Tanzania–Angola) although those genera extending farthest southward have undergone remarkably profuse local speciation. Many odontopygids are well adapted to semi-arid conditions, but many others appear totally restricted to rain forest biotopes. About 13 genera are known from Tanzania, embracing just over 40 species (undescribed material on hand doubles that total). So far two genera, *Hoffmanides* Kraus 1966 and *Xystopyge* Attems 1909, appear to be Tanzanian endemics, both apparently forest-adapted. The first is monotypic with *H. dissutus* (Hoffman) in the Ulugurus, the second with three species in the northeastern corner of Tanzania. None is restricted to montane forest. In general odontopygids do not seem to be especially abundant at higher elevations and known distributions do not suggest any geographic correlations with the Eastern Arc mountains apart from occurrence of *Hoffmanides* in the Ulugurus (of course, this genus may very well be discovered in the Uzungwas and/or Rubehos).

Family Paradoxosomatidae. This, the largest family of Diplopoda, is abundantly represented on all continents except North America and Antarctica. Subsaharan African countries can boast no fewer than 20 genera, and Tanzania alone contains 18 genera (half of them, however, still undescribed). Several genera, such as *Eviulisoma* Silvestri, 1910, *Aklerobunus* Attems, 1931, *Xanthodesmus* and *Ectodesmus* Cook, 1898, are polytypic and fairly widespread; some of them have local species apparently endemic to some of the montane forest isolates. A few genera are endemic to Tanzania, their species being montane obligates.

Suohelisoma Hoffman, 1967, with the single species *ulugurense* Hoffman, seems to be a local derivative of the widespread related taxon *Eviulisoma* known so far only from the Uluguru Mountains (without precise locality but probably near Morogoro). *Nasmodesosoma* Hoffman & Howell is so specialised in gonopod structure that its affinities have not yet been perceived. *Eoseviulisoma* Brolemann, 1920, is known from one species on Mount Kilimanjaro (*E. julinum* Attems) and an apparent congener *E. abnorme* Attems in the Rwenzori. It is interesting that so far (and taking undescribed taxa into account) these several genera do not seem to have indulged in much local speciation as described, e.g. for many spirostreptid and oxydesmid genera.

Family Gomphodesmidae. This family is restricted to subsaharan Africa, from Sierra Leone and The Gambia to Cape Peninsula, a distribution paralleling that of the Odontopygidae. The two groups have further in common a tendency for maximal generic diversity in equatorial Africa with the few southward-ranging genera speciating profusely south of the Zambezi. Again, gomphodesmids represent a great biogeographic potential presently unavailable because of the lack of modern revisions. Only a few genera have so far been monographed, out of the 23 currently recognised. Thirteen genera occur in Tanzania (the largest number for any single country) but unpublished information to hand indicates that an equal number remain to be defined and published, some of them endemics in the eastern mountain arc. What little is known about the local distribution of the named taxa may be briefly summarised.

The nominate genus *Gomphodesmus* is so far monotypic, the type species *G. castaneus* Cook restricted to the East Usambaras. A closely related genus *Usambaranus* (Hoffman & Howell, 1983) is likewise monotypic, *U. stuarti* Hoffman & Howell being known only from the West Usambaras. Presumably differentiation of these genera has been effected or enhanced by the Lwengera Valley; if so, an accelerated evolutionary rate would seem indicated for the taxa involved.

Uluguria inexpectata (Hoffman, 1964) is so far known only from the Uluguru Mountains (exact locality unknown), and both the species and genus are doubtless restricted to those mountains. Related genera occur to the north around Kilimanjaro and eastern Kenya; none so far are known south of the Ulugurus.

Mychodesmus micramma (Cook, 1898) was described only from 'Mpwapwa'. However, recently collected material suggests that this species (and genus) is endemic in the Rubehos, with the type specimen carrying only the name of the original territory (or the nearest large settlement at the time).

Pending completion of a family revision no inferences can be drawn concerning the geographic affinities of the genera mentioned above. It does seem, however, that gomphodesmids follow the general pattern of local speciation associated with climatic fragmentation of former rain forest.

Family Oxydesmidae. This endemic African family is the subject of a recently completed monograph (Hoffman, 1990) and details about its representatives in Tanzania can be presented with some confidence. Two of the recognised subfamilies (Oxydesminae and Plagiodesminae) occur in central and western Africa and enter the eastern countries only marginally in Uganda. However, the third, Orodesminae, is widespread from Ethiopia and Sudan south as far as Mozambique and Malawi, represented in this area by three tribes and 19 genera. Excluding those which are extralimital to the area covered in this book, we have the following taxa: Tribe Orodesmini, *Orodesmus* Cook, 1895; Tribe Mimodesmini, *Morogodesmus* Hoffman, 1967, *Gonepacra* Hoffman, 1986, *Allotocoproctus* Hoffman, 1986; Tribe

Ctenodesmini, *Ctenodesmus* Cook, 1896, *Ceratodesmus* Cook, 1896, *Rhododesmus* Cook, 1896, *Lyodesmus* Cook, 1896, *Iringius* Hoffman, 1967, and *Orodesminus* Hoffman, 1965.

Orodesmus is basically a genus of the Kenya–Somalia region, which reaches Tanzania only marginally along the Kenya border (*O. camelus* Cook is known only from 'Tanga') and is more adapted to semi-arid country than to rain forest.

Three genera of the Mimodesmini, however, appear restricted to the block-fault mountains. Two are so far monotypic: *Morogodesmus restans* Hoffman and *Allocotoproctus stoltzei* Hoffman are known only from the northern Ulugurus; both are relatively specialised within the tribe and the latter is one of the most disjunct species of Oxydesmidae in terms of body-form modifications.

Gonepacra, with seven species (only one of which was described prior to 1986) is a genuine obligate of montane rain forest, occurring in the Uluguru (2 species), Uzungwa (3), Rubeho (1) and Mbarika (1) mountains. The members of this genus differ not only in details of gonopod structure but in external modifications of the body as well, suggesting a fairly early stage in its phylogenetic development. *G. eos* Hoffman of the Ulugurus is closely related to *G. scharffi* Hoffman of the northern Uzungwas; possibly subspecific status better reflects their degree of differentiation although intergradation between the two now seems precluded by lowlands in the Mikumi Region.

Genera of the tribe Ctenodesmini are of particular interest as regards montane distribution upon the exclusion of *Orodesminus*, the two species of which occur in the coastal region from Somalia to Mozambique. *Ctenodesmus* is a small genus (8 species) largely endemic in Kenya whence several species have been found so far only in high mountains of volcanic origin. *C. kibonotanus* (Attems) and *C. basilewskyi* Hoffman are known only from Mount Kilimanjaro; the much larger *C. pectinatus* (Karsch) is confined to lowlands in Tanga Region.

Lyodesmus is the most widespread and speciose (23 taxa now recognised), and probably the most generalised of the ctenodesmine genera. It occurs from eastern Zaïre to Zanzibar Island and southward into Mozambique and Malawi with the majority of the species concentrated in western Tanzania and adjoining parts of Zaïre. In eastern Tanzania the species tend to be widely dispersed and more specialised, perhaps relicts from a former more uniform distribution. Ranging from sea level to nearly 3000 m, the genus has numerous montane forest components. *L. fischeri* appears to be confined to Kilimanjaro and the north Pares, but in what precise biotope remains a little uncertain; its nearest relatives are four species in the mountains bordering on Lake Tanganyika. Two species occur in the Ulugurus, *L. rubidopsis* (Kraus) and *L. kimboza* Hoffman; these are both large and rather disjunct forms, the latter having an even bigger counterpart, *L. dollfusi* (Cook) in the northern Uzungwas. *L. rodgersi* is likewise endemic in the Uzungwas (on Mount Nyumbenito) but in contrast to *dollfusi* is one of the smaller known lyodesmids. *L. ornaticollis* Hoffman and *L. condylostethus* Hoffman appear to be native to the Rubeho Mountains; these species are relatively quite specialised forms both in gonopod structure and in the incipient modifications of anterior metaterga (unusual for this genus). It is a singular fact that no species of *Lyodesmus* occurs within the geographic range of the apparently derivative genus *Ceratodesmus*.

Rhododesmus (2 species) is likewise largely endemic in Tanzania: only the lowland species *R. m. mastophorus* (Gerstaecker) extends northward as far as Malindi, Kenya. (*R. m. priodus* Cook was described from Dar es Salaam but has never been found there subsequently and the type doubtless came from some inland locality not yet discovered.) *R. planus* (Kraus) has an interesting distribution along the Eastern Arc: *R. p. planus* occurs in the Ulugurus (including the isolated Malundwe Hill in Mikumi National Park), *R. p. krausi* Hoffman is found only on Pongwe Mountain in the Wami River plain, and *R. p. flavomarginatus* Hoffman represents the species in the northern Uzungwas (at Mwanihana).

Iringius (5 species) is a rather homogeneous taxon endemic in central and eastern Tanzania but not restricted to montane rain forest. *I. rossi* Hoffman was collected '91 miles northeast of

Iringa', which is presumably in the Ruaha Gorge; *I. r. kisarawensis* Hoffman & Howell occurs in the Pugu Forest near Dar es Salaam, *I. enghoffi* Hoffman and *I. orestes* Hoffman represent the genus in the easternmost Uzungwas, *I. minimus* Hoffman at the westernmost, and *I. rungwe* Hoffman is endemic to the Rungwe area. *I. rossi* may be found in the southern Ulugurus, and probably several additional species in the Uzungwa and Rubeho ranges. In gonopod characters, *Iringius* resembles *Ceratodesmus*, from which it may be derived. As regards peripheral features, the species of *Iringius* are certainly much more specialised, having a strongly modified hypoproct as well as such items as reduced pore formulas, areolated metaterga, and spinose metazonal sides.

The pinnacle of localised oxydesmid speciation occurs in *Ceratodesmus* which has nine species in and around the Usambaras (see Table 6.1). Only one species, *C. cristatus* Cook, is a lowland form (widespread in Tanga Region) but it too ascends to 100 m in the East Usambaras. Interestingly, its closest relative appears to be *C. gracilior* Hoffman, known only from Kanga Mountain in the Ngurus and separated by about 100 km from the nearest locality for *cristatus*. Perhaps this situation reflects more extensive distributions in some earlier pluvial stage. Otherwise, the East Usambaras harbour the nominate subspecies of *Ceratodesmus ansatus* Cook, which has the closely related *C. a. hesperius* in the West Usambaras and *C. a. nguruensis* in the Ngurus. The nucleus of this genus is clearly the West Usambaras, where seven taxa are known to occur, two of them subspecies in different forest isolates (*C. c. coriarius* and *C. c. angusticlavius*). Three species are known at Mazumbai: *C. coriarius*, *C. fraterculus*, and *C. mazumbai*, all strikingly distinct in various external characters of size, coloration and ornamentation. One exceptionally interesting relationship is that between *C. msuyai* Hoffman of the West Usambaras (Ambangulu) and its apparent sister species *C. eburatus*, localised on the relatively minute Pongwe Mountain 100 km to the south. The inference that I draw from the distributions of this species pair, plus *cristatus* and *gracilior* and the subspecies of *ansatus*, is that some continuity of rain forest habitat must have existed

between the Usambaras, Pongwe, and Ngurus up to some time after the differentiation of the taxa mentioned. The existing disjunction of habitat (assuming it to be real and not illusory) may be a relatively recent episode, perhaps from the late Pleistocene.

Almost certainly other members of this complex genus will be found when other forest isolates in the Usambaras (and perhaps Pares) have been sampled. The interactions of syntopic ceratodesmids would appear to be a specially interesting problem. The presence of so many taxa in such a limited area suggests a kind of intense insular speciation process may be in operation as well.

The postulation has been made (Hoffman, 1990) that *Lyodesmus* may represent the ancestral ctenodesmoid stock from which the various more local and specialised ctenodesmine genera have evolved by isolation within the condensing rain forest areas of the block-fault and other massifs in Tanzania. Such genera may be characterised as neoendemics in the sense of Braun-Blanquet (1923).

Summary of mountain faunas

In so far as diplopods are concerned, only three areas can be considered as reasonably well known: the East Usambaras around Amani, the northern Ulugurus, and the easternmost end of the Uzungwas at Mwanihana. Of these three, perhaps only half of the resident faunas have been discovered so far. Many of the taxa already collected are still either undescribed or belong in very poorly understood families. A major difficulty remains in deciding which species are really confined to montane forests or simply have been overlooked at lower elevations and different biotopes. Such a distinction is still a problem for the much better known bird faunas, lending appreciation to its relative magnitude as regards millipedes.

The Usambaras have been fairly well studied around Amani, especially through the efforts of Henrik Enghoff, the only diplopodologist to have worked in East Africa so far. The western subgroup has been sampled especially around

Table 6.1. *Distribution of the endemic montane oxydesmids of Tanzania. The low numbers cited for the Nguru and Rubeho mountains doubtless reflect inadequate sampling to date. Occurrences for the Usambaras, Ulugurus, and Uzungwas are more substantive. Former continuity between the Uluguru and Uzungwa ranges is strongly implied by records for* Gonepacra *and* Rhododesmus, *at the species level in the latter case. Affinity of Rungwe Mountain with the Uzungwas will surely be amplified when its fauna has been more thoroughly collected*

| Species | Mountains | | | | | |
	Usambara	Nguru	Uluguru	Rubeho	Uzungwa	Elsewhere
Gonepacra eos			×			
Gonepacra scharffi					×	
Gonepacra mufindi					×	
Gonepacra kiellandi				×		
Gonepacra lescurei			×			
Gonepacra massagati					×	
Gonepacra muhulu						Mahenge Forest
Morogodesmus restans			×			
Allocotoproctus stoltzei			×			
Lyodesmus rubidopsis			×			
Lyodesmus kimboza			×			
Lyodesmus rodgersi					×	
Lyodesmus condylostethus condylostethus					×	
Lyodesmus condylostethus montanus				×		
Lyodesmus dollfusi					×	Mahenge Forest
Lyodesmus ornaticollis				×		
Rhododesmus planus planus			×			
Rhododesmus planus pongwe						Pongwe Mountain
Rhododesmus planus krausi					×	
Ceratodesmus ansatus ansatus	×E					
Ceratodesmus ansatus hesperus	×W					
Ceratodesmus ansatus nguruensis		×				
Ceratodesmus magamba	×W					
Ceratodesmus coriarius	×W					
Ceratodesmus fraterculus	×W					
Ceratodesmus mazumbai	×W					
Ceratodesmus cristatus	×E					Tanga lowlands
Ceratodesmus gracilior		×				
Ceratodesmus eburatus						Pongwe Mountain
Ceratodesmus msuyai	×W					
Iringius rossi rossi					×	
Iringius rossi kisarawensis						Mafia; Pugu Hills
Iringius minimus					×	
Iringius orestes					×	
Iringius enghoffi					×	
Iringius rungwe						Rungwe Mountain
Totals [genera/species]	1/8	1/2	5/7	2/3	4/11	

Mazumbai, but many of the forest isolates in both ranges are still totally unknown.

Of the 13 or 14 Usambarian genera presumed to be 'montane', at least six are strictly endemic. The two harpagophorid genera, *Obelostreptus* and *Apoctenophora*, are of unknown geographic affinity; they seem to have little in common with the South African taxa and may represent a little known (or largely extinct) Ethiopian fauna. The endemic genus and species *Elythesmus enghoffi* (Cryptodesmidae: Thelydesminae) is known only from Amani, and is interesting in showing similarities with the Liberian genus *Thelydesmus*; related taxa may later be found in Central Africa. The affinities of the two gomphodesmid genera *Gomphodesmus* and *Usambaranus* are not yet established, but presumably lie with some other (probably lowland) regional taxa. These two genera (each monotypic) are of interest in that the first occurs in the eastern segment, the latter in the western, of the Usambaras: vicariation at the generic level paralleling the situation in *Apoctenophora* in which two species represent the genus in the two regions. With respect to oxydesmids, it is singular that no species of *Lyodesmus* have yet been found. This family is exclusively represented by *Ceratodesmus*, with two taxa only in the eastern range, and seven in the western, and every indication that many others will be discovered in unsampled forest isolates. Three species may be found at a single locality; one species (*C. ansatus*) has local races in both the East and West Usambaras and in the Ngurus; subspeciation may occur between adjacent isolates in the West range (e.g. *C. coriarius*); one species pair (*C. msuyai* and *C. eburatus*) links the Usambaras with Pongwe Mountain far to the south. There is some reason to believe that speciation in *Ceratodesmus* was already well advanced before development of the savanna biome condensed the genus back into the mountains where it is now mostly confined. Biogeographically there are few connections between the Usambaran millipede fauna and that of the Ulugurus. One spirostreptid genus (undescribed!) is represented in both ranges as well as the Uzungwas, but it has lowland species as well and is probably not a rain forest obligate.

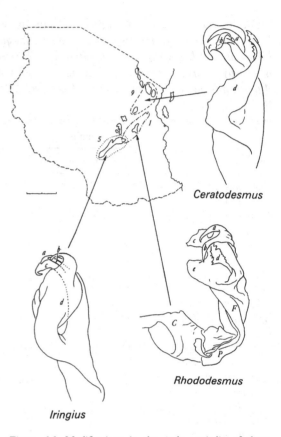

Figure 6.1. Modifications in the male genitalia of three related oxydesmid genera associated with the Eastern Arc Mountains. All four of the distal processes labelled *a*, *b*, *c* and *d* are present, but represented by different configuration. *Ceratodesmus* (upper right), associated with the Usambara and Nguru mountains, has process *b* enlarged and modified into a hoodlike structure and process *c* enlarged and shieldlike. *Rhododesmus* (lower right), occurring in the Uluguru and northern Uzungwa ranges, has process *d* shortened and distally concave with laciniate edges, process *b* elongated and falcate, process *c* reduced in size, and process *e* added. In this drawing of the entire gonopod, *C* indicates the unit homologous to the coxa of the walking legs, *P* identifies the prefemur, and *F* the femur. More distal elements cannot be confidently homologised. *Iringius* (lower left) of the Uzungwa Mountains has process *c* enlarged and partly concealing the bases of the others, which are reduced in size.

All three of the genera have one or more lowland species, but are considered neoendemics of the range indicated, possibly evolving from a widespread common ancestral form. Of the three, *Ceratodesmus* and *Iringius* appear to be most closely related even though neither occurs in the Ulugurus. However, either, both, or some intermediate taxon may be expected to occur in the Rubeho and Ukaguru mountains. Numbers beside the outlined areas indicate how many species of each genus are known from each.

The Uluguru Mountains have been sampled chiefly at the northern end, where accessible from Morogoro, but most of the available material has not yet been studied. Endemism appears to be high: a spirostreptid genus (*Haplogonopus*), and an odontopygid (*Hoffmanides*), two oxydesmid genera (*Allocotoproctus* and *Morogorius*), a gomphodesmid (*Uluguria*), and a paradoxosomatid (*Suohelisoma*) have already been described and yet others are identified (R.L.H., unpublished data). Interestingly, these taxa are all monotypic. Although the oxydesmid genera *Lyodesmus*, *Gonepacra* and *Rhododesmus* occur in the Ulugurus, there is little indication of rampant speciation such as *Ceratodesmus* has produced (although future explorations, particularly in the south, may change that impression). *Rhododesmus planus* has a subspecies on Pongwe Mountain, and another at the easternmost end of the Uzungwas, and sister species of *Gonepacra* occur in the Ulugurus and Uzungwas, suggesting substantial recent connections between the two.

Some elements of the Uluguru fauna occur also in the Mindu Mountains north and west of Morogoro, and at least a local population of *Rhododesmus planus* represents the fauna on Pongwe Mountain (the only place at which a rhododesmid coexists with a *Ceratodesmus*). There are no known connections with the millipedes of the Ngurus or Ukagurus, but of course these ranges are still *terrae incognitae* as regards soil animals.

Proceeding southwest along the mountain arc, knowledge of the Diplopoda regrettably decreases exponentially. For the Uzungwa range, only the vicinity of Mwanihana (above Sanje) has been sampled, and of extant collections, little more than the oxydesmids have been examined. As already noted, these animals imply close former connections with the Ulugurus despite the present lowland savanna barrier between them. There is nothing yet on record in the category of endemic genera, although the oxydesmid genera *Gonepacra* and *Iringius* are nearly so and both seem to have indulged in local speciation (three Uzungwean species in each). There is also some very meagre evidence that the fauna at the western end of the Uzungwas (e.g. around Mufindi) differs from that at Mwanihana, with more influences from the 'Katangan' fauna.

So little is known concerning the millipedes of the Poroto Mountains (in which Mount Rungwe is located), that nothing tangible can be mentioned at this point.

Conclusions

It is almost idle to underscore the point that the Eastern Arc mountains with any remnants of original rain forest represent a natural laboratory of surpassing interest for the study of local evolution and dispersal: this is already known for many plant and animal groups. Millipedes compel attention, however, because of the more intensive degree of endemism induced by their biotope requirements, which for most species are simply the continuous existence of damp broadleaf forest. With rather limited vagility, millipedes naturally respect ecological isolating barriers and local speciation tends to be substantial. For instance, former connections with the forest blocks of the Congo and West Africa are evident only at the family level, there being *no millipede genera in common* between the Tanzanian basement mountain faunas and those of the Congo and Camerounian forests. As noted, differences between the montane faunas even within Tanzania tend to be at the generic level, with few species occurring on more than a single range. This degree of 'fine-tuning' makes the study of millipede evolution particularly intriguing. An additional point is that small soil animals may survive in remnant patches of rain forest too small to support large or even moderate sized vertebrates, so long as enough remains to ensure adequate substrate dampness.

The raw materials for study are in place already. Three major problems remain to impede their utilisation. First and foremost, adequate protection of representative areas must be in effect. Secondly, sampling of the fauna must be undertaken on a comprehensive basis. Thirdly, somebody must be able to work up the materials and translate field collections into published accounts embodying both basic systematics and evolutionary inferences. All three are formidable challenges to face; and, to some extent, the direction of the first and second efforts must depend upon the success of the third. Will necessary support of basic systematic work ever be forthcoming from agencies now enamoured with advanced biochemically based analysis of already known taxa, or will it finally come too late?

References

ATTEMS, C. (1909a). Myriapoda. In *Wissenschaftliche Ergebnisse der Schwedischen Zoologisches Expedition nach dem Kilimandjaro, dem Meru und den Umgebenden Massaisteppen Deutsch- Ostafrikas 1905–1906 unter leitung von Prof. Dr. Yngve Sjoestedt,* Vol. 3, No. 19.

ATTEMS, C. (1909b). Äthiopische Myriopoden. Gesammelte von Prof. O. Neumann und K. v. Erlanger. *Zoologischer Jahrbücher, Abteilung für Systematik* 27, 391–418.

ATTEMS, C. (1914). Afrikanische Spirostreptiden nebst Ueberblick ueber die Spirostreptiden orbis terrarum. *Zoologica (Stuttgart)* 25 (65–66), 1–233.

BRAUN-BLANQUET, J. (1923). *L'Origine et le Developpement des Flores dans le Massif Centrale de France.* Paris.

BROLEMANN, H. W. (1920). Diplopoda. In *Voyage de Ch. Alluaud et R. Jeannel en Afrique Orientale (1911–1912). Resultats scientifiques. Myriapodes,* Vol. III, pp. 49–298.

COOK, O. F. (1895). East African Diplopoda of the suborder Polydesmoidea, collected by Mr. William Astor Chanler. *Proceedings of the United States National Museum* 18, 18–111.

COOK, O. F. (1896). The genera of Oxydesmidae. *Brandtia* 3, 9–12.

COOK, O. F. (1899). African Diplopoda of the family Gomphodesmidae. *Proceedings of the United States National Museum* 21, 677–739.

DEMANGE, J.-M. (1977). Nouveaux Myriapodes de Tanzanie. Description de deux especes nouvelles de Diplopodes. Affinities de quelques genres d'Oxydesmidae. *Bulletin du Museum National d'Histoire Naturelle, Sér. 3,* No. 451, 507–18.

ENGHOFF, H. (1977). Revision of the East African millipede genus *Epibolus* Cook, 1897, with notes on the biology of *E. pulchripes* (Gerstaecker) (Diplopoda: Spirobolida: Pachybolidae). *Entomologica Scandinavica* 8, 1–8.

ENGHOFF, H. & ENGHOFF, I. B. (1976). Notes on myriapods observed and collected in Tanzania and Kenya during the summer 1974. Zoological Museum, Copenhagen, unpublished MS.

HOFFMAN, R. L. (1964). Über einige ostafrikanische Diplopoda Polydesmida der Zoologischen Staatssammlung München. *Opuscula Zoologica (München),* No. 79, pp. 1–10.

HOFFMAN, R. L. (1967). A second contribution to the knowledge of East African orodesmid millipeds. *Revue de zoologie et de botanique africaines* 75, 95–129.

HOFFMAN, R. L. (1990). Myriapoda 4. Family Oxydesmidae. In *Das Tierreich,* lief. 107, pp. 1–515.

HOFFMAN, R. L. & HOWELL, K. M. (1980). Records and descriptions of East African orodesmid millipeds (Polydesmida: Oxydesmidae). *Revue de zoologie et de botanique africaines* 94, 433–56.

HOFFMAN, R. L. & HOWELL, K. M. (1983). *Dendrostreptus,* a new genus for an arboreal Tanzanian millipede, with notes on related forms (Diplopoda: Spirostreptidae). *Revue de zoologie africaine* 97, 625–632.

KRAUS, O. H. (1958). Myriapoden aus Ostafrika (Tanganyika Territory). *Veröffentlichen der Uberseemuseum in Bremen, Ser. A* 3, 1–16.

RIBAUT, H. (1907). Voyage de M. Maurice de Rothschild en Ethopie et dans l'Afrique orientale. *Annales de la Société Entomologique de France* 76, 499–514.

VERHOEFF, K. W. (1941). Studien an äthopischen Diplopoden. *Jenaische Zeitschrift für Naturwissenschaften* 73, 231–74.

7 The Linyphiid spider fauna (Araneae: Linyphiidae) of mountain forests in the Eastern Arc mountains

Introduction

The spider fauna of East Africa is not particularly well known, but one group of spiders, the sheet-web weavers or dwarf spiders (family Linyphiidae), is much better known than any other group. This is quite strange since these spiders are small (most spiders are less than 3 mm in total length) and do not have any remarkable body colours. The family Linyphiidae is one of the largest spider families (second only to the jumping spiders, Salticidae) with some 3600 species described in 403 genera (Platnick, 1989).

The common English names for these spiders (dwarf or sheet-web spiders) are not very appropriate, since linyphiids are not smaller than spiders in several other families and sheet webs resembling those of linyphiids are also built by members of other spider families such as the Theridiidae, Cyatholipidae, Diguetidae, Pholcidae, Araneidae and Uloboridae (Eberhard, 1990). Furthermore, sheet webs are built only by some linyphiids, particularly representatives of the subfamily Linyphiinae.

An impressive amount of information on the East African linyphiid fauna has been published during the last 30 years, especially on the montane fauna. In East Africa, many mountains have now been explored at least once and this has revealed a relatively rich linyphiid fauna. However, 'rich' should be understood in an African context, as the diversity is low if one compares the number of species on each mountain in East Africa with the number of species on mountains in the northern hemisphere. This chapter describes the linyphiid fauna of the Eastern Arc mountain forests (Figures 7.1 and 7.2). Information comes from my own research (Scharff, 1989, 1990a, b, c, 1992) and data published elsewhere.

Forest faunas are difficult to define ecologically, since it is difficult to make a clear ecological distinction between forest and non-forest species, as species of more open habitats and species with broad habitat preferences (transgressor species) intermingle with true forest species. An example of such a transgressor species is *Metaleptyphantes perexiguus* (Simon & Fage, 1922), which is widely distributed in the semi-arid African bushlands but also extends its range into the African humid rain forests. In this chapter, 'forest species' are defined as any species recorded from forests and 'true forest species' as species which have been recorded only from forests, forest edges or gallery forests (= riverine forest), and therefore are thought to be dependent on the forest habitat for survival (see also Scharff, 1992).

The Linyphiid fauna of the Eastern Arc montane forests

From a zoogeographical point of view, areas should be defined by their monophyletic taxa and such a situation does not exist for linyphiids within the geographical boundaries defined in this book. If the boundaries are extended to include the Nyika Plateau and Viphya Mountains, in

Figure 7.1. View over Mwanihana Forest, Uzungwa Mountains.

Figure 7.2. Sanje Falls. Mwanihana Forest, Uzungwa Mountains.

Malawi, the Eastern Arc mountains would be defined by the monophyletic taxon *Ophrynia* which includes 11 species, distributed from the Usambara Mountains in northeastern Tanzania to the Viphya Mountains in Malawi (see Figure 7.3). This geographical definition of the Eastern Arc mountains is very similar to 'The Tanganyika–Nyasa Montane Forest Group' (as defined by Moreau, 1966) based on the distribution of forest birds. The linyphiid fauna of the areas covered in this book is, however, remarkable in two ways:

1. The high number of endemic species on the individual mountains, much higher than on other mountains in central East Africa (northern Tanzania, Kenya and eastern Uganda).

Figure 7.3. Total distribution of the genus *Ophrynia*. 1. *O. uncata* (East and West Usambara) 2. *O. rostrata* (East Usambara) 3. *O. galeata* (Uluguru Mts) 4. *O. infecta* (Uluguru Mts) 5. *O. revelatrix* (Uluguru Mts) 6. *O. summicola* (Uluguru Mts) 7. *O. insulana* (Uzungwa Mts, Mwanihana) 8. *O. perspicuua* (Uzungwa Mts, Mwanihana) 9. *O. juguma* (Uzungwa Mts, Chita) 10. *O. truncatula* (Uzungwa Mts, Chita) 11. *O. superciliosa* (Viphya Mts).

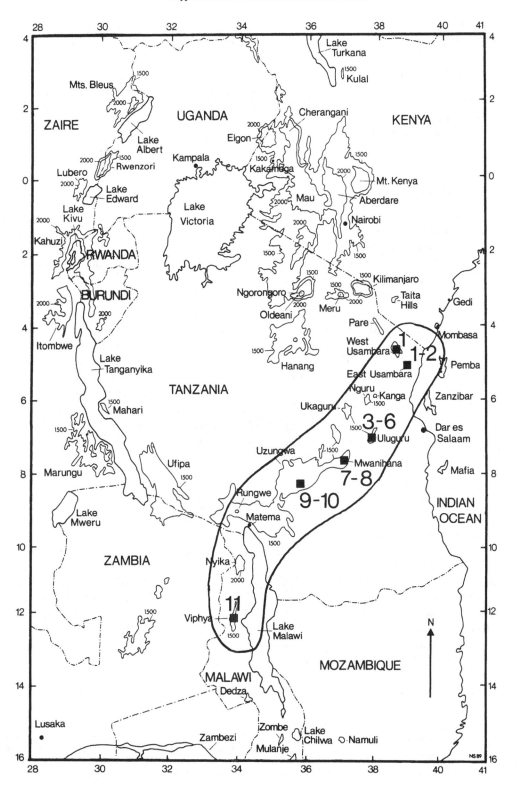

2. The total lack of representatives of genera typical for the mountains in central East Africa and the Albertine Rift (*Trachyneta*, *Afroneta* and *Microcyba*).

Each mountain is thus well defined by a number of endemic species, but the group itself (the Eastern Arc) is artificial (paraphyletic as regards spiders).

Historical background

Even though the investigation of the African linyphiid fauna started in the last century, only 56 species had been discovered and described up to 1960 (Russell-Smith, 1981a). Nearly 30 years later, a total of 358 linyphiid species had been recorded from Africa south of the Sahara (Scharff, 1990a) and many more will no doubt be found in the future. A more detailed historical account for the period before 1960 is given by Holm (1962) and van Helsdingen (1969). Information covering the period after 1960, together with more detailed information on the total East African linyphiid fauna, is given in Scharff (1992).

Our present knowledge and the increased interest in East African linyphiids are based primarily on two papers (Denis, 1962; Holm, 1962), both of which deal with the subfamily Erigoninae (then considered to be a family). Denis (1962) reviewed the erigonines then known from Africa, and Holm (1962) presented a preliminary discussion of the zoogeography of the East African erigonines. Based on a large amount of new material, Holm could show that the 'individual mountains carry numerous endemic taxa, and that these taxa often consist of closely related species belonging to the same genus, and replacing each other on the different mountains' (Holm, 1962: p. 194). He also pointed out that 'endemism increases with height, depending obviously upon the degree of isolation which is greater in the higher, and smaller in the lower zones' (Holm, 1962: p. 195).

With regard to the lower zones of montane forests, Holm found that the few species then known tended to occupy several mountains: 'on the whole we can reckon that the species which occur both in the rain forest on the mountain and in gallery forest along the rivers are widely distributed over the region, and probably occur upon several of the mountains' (Holm, 1962: p. 195). Thus, the fauna in the montane forest zone was not considered sufficiently isolated, or its ability for long-distance dispersal was considered so great, that contact between the different forest areas was envisaged, with a shared fauna on the different mountains as a result.

The last observation was in good accordance with the knowledge at that time about the avifauna of the areas investigated by Holm (1962). These areas have almost no endemic species and most of the birds are shared with other montane forest areas in Kenya, Uganda and eastern Zaïre (Moreau, 1966).

As regards the higher altitudes, above the forest zone, Holm pointed out that the degree of endemism was much higher. The presence of many endemic linyphiids at high altitudes (Holm, 1962) was quite surprising, as these spiders are small, agile and supposed to be good aeronauts (ballooners). Thus Holm's endemic species could easily be rejected with the argument that the proposed endemic species would be found on other East African mountains which had not yet been investigated in 1962.

Many new areas were investigated for linyphiids during the following decades (Figure 7.4) and these collections, together with other information on the environmental history of East Africa, have made it possible to test some of Holm's hypotheses.

Even though the Eastern Arc mountains are well known for their many endemic species among birds, butterflies and reptiles (Loveridge, 1933; Carcasson, 1964; Moreau, 1966), entomological exploration started very late. Apart from isolated descriptions of single species collected by travelling naturalists, no major effort had been made to collect and describe the arthropod fauna of the Eastern Arc mountains until 1957. The first

Figure 7.4. Map of eastern Africa. The stars represent areas that have been investigated for spiders of the family Linyphiidae. Stars within the encircled area represent forest areas investigated by the author.

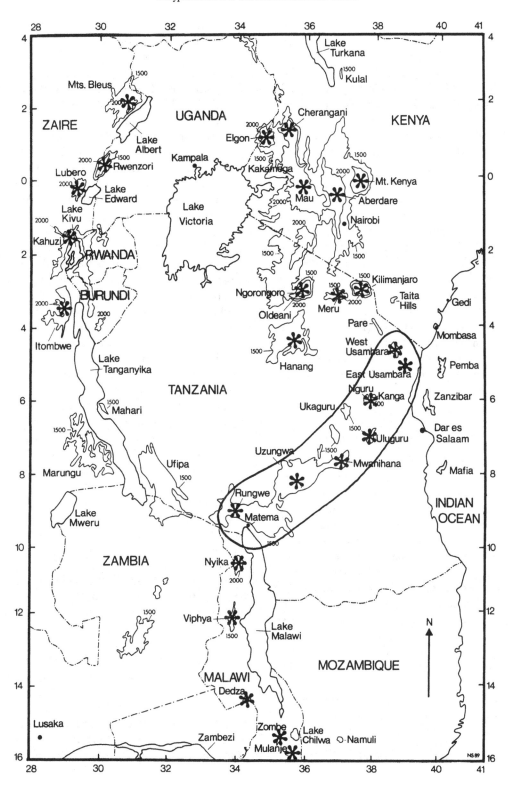

major entomological survey was carried out on the Uluguru Mountains by Belgian teams of entomologists in 1957 (Basilewsky & Leleup, 1960) and again in 1971 (Berger, Leleup & Debecker, 1975). Since 1974, the Zoological Museum in Copenhagen, Denmark, has carried out entomological surveys in the Usambara Mountains (1974, 1975, 1976, 1977 and 1979), Uluguru Mountains (1981, 1982 and 1984), Mount Rungwe (1980), Uzungwa Mountains (1981, 1982 and 1984) and Mount Kanga (1984) (Scharff, 1990c).

Before 1970, only three species of linyphiids were known from the Eastern Arc, *Neriene kibonotensis* (Tullgren, 1910), *Ceratinopsis benoiti* (Holm, 1968) and *Microlinyphia sterilis* (Pavesi, 1883), all from the Ulugurus and all collected by the Belgian entomological expedition of 1957. The linyphiids from this material were described and identified by Holm (1968) and van Helsdingen (1969, 1970). Since 1986, another 58 species have been described and recorded from the Eastern Arc mountains (Jocqué & Scharff, 1986; Scharff, 1990b).

Diversity and distributions

The mountain linyphiid fauna of Tanzania is without doubt more diverse than the 97 species presently reported. Of these, 61 species are known from the Eastern Arc mountains and no fewer than 50 of these are endemic (82%). Several other Tanzanian mountains have not yet been investigated and some have been only very superficially explored. While additional collecting in these areas will probably reveal many new species, I believe that those areas investigated in the Uluguru and Uzungwa mountains have been so well collected that the majority of the linyphiid species from these areas have been found (see discussion). All Eastern Arc species are listed in Table 7.1.

Only 11 species (18%) have a distribution that also covers areas outside the Eastern Arc. They can be divided as follows.

1. **Widespread African species**: 7 species.

Neriene kibonotensis, Metaleptyphantes perexiguus, Meioneta habra, Microlinyphia sterilis, Lepthyphantes extensus, Metaleptyphantes clavator, Bursellia setifera.

2. **Widespread species also found outside Africa**: 1 species. *Erigone prominens* (Africa, Oman, Japan and New Zealand).
3. **African species with disjunct distribution**: 3 species. (a) **Wide disjunctions (East and West Africa):** *Bursellia setifera, Proelauna humicola.* (b) **Close disjunctions (within East Africa):** *Araeoncus victorianyanzae, Metaleptyphantes ovatus.*

If the linyphiid fauna of the Eastern Arc mountains is compared with that of the nearby mountain areas in the north and south, and the widely distributed species mentioned above (1 and 2) are excluded, only two species (*Araeoncus victorianyanzae* and *Metaleptyphantes ovatus*) are shared with the northern areas (Kenya) and one species (*Proelauna humicola*) is shared with areas in the south (Malawi).

Nine of the 19 genera recorded from the Eastern Arc mountains are widespread genera that are also well known from the Palaearctic Region, some even from the Nearctic Region. Some of the East African species described in this group of genera are taxonomically very closely related to Palaearctic species (Holm, 1962; van Helsdingen, 1969). Thus, 14 of the 16 East African species of *Walckenaeria* belong to the peculiar looking species group (subgenus) *Tigellinus* Simon, 1884 (*sensu* Wunderlich, 1972). Species belonging to this group of spiders have a dorsal lobe on the carapace carrying the posterior median eyes. The shape of the lobe, combined with the position of the eyes and spines, makes it look very much like the head of a dog with whiskers (Scharff, 1990b, figs 214 and 223). The African species form a clearly defined group of species most closely related to the European *Walckenaeria furcillata* (Menge, 1869). *Walckenaeria furcillata* is widely distributed in Europe, but it is the only known Palaearctic representative of the species group *Tigellinus* (Holm, 1984).

Table 7.1. *Distribution of the forest linyphiid fauna of the Eastern Arc mountains*

Species	Ru	Uz	Ul	Us	Other distribution
Araeoncus victorianyanzae Berland, 1936		×	×		Kenya 1000–2950 m (F)
Bursellia paghi Jocqué & Scharff, 1986	×				Endemic 1800–1900 m (F)
Bursellia setifera (Denis, 1962)		×			East and West Africa 1100–2300 m (F, WL)
Callitrichia afromontana Scharff, 1990		×			Endemic 1800–1850 m (F)
Callitrichia criniger Scharff, 1990		×			Endemic 1500–1600 m (F)
(*) *Callitrichia mira* (Jocqué & Scharff, 1986)				×	Endemic 1000 m (F)
(*) *Callitrichia pileata* (Jocqué & Scharff, 1986)	×				Endemic 1900 m (F)
(*) *Callitrichia pilosa* (Jocqué & Scharff, 1986)				×	Endemic 1000 m (F)
Callitrichia sellafrontis Scharff, 1990		×			Endemic 1500–1650 m (F)
(*) *Callitrichia simplex* (Jocqué & Scharff, 1986)			×		Endemic 250–1400 m (F)
Ceratinopsis benoiti (Holm, 1968)		×	×		Endemic 600–1300 m (F)
Elgonia annemetteae Scharff, 1990		×			Endemic 1300–1650 m (F)
Elgonia annulata (Jocqué & Scharff, 1986)			×		Endemic 1100–2138 m (F)
Elgonia basalis (Jocqué & Scharff, 1986)		×		×	Endemic 1400–1800 m (F)
Elgonia falciformis Scharff, 1990		×			Endemic 750–1050 m (F)
Elgonia perturbatrix (Jocqué & Scharff, 1986)				×	Endemic 1000 m (F)
Elgonia projecta (Jocqué & Scharff, 1986)		×			Endemic 1300–2138 m (F)
Elgonia rungwensis (Jocqué & Scharff, 1986)	×				Endemic 1800–1900 m (F)
Elgonia stoltzei (Jocqué & Scharff, 1986)				×	Endemic 1600–1900 m (F)
Erigone prominens Bösenberg & Strand, 1906				×	West Africa, Japan, New Zealand, Oman 500–2200 m (F, GR)
Lepthyphantes bakeri Scharff, 1990		×			Endemic 1400–1650 m (F)
Lepthyphantes chita Scharff, 1990		×			Endemic 1300–1500 m (F)
Lepthyphantes dolichoskeles Scharff, 1990		×			Endemic 1500 m (F)

Table 7.1. (*cont.*)

Species	Ru	Uz	Ul	Us	Other distribution
Lepthyphantes extensus Locket, 1968	×	×			Africa, Madagascar, Comoro Is., St Helena Is. 1650–1800 m (F)
Lepthyphantes howelli Jocqué & Scharff, 1986	×				Endemic 1900 m (F)
Lepthyphantes msuyai Scharff, 1990		×			Endemic 1400–1650 m (F)
Lepthyphantes pennatus Scharff, 1990		×			Endemic 600 m (F)
Lepthyphantes phialoides Scharff, 1990		×			Endemic 750–1650 m (F)
Lepthyphantes stramencola Scharff, 1990		×			Endemic 1800–1850 m (F)
Locketidium stuarti Scharff, 1990		×			Endemic 1050 m (F)
Mecynidis antiqua Jocqué & Scharff, 1986			×		Endemic 1600–2400 m (F, GR)
Mecynidis ascia Scharff, 1990		×			Endemic 1400 m (F)
Mecynidis scutata Jocqué & Scharff, 1986	×	×			Endemic 1050–1900 m (F)
Mecynidis spiralis Jocqué & Scharff, 1986			×		Endemic 250 m (F)
Meioneta habra Locket, 1968		×		×	Kenya, West Africa, South Africa >600 m (F, C)
Metaleptyphantes clavator Locket, 1968				×	Angola, Zaïre 600–2100 m (F)
Metaleptyphantes ovatus Scharff, 1990		×			Kenya, Tanzania 300–2300 m (F)
Metaleptyphantes perexiguus (Simon & Fage, 1922)		×	×		Africa, Comoro Is. 30–1600 m (F, GF, GR, WL, C)
Microlinyphia sterilis (Pavesi, 1883)		×	×	×	Africa >1200 m (F, GR)
Neriene kibonotensis (Tullgren, 1910)	×	×	×	×	Africa 20–3500 m (F, GR, GF)
Oedothorax usitatus Jocqué & Scharff, 1986	×				Endemic 1600–1800 m (F, GF)
Ophrynia galeata Jocqué & Scharff, 1986			×		Endemic 1100–1900 m (F)
Ophrynia infecta Jocqué & Scharff, 1986			×		Endemic 2138 m (F)
Ophrynia insulana Scharff, 1990		×			Endemic 600–1850 m (F)
Ophrynia juguma Scharff, 1990		×			Endemic 1050–1650 m (F)
Ophrynia perspicuua Scharff, 1990		×			Endemic 1250 m (F)

Table 7.1. (*cont.*)

Species	Ru	Uz	Ul	Us	Other distribution
Ophrynia revelatrix Jocqué & Scharff, 1986			×		Endemic 1400–1800 m (F)
Ophrynia rostrata Jocqué & Scharff, 1986				×	Endemic 900–1000 m (F)
Ophrynia summicola Jocqué & Scharff, 1986			×		Endemic 2138 m (F)
Ophrynia truncatula Scharff, 1990		×			Endemic 1300 m (F)
Ophrynia uncata Jocqué & Scharff, 1986				×	Endemic 900–1900 m (F)
Pelecopsis papillii Scharff, 1990		×			Endemic 1500–1650 m (F)
Pelecopsis sanje Scharff, 1990		×			Endemic 1650–1850 m (F)
Proelauna humicola (Miller, 1970)		×			Malawi, Angola 500–1100 m (F, WL)
(*) *Toschia virgo* Jocqué & Scharff, 1986			×		Endemic 2100 m (F)
Ulugurella longimana Jocqué & Scharff, 1986			×		Endemic 1600–1900 m (F)
Walckenaeria allopatriae Jocqué & Scharff, 1986			×		Endemic 1600 m (F)
Walckenaeria gologolensis Scharff, 1990		×			Endemic 1650–1850 m (F)
Walckenaeria kigogensis Scharff, 1990		×			Endemic 1700–1900 m (F)
Walckenaeria tanzaniensis Jocqué & Scharff, 1986	×				Endemic 1900 m (F)
Walckenaeria uzungwensis Scharff, 1990		×			Endemic 1650 m (F)

Note: Abbreviations: F, forest; GR, grassland; GF, gallery forest; WL, woodland; C, cultivated land; Uz, Uzungwa Mts; Ul, Uluguru Mts; Us, Usambara Mts; Ru, Mt Rungwe.

(*) According to Jocqué (1983), *Atypena* Simon, 1894 should be a senior synonym of *Callitrichia* Fage, 1936, and Wunderlich (1978) has suggested that *Oedothorax* Bertkau, 1883 should be a senior synonym of *Toschia* di Caporiacco, 1949 and *Callitrichia* Fage, 1936. I prefer to keep the species in the traditional genera until revisions of the genera in question has been carried out. The *Atypena* species described by Jocqué & Scharff (1986) are therefore referred to *Callitrichia*: *Callitrichia mira* (Jocqué & Scharff, 1986); *Callitrichia pileata* (Jocqué & Scharff, 1986); *Callitrichia pilosa* (Jocqué & Scharff, 1986); *Callitrichia simplex* (Jocqué & Scharff, 1986).

Nine genera are endemic to Africa, of which two (*Ulugurella* and *Ophrynia*) are still known only from East Africa and *Ulugurella* only from the Eastern Arc mountains.

One genus, *Metaleptyphantes*, might have a much larger distribution. Members of this genus have also been recorded from Sri Lanka (van Helsdingen, 1985). Very little has been published on linyphiids in other parts of the southern hemisphere (Oriental and Neotropical regions) and the taxonomical affinity of the endemic African genera is therefore unknown. Affinities between

Madagascar and East Africa have been suggested for spiders (Griswold, 1990: p. 199; 1991), but only four species of linyphiids (*Thyreobaeus scutiger* Simon, 1888; *Tmeticides araneiformis* Strand, 1907; *Thapsagus pulcher* Simon, 1894; *Lepthyphantes extensus* Locket, 1968) are known from Madagascar. The first three species represent endemic, monotypic genera. The fourth is a widespread species known also from Africa south of the Sahara and from the Comoro Islands.

Although nearly half of the genera in the Eastern Arc are shared with the northern hemisphere, it is the endemic African genera which dominate in numbers of species. The Eastern Arc shares 47% of its genera with the Holarctic, but these genera hold only 36% of the species. The majority of the Eastern Arc mountain linyphiids (64%) belong to the endemic African genera.

Three of the most species-rich genera (*Ophrynia*, *Elgonia* and *Callitrichia*) in the Eastern Arc mountain forests are endemic to Africa and make up nearly 42% of the total linyphiid fauna. Each of these genera is represented by closely related species that replace one another on the individual Eastern Arc mountains. This pattern has been confirmed when new forest areas in the Eastern Arc have been explored. Thus, two new isolated forest areas in the Uzungwa Mountains (Uzungwa Scarp Forest Reserve and Kigogo Forest Reserve) were explored in 1984. New representatives of both *Ophrynia* and *Callitrichia* were found there. The distribution of the genus *Ophrynia* illustrates this kind of distribution pattern (Figure 7.3).

The total lack of representatives of the genera *Afroneta*, *Trachyneta* and *Microcyba* in the Eastern Arc mountains is peculiar, as species of these genera are well represented in the neighbouring montane forest groups of Kenya and eastern Zaïre. The lack of high altitudes (above 3000 m) in the Eastern Arc mountains could be an explanation, since most of these species occur at high altitudes in the alpine habitats above the montane forest belt. This cannot, however, be the whole explanation, since representatives of these genera are also found in the montane forests at lower altitudes (below 3000 m). The problem will be further discussed later in this chapter.

Endemism of the montane Linyphiids of the Eastern Arc mountains

From Table 7.2 it can be seen that although only one of the genera in the Eastern Arc Montane Forest Group is a possible endemic, the majority being either endemic to Africa or shared with the northern hemisphere, a very high proportion of the species are endemic.

If we look at the total linyphiid fauna of each mountain in the Eastern Arc, endemism is high on all mountains but seems to be highest on the Uluguru and Uzungwa mountains (Table 7.3). Interestingly, these are also the best investigated mountains in the Eastern Arc. Thus, in contrast to what might have been expected, further fieldwork in the Uzungwa and Uluguru mountains has revealed a higher proportion of endemic species – not just new range extensions of species from other areas.

If the linyphiid fauna of the best investigated mountains in the Eastern Arc is compared with those of the best investigated mountains in Kenya (belonging to the Central East Africa Forest Group), there is no big difference in the number of true forest species from single forest areas on

Table 7.2. *Endemism of genera and species of linyphiid spiders in the Eastern Arc mountains*

Taxa	Total number of taxa	Number of endemic taxa	Percentage endemism
Genera	20	1	5
Species	61	50	82

Table 7.3. *Species richness and endemism of linyphiid spiders on selected mountains in the Eastern Arc mountains*

Mountain	Number of species	Number of endemics	Percentage endemism
Usambara	12	7	58
Uluguru	17	12	70
Uzungwa	35	25	71

Source: Scharff (1992).

Table 7.4. *Species richness and endemism of 'true forest species' of linyphiid spiders within four forest areas in East Africa*

Site	Number of species	Number of endemics	Percentage endemism
Uluguru Mts	14	12	86
Uzungwa Mts	31	25	81
Mt Kenya	15	5	33
Mt Elgon	16	4	25

Note: The high number of forest species recorded from the Uzungwa Mts represent data from several forest areas.
Source: Scharff (1992).

each mountain, but there are big differences in the degree of endemism – the moist forest areas of individual mountains of the Eastern Arc have a much higher degree of endemism (Table 7.4) (Scharff, 1992).

Altitudinal distribution of species

Very little is known about the altitudinal distribution of East African linyphiid species. Studies carried out on Mount Kenya in Kenya have shown that the linyphiid fauna of the montane forest area on this mountain is very different from that of the higher Afroalpine area (Holm, 1962; Bosmans, 1977, 1979), but very little information is available from the East African lowland areas. Thus, nothing is known about the difference, if any, between the fauna of lowland and highland forests. A further complication is the fact that there are only a few areas left in East Africa where studies of forest communities can be carried out in a continuum from lowland to highland forest habitats, since most areas have lost their forest cover below 1500–2000 m during the last 100 years. One of the very few areas where such a study can still be carried out is in the Eastern Arc mountains, where the eastern side of the Uzungwa Mountains and Mount Kanga carries a continuous forest cover from about 300 m to 2000–2500 m. Collection along altitudinal gradients in the Uluguru and Uzungwa moun-

tains in the period 1980–4 revealed the following patterns (Tables 7.5 and 7.7).

In the Mwanihana Forest Reserve, where collecting was carried out at 100 m intervals along an altitudinal gradient from 300 to 1850 m, it was not possible to show any dividing line between a lowland and highland fauna. Thus, even though there is a clear difference between the species composition of the linyphiid fauna of the lowest and highest parts of the forest, it is not possible to show any critical altitude separating lowland forest from highland forest.

An intermediate rain forest zone, where species from higher altitudes have their lower limits and where species from lower altitudes have their upper limits have been suggested by ornithologists (Moreau, 1935) for the Usambara Mountains, between 750 and 1350 m. A similar intermediate forest zone has been suggested for the eastern part of the Uluguru Mountains between 800 and 1500 m, by botanists (Pocs, 1976) and for the Uzungwa Mountains between 750 and 1250 m (Rodgers & Homewood, 1982). From Tables 7.5–7.7 it can be seen that such a zone of mixture does also exist for the linyphiid fauna, but the extension of the zone differs between species and also between the two mountains investigated (Uzungwa and Uluguru). Furthermore, it is not possible to find any species restricted to an intermediate rain forest zone; the linyphiid fauna changes gradually.

An interesting case of altitudinal distribution is shown by species of the genus *Ophrynia* on the Uluguru Mountains (Table 7.5). A remarkably high number of *Ophrynia* species have been recorded from this mountain and the species do not show an even distribution. Two species (*O. summicola* and *O. infecta*) are restricted to the ridge top (elfin forest) and have not been found below 2100 m. Another species (*O. galeata*) is found above 1800 m on the western side of the mountain and down to 1100 m on the eastern side (near the present man-made forest edge on the eastern side). The fourth species (*O. revelatrix*) has been recorded only from the western side where it occurs between 1400 and 1800 m (a man-made forest edge has been created on the western side, below Lupanga Peak, at 1400 m).

Table 7.5. *Distribution of species in the Uluguru Mountains, Tanzania*

Species	Altitude (m)															
	1400	1500	1600	1700	1800	1900	2138	1600	1300	1200	1100	250	1700	2400	2100	1700
Araeoncus victorianyanzae															×	
Callitrichia simplex	×											×				
Ceratinopsis benoiti													×			
Elgonia annulata					×		×	×	×		×					
Elgonia projecta					×			×	×							
Mecynidis antiqua							×							×	×	
Mecynidis spiralis											×					
Metaleptyphantes perexiguus							×									
Microlinyphia sterilis						×								×		×
Neriene kibonotensis												×				
Ophrynia galeata					×		×		×		×			×	×	
Ophrynia infecta						×										
Ophrynia revelatrix	×		×		×											
Ophrynia summicola							×									
Toschia virgo															×	
Ulugurella longimana					×	×	×	×							×	
Walckenaeria allopatriae		×							×							

Note: ×, records; - -, anticipated.

Key to altitudes: from left to right: 1400–2138–1100, Lupanga (northern Uluguru); 250, Kimboza (eastern foothill of northern Uluguru); 1700–1700, Lukwangule (southern Uluguru).

Table 7.6. *Distribution of species in the Usambara Mountains*

Species	Altitude (m)			
	East		West	
	900	1000	1600	1900
Callitrichia mira		×		
Callitrichia pilosa		×		
Elgonia basalis			×	
Elgonia perturbatrix		×		
Elgonia stoltzei			×	×
Erigone prominens		×		
Meioneta habra		×		
Metaleptyphantes clavator		×		
Microlinyphia sterilis			×	
Neriene kibonotensis		×	×	
Ophrynia rostrata	×	×		
Ophrynia uncata	×		×	×

Note: ×, records; - -, anticipated.

Table 7.7. *Distribution of species in the Mwanihana Forest Reserve, Uzungwa Mountains*

Species	Altitude (m)													
	300	500	600	700	800	900	1000	1100	1250	1400	1650	1700	1800	1850
Araeoncus victorianyanzae							×							
Callitrichia afromontana													×	×
Ceratinopsis benoiti			×	--	--	×								
Elgonia basalis							×	×	--	×	×	×	×	×
Lepthyphantes bakeri										×	×			
Lepthyphantes extensus											×			
Lepthyphantes pennatus			×											
Lepthyphantes stramencola													×	×
Locketidium stuarti			×	×	×									
Mecynidis ascia										×				
Meioneta habra			×											
Metaleptyphantes ovatus	×	×	×	×										
Metaleptyphantes perexiguus	×	×	×	×	×	×	×							
Microlinyphia sterilis											×			
Neriene kibonotensis	×	×	×	×	×	×	×	×	×	×	×	×	×	×
Ophrynia insulana		×	×	×	×	×	×	×	×	×	×	×	×	×
Ophrynia perspicuua									×					
Pelecopsis sanje											×	×	×	×
Proelauna humicola				×	×	×	×							
Walckenaeria gologolensis													×	×

Note: ×, records; --, anticipated.
Source: Scharff (1990b).

Such irregular distributions, with differences in altitudinal distribution between the eastern and western sides of the mountain, are in good agreement with the distribution of other East African forest organisms (Hamilton, 1976, 1982) and is probably correlated with the uneven distribution of rainfall over the mountain. The eastern and upper parts of the western side of Uluguru are much wetter than the lower parts of the western side.

Discussion

Diversity

The number of forest species in the Eastern Arc mountains seems low if one judges from the mountains investigated until now. It is low compared with forest linyphiid faunas in the Mediterranean mountains, where a single mountain locality (in a huge mountain range) can hold more species than the total known Eastern Arc mountain linyphiid fauna (Bosmans, Maelfait & De Kimpe, 1986).

The apparently low numbers of linyphiids in the Eastern Arc mountains are in accordance with findings from other African habitats (Bosmans, 1977, 1979; Jocqué, 1981a) and could indicate a low relative number of linyphiid species in Africa. It is important here to emphasise that although this is apparently the case for the linyphiids, other spider taxa show different patterns. Spiders of the families Lycosidae, Salticidae and Gnaphosidae seem to be much more diverse in the tropics (Russell-Smith, 1981b; Jocqué, 1984).

A comparison of the total number of species on well investigated mountains in the Eastern Arc and other parts of eastern Africa reveals very different numbers. This is obviously related to collecting effort, as some areas are better investigated than others, but is also a result of the size and diversity of habitats, ecological conditions and environmental history. Thus, Mount Kenya is a much higher mountain than the Uluguru Mountains and has more different habitats (among others, a well-developed *Erica* and Afroalpine zone). It is therefore not strange that 41 linyphiid species have been recorded from Mount Kenya, while only 17 have been recorded from the Uluguru Mountains. Interestingly, the number of 'true forest species' is nearly the same on the two mountains, with 15 species on Mount Kenya and 14 species on Uluguru Mountains.

Endemism

Perhaps the most remarkable thing about the linyphiid fauna of the Eastern Arc is the very high number of endemic species. This high degree of endemism could be refuted by the argument that the assumed endemic species would be found on other mountains not yet investigated, or that the number of endemic species would be reduced by taxonomic synonymy. However, many mountains have been investigated since the work of Holm (1962), including the Eastern Arc mountains and mountains in Malawi, and this has not changed the fact that some mountains, most notably those of the Eastern Arc, harbour considerable numbers of endemic species. The reason for the high degree of endemism on the Eastern Arc mountains, compared with that of other mountains in Kenya and Tanzania, could be a result of different environmental history, i.e. a difference in the geological and climatological history of these areas which is reflected in the present-day distribution of flora and fauna.

A number of authors have suggested that the high degree of endemism in the Eastern Arc mountains is a result of a long-established forest cover, which has probably changed in extent but always been present and of a rather moist type (Carcasson, 1964; Moreau, 1966; Hamilton,

1976; Schiøtz, 1981). These forests have thus acted as refuge areas for forest species during dry climatic periods in the past. Repeated extensions and contractions of the forest cover, caused by climatic changes, enabled forest faunas of different mountains to make contact in the forest extension phase and caused isolation in the forest contraction phase. If isolation was long enough, speciation would be expected. The genus *Ophrynia* could be taken as an example of this hypothesis (Figure 7.3). Such a distribution could be the result of an ancestral, once widespread population of a *Ophrynia* living in forest which had a larger distribution in East Africa than that at present, bearing in mind the extensive forest destruction during the last 100–200 years. When climatic conditions changed, the forest cover shrank to the small disjunct forest 'refuge' areas we observe today. The ancestral population of spiders was thus fragmented, and because of prolonged reproductive isolation, new species evolved in each refuge area. In this way, a number of closely related species would tend to replace each other geographically.

This pattern could have changed during later forest extensions/contractions where species could perhaps spread to new areas, or to areas which were already occupied by other species. This kind of range extension, where species spread throughout suitable habitats that acted as corridors or bridges, should be distinguished from dispersal, where a species actively moves across a barrier of unsuitable habitat(s). Dispersal cannot be ruled out as a factor responsible for some patterns observed, and could blur the picture of distributions caused by geographical segregation, but both factors should be used with care in the interpretation of the distributions observed.

How the distribution and diversity of the East African linyphiid fauna fit into the theory of 'refuges' cannot be determined by just looking at the data from the Eastern Arc mountains and the Central East Africa Forest Group. It must await a total phylogenetic analysis of monophyletic groups with representatives in these areas and anywhere else (also if they occur outside postulated 'refugia'). These results should then be tested against information on the environmental history (geology, climatic and vegetation history) of the areas. Only then will it be possible to construct zoogeographical hypotheses that are not just speculation.

It should be noted that the theory of the Eastern Arc forest refuges is based on quaternary climatic investigations (including palynological surveys) of the high mountains of East Africa (Mount Kilimanjaro, Mount Kenya, Mount Elgon, and the Rwenzori Mountains) and distributional data of fauna and flora. No study of the vegetational history of the Eastern Arc mountains, to document past composition of vegetation on the individual mountains, has been carried out, and this should have high priority in the future. Valuable bogs exist on both the Ulugurus and on Mount Rungwe.

Linyphiids are able to disperse passively by the wind and it is tempting to use this ability to explain distribution patterns. It is nevertheless difficult to explain the distribution of the Eastern Arc linyphiids as a result of ballooning. The existence of many endemic species within certain genera does not support a theory that postulates frequent and easy contact between closely set mountains. Thus, although the Uluguru Mountains are situated only 250 km south of the Usambara Mountains, they do not share a single true forest species. It seems that even though the distance between the forests is not too great to prevent active dispersal, the habitats lying between the forests act as strong barriers.

According to previous studies (Holm, 1962; Bosmans, 1977, 1979), the linyphiids in the montane forest belt and the gallery forests of the East African mountains should be widely distributed. The distribution of linyphiids in the montane forests of the much lower mountains of the Eastern Arc is, however, very different. There are very few widely distributed linyphiid species and most of these are well-known African transgressor species occurring in a wide range of habitats.

Historical factors can be used to explain why the Eastern Arc mountains do not have any representatives of the genera *Afroneta*, *Trachyneta* and *Microcyba*, even though these mountains should be high enough to harbour these species.

Representatives of these genera do occur on nearby mountains in Zaïre, Kenya and Tanzania. First of all, the fauna of the Eastern Arc could have another origin which did not include these genera or their ancestors, and representatives of these genera have not been able to colonise the Eastern Arc since, even though conditions for colonisation may have been favourable during periods in the past.

Several studies in both tropical and temperate regions have shown that the lower altitudinal limit of montane species varies with the size of the mountain. Thus species which occur on high mountains often descend to altitudes that are lower than the summit of surrounding mountains, from which they are absent (White, 1978). This phenomenon has been explained by various authors as a result of wind exposure, different temperatures at the same altitude on big and small mountains, or exposure to fog. Historical reasons have also been suggested. According to the latter, the montane organisms on the smaller mountains became extinct during a warmer period in the past, whereas they could find refuge in higher altitudes of the taller mountains (White, 1978).

Conclusion

We do not yet know the total Eastern Arc linyphiid fauna. There are still mountains that have never been investigated. Of these, those of importance for the delimitation and distinction of the Eastern Arc Mountain Forest linyphiid fauna should be especially targeted (Pare, Rubeho, Nguru, Kanga and Ukaguru mountains in Tanzania and Taita Hills in Kenya).

As for the linyphiids, and maybe many other groups, delimitation of faunas based on autecological studies of single species does not work. As an alternative we should try to work out the phylogenetic relationships among the genera with endemic representatives on individual Eastern Arc mountains. Through this approach, faunas will be delimited by their pattern of sister-group disjunctions (Griswold, 1991: p. 75).

Based on our present knowledge, the individual mountains in the Eastern Arc appear to harbour a distinct linyphiid fauna with high numbers of endemic forest species, much higher than have been found on any other East African mountains investigated. No species are restricted to a range covering only the Eastern Arc.

The Usambara, Uluguru, and Uzungwa mountains appear to harbour very closely related linyphiid species that replace one another on the individual mountains. Each mountain thus has its own species composition at both low and high altitudes, but it is not possible to draw any critical altitudinal line between a lowland and highland linyphiid fauna or to define a particular intermediate forest linyphiid fauna. There is a gradual change in the species composition from lowland to highland and it is therefore important to preserve the whole continuum of habitats, when and if conservation of these unique forests should be considered.

This zoogeographical pattern indicates that the linyphiid faunas of the Usambara, Uluguru, and Uzungwa mountains have been effectively isolated from each other and from other forest areas for a long period of time. The high degree of taxonomical isolation and geographical replacement of closely related species within certain genera of linyphiids point towards a zoogeographical indicator group of great value.

Summary

Based on present knowledge, and data from the best investigated mountains in Africa, the geologically old East African mountains (mountains in the Albertine Rift and in the Eastern Arc) harbour a more distinct linyphiid fauna, with higher numbers of endemic species, than can be found on the geologically younger East African mountains such as Mount Kilimanjaro, Mount Kenya and Mount Meru. The number of endemic species is high on each of the geographically close, old coastal basement mountains in Tanzania (the Eastern Arc mountains), despite the fact that some linyphiids are known for their strong ability to disperse passively with the help of wind. Other East African mountain areas, which are also geographically close, hold a much larger proportion of widespread species. However, when one compares the total species richness of true forest linyphiids on the different mountains, the geologically old mountains do not show particularly high species richness.

In the northern hemisphere, linyphiids are known for their ability to disperse by ballooning. For unknown reasons, linyphiids move upwards from the lower vegetation to the tips of grasses and branches where they assume a tiptoe position in which the abdomen of the spider is raised while the spider simultaneously pulls out silk from its spinnerets. The silk thread is caught by the wind and the spider is carried away with it, using it as a kind of glider. Air samples taken up to several thousand metres above the ground have shown that spiders make up an important part of the airborne animals (aerial plankton) carried long distances by the wind (Greenstone, 1990). Ballooning has been used to explain the wide distribution of spider species and the occurrence of mainland spider species on isolated oceanic islands (Bristowe, 1939, 1941; Gertsch, 1949; Foelix, 1982).

Linyphiid spiders are found in both lowland and highland forests. Preliminary results from collecting along altitudinal gradients in the Uzungwa Mountains show differences between lowland and highland forest faunas, but the changes are gradual and no critical borderline can be identified (Scharff, 1992).

Investigations carried out on the East African mountains have also revealed a number of genera each with closely related species replacing one another on certain mountains. Each of these mountains has its own representative(s) of particular genera and the distributional ranges of species belonging to these genera are limited to certain geographical areas (Holm, 1962; Jocqué & Scharff, 1986; Scharff, 1990b, 1992). Although only scattered information is available on the East African lowlands (all kinds of habitats), and there are still many unexplored areas in the highlands, general patterns can now be presented and discussed.

Acknowledgements

I would like to extend my sincere thanks to Henrik Enghoff, Konrad Thaler, Bent Muus, Rudy Jocqué, Jon Lovett, Sam Wasser, Kim

Howell and Charles E. Griswold for reading and commenting earlier versions of this paper. I am particularly grateful to the National Research Council in Tanzania for permission to carry out research in the Tanzanian forest reserves. The work was supported by grants from the Danish Natural Science Research Council (grants nos. 11–6542 and 11–7440).

References

BASILEWSKY, P. & LELEUP, N. (1960). Mission Zoologique de l'I.R.S.A.C. en Afrique orientale (P. Basilewsky et N. Leleup, 1957). I. Enumération et description des stations prospectées. *Annales du Musée Royal du Congo Belge (Tervuren)* **81**, 11–26.

BERGER, L., LELEUP, N. & DEBECKER, J. (1975). Mission entomologique du Musée Royal de l'Afrique Centrale aux Monts Uluguru, Tanzanie (L. Berger, N. Leleup et J. Debecker, V–VIII 1971). I. Introduction. *Revue de Zoologie africaine* **89**(3), 673–80.

BOSMANS, R. (1977). Spiders of the subfamily Erigoninae from Mt. Kenya. Scientific report of the Belgian Mt. Kenya Bio-Expedition, No. 3. *Revue de Zoologie africaine* **91**(2), 449–72.

BOSMANS, R. (1979). Spiders of the subfamily Linyphiinae from Mt. Kenya (Araneae; Linyphiidae). Scientific report on the Belgian Bio-Expedition 1975, No. 17. *Revue de Zoologie africaine* **93**(1), 53–100.

BOSMANS, R., MAELFAIT, J.-P. & DE KIMPE, A. (1986). Analysis of the spider communities in an altitudinal gradient in the French and Spanish Pyrenees. *Bulletin, British Arachnological Society* **7**(3), 69–76.

BRISTOWE, W. S. (1939). *The Comity of Spiders I.* London: Ray Society.

BRISTOWE, W. S. (1941). *The Comity of Spiders II.* London: Ray Society.

CARCASSON, R. H. (1964). A preliminary survey of the zoogeography of African butterflies. *East African Wildlife Journal* **2**, 122–57.

DENIS, J. (1962). Notes sur les Erigonides. XX. Erigonides d'Afrique orientale avec quelques remarques sur les Erigonides ethiopiens. *Revue de Zoologie et de botanique africaines* **65**, 169–204.

EBERHARD, W. G. (1990). Function and phylogeny of spider webs. *Annual Review of Ecology and Systematics* **21**, 341–72.

FOELIX, R. F. (1982). *Biology of Spiders*. Cambridge, MA: Harvard University Press.

GERTSCH, W. J. (1949). *American Spiders*. New York: Van Nostrand.

GREENSTONE, M. H. (1990). Meteorological determinants of spider ballooning: the roles of thermals vs. the vertical windspeed gradient in becoming airborne. *Oecologia* **84**, 164–8.

GRISWOLD, C. E. (1990). A revision and phylogenetic analysis of the spider subfamily Phyxelidinae (Araneae, Amaurobiidae). *Bulletin of the American Museum of Natural History* **196**, 1–206.

GRISWOLD, C. E. (1991). Cladistic biogeography of afromontane spiders. *Australian Systematic Botany* **4**(1), 73–89.

HAMILTON, A. C. (1976). The significance of patterns of distribution shown by forest plants and animals in tropical Africa for the reconstruction of upper Pleistocene palaeo-environments: a review. *Palaeoecology of Africa* **9**, 63–97.

HAMILTON, A. C. (1982). *Environmental History of East Africa*. London: Academic Press.

HOLM, Å. (1962). The spider fauna of the East African mountains. *Zoologiska Bidrag från Uppsala* **35**, 19–204.

HOLM, Å. (1968). Spiders of the families Erigonidae and Linyphiidae from East and Central Africa. *Annales, Musée Royal de l'Afrique Centrale, Sciences Zoologiques* **171**, 1–49.

HOLM, Å. (1984). The spider fauna of the East African mountains Part II. The genus *Walckenaeria* Blackwall. *Zoologica Scripta* **13**, 135–53.

JOCQUÉ, R. (1981a). Erigonid spiders from Malawi (Araneida; Linyphiidae). *Revue de Zoologie africaine* **95**(2), 470–92.

JOCQUÉ, R. (1981b). Some linyphiids from Kenya with the description of *Locketidium* n. gen. (Araneida; Linyphiidae). *Revue de Zoologie africaine* **95**(3), 557–69.

JOCQUÉ, R. (1983). Sur la synonymie de *Callitrichia* Fage et *Atypena* Simon avec la redescription de quelques espèces paleotropicales (Araneae, Linyphiidae). *Bulletin du Museum national d'Histoire naturelle, sér. 4*, Section A, No. 2, **5**, 235–45.

JOCQUÉ, R. (1984). Considerations concernant l'abondance relative des araignées errantes et des araignées à toile vivant au niveau du sol. *Revue Arachnologique* **5**(4), 193–204.

JOCQUÉ, R. & SCHARFF, N. (1986). Spiders (Araneae) of the family Linyphiidae from the Tanzanian mountain areas Usambara, Uluguru and

Rungwe. *Annales du Musée royal de l'Afrique Centrale, Sciences Zoologiques* **248**, 1–61.

LOVERIDGE, A. (1933). Reports on the scientific results of an expedition to the southwestern highlands of Tanganyika Territory. I. Introduction and zoogeography. *Bulletin of the Museum of Comparative Zoology* **75**, 1–43.

MOREAU, R. E. (1935). A synecological study of Usambara, Tanganyika Territory, with particular reference to birds. *Journal of Ecology* **23**, 1–43.

MOREAU, R. E. (1966). *The Bird Faunas of Africa and its Islands.* London: Academic Press.

PLATNICK, N. I. (1989). *Advances in Spider Taxonomy 1981–1987.* Manchester: Manchester University Press.

PÓCS, T. (1976). Vegetation mapping in the Uluguru Mountains (Tanzania; East Africa). *Boissiera* **24**, 477–98.

RODGERS, W. A. & HOMEWOOD, K. M. (1982). Biological values and conservation prospects for the forests and primate populations of the Uzungwa Mountains, Tanzania. *Biological Conservation* **24**, 285–304.

RUSSELL-SMITH, A. (1981a). Notes on African Linyphiidae. *Newsletter, British Arachnological Society* **30**, 4–6.

RUSSELL-SMITH, A. (1981b). Seasonal activity and diversity of ground-living spiders in two African savanna habitats. *Bulletin, British Arachnological Society* **5**(4), 145–54.

SCHARFF, N. (1989). New species and records of afrotropical Linyphiidae (Araneae). *Bulletin, British Arachnological Society* **8**(1), 13–20.

SCHARFF, N. (1990a). A catalogue of African Linyphiidae (Araneae). *Steenstrupia* **16**(8), 117–52.

SCHARFF, N. (1990b). Spiders of the family Linyphiidae from the Uzungwa mountains, Tanzania (Araneae). *Entomologica Scandinavica*, Supplement **36**, 1–95.

SCHARFF, N. (1990c). The Zoological exploration of the moist forests of Tanzania carried out by the University of Copenhagen in the period 1968–1988. In *Research for Conservation of Tanzanian Catchment Forests.* Proceedings from a workshop held in Morogoro, Tanzania, 13–17 March 1989, pp. 1–176. ed. I. Hedberg and E. Persson. Uppsala: Uppsala University Press.

SCHARFF, N. (1992). The linyphiid fauna of Eastern Africa (Araneae, Linyphiidae) – distributional patterns, diversity and endemism. *Biological Journal of the Linnean Society* **45**, 117–54.

SCHIØTZ, A. (1981). Short communication – the amphibia in the forest basement hills of Tanzania: a biogeographical indicator group. *African Journal of Ecology* **19**, 205–7.

VAN HELSDINGEN, P. J. (1969). A reclassification of the species of *Linyphia* Latreille based on the functioning of the genitalia (Araneida; Linyphiidae) I. *Zoologische Verhandelingen (Leiden)* **105**, 1–303.

VAN HELSDINGEN, P. J. (1970). A reclassification of the species of *Linyphia* based on the functioning of the genitalia (Araneida, Linyphiidae) II. *Zoologische Verhandelingen (Leiden)* **111**, 1–86.

VAN HELSDINGEN, P. J. (1985). Araneae: Linyphiidae of Sri Lanka, with a note on Erigonidae. *Entomologica Scandinavica*, Supplement **30**, 13–30.

WHITE, F. (1978). The Afromontane Region. In *Biogeography and Ecology of Southern Africa*, ed. W. J. A. Werger, pp. 463–513. The Hague: Junk.

WUNDERLICH, J. (1972). Zur kenntnis der Gattung *Walckenaeria* Blackwall, 1883 unter besonderer Berucksichtigung der europaischen Subgenera und Arten. *Zoologische Beiträge* (N.F.) **18**, 371–427.

WUNDERLICH, J. (1978). Zur kenntnis der Gattung *Oedothorax* Bertkau 1883, *Callitrichia* Fage 1936 und *Toschia* Caporiacco 1949. *Senckenbergiana Biologica* **58** (3/4), 257–60.

8 The montane butterflies of the eastern Afrotropics

R. DE JONG AND T. C. E. CONGDON

Introduction

In the eastern half of Africa the forests are mainly restricted to mountains and surrounded by savanna or even semi-desert. They have very appropriately been compared with an archipelago by White (1981). Their isolation and the high degree of endemism of flora and fauna raise questions about the evolutionary and geographic history of the forests as a whole and of the species living there. As with real islands the central questions are:

1. From the geographic point of view, have these islands always been isolated or is there a history of interconnections?
2. From the biological point of view, how are the species distributed and related, and where do their sister species live?

Starting from an allopatric speciation model there is a causative correlation between the two questions in such a way that the geographic history of the islands must have influenced the evolution of the species living in the islands. Biogeography is concerned with this correlation. To put it in a simple way, the biogeographic question is:

3. Have the species of a particular island originated on the spot (and if so, how about the ancestors) or are they colonists from elsewhere (either by jump dispersal or following a range expansion of the habitat), or a mixture of both (and if so, what are the proportions)?

With real islands the ecological difference between the island and its surroundings (the sea) is so extreme that an origin of the great majority of the terrestrial island organisms from the sea is most unlikely if not out of question. This may be different for ecological islands like forests surrounded by open formations. Here the ecological difference is, in principle, bridgeable. In a genus like *Spialia* (Hesperiidae) with 26 mainly African species only one of which (*S. ploetzi* (Aurivillius)) is a forest species, a change in ecological requirements must have taken place, in this case an adaptation to forest environment by an open formation species (de Jong, 1978). Similarly, and maybe even more instructive, is the genus *Colotis* which has 42 species in Africa, all living in woodland and open, often arid, habitats except *C. elgonensis* which is restricted to highland forest. Indeed, in any monophyletic group of species, some of which live in open formations and some other in forests such a change must be postulated. Thus with ecological islands we must also deal with the question:

4. What is the ecological origin of the constituent species?

In this chapter we shall address these questions by way of the butterflies of the mountain forests of the eastern part of Africa, roughly east of Long. 30° E. Forests are not the only type of natural habitat at higher elevations. There may be marshy areas and large stretches of grassland within the forest belt, and where the mountains are high enough there is a treeless ericaceous belt or even an alpine belt (see Hedberg, 1986). The butterflies of these habitats may have a different history from the forest species. They will be briefly dealt with as an illustration of the complex and dynamic character of the butterfly fauna of the African mountains, but their numbers are low if compared with the rich forest fauna.

The area covered in this chapter is the eastern half of Africa, east of the Congo basin, and from the highlands of Ethiopia in the north to the Zambezi River in the south. We have termed this area 'Eastern Afrotropics' (abbreviated EAT), so as to avoid confusion with the term 'Eastern Africa' which is used in this book to cover a more restricted area. The mountains in this area are complex and consist of ranges of widely different geological origin and age. It would not be possible to deal adequately with all mountains in a single chapter. Therefore we shall make a selection and leave a number of mountains such as Marsabit, Kulal and Karisia Hills in N Kenya and the Taita Hills in SE Kenya out of consideration, except where they may be of assistance in a discussion. Some other, complex mountain ranges are taken as a single range here for the sake of convenience, e.g. 'Ethiopia' (see Hamilton, 1982, fig. 53 for the disjunct nature of the mountain forest in Ethiopia). We do not think this detrimental to our conclusions.

Old mountains are found in the so-called Eastern Arc. These mountains run in a broad arc through eastern Tanzania from the Pare Mountains in the north near the Kenya border through the Usambaras, Nguu, Nguru, Ukaguru, Rubeho, Uluguru and Uzungwa to the Southern Highlands (Figure 8.1). The mountain range continues southward into Malawi (Nyika Plateau and further south). To the north, the Taita Hills in Kenya belong to the same formation. They consist of ancient basement and crystalline rocks at least 1 600 million years old. They were, however, not uplifted all at the same time. Some areas like the Usambara and Uluguru mountains are supposed to be of Jurassic age, while the Nguru and Uzungwa mountains were the result of tilting of huge blocks at about the end of the Miocene (around 12 Myr BP) (Moreau, 1966). Other, higher mountains in East Africa came into being during the Pliocene when the enormous

volcanoes Kilimanjaro, Mount Kenya and Mount Elgon began to pile up and the gigantic block Rwenzori started to rise. In the Pleistocene also, the East African surface was not at rest. Rwenzori continued to rise, the Virunga Volcanoes northeast of Lake Kivu and the Rungwe volcano at the head of Lake Malawi (in contact with the much older Southern Highlands), and the volcanoes Kulal and Marsabit in north Kenya were built up.

The complex geological history of the eastern half of Africa, combined with dramatic climatic changes in the Pleistocene and coinciding with the northern Ice Ages, renders the montane habitats in EAT a potentially most interesting subject for biogeographic analysis.

The mountains of EAT have been very differently studied regarding their butterfly populations. Many distributional records of remote mountain areas in west Tanzania (Ufipa, Mpanda, Kigoma) and south and east Tanzania (Eastern Arc mountains) have become available only in recent years thanks to the indefatigable efforts of J. Kielland.

Methods

To find an answer to the question of the origin and differentiation of the montane butterfly fauna we shall describe several possible scenarios, predict distribution patterns on the basis of these scenarios, and test the predictions against the actual patterns found. An important tool would be cladistic biogeography, in which similar sequential speciation patterns in different groups are used to frame hypotheses about the sequence of connections and disconnections of areas. Prerequisite for the application of this tool is the availability of well-supported phylogenies. These are extremely scarce for montane groups in EAT. In the few instances that phylogenies of montane butterflies are available, there is, moreover, so much distributional overlap that they are not very helpful in this respect, but they are helpful in others, for example by indicating differentiation within Eastern Africa and change of habitat preference. Vigorous as cladistic biogeography may be, it also has its limitations. For instance, it starts

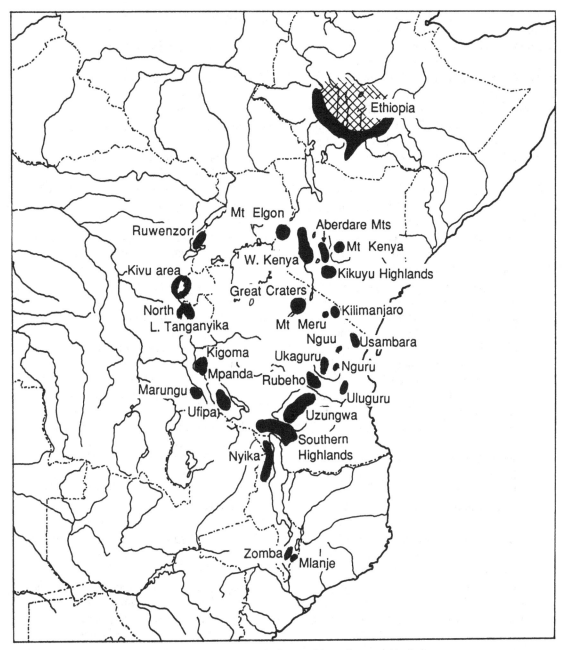

Figure 8.1. Mountain areas in the Eastern Afrotropics as used in the text.

from the idea that by far the most important mode of speciation is allopatric speciation. This claim is not substantiated. It may be true, but it is unwise to exclude other modes of speciation from the start, especially parapatric speciation involving habitat change. If it occurred, it must have resulted in a different distribution pattern.

As in cladistic biogeography we shall define areas of endemism. An area of endemism is, in principle, every area to which a taxon is restricted.

The idea is that a concentration of endemics in a single area is indicative of a prolonged period of isolation. It is only a descriptive term and should not be confused with area of origin, which is an explanation of a pattern found (of course, the two may coincide).

There are so few butterfly species above the timber-line that they will be dealt with in a single paragraph. For the other butterflies we shall proceed along the following lines.

1. Three kinds of habitats are distinguished: (a) Afroalpine (above the timber-line), (b) montane grassland, (c) montane forest. Each will be dealt with separately since they may have different histories of expansion and fragmentation.

2. A description is given of possible scenarios and distribution patterns are predicted based on these scenarios.

3. The present distribution of montane habitats is well known. On this basis a number of potential (not necessary all) areas of endemism are distinguished.

4. Authors may differ as to the taxonomic rank of isolates, i.e. whether a population is a subspecies of a more widely distributed species or if it constitutes a species of its own. We have generally followed Carcasson (1981), unless otherwise stated.

5. The distribution of all species and subspecies restricted to montane habitats in EAT is listed and actual areas of endemism are identified. These areas may consist of one or more of the potential areas of endemism. Areas of endemism are grouped into larger areas of endemism in such a way that the new area has more endemics than the sum of the endemics of the constituent areas. Subspecies have been included since they can break up otherwise uninformative wide distributions. If in a grouping of areas all subspecies of a species are included, that species is another endemic of the combined area and must be added to the further endemics of the combined area.

The same holds for monophyletic groups consisting of endemic species, but as said before, this kind of information is very scarce.

6. For the delimitation of the areas of endemism, only taxa that are endemic to EAT have been taken. Other taxa are of interest in showing relations to other areas, but they cannot be used to delimit a possible area of endemism within the boundaries of EAT.

7. The predictions given above are tested against the actual distributions. Where possible phylogenetic data will be used.

8. New hypotheses will be framed if needed to explain actual distribution patterns. In these cases the kind of evidence needed for testing will be indicated.

9. Area relationship as conceived here is a descriptive term and not explanatory as in cladistic biogeography; two areas are considered related if a taxon (be it an extant taxon or an ancestor) occurs in and is restricted to both; how the relationship originated is a matter of explanation (process), not of definition (pattern).

10. Even though phylogenetic evidence may be lacking, genera are taken to be monophyletic groups.

11. If one species of a genus occurs in area A and all other species in area B, and if there is no evidence of the single species being the sister species of the rest of the genus, the species in area A, or rather its ancestor, is supposed to have originated from area B.

12. Being restricted to a montane habitat is considered a specialisation, i.e. an apomorphy if the sister species is not restricted to this kind of habitat.

Above the timber-line

The upper limit of the montane forest belt in EAT is between 2800 and 3300 m a.s.l. It is usually followed by a belt dominated by Ericaceae or moorland up to 3500–4100 m (Afroalpine moorland: Moreau, 1966; étage afro-subalpin:

Bernardi, 1979). Still higher, the Ericaceae are replaced by an alpine vegetation and eventually by bare rock (étage afro-alpin: Bernardi, 1979). In EAT these belts are found only on the young high mountains, mainly of volcanic origin, like Mount Elgon, the Aberdares, Mount Kenya, Kilimanjaro and Mount Meru. Since the mountains of the Eastern Arc are not high enough to support these belts, the treatment of the other mountains here seems out of place; they are included only for the sake of completeness. Moreover, it must be kept in mind that at the height of the last glaciation *c.* 22 000–14 000 BP, when the vegetational belts were displaced downwards by about 1000 m (Hamilton, 1982, 1989), this type of vegetation must have been much more widespread, and part of the Eastern Arc mountains may have acted as stepping stones for dispersal of organisms adapted to this environment. For the butterflies, however, this is irrelevant since there are no species restricted to this type of habitat in Africa. The butterflies that do occur here are from lower altitudes. Coe (1967) reported that they can be numerous in the alpine zone of Mount Kenya where they may even be found flying above the glaciers. According to Coe these butterflies are brought there by air currents during the day. At night they die from the intense cold and fall into the ice. Next morning the body and wings absorb the sun's heat which melts the ice below the body, and the insect slowly sinks into a small pit. The diversity of these butterflies is apparently very low. Coe mentioned four species only:

Colias electo (Linnaeus, 1763) – a widespread pierid occurring from southwestern Arabia and northern Ethiopia to South Africa, especially in montane grasslands; in South Africa it is one of the commonest and most widespread species (Dickson, 1978) (see also below);

Vanessa cardui (Linnaeus, 1758) – a strongly migratory nymphalid, probably the most widespread butterfly of the world occurring throughout Africa, the warmer parts of Europe and temperate Asia (in spring migrating northward as far as Iceland and northern Scandinavia), most

of the Oriental Region, throughout North America, and as a straggler in Western Australia.

Harpendyreus aequatorialis (Sharpe, 1891) – a lycaenid of the montane grasslands of Kenya and northern Tanzania (see also below);

Zizula hylax (Fabricius, 1775) – a small but very widespread lycaenid occurring throughout the Afrotropical and Oriental regions.

The absence of true alpine butterflies in Africa is surprising in view of the relatively high diversity of alpine butterflies in the Holarctic. In the higher altitudes of the Andes, Descimon (1986) also observed a poor butterfly fauna compared with the Holarctic in spite of the large extent of apparently suitable habitats. He also remarked on the poor butterfly diversity in Africa, but the species he mentioned mainly belong to the montane grassland. He explained the poverty in butterflies of the African oreal regions by 'the youth, the small size and the isolation of these montane islands'. Indeed, the alpine biome in Africa may be not older than 1–2 Myr BP, when the great volcanoes arose, and certainly its isolation from the much older and richer Palaearctic alpine biome was and is so severe that one cannot imagine even a very long distance dispersal. On the other hand, its age was great enough for a highly specialised and endemic flora to develop (see, for example, Hedberg, 1986). The key factor for the species poverty in butterflies may be ecological rather than historical, such as absence of suitable food plants. So far, however, this is speculation.

Montane grasslands

General

Montane non-forest habitats have attracted much less attention than the Afroalpine and montane forest habitats. One reason may be that the habitat is poor in characteristic species. It is not a uniform habitat, varying from woodland to open grassland. We shall confine ourselves to the grassland since this habitat seems to have more characteristic species than the woodlands where most species are the same as at lower elevations.

The montane grassland is difficult to delimit exactly. Not only is the lower limit often obscure, but the grassland may also gradually merge into open woodland and other types of open vegetation. Near Maralal (N Kenya), for example, there are wide stretches of grassland along the eastern rim of the Rift Valley, at over 2000 m. This grassland is gradually replaced locally by more or less open *Juniperus* and *Podocarpus* forest, and south of Maralal it descends to lower elevations and changes into *Acacia* savanna. As a consequence areas of montane grassland are usually not much isolated from each other and are often interconnected by other types of open formations. Moreover, although part of the montane grasslands must be natural, their extent has been greatly enlarged by human activities (Kielland, 1978; Hamilton, 1982) thus increasing the possibility of interchange for the species restricted to this habitat. It is therefore not surprising that endemism in this habitat is rather low in EAT. The situation is different in South Africa where there is a better developed grassland fauna with many endemic species, especially in the Lycaenidae; for example, the genus *Thestor* numbers 23 species which are all restricted to the Cape Province except for two species which reach Transvaal.

An interesting ecological feature of the grasslands is that they are often burnt at the start of the dry season. Although this may often or even usually be attributable to human interference natural fires also occur. A number of plant species depend on the fires and do not appear if the grass remains unburnt. Butterflies dependent on these herbs (and thus also on the fires), like *Lepidochrysops* species (Lycaenidae), tend to have a very short season in August and September. At this time of the year night temperatures can fall below freezing point. Windswept, fire-blackened stubble in the coldest, driest time of the year can be an inhospitable habitat.

It is not always easy to decide if a species is characteristic for grassland. Some of the species selected may not be strictly grassland species; for example, the *Issoria* species, the larvae of which

live on *Viola*, usually occur not far from the forest. Similarly one can argue about the inclusion or exclusion of *Aloeides* species. A few of the 43 species of this essentially South African genus occur in grasslands in EAT. *A. conradsi*, for instance, can be common in montane grassland near Maralal (N Kenya) at 2000 m (R. de J., personal observations). Since the species concerned can also occur at lower levels and may be found in open woodland as well, they have been left out. Inclusion of such species would obscure the history of the strictly montane species.

While selecting potential areas of endemism we have chosen the same mountains or mountain ranges for the grassland and the forest butterflies to facilitate comparison between the two. Not all mountains, however, carry montane grassland. It is, for instance, absent from Usambara, Nguu and Nguru. It is, therefore, not surprising that no endemic grassland butterflies are found there.

Scenarios and predicted patterns

Since the present extent of the montane grasslands has been so strongly enlarged by human activities it is uncertain if the grasslands ever were more extensive than at present even when montane conditions were more widespread. This makes it difficult to frame hypotheses. In contrast to the montane forest, there was (and is) no reservoir of grassland species to the west. The only area in Africa with a well-developed montane grassland fauna is South Africa. So as a start we may consider the following options:

1. South African connection. The montane grassland butterflies of EAT originated from a South African fauna that once expanded its range to the north and later became isolated there. This leads to a recurrent pattern of sister-group relationships between taxa in EAT and South Africa, and of taxa (be it subspecies, species, or monophyletic groups) covering all or most of the available habitats in EAT.

2. South African connection. The South African species expanded northward by

jump dispersal. Again sister-group relationships with South African taxa, and further a concentration of taxa in the southern areas.

3. Origin in EAT. The sister species are to be found in EAT at lower levels and/or in other habitats (possibly mainly woodland since the change in habitat preference is then relatively small). No special geographic pattern of phylogenetic relationships to be expected. See also the discussion below.

Since human occupancy and activities in the area are probably too recent for having created opportunities for speciation (the effect may even be the other way round since grasslands have become more extensive and less isolated), the presence of taxa endemic to small areas can be taken as an indication of the age of the grassland there predating human interference.

Notes on the genera

In Appendix 8.1 we have listed all species and subspecies that in the mountain ranges from Ethiopia to east Zaïre and Malawi are restricted to montane grassland. Data have mainly been taken from d'Abrera (1980), Berger (1981), Bernardi (1980), Carcasson (1981), Carpenter (1935), Gifford (1965), Kielland (1978), van Someren (1957), and unpublished observations (especially T.C.E.C.). In some cases authors do not agree about how restricted a species is to a particular habitat, but we think this Appendix gives a fair idea of the species and subspecies restricted to montane grassland. Some of the areas are under-collected and a number of taxa are undoubtedly more widely distributed than given in the table. The following notes may be added.

Metisella (Hesperiidae). This strictly Afrotropical genus of about 26 species is at present under revision (R. de Jong, unpublished data). *M. carsoni* belongs to a group of six species that are generally found in open habitats and grasslands, from East Africa to Nigeria and Angola.

Kedestes (Hesperiidae). With 18 species all over Africa, in woodland, open habitats and grassland. In EAT there are eight species, but only one (*barberae*) is restricted here to montane grassland; its further distribution is Zimbabwe to Cape Province.

Colias (Pieridae). About 70 species, almost entirely Holarctic. In the Old World two species occur south of the Palaearctic. One, *C. erate*, with an otherwise Palaearctic distribution, reaches southern India and Ethiopia (and Sudan); the other, *C. electo*, is an Afrotropical endemic that is distributed in mountainous areas from southern Arabia through Ethiopia, East and southeastern Africa to Angola and South Africa, with an isolated population in Cameroun. It is so similar to the Palaearctic (mainly Mediterranean) *C. croceus* that several authors have considered the two conspecific. The reason for keeping them specifically separate is not clear. *C. croceus* has been found as far south as Tibesti in the central Sahara (Bernardi, 1962). It is a well-known migrant in Europe. Migratory behaviour of *C. electo* has not been reported.

Pieris (Pieridae). Opinions differ widely as to the delimitation of this genus. It may have as many as 30 or as few as 15 species. In current usage it is always considered Holarctic with slight penetrations to the south. *P. brassicoides* is the only representative in the Afrotropical region. It is supposed to form a monophyletic group with *P. brassicae* (Europe to Central Asia) and *P. deota* (Pamir, Kashmir) (Bernardi, 1947; Robbins & Henson, 1986).

Capys (Lycaenidae). According to Carcasson (1981) the genus has five species, two of which are confined to higher elevations in South Africa. Of the remaining species *brunneus* is listed from highland grasslands of southern Tanzania and Malawi, but it has not been found in Malawi since 1913 (Gifford, 1965), and Kielland (1978) lists only a single male from woodland in Kigoma (W Tanzania). Therefore *brunneus* is not listed in Appendix 8.1. Carcasson divides the highland grassland species *disjunctus* into three subspecies:

disjunctus from South Africa to Zimbabwe, *connexivus* from Zimbabwe to southern Tanzania and Angola, and *bamendanus* from Cameroun. According to Kielland (1978), *disjunctus* and *connexivus* are separate species, the former occurring in open grassland, the latter in woodland and grassland. We follow Kielland here, and since *connexivus* is not exclusively a grassland species it is not included in Appendix 8.1. It leaves the exact relationship of *bamendanus* open, although on the basis of distribution only a close relationship with *connexivus* is most probable.

Harpendyreus (Lycaenidae). Of the 15 species of this genus 12 are restricted to the montane grasslands of the study area, and three occur in South Africa, in (highland) grassland and (one species) in open habitats.

Actizera (Lycaenidae). Three species: one endemic in Madagascar, one in open habitats from East to South Africa, in West Africa and Madagascar, and *stellata* in highland grassland in eastern Africa and, disjunctly, in the Cape Province.

Eicochrysops (Lycaenidae). The two taxa listed in Appendix 8.1 are considered subspecies and conspecific with the South African *messapus* (nominotypical subspecies) and *mahallakoaena* (South Africa to Tanzania). However, Kielland (1978) lists *nandianus* and *mahallakoaena* as separate species both occurring in southwestern Tanzania. This classification is followed here. Since *mahallakoaena* occurs in grassland (Kielland, 1978) as well as in deciduous woodland (Gifford, 1965) it is not included in Appendix 8.1.

Euchrysops (Lycaenidae). The genus is, with 23 species, distributed throughout the Afrotropical region, Madagascar, Comoro Islands, and in Arabia. The species are found in arid woodland, open habitats, grassland and highland grassland. Apart from the three montane grassland species listed in Appendix 8.1 there are nine species of the genus in the area.

Lepidochrysops (Lycaenidae). A large genus of over

100 species throughout the Afrotropical region, Madagascar and in Arabia, flying in woodland, arid woodland, open habitats, and (highland) grassland. Apart from the six species listed in Appendix 8.1 there are some 20 more species in the area.

Lycaena (Lycaenidae). In its broadest sense the genus comprises about 30 Palaearctic and 16 Nearctic species, but usually it is split into a number of genera. In its restricted sense the genus comprises about nine Palaearctic, one Nearctic and one Holarctic species. Most remarkable is the occurrence in New Zealand of three species which differ only slightly from Palaearctic relatives. From the Afrotropical region four species are reported: *phlaeas* (Holarctic, one of the most adaptable butterflies, from the desert oases in Iraq to the shores of the Arctic Ocean), the East African *abbotii* (could as well be a subspecies of *phlaeas*, according to Stempffer, 1967), and the South African *orus* and *clarki*.

Antanartia (Nymphalidae). One species in Madagascar, Réunion and Mauritius, one (*abyssinica*) in the highland grasslands of EAT, and four in forests all over Africa (and in Madagascar). Of the latter, three occur in the study area.

Argyreus (Nymphalidae). A monotypic genus. *A. hyperbius* occurs widely in the Oriental region, from Sri Lanka and India to Java and the Philippines. Some authors place it in the Palaearctic genus *Argynnis*, the delimitation of which is still a matter of debate.

Issoria (Nymphalidae). In the Afrotropical region there are three species. All three are represented in EAT: *hanningtoni* is an endemic, *smaragdifera* also occurs slightly further south (E Zimbabwe), *baumanni* also occurs further west (border region Zaïre/Zambia; isolated localities in Zaïre; Cameroun). The species are found in grasslands but usually near forest. The genus is represented in the Palaearctic by three species, two of which are restricted to Asiatic mountains. As delimited by Warren (1956) the genus also comprises five South American species (higher parts of the

Bolivian and Peruvian Andes, south to Tierra del Fuego), but according to Shirozu & Saigusa (1973) the South American species belong to a separate genus, *Yramea*, that is not even closely related to *Issoria*.

Neocoenyra (Nymphalidae). There are 12 species distributed from Ethiopia to Angola and Malawi, along forest margins, in woodland, grassland and arid habitats. Apart from the four species restricted to the grasslands in the Eastern Arc mountains there are seven other species in the area covered here.

Distinction and coherence of areas of endemism

The list of potential areas of endemism and their endemicity is given in Table 8.1. Two points are immediately apparent: (i) 13 of the 27 potential areas of endemism have no endemics at all (and thus are not areas of endemism, at least for grassland butterflies), and (ii) the endemics are rather evenly distributed over the remaining areas, without concentrations except in Ethiopia. There are 26 taxa (50%) confined to a single area. The remaining 26 taxa occur in two or more areas and therefore can in principle be used to define larger areas of endemism (that may include areas

Table 8.1. *Distribution of butterflies (species and subspecies) restricted to montane grassland over potential areas of endemism*

	N	$\%$	N(end)	$\%$(end$N$)	$\%$(end$T$)
1 Ethiopia	8	15.38	8	100.00	15.38
2 Mt Kenya	6	11.54	0	0	0
3 Aberdare Mts	5	9.62	0	0	0
4 Kikuyu Highlands	8	15.38	1	12.50	1.92
5 W Kenya	8	15.38	0	0	0
6 Mt Elgon	9	17.31	1	11.11	1.92
7 Kilimanjaro	7	13.46	1	14.29	1.92
8 Mt Meru	3	5.77	1	33.33	1.92
9 Great Craters	8	15.38	0	0	0
10 Usambara	1	1.92	0	0	0
11 Nguu	1	1.92	0	0	0
12 Nguru	1	1.92	0	0	0
13 Uluguru	3	5.77	1	33.33	1.92
14 Ukaguru	2	3.85	0	0	0
15 Rubeho	2	3.85	0	0	0
16 Uzungwa	7	13.46	2	28.57	3.85
17 Southern Highlands	5	9.62	1	20.00	1.92
18 Nyika	4	7.69	1	25.00	1.92
19 Zomba	1	1.92	0	0	0
20 Mlanje	0	0.00	0	0	0
21 Ufipa	7	13.46	1	14.29	1.92
22 Mpanda	7	13.46	2	28.57	3.85
23 Kigoma	3	5.77	0	0	0
24 Marungu	3	5.77	1	33.33	1.92
25 North L. Tanganyika	6	11.54	0	0	0
26 Kivu area	10	19.23	3	30.00	5.77
27 Rwenzori	5	9.62	2	40.00	3.85

Note: N, number of taxa per area; *N*(end), number of taxa endemic to area; %, *N* in relation to total number of taxa; %(end*N*), percentage of *N* endemic to area; %(end*T*), percentage of total number of taxa endemic to area.

without endemics of their own). In practice, however, these 26 taxa are distributed in such a way that the distinction of larger areas of endemism is problematic. This may result partly from undercollecting or local extinction, but may equally well reflect the real situation, i.e. that within EAT there are no clear subdivisions with a montane grassland fauna of their own, with the exception of Ethiopia. In the following discussion the areas are grouped on the basis of their proximity and/or geological history.

Ethiopia

All eight taxa are endemic. With the exception of *Eicochrysops sebagadis* they are subspecies of more widely spread species. In two cases the related subspecies occur outside Africa; they will be dealt with below (External geographic connections, p. 143). Of the five remaining taxa the related subspecies occur in Kenya (*Antanartia abyssinica*, *Euchrysops mauensis*), northern Tanzania (*Pieris brassicoides*, *Antanartia abyssinica*), east Zaïre/west Uganda (*Antanartia abyssinica*, *Euchrysops mauensis*, *Lycaena phlaeas*), and western Tanzania (*Euchrysops mauensis*), while *Colias electo* is too widespread (even reaching South Africa) to be informative. What strikes here is the absence of a relationship with the Eastern Arc mountains.

Kenya

There is one endemic of Mount Kenya plus the Aberdares (*Lepidochrysops elgonae moyo*) and one of the Nandi–Cherangani area plus Mount Elgon (*L. e. elgonae*). The two subspecies together unite the central and west Kenyan mountains, i.e. the mountains east and west of the Rift Valley. There is one other such taxon, namely *Issoria h. hanningtoni*. No further areas of endemism can be distinguished. See also the next area.

North Tanzania

There are four taxa restricted to the area Kilimanjaro–Mount Meru–District of Great Craters. None occur in all three areas, but in one taxon (*Harpendyreus aequatorialis vulcanica*) it is the intervening area (Mount Meru) that is missing. Undercollecting seems a probable cause so that we may consider this taxon an endemic of north Tanzania

as a whole. There are three taxa linking north Tanzania to Kenya (*Antanartia abyssinica jacksoni*, *Euchrysops nandensis*, *Euchrysops c. crawshayi*). None of these occur in more than four subareas, but all three are found in the Kikuyu Highlands in central Kenya.

Eastern Arc mountains

Seven taxa occur only here, but there are no taxa linking all areas together. *Neocoenyra heckmanni* indicates a coherence of the areas from Ukaguru to the Southern Highlands. Similarly, *N. jordani* links three more northern areas. If the two species proved to be sister species this would be the only argument for uniting most of the areas of the Eastern Arc Mountains, but so far there are no reasons for considering this mountain range (or the greater part of it) a single area of endemism.

Malawi

From the geographic point of view there is no reason to suppose that the three areas would show a special coherence. Indeed they do not, nor is there any special relationship with the Eastern Arc mountains or even with the Southern Highlands alone.

West Tanzania

As with the areas in Malawi the mountain ranges in west Tanzania cannot be defined as a single area. There are exclusive connections with the next range (*Capys c. catharus*, *Harpendyreus major*), although not including all areas, and a similar connection is extended to west Kenya (*Euchrysops m. mauensis*).

Western Border Range

We use this name for the mountain ranges bordering the study area as we have chosen it, to the west, i.e. from Lake Tanganyika to Lake Mobuto. Although relatively rich in endemics (10 taxa), there are no endemics found in all four areas and four taxa are known from a single area only. Three species (*Harpendyreus aequatorialis*, *H. marungensis*, *Euchrysops crawshayi*) link (part of) the Western Border Range to (part of) Kenya–northern Tanzania, but there are no taxa restricted to these areas and occurring in all parts.

Table 8.2. *Distribution of montane grassland butterflies (species and subspecies) over combined mountain areas (for explanation, see Table 8.1)*

	N	$\%$	N(end)	$\%$(end$N$)	$\%$(end$T$)
1 Ethiopia	8	15.38	8	100.00	15.38
2 E Tanzania	11	21.15	7	63.64	13.46
3 Malawi	4	7.69	1	25.00	1.92
4 W Tanzania	11	21.15	3	27.27	5.77
5 W Range	15	28.85	10	66.67	19.23
6 Kenya–N Tanzania	19	36.54	12	63.16	23.08

Summary

It can be concluded that the 58 taxa of montane grassland butterflies do not show much of a geographic pattern. A cautious subdivision of the montane grasslands could be (see also Table 8.2): (i) Ethiopia, (ii) Kenya–north Tanzania, (iii) east Tanzania, (iv) Western Border Range, with connections between the four. The number of taxa supporting such a subdivision is pretty low, but these ranges show more internal and inter-range coherence than the mountains further south in Tanzania and Malawi. Also note the absence of widespread endemics. Widespread taxa like *Colias electo*, *Actizera stellata* and the three *Issoria* species together are not endemic to EAT. Even if a proportion of the species listed in Appendix 8.1 turned out to form monophyletic groups within EAT (possible in all genera with at least two species represented), the coherence would hardly be stronger.

External geographic connections

Geographic connections are in all directions, to the north, west and south. They have been summarised in Table 8.3. The connections are, however, of a different kind.

Northern connections

Five of the 15 genera are represented outside Africa: *Colias*, *Pieris*, *Argyreus*, *Issoria* and *Lycaena*. Except for *Issoria* the African species are either the same as outside Africa (*Colias erate*, *Argyreus hyperbius*, *Lycaena phlaeas*) or they are very closely related to non-African species. These species or their ancestors are supposed to have originated

Table 8.3. *Geographical connections in montane grassland butterflies expressed as the number of shared species or sister species*

Area	1	2	3	4	5
Inside area	3	12	9	10	4
Northern	5	4	4	3	1
Western	1	1	1	2	1
Southern	2	2	4	2	–
Total	9	17	15	14	5

Note: 1, Ethiopia; 2, Kenya–N Tanzania; 3, Eastern Arc mountains–Malawi; 4, E Zaïre to NW Tanzania; 5, Rwenzori.

outside Africa (see Methods, p. 134). The case of *Issoria* is more difficult to interpret: there are two opposing views. Either the African species have a Palaearctic ancestry (de Jong, 1976), or their occurrence in Africa is a relict distribution of Gondwana origin (Bernardi, 1980). It seems fashionable to interpret the occurrence of present-day taxa in areas that over 80 Myr BP formed part of Gondwana as an indication of Gondwana origin. However, at least in this case there are good arguments against it. First, there are no indications that the genus *Issoria* existed 80 Myr BP. The oldest butterfly fossils known are about 40 Myr old and belong to extinct genera of the Papilionidae. Secondly, it would be very strange for a Gondwana element to be mainly restricted to mountains that are at most 2 Myr old. Thirdly, there is a rich development (11 genera) of the Argynnina, the subtribe *Issoria* belongs to, in the Holarctic; there is no evidence that the

African species form the sister group of the Holarctic or Palaearctic *Argynnina* as a group. Thus a Palaearctic ancestry of *Issoria* seems most likely.

There are nine species with a northern (or northeastern) ancestry, i.e. more than one fifth (21.4%) of all species in the area that are restricted to montane grassland. In three species the African populations are considered conspecific with populations outside Africa. In the other species the ancestry is more remote. Not surprisingly, the northern element is strongest in Ethiopia where two species reach their southern limit in Africa (*Colias erate*, *Argyreus hyperbius*), and where three more species with northern ancestry occur, in total more than half of the montane grassland species in Ethiopia.

Western connections

There are three instances of a connection between EAT and Cameroun: *Colias electo*, *Issoria baumanni* and *Capys disjunctus* (through *connexivus* and *bamendanus*). In these cases the complete absence of relatives in Cameroun points to an East African origin of the Cameroun populations. It is interesting to note that the first two species are known from a few localities in Zaïre west of the area under study: they may have a wider habitat tolerance than the other species. The case of the *Capys* species is less clear since the exact relationship between *disjunctus*, *connexivus* and *bamendanus* is uncertain.

Southern connections

Here we consider only connections with South Africa, so we leave aside the case of *Issoria smaragdifera* which occurs just south of the area under study in eastern Zimbabwe. There are remarkably few such connections (four only; 10% of the montane grassland species of EAT): *Kedestes barberae*, *Colias electo*, *Capys disjunctus* and *Actizera stellata*. In all cases it is the same species occurring in EAT as well as in South Africa, no sister group situations being known. Since the species concerned have relatives in EAT (except *Colias electo*, which has a northern origin), there is no reason to suppose that the species were orig-

inally South African and expanded their range to the north.

In some instances the external connections in different directions concern the same species. For the majority of the species (30; 71.5%) there is no evidence of a non-African ancestry nor of an ancestry outside EAT.

Ecological relationships

This section examines if species whose geographic ancestry is obscure may have originated on the spot from species with other habitat preferences. Therefore we shall restrict ourselves to the genera whose species have an African ancestry.

Metisella. Most *Metisella* species live in forested areas, but an apparently monophyletic group of six species is found in open habitats. Only one of these, *M. carsoni*, is restricted to montane grasslands, so its habitat preference seems derived. Its sister species is, however, still unknown. It could be the sympatric *formosa* or an allopatric species.

Kedestes. Since there are no further species restricted to montane grassland among the 18 species of the genus, the preference for montane grassland in *barberae* seems derived. Since its range is so wide and potential sister species are found in South as well as in East Africa, it is of little help in elucidating *barberae*'s origin.

Capys. The species are found in montane grasslands as well as in woodland. A switch in habitat preference is not needed to understand the present situation, i.e. the ancestors of the two species in Appendix 8.1 could also have lived in montane grasslands.

Harpendyreus. Only one of the three species in South Africa occurs in open habitats in general, all 14 other species (including all species in EAT) live in montane grassland. So also the ancestors of the species in EAT probably preferred this kind of habitat.

Actizera. The only other continental African species overlaps the distribution range of *stellata* over the whole area from East Africa to South Africa. No clue exists as to what could have been the original habitat preference of the genus.

Eicochrysops. In view of the range in habitat preferences, the occurrence of six more species in the study area and the possibility that the only two taxa that are confined to montane grassland are conspecific with each other as well as with taxa with a wider ecological tolerance, the preference for montane grassland is most likely a derived character in the genus.

Euchrysops. Since preference for montane grassland occurs in only three of the 23 species, and nine more species of the genus occur in EAT, a shift in habitat preference towards montane grassland is not unlikely.

Lepidochrysops. As for *Euchrysops*, apart for the numbers being higher.

Antanartia. Since *abyssinica* is the only montane grassland species and three other species of the genus occur in the area it seems likely that the habitat preference is a derived condition.

Neocoenyra. The montane grassland species are restricted to the Eastern Arc mountains. The eight other species of the genus occur in a variety of other habitats at lower elevations. The related genus *Pseudonympha*, which is restricted to southern Africa, has a number of montane grassland species apart from species in other open habitats. Even if the two genera were sister groups there is no reason for considering the preference for montane grassland a primitive condition in *Neocoenyra*.

Discussion

The two patterns above were predicted based on the alternative hypotheses that the montane grassland fauna of EAT would either be a derivative of the South African grassland fauna, or be a local development from a fauna or faunas with different ecological requirements. The pattern found contradicts the hypothesised South African origin: there are no recurrent patterns of sister-group relationships between EAT and South Africa, there are no taxa at the same time widespread in and endemic to EAT, and there is no concentration of montane grassland taxa in the southern areas in EAT. The few South African connections found can indicate only limited jump dispersal, possibly by species with a slightly wider ecological tolerance than most grassland species. Interestingly, the pattern(s) found rather indicate a geographic connection in the opposite direction – from the north. Even though only a limited number of taxa are involved we find here the pattern predicted for a South African connection under the jump dispersal model: same species as or sister-group relationships with taxa outside Africa, and the highest concentration in the north (Ethiopia).

Almost three quarters of the species do not show close relationships with taxa outside EAT. Even though their sister-group relationships are badly known, we can safely assume that in most cases the sister group is to be found in EAT. These species agree with, or at least do not contradict, the predictions of the internal differentiation model in which it is supposed that species originated on the spot from species with a different habitat preference, and later expanded their range according to their dispersal powers, leading to a mosaic of small, isolated ranges and differently overlapping larger ranges. Under this model coherence of areas of endemism is a function of dispersal rather than of former connections, since the latter would have led to a much less 'chaotic' pattern. The model will be further discussed after the forest species have been dealt with.

As stated above, the connection with the Cameroun highlands is apparently attributable to a few species dispersing to the west. Since so few species are involved and these species, because of their also otherwise wide distributions, possibly have a larger ecological tolerance than the other montane grassland species, they cannot be indicative of a former continuous belt of montane grassland conditions stretching from EAT to Cameroun.

Conclusion

The montane grasslands of EAT are populated by butterfly species of widely different origins. A proportion of the species occur at lower elevations and/or in other habitats as well. The species that are more or less restricted to this kind of habitat have mainly originated on the spot from species with other habitat preferences. A much smaller proportion of the fauna (20–28%) originates from an influx from outside, particularly from the north (Palaearctic and Oriental regions).

Forests

General

The butterfly fauna of Africa numbers well over 3200 species (Carcasson, 1981). Of these about 56% are restricted to forests. The forest species are not evenly distributed over the forest area: by far the greatest diversity is found in the large central forest block. In view of its size this is not remarkable, but it is not only area that determines the number of species: perhaps more important are the vicissitudes in the extent of the forest over geological time and the diversity in the types of forest. By far the majority of forest species are restricted to low and/or medium levels. Although a number of these species may ascend to higher levels, the butterfly fauna usually changes with increasing altitude and species restricted to higher levels appear. In EAT there are forests at all altitudinal levels, although most forests at low and medium levels have been destroyed in historical times by human intervention, strengthening the isolation of the montane forests. Forests at low and medium elevation are found in east Zaïre, Uganda, west Tanzania, east Tanzania (Usambara, Nguu, Nguru, Uluguru, Uzungwa; patches along the Indian Ocean coast), Kenya (patches along the coast and at medium levels in west Kenya) and Malawi. We are concerned here with the montane forest butterfly fauna only. In EAT, where by far most of the montane forest species of Africa occur, they number 177. Although this number is low relative to the number of lowland species, the disjunctions and the many endemics

restricted to single mountains renders this fauna of disproportional interest.

It is not clear why there are many fewer montane forest species than lowland forest species. One reason could be that the area of the montane forest is much smaller. This may be so, but on the other hand the area is strongly split up, offering more opportunities for speciation events. Another reason for the relatively low number could be the young age of the montane forest. This is, however, very uncertain. Part of the Eastern Arc mountains is supposed to have been uplifted during Jurassic times (Moreau, 1966). If true, these early mountains could have carried forests at higher altitudes at a time when the butterflies as a group did not even exist. The extent, however, could have been very limited, and most of the present area of montane forest is certainly not older than a few million years. Still another reason could be that the montane forest fauna suffered much from climatic vicissitudes and could not be supplied from the rich lowland forest to the west by sheer distance (there were, and still are, patches of lowland forest along the east coast, but they are rather poor in species). This explanation is unlikely since an important refugial area of lowland forest during extreme contraction of the forest is supposed to have been at the eastern rim of the Congo basin against (or even in) the mountains fringing the great lakes from Lake Tanganyika to Lake Mobuto (Hamilton, 1981, 1982; Crowe & Crowe, 1982). This mountainous area (to be called the Western Border Range, below) is not even the area richest in montane forest butterflies. Maybe an explanation of the relatively low number of montane forest butterflies is quite simple and should be sought in an ecologically less diversified environment with stronger constraints (e.g. lower temperature).

Scenarios and predicted patterns

Unless the butterflies as a group originated in the highland forests of EAT (which is most unlikely, if only because of the fact that the butterflies as a group already existed long before the greater part of EAT was uplifted), the montane forest butterflies of eastern Africa must have come from elsewhere. There are three options: from other mountain ranges, from lowland forest, or from other habitats. These options lead to the following scenarios, with deduced predictions regarding distribution pattern.

Origin from other mountain ranges

No continuous forest connection with areas outside Africa is known in Tertiary or Quaternary times. This leaves only other African mountains as possible source areas. It has been stated (Moreau, 1963, 1966; Carcasson, 1964; van Zinderen Bakker, 1967) that the montane forest belt was depressed so far that a contact was established between the Cameroun highlands (and perhaps mountain ranges in Angola) and the mountains in EAT. Such a contact was assumed to have existed during the Pleistocene at the time of the northern Ice Ages. If there had been any substantial contact the predicted pattern should show a fauna that in the Cameroun highlands and the mountains in EAT must be similar, with shared species or at most with sister species. Relatives of the shared taxa should be found in or around the source area. Since in this scenario the montane forests of EAT were also in mutual contact, the distribution of the species shared with Cameroun should be more or less even over the mountains in EAT.

Fragmentation of a once continuous forest belt across the African continent

Several authors (e.g. Carcasson, 1964) have projected such a belt. There is certainly good evidence for dramatic climatic changes even or especially in the near past (Pleistocene) (summary in Hamilton, 1982). It seems, however, that these projections are more based on assumptions about what would have happened to the forest during colder or wetter times, than on geological or palaeontological evidence. If such a belt ever existed, and if it existed long enough, a more or less uniform forest fauna must have occurred across Africa, although relief may have caused local variations. After fragmentation the forest fauna was trapped in EAT in refugia along the coast and in the mountains. The refugial areas were not extremely small, but may have been much smaller during periods with a drier climate. The age of the belt is very important here, either before or after the Plio–Pleistocene mountain building.

(a) Forest belt before the Plio–Pleistocene orogenic phase. The mountains present were mainly the Eastern Arc mountains. In these mountains the montane forest fauna originated and from here the fauna dispersed at a much later date over the other, more recent montane forests. The pattern is: highest concentration in the Eastern Arc mountains, the numbers decreasing with increasing distance from these mountains; and a twin relationship between taxa of the Eastern Arc mountains and of the Central African lowland forest block (which extends eastward as far as west Kenya). Other patterns, e.g. a twin relationship between the mountains of EAT (or at least a large part including the Eastern Arc mountains) and the lowland forest block that could arise if a species later extended its range from the Eastern Arc mountains over other mountains as well, are difficult to distinguish from the next scenario and therefore not fit to falsify the present scenario.

(b) Forest belt after the Plio–Pleistocene orogenic phase. In this case there must have been a more or less uniform fauna throughout the montane forests. Undoubtedly part of the fauna became extinct when the forests retreated to higher elevations and distribution areas of the species became fragmented. Depending on the occurrence of speciation as a consequence of the fragmentation of the habitat the predicted pattern consists of groups of closely related, allopatric species without clear hierarchical relationships (they all arose 'at the same time'), or of species widespread in the montane forests of EAT (and maybe differentiated here in a number of subspecies) and also occurring in the Central African forest block. Further, it may be expected that there are no large differences between the areas in numbers of taxa and numbers of endemics, unless there is evidence of large-scale local extinction (see below).

Ecological differentiation

Ecologically the montane forest butterflies can have originated from three sources: from lowland forest butterflies (in a broad sense, not ascending over 1500 m); from butterflies living in open habitats (woodland to treeless habitats); and from other montane forest butterflies. Scenarios for an origin from other mountain areas and for fragmentation of a continuous forest belt with adaptation from lowland to highland conditions have just been dealt with. They both have a strong geographical component. Here we are concerned with the possibility that the montane forest butterflies arose within EAT from butterflies living in lowland forest or in non-forest habitats. It has, of course, also a geographical component since different habitats cannot occupy exactly the same place, but they form a mosaic, a patchwork within EAT. Consequently, twin relationships can be expected within EAT and between different habitats, and not between EAT and some area outside. In the case of origin from lowland forest butterflies the difference with the fragmentation scenarios is that it is irrelevant whether or not there ever was a continuous forest belt. What counts is the presence of an ancestor at lower elevations in EAT, and how this ancestor came there is unimportant. To come into consideration for this scenario a montane forest species must have a sister species in forests at lower elevations but in the same area. Of course, if such situations exist they can also be explained by a secondary dispersal of the lowland forest species into EAT after fragmentation of a continuous belt; but apart from this being an extra assumption, why should a lowland forest species do so if it could not survive there before without adapting itself to montane forest?

It seems unlikely that the old mountains of the Eastern Arc, if they carried montane forest from an early date, had no butterfly fauna at all. There was plenty of time for this fauna to develop from the fauna around, apart from the possibility of a contact with forests to the west. If so, the Eastern Arc mountains could be expected to have the highest number of montane forest butterflies. Later this fauna could have spread to other mountains where in the meantime also montane forest species could have evolved from species with other habitat preference. As a result a mosaic of partly overlapping ranges can be expected

without a geographic pattern of phylogenetic relationships. Whatever its origin, once a montane forest species had come into being it could follow its own course. According to opportunities offered it could disperse through and differentiate in the montane habitats in EAT, possibly obscuring its origin. Thus, we must consider the following scenario.

Internal development

If the montane forests of EAT ever were continuous, irrespective of a possible contact with lowland forest or montane forest in Cameroun, they must have had a more or less uniform fauna. In that case we may expect to find widespread species as well as monophyletic groups of allopatric species together covering most or all of the mountain ranges. A complicating factor could be large-scale local extinction. Several authors (e.g. Livingstone, 1975; van Zinderen Bakker, 1982) maintain that during the height of the last Glacial (about 18 000 BP) the climate was not only cooler but also much drier and most of EAT was covered with open woodland or arid vegetation. Forests were found only in the wettest areas such as the Eastern Arc mountains and the Western Border Range. If this really happened there should be a high concentration of endemics in these refugial areas, and few or no endemics in the other mountain ranges. If the montane forest fauna was uniform before it became extinct except in the refugia, there should be either many shared species or many sister relationships between the Eastern Arc mountains and the Western Border Range.

The scenarios are not mutually exclusive: all could have occurred separated in time, or some even simultaneously but in different parts of the area.

Notes on the genera

All species and subspecies of butterflies that in EAT are restricted to montane forest are listed in Appendix 8.2. Data sources are the same as for the montane grassland butterflies. Additional references are mentioned below.

Celaenorrhinus (Hesperiidae). A circumtropical genus (except for New Guinea and Australia) with about 30 species in the Afrotropical region. The species are generally found in forests where the butterflies often settle with wings spread on the underside of leaves of herbs and shrubs. Apart from the listed species there are six more species in the study area. The African species fall into two groups, a white-spotted and a yellow-spotted group. The latter group (of the listed species *kivuensis* belongs to this group) seems to be monophyletic (de Jong, 1982). It occurs throughout the Afrotropical region wherever habitat is appropriate. White-spotted species do not occur in the southern part of the continent (Zimbabwe and further south) but there are two species in Madagascar, where the yellow-spotted species are absent. The genus as a whole is subject to a long-term study (de Jong, unpublished data).

Abantis (Hesperiidae). An Afrotropical genus of 18 species, flying throughout Africa in open habitats, woodlands and forests. *A. meru* is the only montane forest species. It was described by Evans as a subspecies of *A. paradisea*, a woodland species occurring from Somalia to South Africa, but raised to full species status by Carcasson (1981). Because of its similarity to *paradisea* it could be its sister species, if not simply a high altitude subspecies. There are five more species in the area.

Metisella (Hesperiidae). This Afrotropical genus is subject to a monographic study (R. de Jong, unpublished data). The arrangement in Appendix 8.2 has not been published before and the arguments will be given in a future publication, as well as the descriptions of the undescribed species. The genus numbers about 26 species occurring from Kenya and Somalia west to Cameroun and south to South Africa, in arid places, grasslands, woodlands, forests. The highland forest species fall into several monophyletic groups.

Ampittia (Hesperiidae). Two African species. Apart from the endemic of the Eastern Arc mountains, *parva*, there is a woodland species in

Central, East and Southeast Africa. Six species occur from India to Java and the Amur region.

Gorgyra (Hesperiidae). A strictly African genus, 19 species mainly occurring in forests in West and Central Africa, with a few species in EAT (going south as far as Malawi and Mozambique). In addition to *bibulus* (which according to Carcasson (1981) is also found in 'Nigeria' – this could be Cameroun), six more species occur at lower elevations in the study area.

Ceratrichia (Hesperiidae). An African genus of 16 mainly West African forest species. In addition to *bonga*, the only representative of the genus in Tanzania, eight more species occur in the study area, at lower elevations and only in the area Zaïre–Uganda–Kenya.

Chondrolepis (Hesperiidae). An African genus with seven species, five of which are restricted to the highland forests of EAT. Of the two species not listed in Appendix 8.2, *nero* is restricted to the highlands of Cameroun, while *niveicornis* (its sister species) has a wide distribution from Cameroun to Ethiopia and south to Angola and Zimbabwe. The genus was revised by de Jong (1986). Figure 8.2 gives the phylogenetic tree. Note that in one of the six sister group situations there is allopatry of the sister groups.

Zenonia (Hesperiidae). An African genus of three species only: *zeno*, commonly found in woodlands all over the Afrotropical region; *anax*, in forests in Tanzania, Malawi and south Zaïre; and *crasta* (see Appendix 8.2). The last two species are allopatric, but it is not certain that they are sister species.

Papilio (Papilionidae). A huge, cosmopolitan genus of *c.* 220 species, occurring from desert oases to tundra and to rain forest. The genus is sometimes divided into a number of genera but according to Miller (1987) these so-called genera are not convincingly monophyletic. In the Afrotropical region 45 species are known. Accepting Hancock's (1983) subdivision the 15 species of Appendix 8.2 fall into the following species groups:

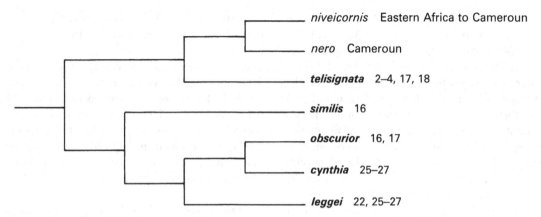

niveicornis Eastern Africa to Cameroun

nero Cameroun

telisignata 2–4, 17, 18

similis 16

obscurior 16, 17

cynthia 25–27

leggei 22, 25–27

Figure 8.2. Phylogeny and distribution of the genus *Chondrolepis*, after de Jong (1986). Names in bold refer to Afromontane forest species. Numbers refer to the areas listed in Table 8.1.

leucotaenia group, *leucotaenia*; sole species of the group; sister group of the *delalandei* (Madagascar), *phorcas* (throughout Africa) and *hesperus* groups combined;

hesperus group, *nobilis*, *pelodorus*; three more species in the group, in forests, one of which also in the study area at lower elevations;

nireus group, *mackinnoni* to *thuraui*; nine more species in the group; *charopus* also in Cameroun (nominotypical subspecies); the group was studied by Hancock (1984); for the phylogenetic relationships and distributions, see Figure 8.3;

rex group, *rex*; sole species of the group, also represented in Cameroun by a separated subspecies; sister group of the *cynorta* group;

cynorta group, *echerioides* through *sjoestedti*; three more species in the group, in forests at lower elevations, two also in the study area. *P. echerioides* extends southward in two subspecies to Zimbabwe and Natal/Transvaal.

Colotis (Pieridae). A mainly African genus with 42 species in the Afrotropical region, one of which also occurs in North Africa and occasionally in southern Spain. Three other African species extend into India where four more species occur.

All species are found in woodland and open habitats, often quite arid. *C. elgonensis* is the only species of the genus found in (and restricted to) highland forest, not only in EAT but also in Cameroun. In the study area 30 more species occur.

Belenois (Pieridae). Indo-African, with 23 species in the Afrotropical region, one of which occurs in India as well. The species are found in open habitats, woodland and forest. In addition to the species in Appendix 8.2, 11 more species occur in the study area.

Mylothris (Pieridae). An African genus of 31 woodland and forest species. In addition to the species in Appendix 8.2, 12 more species occur in the study area. Carcasson (1981) treats *leonora* as subspecies of *M. crawshayi*; this is followed here. *M. sagala* extends south to east Zimbabwe (separate subspecies), and also occurs in Cameroun (also separate subspecies).

Alaena (Lycaenidae). The genus is strictly Afrotropical and belongs to an endemic African subfamily, Lipteninae. It numbers 22 species in a wide range of habitats from arid open conditions to forest. It does not occur in West Africa. In addition to the four species in Appendix 8.2, 10 more species occur in the study area.

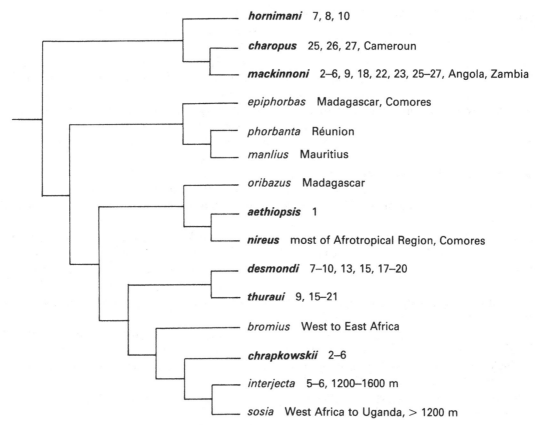

Figure 8.3. Phylogeny and distribution of species of the *Papilio nireus* group, after Hancock (1984). Names in bold refer to Afromontane forest species. Numbers refer to the areas listed in Table 8.1.

Ornipholidotos (Lycaenidae). Another genus of the Lipteninae. Its 29 species are entirely confined to forest. Most species are found in Central Africa. In East Africa the genus occurs in Uganda (9 species), Kenya (and adjoining Somalia; 2 species) and Tanzania (the same two species as in Kenya, plus the one in Appendix 8.2). One of the species of Kenya and Tanzania is in East Africa found mainly in the lowland forests near the coast and it extends to the south as far as Natal; no other species occur south of Tanzania.

Mimacraea (Lycaenidae). Still another genus of the Lipteninae, with 18 species in forests of West, Central and East Africa; one species lives in woodland in East Africa and extends southward to Zimbabwe. In addition to *M. gelinia* six more species occur in the study area, but none in the

Usambara Mountains or elsewhere in east Tanzania.

Baliochila (Lycaenidae). Still another genus of the Lipteninae, with 17 species distributed in woodlands and forests of EAT, from Kenya to Natal. No records from Central or West Africa, except for a single species from Liberia. In addition to the species in Appendix 8.2, 12 more species have been recorded from the study area.

Aslauga (Lycaenidae). This and the remaining genera of the Lycaenidae belong to subfamilies that are not restricted to Africa. *Aslauga* is mainly West African, with eight species in lowland forest and one in woodland, the latter having a wider distribution, from Gabon to Kenya and south to the Cape Province (not certain if this is all one species). *A. orientalis* is the only species known from the Usambara Mountains.

Spindasis (Lycaenidae). An Afro-Oriental genus with 26 species in Africa. Mainly found in woodlands, but a few species also in open arid habitats and in forests. In addition to the single species in Appendix 8.2, 11 more species occur in the study area.

Epamera (Lycaenidae). This genus and the next two genera were considered subgenera of the large African genus *Iolaus* (over 100 species) by Carcasson (1981). *Epamera* numbers some 54 species all over Africa in open habitats, arid woodlands, woodlands, and forests, from lowland to highland. *E. bansana* occurs in Cameroun as well (as a separate subspecies). In addition to the species listed in Appendix 8.2, about 25 more species have been recorded from the study area.

Etesiolaus (Lycaenidae). So far a monotypic genus, the single species known being distributed in forests from West Africa to west Kenya and the Usambara Mountains. The yet undescribed species recorded here is, thus, the second species from the Usambara Mountains.

Iolaphilus (Lycaenidae). The genus numbers 25 species, in forest and woodland all over Africa. In addition to the two species in Appendix 8.2, 13 more species occur in the study area.

Hypolycaena (Lycaenidae). A widespread genus in the Old World tropics, extending eastward as far as the Bismarck Archipelago and Australia. In Africa 19 species are known, flying in woodland and forest. In addition to *jacksoni*, 12 more species occur in the study area.

Pilodeudorix (Lycaenidae). The genus, together with the next and a number of other genera, was placed as a subgenus of *Deudorix* by Carcasson (1981). In this sense, *Deudorix* is widespread in the Old World tropics, extending to the southwest Pacific islands. *Pilodeudorix* is restricted to Africa. It numbers 11 species, in woodland and forest all over Africa. Apart from the species listed in

Appendix 8.2, four more species occur in the study area.

Virachola (Lycaenidae). The genus is well developed in Africa with 29 species; a few species occur in the Oriental and Australian regions. The African species are found in arid and open habitats, woodland and forest throughout the region. Apart from the three species in Appendix 8.2, 17 species occur in the study area.

Anthene (Lycaenidae). The genus is widely distributed throughout the tropics of the Old World, but best represented in Africa where 83 species occur, in all kinds of habitats. The high number of undescribed species in the Eastern Arc mountains alone indicates that the knowledge of this genus is far from complete. In the study area 42 species occur in addition to the 10 listed in Appendix 8.2.

Uranothauma (Lycaenidae). The genus is restricted to Africa. So far 18 species are known, five of which were discovered recently in the Eastern Arc mountains. All species occur in forest, especially highland forest, but a few widespread species are also found in woodland. In addition to the species listed in Appendix 8.2, four more species occur in the study area.

Castalius (Lycaenidae). An African genus of 10 species, found mainly in woodland in eastern and southern Africa but two species are also known from West Africa. *C. margaritaceus* is probably represented in Cameroun by a separate subspecies. With the exception of two species all species occur in the study area.

Abisara (Lycaenidae). About 20 species of which 11 occur in the Afrotropical Region and the rest in the Oriental Region. They are forest butterflies, widespread in Africa but not found south of Angola and Malawi. In addition to *delicata* (which is also found in Cameroun), three more species occur only in the study area.

Charaxes (Nymphalidae). A large genus of the Old World tropics, with some 150 species in the Afrotropical Region (one species of which extends into the Mediterranean) (Henning, 1988). The species are found throughout Africa in woodland and forest. Although very popular with collectors and subject of numerous papers and several books, *Charaxes* species are not well known as regards their phylogenetic relationships. About 62 species are known from the study area in addition to those listed in Appendix 8.2. *C. xiphares* and *acuminatus* are also found in Cameroun, as separate subspecies.

Cymothoe (Nymphalidae). An African genus of 71 species, exclusively in forest at lower elevations except for the montane forest species listed in Appendix 8.2. Apart from these highland species 13 more species occur in the study area.

Kumothales (Nymphalidae). A monotypic genus only known from eastern Zaïre and southwestern Uganda.

Pseudathyma (Nymphalidae). A small genus of seven forest species found in West, Central and East Africa. All but one occur in the study area in Uganda, W Kenya, W Tanzania. *P. uluguru* is the only species in the Eastern Arc mountains.

Pseudacraea (Nymphalidae). An African genus of 13 species found in woodland and forest throughout Africa. Eight species are found in the study area, in addition to *deludens*.

Neptis (Nymphalidae). A large genus found in the Afrotropical, Palaearctic, Oriental and Australian regions. In Africa 58 species are known. They live in woodland and forest; six are found throughout Africa. About 33 species occur in the study area in addition to those listed in Appendix 8.2. The listed species are restricted to EAT, with the exception of *occidentalis*, which is represented in Cameroun by a separate subspecies.

Sallya (Nymphalidae). This genus is found throughout the Afrotropical region and numbers 14 species in woodland and forest. There are nine species in the study area in addition to *pseudotrimeni*.

Hypolimnas (Nymphalidae). The genus, and some of the species, are widely distributed in the Old World tropics. In Africa 13 species occur, mainly in lowland forest but two species (which also extend throughout the Oriental and Australian regions) are also found in open habitats. There are eight species in the study area in addition to *antevorta*.

Precis (Nymphalidae). There is a confusing use of the names *Precis* and *Junonia* in the literature. Together they are found almost world-wide with the exception of the greater part of the Palaearctic. Carcasson (1981) distinguished between the two for the African species. We shall not enter into the debate but simply follow Carcasson here. The confusion makes, however, the extension of *Precis sensu* Carcasson outside Africa uncertain. The genus numbers 15 species in Africa, flying in arid habitats, marshy areas, woodland and forest. Twelve species are found in the study area in addition to *milonia*, which is also found in Cameroun (represented by a separate subspecies).

Antanartia (Nymphalidae). An African genus of six species, distributed throughout Africa in forest and one species in highland grassland near forest (see under Montane grasslands, p. 140). The two species listed in Appendix 8.2 are not restricted to eastern Africa; *schaeneia* extends through the mountains of east Zimbabwe to South Africa (separate subspecies), while *dimorphica*, apart from South Africa (same subspecies as in eastern Africa; not further south than north Transvaal), is also found in Cameroun (separate subspecies), and even in Grand Comoro. In addition to these species, two more species occur in the study area; the two remaining species of the genus are restricted to Madagascar, Mauritius and Réunion. The genus was monographed by Howarth (1966).

Bematistes (Nymphalidae). An African genus with 23 species in forest (mainly lowland) throughout Africa. There are 14 species in the study area in addition to the three listed in Appendix 8.2. *B. obliqua* is represented in Cameroun by a separate subspecies (the nominotypical one).

Acraea (Nymphalidae). A large genus with 155 species in Africa (one of which extends to India), a few in the Oriental region and one species flying as far east as Australia and Fiji. The species live in a variety of habitats: arid areas, marshy areas, grassland, woodland and forest. The 11 species listed in Appendix 8.2 are restricted to the study area except *kraka* which is found in Cameroun as well. Many more species occur in the study area.

Gnophodes (Nymphalidae). A small genus of three forest species only. Of the two species not in Appendix 8.2, one (*betsimena*) is found throughout Africa, the other (*chelys*) occurs from West Africa to west Kenya.

Aphysoneura (Nymphalidae). A monotypic genus with unknown relationships. Outside the study area it has been found in east Zimbabwe and in Cameroun.

Bicyclus (Nymphalidae). A large African genus with about 80 species, mainly in forest but also in woodland. At least 36 species occur in the study area in addition to those listed in Appendix 8.2. The genus was monographed by Condamin (1973). His diagram of relationships between species groups (not cladistic) served as a basis for Figure 8.4.

Ypthima (Nymphalidae). The genus numbers 15 species in Africa, while there are many more species in the Oriental region and also a few in the southern and eastern Palaearctic. One common African species, *asterope*, extends to India. The African species are generally found in open habitats and grassy places, but may also occur in woodland. *Y. albida* is the only African species restricted to forest; outside the study area, it is represented in Cameroun by a separate subspecies. In the study area six more species occur.

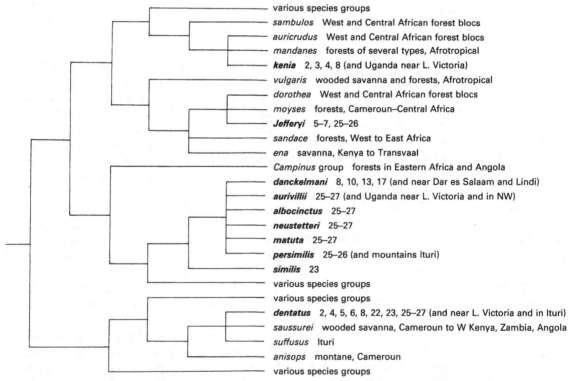

Figure 8.4. Possible phylogeny and distribution of species of *Bicyclus*, after Condamin (1973). Names in bold refer to Afromontane forest species. Numbers refer to the areas listed in Table 8.4.

Amauris (Nymphalidae). This African genus numbers 15 species (Ackery & Vane-Wright, 1984) generally found in forests. In addition to *ellioti*, nine more species occur in the study area. Ackery & Vane-Wright (1984) studied the phylogeny of the genus. For our purpose the results are not very helpful, *ellioti* being the sister species of a group of 12 species.

Distinction and coherence of areas of endemism

The list of potential areas of endemism and their endemicity is given in Table 8.4. There are striking differences with the montane grassland butterflies (Table 8.1): there are fewer potential areas without endemics, the endemics are not evenly distributed over the areas with endemics, and the highest concentration of endemics in a single area is found in the Usambara Mountains where no endemic montane grassland species occur.

The endemics restricted to a single area account for 106 (i.e. 41%) of the 259 taxa listed in Appendix 8.2. Thus, 59% of the taxa are found in two or more areas. This percentage is slightly higher than for grassland butterflies, and since the total number of taxa is 4.5 times as high there are many more forest than grassland taxa indicating coherence between areas.

Table 8.5 gives eight areas formed by combinations according to the steps mentioned under Methods (p. 134). Two combinations, namely, the Eastern Arc mountains and what we have called the Western Border Range (from the Marungu Plateau to the west side of Lake Mobuto – for-

Table 8.4. *Distribution of butterflies (species and subspecies) restricted to montane forests over potential areas of endemism (for explanation, see Table 8.1)*

	N	%	N(end)	%(end$N$)	%(end$T$)
1 Ethiopia	13	5.04	13	100.00	5.04
2 Mt Kenya	29	11.24	2	6.90	0.78
3 Aberdare Mts	22	8.53	0	0	0
4 Kikuyu Highlands	33	12.79	3	9.09	1.16
5 W Kenya	38	14.73	0	0	0
6 Mt Elgon	38	14.73	3	7.89	1.16
7 Kilimanjaro	23	8.91	4	17.39	1.55
8 Mt Meru	14	5.43	0	0	0
9 Great Craters	22	8.53	2	9.09	0.78
10 Usambara	37	14.34	15	40.54	5.81
11 Nguu	4	1.55	0	0	0
12 Nguru	27	10.47	8	29.63	3.10
13 Uluguru	37	14.34	10	27.03	3.88
14 Ukaguru	10	3.88	0	0	0
15 Rubeho	18	6.98	1	5.56	0.39
16 Uzungwa	45	17.44	8	17.78	3.10
17 Southern Highlands	30	11.63	0	0	0
18 Nyika	28	10.85	9	32.14	3.49
19 Zomba	14	5.43	1	7.14	0.39
20 Mlanje	20	7.75	4	20.00	1.55
21 Ufipa	14	5.43	1	7.14	0.39
22 Mpanda	20	7.75	1	5.00	0.39
23 Kigoma	22	8.53	2	9.09	0.78
24 Marungu	5	1.94	0	0	0
25 North L. Tanganyika	34	13.18	2	5.88	0.78
26 Kivu area	60	23.26	12	20.00	4.65
27 Rwenzori	46	17.83	5	10.87	1.94

Table 8.5. *Distribution of montane forest butterfly species over combined mountain areas, and occurrence in Cameroun (for explanation, see Table 8.1)*

	N	%	N(end)	%(end$N$)	%(end$T$)	Shared with Cameroun
1 Ethiopia	10	5.68	4	40.00	2.27	3
2 C Kenya	36	20.45	4	11.11	2.27	8
3 W Kenya	42	23.86	4	9.52	2.27	12
4 N Tanzania	35	19.89	4	11.43	2.27	6
5 E Tanzania	89	50.57	56	62.92	31.82	6
6 Malawi	37	21.02	8	21.62	4.55	6
7 W Tanzania	35	19.89	2	5.71	1.14	9
8 W Range	67	38.07	30	44.78	17.05	15

merly Lake Albert), stand out because of their high numbers of endemics.

(a) In the Eastern Arc mountains 36 of the 56 endemic species occur in one of the areas, leaving only 20 species that are distributed over two or more areas. These mountains form a chain in such a way that absence of a taxon in an intervening area is likely to be the result of undercollecting or extinction and does not indicate a historical gap between the areas. Thus occurrence of a species in two areas separated by two other areas indicates coherence of all four areas rather than a special connection between two. Seen in this light and focusing on the 56 species that are restricted to the Eastern Arc mountains, only two species (*Papilio fuelleborni* and *Uranothauma williamsi*) can be said to indicate the coherence of all eight areas, although the former is not known from one area and the latter from two. Two more species (*Ampittia parva* and *Epamera dubiosa*) connect seven areas, from the Usambara Mountains to Uzungwa. In comparison with the number of endemics in the Usambara Mountains (12 species, 15 taxa if subspecies are included), Nguru (7, 8), Uluguru (8, 10) and Uzungwa (8, 8) these numbers are too low to suppose that the Eastern Arc mountains ever formed a single area of evolution (i.e. a single isolated area). It rather seems that the four areas just mentioned were each, so to say, nuclei of speciation from where some of the new species spread to nearby areas.

It is unlikely that none of the species that orig-

inated in the Eastern Arc mountains and expanded their ranges there ever reached other areas to the north (Kilimanjaro, Mount Meru, District of Great Craters, together listed as north Tanzania in Table 8.5) and to the south (Nyika Plateau). There are exclusive connections between these areas and the Eastern Arc mountains. Three species (*Papilio hornimani*, *Neptis aurivillii* and *Bicyclus danckelmani*) are found only in north Tanzania and the Eastern Arc mountains. Similarly, four species (*Metisella decipiens*, *M. nyika*, *Mylothris crawshayi* and *Uranothauma cuneatum*), and two subspecies (*Papilio desmondi usambarensis* and *Charaxes ansorgei levicki*) are found only in Nyika Plateau and the Eastern Arc mountains. We do not suggest that all these taxa originated in the Eastern Arc mountains: they merely indicate that at their northern and southern ends the Eastern Arc mountains had contacts with other areas. Most of the taxa concerned have a rather wide distribution in the Eastern Arc mountains and are apparently rather good dispersants.

(b) The Western Border Range encompasses the following areas of Table 8.4: 24 (Marungu Plateau), 25 (north end of Lake Tanganyika), 26 (Kivu area), and 27 (Rwenzori to Lake Mobuto). None of the taxa endemic to the Western Border Range occurs on the Marungu Plateau, so we leave this area out of further consideration and restrict the Western Border Range to the mountains between the northern end of Lake

Tanganyika and the west side of Lake Mobuto. Of the 30 species restricted to this combined area 12 occur in one area only, mainly in the Kivu area (9). Thus 18 species are found in two or three areas, most of which (11) occur in all three areas. These figures illustrate the strong coherence of these areas and at the same time show the important position of the Kivu area to which not only 30% of the endemics of the Western Border Range are restricted but where only three endemics of this combined area are missing. It is not clear why there are many more endemics in the Kivu area than in Rwenzori, unless the former offers better opportunities for speciation by geographic isolation, being more extensive and with more mountain tops.

Not surprisingly, there are connections between the Western Border Region and the Mount Elgon/west Kenya area, four species being exclusively distributed in both areas. All these species occur in the Kivu area. Similarly there are two species that only occur in west Tanzania and the Western Border Range. In view of the large number of species and the high endemicity in the Western Border Range, the exclusive connections to the northeast and to the southeast must be regarded the result of range expansion rather than the combinations Western Border Range/west Tanzania or Western Border Range/west Kenya ever having been single areas of speciation.

(c) The six remaining combined areas of Table 8.5 have few endemics only and cannot be considered important centres of diversification. Two of the combined areas, west Tanzania (consisting of the areas Ufipa, Mpanda and Kigoma), and Malawi (with Nyika Plateau, Zomba and Mlanje), have been distinguished only for the sake of convenience; they lack characterising species (i.e. endemics that occur in all subareas). Further north there is more coherence although also there the numbers of endemics are low. Ethiopia, with four endemics and six widespread species, stands slightly apart since it cannot be combined with any particular area. The remaining three combined areas, west Kenya, central Kenya and north Tanzania, share 40% of their species, but only one species, *Papilio chrapkowskii*, occurs

exclusively in all three areas, so there is little reason for combining the three. Also any of the three possible combinations of two of the three areas is not significant. The combination central Kenya/west Kenya, for instance, would be characterised by two out of the 10 endemics found in the two areas (i.e. eight are confined to one area only), the combination central Kenya/north Tanzania by one out of nine. How could we (and why should we) select one of these combinations?

Summarising, as far as we can speak of pattern here the following pattern emerges. In EAT there are 259 taxa (177 species) restricted to highland forest. Of these, 255 (157 species) do not occur outside EAT. The endemics are not uniformly distributed: 106 (64 species) are found in one of the initially recognised areas of endemism with highest concentrations in the Usambara Mountains (15, 12), Nguru (8, 7), Uluguru (10, 8), Uzungwa (8, 8) and Kivu area (12, 9). There are seven potential areas of endemism without endemics even at the subspecies level (10, if we only consider the species level). Although they can no longer be considered areas of endemism most of them can be combined with other areas into larger areas of endemism. There are no endemic species that occur in more than 19 areas, and only nine occur in more than 10 areas. Thus, the mountains in EAT as a whole cannot be considered an area of endemism. Combination of areas into larger areas of endemism (i.e. characterised by endemics occurring in all subareas) yields two larger areas with high concentrations of endemics, namely, the Eastern Arc mountains and the Western Border Range. Other combinations are of minor importance as far as endemism is concerned, and seem to consist mainly of rather loose sets of individual mountains.

External geographic connections

The connections with areas outside EAT are different from those found in the grassland butterflies.

Oriental connection

In contrast to the grassland butterflies, the forest

butterflies have no connections with areas outside Africa at the species level. This is not surprising since as far as we know there has never been a forest connection between the Oriental and Afrotropical regions (unless we go back to Gondwana times). At the genus level, however, there are several connections involving 17 of the 46 genera. None of these genera is restricted to forest (i.e. some of the species live in woodland or still more open habitats) and it seems likely that the contact came about through species with preference for more open habitats. It implies either that the African forest species of these genera originated from species with another habitat preference, or that the habitat preference was originally for forests and the Oriental contact could come into being only when a species arose with a changed habitat preference. Anyway, the importance of the Oriental connection for the butterflies of the EAT montane forests of EAT seems to be negligible.

An exception to the above may be the genus *Celaenorrhinus*. The species are generally forest species. It is the only genus of the Hesperiidae with a pantropical distribution. The possibility that this distribution is very old must be considered.

Western connection

West of the Western Border Range there are several mountain areas high enough to carry highland forest, in south Zaïre/north Zambia, Angola and Cameroun. We are concerned only with the third area here: the montane fauna of Angola is badly known, and the mountains of south Zaïre/north Zambia are relatively close. There are 17 species (10%) of the montane forest fauna in East Africa that are represented in Cameroun (by a separate subspecies). The shared species are not evenly distributed in EAT, the largest number being found in the Western Border Range (see Table 8.5). In Cameroun the lowland forest fauna ascends to unusual heights in the mountains because of the high humidity; the strictly montane forest fauna is poor in species (Carcasson, 1964), and these species are apparently the ones that are shared with EAT.

Southern connection

Among the montane forest species of EAT there are only three that are also represented in South Africa. One of these, *Antanartia dimorphica*, is moreover also found in Cameroun and is thus a rather widespread species. Another species of this genus in East Africa, *A. schaeneia*, is also found in South Africa, where still another species occurs (*A. hippomene*). Even if we knew the exact relationships in this genus, the large distributional overlaps would hamper a geographic analysis of the speciation events. The third species also found in South Africa is *Papilio echerioides*. This species belongs to the *cynorta* group (Hancock, 1984) the seven species of which are found from West to East Africa and south to Angola and Malawi, only *echerioides* going further south and reaching Transvaal and Natal. We can hardly avoid the conclusion that this species is a northern intruder in South Africa.

Ecological relationships

In cases where phylogenetic data are available we can test whether the sister species (or sister group) of a montane forest species has the same habitat preference. As stated above such information is scarce, but the data available are interesting enough to mention here.

All five *Chondrolepis* species listed form a monophyletic group (de Jong, 1986). Thus, the speciation events leading to these species did not involve habitat change. The sister group of the five species has a slightly wider habitat preference, also occurring at much lower elevations, down to 800 m (and, not surprisingly, a wider distribution reaching Cameroun). Although there is altitudinal overlap the strictly montane forest species must have originated from a species generally living at lower elevations. The *danckelmani* species group of the genus *Bicyclus* is supposed to be monophyletic (Condamin, 1973). It consists of six montane forest species in EAT (all listed in Appendix 8.2). The evolution of the group did not need to involve a habitat change. The sister group of the *danckelmani* group, the *similis* group (with a single species), is also restricted to montane forest. The sister group of the similis and *danckelmani* groups combined, the *sanaos* group, consists of two species of lowland forest distributed from West Africa to west Kenya. In the three remaining *Bicyclus* species, each belonging to another species group, we must suppose a habitat change, either from forest or from woodland to montane forest, during their evolution. The *nireus* species group of the genus *Papilio* shows a similar pattern, i.e. there are some species whose sister species lives in montane forest (*hornimani–charopus–mackinnoni, desmondi–thuraui*), and some other species (*aethiopsis, chrapkowskii*) whose sister species flies at lower elevations, either in forest or (also) in woodland.

Anticipating a study on the phylogeny of the genus *Metisella* it can be said that the 13 species listed in Appendix 8.2 fall into a number of species groups which are entirely restricted to montane forests. Since the genus is still under study details cannot be given. Also in the genus *Uranothauma*, with 12 of the 18 known species restricted to the montane forests of EAT, diversification apparently took place largely within the montane forest area. In all other genera either the situation is not clear or a habitat switch was involved. In a number of genera the switch is clear even without knowing the sister species, since there are no other congeneric montane forest species. This is particularly the case in the genus *Colotis*, where *elgonensis* is the only forest species while the 41 other species in Africa live in open and often dry habitats. Other apparent switches in habitat preference are found in *Abantis meru, Ampittia parva, Bicyclus dentatus* and *Ypthima albida*. In these cases a switch must have been made from woodland to montane forest since there are no close relatives living in forests. In the other genera the situation is less clear, but since always, except in monotypic genera, different habitat preferences occur in the genera (of the 46 genera represented, 13 are restricted to forest, including lowland forest, the remaining genera also occur in woodland and/or open habitats), at least the change from low or medium to high levels must often have been made. Clearly phylogenetic data would be most helpful here to exactly determine sister species and the ecological switches made.

Discussion

The Cameroun connection

The predicted pattern in case of a continuous connection of montane forest biota between EAT and Cameroun shows a fauna that is similar at both ends, with shared species or, in view of the short time lapse, at most sister species. The pattern found is different. Although there are shared species, the Cameroun fauna is poor in comparison with the fauna in EAT. A continuous connection would explain the shared species, but not the absence in Cameroun of the other 90% of the montane forest fauna of EAT. Moreover, as already stated above when describing the scenario, if there was a connection between Cameroun and EAT, the montane forests in EAT would also be interconnected, leading to a more or less uniform fauna. In that case the species shared with Cameroun should be rather evenly distributed over the montane forests of EAT. This is not the case (see Table 8.5). The pattern does, however, agree with the idea that there never was a continuous connection, but that a number of montane species with a wider ecological tolerance could live at lower altitudes at a time when the fauna there was unsaturated and competition was low (Hedberg, 1969; Diamond & Hamilton, 1980; Hamilton, 1982; de Jong, 1986). The latter situation could have existed when the forest after a period of restriction to refugia rapidly expanded its area.

It is tempting to conclude that Cameroun has no montane forest butterflies of its own but received them from EAT. Actually, however, there is little evidence for or against this. The shared species could have arisen in Cameroun and migrated to the east. This would explain why in EAT the shared species are mainly found in the western areas (see Table 8.5), but the uneven distribution of the shared species over the mountains of EAT could also have quite a different origin, i.e. because the montane forest fauna never was evenly distributed in EAT (see below). What militates against a Cameroun origin is that it is difficult to understand why a poor fauna could successfully penetrate into a much richer fauna and not vice versa. So far an eastern origin of the montane forest butterflies of Cameroun seems most probable, but we should await phylogenetic studies of the species concerned before drawing conclusions.

Fragmentation of a Tertiary forest belt

In this scenario the Eastern Arc mountains should have the highest concentration of montane forest butterflies. This is actually found, but the decrease in numbers with increasing distance from the Eastern Arc mountains, which is a further prediction, is not found at all (see Table 8.5). The other prediction of this scenario is a twin relationship between taxa restricted to the Eastern Arc mountains and taxa in the Central African lowland forest block. In the cases in which phylogenetic data are available (*Metisella, Chondrolepis, Papilio, Bicyclus, Amauris*) this pattern is not found. Even though exact knowledge of the phylogeny of the other butterflies is very limited or absent, a remark can be made since we do know the relatives (i.e. congeners) and their distributions. In those genera in which all species are restricted to lowland forest except for one or more species in the montane forests of the Eastern Arc mountains, we may safely assume that this pattern occurs. However, the only such genus is *Ceratrichia*. Since there are 28 genera with endemic species in the Eastern Arc mountains, nine of which are restricted to forests, it cannot be said that there is convincing evidence for the existence of a Tertiary forest belt across the African continent. This agrees with what was found about the prediction of decreasing numbers just mentioned. It does not mean that a continuous forest belt never existed in the Tertiary, only that so far there is no evidence for it. If the belt ever existed the traces have probably been obscured or wiped out by later developments. This is not surprising for events so remote in time, certainly in the light of the dramatic climatic changes in the Pleistocene that must have upset all ecosystems in the area.

Fragmentation of a Pleistocene forest belt

None of the patterns predicted by this scenario are actually found. There are no groups of closely related allopatric species, together covering most of the mountain areas. Admittedly the species of the *Papilio nireus* group in the montane forests together cover all eight combined areas listed in Table 8.5, but apart from the fact that there is much distributional overlap, all species together or the monophyletic subgroup *mackinnoni–charopus–hornimani* which occurs in seven of the eight areas, do not form a monophyletic group with an allopatric lowland species further west (see Figure 8.3). Similarly the *Charaxes* species listed together cover most of the study area, but they belong to a number of widely different species groups (Henning, 1988). The montane *Uranothauma* species combined also occur in seven of the eight combined areas, but their interrelationships are unknown and there is no good lowland candidate in the genus to form a monophyletic group with the montane species. The *Chondrolepis* species are closely related, forming a monophyletic group, but they are not allopatric and the sister group has a wide distribution in EAT. A twin relationship between montane forest species and lowland forest species to the west can be found in a few instances only. *Ceratrichia bonga* is the only species of the genus in Tanzania, other species living further west, so even though the phylogeny of the genus is not known the sister species of *bonga* must be sought for further west. *Bicyclus kenia* is found in central Kenya/north Tanzania, its sister species or sister group (*mandanes* and/or *auricrudus*) lives from West Africa to west Kenya. *Bicyclus jefferyi* occurs from east Zaïre to Kenya/north Tanzania, its sister taxon (*dorothea* and/or *moyses*) is found in the West and Central African forest blocks. For the *Bicyclus* species, see Figure 8.4.

There are no widespread montane forest species that at the same time occur in the Central African forest block as well. Of the only five species occurring in seven or all eight combined areas (Table 8.5), two are restricted to the study area, two occur in the highlands of Cameroun as well and one is also found in the montane forests of South Africa. Further, it is clear from Tables 8.4 and 8.5 and the remarks above (Distinction and coherence of areas of endemism, p. 155) that the numbers of species and of endemics are very unevenly distributed.

Thus, the patterns found do not support the idea that the butterflies of the montane forests are simply lowland forest species trapped in the mountains when the forests became isolated there. In this connection it is significant to note that species-rich forest genera like *Telipna* (27 species), *Micropentila* (31), *Pentila* (36), *Liptena* (70), *Epitola* (82), *Euriphene* (64) and *Euphedra* (137) have no species at all in the montane forests of EAT. It seems unlikely that a continuous forest belt would have left no traces of the occurrence of these genera in EAT.

Ecological differentiation

Starting from the viewpoint that preference for the montane habitat is a derived condition, the montane forest butterflies either arose from lowland forest butterflies or from butterflies living in non-forest habitats. The possibility of the montane forest butterflies originally being lowland butterflies from the west trapped in mountain refugia has just been dealt with. These are fragmentation scenarios. Here we are concerned with a possible adaptation from a non-forest habitat or a lowland forest habitat to montane forest without geographical isolation. If this really occurred the following patterns are expected (see description of the scenario, above):

1. Twin relationships between montane forest and neighbouring lowland forest or open habitat/woodland within EAT.
2. No recurrent geographic pattern of phylogenetic relationships.
3. Highest concentration of endemics in Eastern Arc mountains, unless the species originated here could later spread through the other mountain ranges (see Internal development, p. 148).

Direct evidence of pattern (1) is rare since we know so little about the phylogeny of the species concerned, but there is also indirect evidence (e.g. because the species is the only one of the genus living in montane forest, and all candidate sister taxa occur nearby at lower levels and/or in other habitats). Thus, adaptation from lower to higher altitude forest must be assumed for *Gorgyra bibulus*, the *Chondrolepis* species combined (see

Figure 8.2), *Papilio aethiopsis*, *P. chrapkowskii* (see Figure 8.3), *Etesiolaus* sp., and *Gnophodes grogani*. Adaptation from another habitat, most probably woodland, must be assumed for *Abantis meru*, *Ampittia parva*, *Colotis elgonensis* and *Ypthima albida*. In the other 37 genera such habitat switches may also have occurred but the phylogeny is unknown and there are possibilities for twin relationships that do not agree with the present pattern. The pattern could as well be common, but we must await phylogenetic studies. For the same reason we cannot say much about prediction (2). Prediction (3), however, the high concentration of endemics in the Eastern Arc mountains, absolutely agrees with the pattern found. It is remarkable that there is one other combined area, the Western Border Range, with a high concentration of endemics. Since this mountain range is much younger than the Eastern Arc mountains, and not older than other mountain ranges in EAT, its high endemicity (although lower than in the Eastern Arc mountains) must have a cause other than antiquity of the habitat. The Western Border Range differs from other mountain ranges in EAT in the close proximity of a large reservoir of lowland forest species (and thus, candidate highland forest species), i.e. the Central African forest block. It could be an explanation for the high endemicity in the Western Border Range relative to the other young mountains in EAT. This idea would be supported if we were to find that the endemics of the Western Border Range each had sister species in the lowland forest nearby. Another possible explanation will be given in the following paragraph.

Internal development

Here we must test two possible scenarios that may both have occurred but not simultaneously: (i) the montane forests once were continuous irrespective of possible contacts with forests to the west, and (ii) there has been large-scale local and differential extinction caused by desiccation. The relevant patterns are: (i) widespread species or widespread monophyletic groups of allopatric species; (ii) highest concentration of endemics and of species in general in wetter areas, i.e. the

Eastern Arc mountains and the Western Border Range. A combination of both scenarios leads to a pattern of either many shared species or many sister species in both the Eastern Arc mountains and the Western Border Range. As already mentioned there are few widespread species: only five species (2.8%) occur in seven or all eight combined areas (Table 8.5). No monophyletic groups of allopatric species are known, but there are some monophyletic groups the species of which together cover most of the mountain ranges, like *Papilio hornimani–charopus–mackinnoni* and the five *Chondrolepis* species. Their number, however, hardly reinforces the pattern of widespread taxa. Equally, there is no evidence for a strong similarity between the Eastern Arc mountains and the Western Border Range. Of the 146 species occurring in one or both of these ranges only 10 (7%) are common to both. The only known twin relationship between the Eastern Arc mountains (in this case with north Tanzania) and the Western Border Range is the *Bicyclus danckelmani* group. Even though the exact relationships in this group are not known the distribution of the species (Figure 8.4) shows that an Eastern Arc mountains–Western Border Range twin relationship must occur here.

On the basis of the patterns found we can only conclude that there is no evidence of a former continuous montane forest habitat. It has even proved difficult to combine areas into larger units, this usually being based on the distribution of a few species only. It points rather to a dispersal than to a fragmentation scenario. Pattern (ii) on the other hand agrees well with what is really found: high concentrations of species in the Eastern Arc mountains and the Western Border Range, and far fewer elsewhere. It is particularly obvious in the endemics (see Table 8.5). Consequently we cannot refute the hypothesis of a large-scale extinction of montane forest species in many mountains of EAT.

Conclusion

The butterflies that are more or less restricted to the montane forests of EAT do not bear evidence of a Tertiary forest belt across Africa, nor of such a belt in the Quaternary. If such a belt ever existed, the traces have been wiped out. Similarly, there is no evidence for a former continuous montane forest belt from East Africa to Cameroun. The rather small proportion of taxa of the montane forests of EAT also occurring in Cameroun can equally well be explained by range extensions through unsaturated forests at low and medium elevations. The majority if not all of the Afromontane butterflies in EAT did not originate from colonisation from elsewhere but by differentiation and switches in habitat preference within EAT. There is no evidence that the switches were attributable to or accompanied by geographic isolation. The age of the Eastern Arc mountains is reflected in their high diversity. Once adapted to the montane forests the species more or less spread over the mountain ranges. Locally the spread may have been facilitated by temporary extensions of the forest, but the butterflies do not indicate a repeated vicariance scenario. It rather seems that dispersal and local extinctions have led to the present distribution pattern.

Summary

The distribution of the butterflies more or less restricted to montane habitats in the Eastern Afrotropics (EAT; roughly east of Long. 30° E) is analysed. Two habitats are investigated in some detail: montane grasslands and montane forests. There are no butterflies characteristic of the treeless Afroalpine zone. The few butterflies that are found there have come from lower elevations, possibly using air currents. It is not clear why there are no Afroalpine butterflies: there may be an ecological rather than a historical origin.

It is concluded that a small proportion (1/4 to 1/5) of the montane grassland butterflies originated from dispersal from the Palaearctic and Oriental regions. By far the greater part, however, originated in the Eastern Afrotropics from species with other habitat preferences. The montane forest fauna of the area has no affinity with montane forest faunas elsewhere except for Cameroun. The similarity (limited, as far as the Eastern Afrotropics are concerned) between the montane forest faunas of Cameroun and the eastern part of

Africa is ascribed to dispersal from the east. Most montane forest species in the Eastern Afrotropics evolved from other species in the same general area but with a different habitat preference, while a small proportion arose by further differentiation of montane forest species in the Eastern Afrotropics. It implies that for the development of the montane grassland fauna and particularly for the montane forest fauna, parapatric speciation (or maybe sympatric speciation) has been of paramount importance, allopatric speciation (i.e. by geographic isolation) playing only a minor role. Evidence for vicariance events is very weak. If the montane forests were ever more coherent than today, the traces have been wiped out, at least in the butterflies.

The ideas put forward here must await further support (or rejection) from phylogenetic studies, which are badly needed for most of the butterflies concerned.

References

ACKERY, P. R. & VANE-WRIGHT, R. I. (1984). *Milkweed Butterflies*. London: British Museum (Natural History).

BERGER, L. A. (1981). *Les Papillons du Zaïre*. Brussels: Weissenbruch.

BERNARDI, G. (1947). Révision de la classification des espèces holarctiques des genres *Pieris* Schr. et *Pontia* Fabr. *Miscell. Ent.* **44**, 65–80.

BERNARDI, G. (1962). Missions Ph. Bruneau de Miré au Tibesti: Lépidoptères Pieridae, Nymphalidae et Danaidae. *Bulletin de l'Institut Français d'Afrique Noire* **24A**, 813–51.

BERNARDI, G. (1979). Contribution à la biogéographie des montagnes africaines: I. Généralités. *Compte rendu des séances de la Société de Biogéographie* **479**, 5–28.

BERNARDI, G. (1980). Contribution à la biogéographie des montagnes africaines: II. Le genre *Issoria* Hübner (Lepidoptera, Nymphalidae). *Compte rendu des séances de la Société de Biogéographie* **487**, 189–99.

CARCASSON, R. H. (1964). A preliminary survey of the zoogeography of the African butterflies. *East African Wildlife Journal* **2**, 122–57.

CARCASSON, R. H. (1981). *Collins Handguide to the Butterflies of Africa*. London: Collins.

CARPENTER, G. D. H. (1935). The Rhopalocera of Abyssinia, a faunistic study. *Transactions of the Royal Entomological Society of London* **83**, 313–447.

COE, M. J. (1967). *The Ecology of the Alpine Zone of Mount Kenya*. The Hague: Junk.

CONDAMIN, M. (1973). Monographie du genre *Bicyclus* (Lepidoptera, Satyridae). *Mémoires de l'Institut fondamental d'Afrique Noire* No. 88, Dakar: IFAN.

CROWE, T. M. & CROWE, A. A. (1982). Patterns of distribution, diversity and endemism in Afrotropical birds. *Journal of Zoology (London)* **198**, 417–42.

D'ABRERA, B. (1980). *Butterflies of the Afrotropical Region*. Melbourne: Lansdowne Editions.

DE JONG, R. (1976). Affinities between the West Palaearctic and Ethiopian butterfly faunas. *Tijdschrift voor Entomologie* **119**, 165–215.

DE JONG, R. (1978). Monograph of the genus *Spialia* Swinhoe (Lepidoptera, Hesperiidae). *Tijdschrift voor Entomologie* **121**, 23–146.

DE JONG, R. (1982). Secondary sexual characters in Celaenorrhinus and the delimitation of the genus (Lepidoptera, Hesperiidae). *Journal of Natural History* **16**, 695–705.

DE JONG, R. (1986). Systematics, phylogeny and biogeography of the chiefly Afromontane genus *Chondrolepis* Mabille (Lepidoptera: Hesperiidae). *Zoologische Verhandelingen* (Leiden) **231**, 1–40.

DESCIMON, H. (1986). Origins of lepidopteran faunas in the High Andes. In *High Altitude Tropical Biogeography*, ed. F. Villeumier and M. Monasterio, pp. 500–32. Oxford: Oxford University Press.

DIAMOND, A. W. & HAMILTON, A. C. (1980). The distribution of forest passerine birds and Quaternary climatic change in tropical Africa. *Journal of Zoology (London)* **191**, 379–402.

DICKSON, C. G. C. (1978). *Pennington's Butterflies of Southern Africa*. Johannesburg: Ad. Donker.

GIFFORD, D. (1965). *A List of the Butterflies of Malawi*. Blantyre: The Society of Malawi.

HAMILTON, A. C. (1981). The Quaternary history of African forests: its relevance to conservation. *African Journal of Ecology* **19**, 1–6.

HAMILTON, A. C. (1982). *Environmental History of East Africa*. London: Academic Press.

HAMILTON, A. C. (1989). African forests. In *Tropical Rain Forest Ecosystems. Ecosystems of the World*, Vol. 14B, ed. H. Lieth and M. J. A. Werger, pp. 155–82. New York: Elsevier.

HANCOCK, D. L. (1983). Classification of the Papilionidae (Lepidoptera): a phylogenetic approach. *Smithersia* **2**, 1–48.

HANCOCK, D. L. (1984). The *Princeps nireus* group

of Swallowtails (Lepidoptera: Papilionidae). Systematics, phylogeny and biogeography. *Arnoldia Zimbabwe* 9, 181–215.

HEDBERG, O. (1969). Evolution and speciation in tropical high mountain flora. *Biological Journal of the Linnean Society* 1, 135–48.

HEDBERG, O. (1986). Origins of the Afroalpine flora. In *High Altitude Tropical Biogeography*, ed. F. Vuilleumier and M. Monasterio, pp. 443–68. Oxford: Oxford University Press.

HENNING, S. F. (1988). *The Charaxinae butterflies of Africa*. Johannesburg: Aloe Books.

HOWARTH, G. (1966). Revisional notes on the genus *Antanartia* (Lepidoptera: Nymphalidae). *Bulletin of the British Museum (Natural History), Entomology (London)* 18, 21–43.

KIELLAND, J. (1978). A provisional checklist of the Rhopalocera of the eastern side of Lake Tanganyika. *Tijdschrift voor Entomologie* 121, 147–237.

KIELLAND, J. (1990). *Butterflies of Tanzania*. Melbourne: Hill House.

LIVINGSTONE, D. A. (1975). Late Quaternary climatic change in Africa. *Annual Review of Ecology and Systematics* 6, 249–80.

MILLER, J. S. (1987). Host–plant relationships in the Papilionidae (Lepidoptera): parallel cladogenesis or colonization? *Cladistics* 3, 105–20.

MOREAU, R. E. (1963). Vicissitudes of the African biomes in the late Pleistocene. *Proceedings of the Zoological Society of London* 41, 395–421.

MOREAU, R. E. (1966). *The Bird Faunas of Africa and its Islands*. New York: Academic Press.

ROBBINS, R. K. & HENSON, P. M. (1986). Why *Pieris rapae* is a better name than *Artogeia rapae* (Pieridae). *Journal of the Lepidopteran Society* 40, 79–92.

SHIROZU, T. & SAIGUSA, T. (1973). A generic classification of the genus *Argynnis* and its allied genera (Lepidoptera: Nymphalidae). *Sieboldia* 4, 99–114.

STEMPFFER, H. (1967). The genera of the African Lycaenidae (Lepidoptera: Rhopalocera). *Bulletin of the British Museum (Natural History)*, Supplement No. 10, pp. 1–322.

VAN SOMEREN, V. G. L. (1957). Revisional notes on *Lepidochrysops* (Lycaenidae: Lampidinae) of Kenya and Uganda, with descriptions of new species and sub-species. *Journal of the Entomological Society of South Africa* 20, 58–78.

VAN ZINDEREN BAKKER, E. M. (1967). Upper Pleistocene and Holocene stratigraphy and ecology on the basis of vegetation changes in sub-Saharan Africa. In *Background to Evolution in Africa*, ed. W. W. Bishop and S. Desmond Clark, pp. 125–47. Chicago: University of Chicago Press.

VAN ZINDEREN BAKKER, E. M. (1982). African palaeoenvironments 18 000 yrs BP. *Palaeoecology of Africa and of the Surrounding Islands and Antarctica* 15, 77–99.

WARREN, B. C. S. (1956). A review of the classification of the subfamily Argynninae (Lepidoptera: Nymphalidae). Part 2. Definition of the Asiatic genera. *Transaction of the Royal Entomological Society of London* 107, 381–92.

WHITE, F. (1981). The history of the Afromontane archipelago and the scientific need for its conservation. *African Journal of Ecology* 19, 33–54.

Addendum

Recently the book *Butterflies of Tanzania* by J. Kielland appeared, with beautiful illustrations and interesting notes with each species on distribution and ecology. Unfortunately it was received too late by the authors to check the data in the book against the data in this chapter.

Appendix 8.1 *Distribution of species and subspecies of butterflies restricted to montane grassland in the Eastern Afrotropics. X = presence; O = absence. For explanation of the areas (1–27), see Table 8.1.*

Species/subspecies													Area number														
	1	2	3	4	5	6	7	8	9	10	11	12	13	14	15	16	17	18	19	20	21	22	23	24	25	26	27
Colias erate marmoana	X	O	O	O	O	O	O	O	O	O	O	O	O	O	O	O	O	O	O	O	O	O	O	O	O	O	O
Colias electo meneliki	X	O	O	O	O	O	O	O	O	O	O	O	O	O	O	O	O	O	O	O	O	O	O	O	O	O	O
Colias electo pseudohecate	O	X	X	X	X	X	X	O	X	X	O	O	X	X	X	X	X	X	X	X	X	X	X	O	O	X	O
Colias electo hecate	X	O	O	O	O	O	O	O	O	O	O	O	O	O	O	O	O	O	O	X	O	O	O	O	O	X	O
Pieris brassicoides brassicoides	X	O	O	O	O	O	O	O	O	O	O	O	O	O	O	O	O	O	O	O	O	O	O	O	O	O	O
Pieris brassicoides marghanita	O	O	O	O	O	O	O	X	X	O	O	O	O	O	O	O	O	O	O	O	O	O	O	O	O	O	O
Antanartia abyssinica abyssinica	X	O	O	O	O	X	X	O	O	O	O	O	O	O	O	O	O	O	O	O	O	O	O	O	O	O	O
Antanartia abyssinica jacksoni	O	X	X	X	X	X	X	X	X	X	X	O	O	O	X	X	X	X	X	X	X	X	X	X	X	X	X
Antanartia abyssinica vansomereni	O	O	O	O	O	O	O	O	O	O	O	O	O	O	O	O	O	O	O	O	O	O	O	O	O	O	O
Argyreus hyperbius neuma	X	O	O	O	O	O	O	O	O	O	O	O	O	O	O	O	X	X	O	X	O	O	O	O	O	O	O
Issoria smaragdifera smaragdifera	O	O	O	O	O	O	O	O	O	O	O	O	O	O	O	X	O	O	O	O	O	O	O	O	O	O	X
Issoria smaragdifera reducta	O	X	X	X	X	X	X	O	X	O	O	O	O	O	O	O	O	O	O	O	O	O	O	O	O	O	O
Issoria hanningtoni hanningtoni	O	O	O	O	O	O	O	O	X	O	O	O	O	O	O	O	O	O	O	O	O	X	X	X	O	O	O
Issoria hanningtoni jeanneli	O	O	O	O	O	O	O	O	O	O	O	O	O	O	O	O	O	O	O	O	O	X	X	X	X	X	X
Issoria baumanni baumanni	O	O	O	O	O	O	O	O	O	O	O	O	O	O	O	O	O	O	O	O	O	O	O	X	X	X	O
Issoria baumanni excelsior	O	O	O	O	O	O	O	O	O	O	O	O	O	O	O	O	O	O	O	O	X	X	O	X	O	O	O
Issoria baumanni katangae	O	O	O	O	O	O	O	O	O	O	X	X	X	O	O	X	X	O	O	O	O	O	O	O	O	O	O
Neocoenyra heckmanni	O	O	O	O	O	O	O	O	O	O	X	X	X	O	X	X	X	O	O	X	O	O	O	O	O	O	O
Neocoenyra jordani	O	O	O	O	O	O	O	O	O	O	O	O	O	O	O	O	O	O	O	O	O	X	X	O	X	O	O
Neocoenyra fuelleborni	O	O	O	O	X	X	O	O	X	O	O	O	O	O	O	O	O	O	O	O	O	O	O	O	O	O	O
Neocoenyra species	O	O	O	O	O	O	O	O	O	O	O	O	O	O	O	X	X	O	O	O	O	X	X	O	X	O	O
Capys disjunctus	O	O	O	O	O	X	O	O	O	O	O	O	O	O	O	O	O	O	O	O	O	O	O	O	O	O	O
Capys catharus catharus	O	X	X	X	X	O	O	O	O	O	O	O	O	O	O	O	O	O	O	O	X	X	X	X	X	X	X
Capys catharus rileyi	O	O	X	X	O	O	X	O	X	O	O	O	O	O	O	O	O	O	O	O	O	O	O	O	O	O	O
Harpendyreus aequatorialis aequator	O	O	O	O	O	O	O	O	O	O	O	O	O	O	O	O	O	O	O	O	O	O	O	O	O	O	O
Harpendyreus aequatorialis vulcanica	O	O	O	O	O	O	O	O	O	O	O	O	O	O	O	O	O	O	O	O	O	O	O	O	O	O	O
Harpendyreus meruanus	O	O	O	O	O	O	O	O	O	O	O	O	O	O	O	O	O	O	O	O	O	O	O	O	O	O	O
Harpendyreus boma	O	O	O	O	O	O	O	O	O	O	O	O	O	X	O	X	X	O	O	O	O	O	O	X	O	X	O
Harpendyreus juno	O	O	O	O	O	O	O	O	O	O	O	O	X	O	O	O	O	O	O	O	X	O	O	O	O	O	O
Harpendyreus marungensis marungensis	O	O	O	O	O	O	O	O	O	O	O	O	O	O	O	O	O	O	O	O	O	O	O	X	X	X	O
Harpendyreus marungensis wollastoni	O	O	O	O	O	O	O	O	O	O	O	O	O	O	O	O	O	O	O	O	O	O	O	O	O	X	O
Harpendyreus bergeri	O	O	O	O	O	O	O	O	O	O	O	O	O	O	O	O	O	O	O	O	X	X	X	X	X	X	X
Harpendyreus major	O	O	O	O	O	O	O	O	O	O	O	O	O	O	O	O	O	O	O	O	O	O	O	O	X	O	O
Harpendyreus marlieri	O	O	O	O	O	O	O	O	O	O	O	O	O	O	O	O	O	O	O	O	O	O	O	O	X	X	O
Harpendyreus hazelae	O	O	O	O	O	O	O	O	O	O	O	O	O	O	O	O	O	O	X	O	O	O	O	O	O	X	O
Harpendyreus kisaba	O	O	O	O	O	O	O	O	O	O	O	O	O	O	O	O	O	O	O	O	O	O	O	O	O	X	O
Harpendyreus argenteostriata	O	O	O	O	O	O	O	O	O	O	O	O	O	O	O	O	O	O	O	O	X	O	O	O	O	O	X
Harpendyreus reginaldi	O	O	O	O	O	O	O	O	O	O	O	O	O	O	O	O	O	X	O	O	O	O	O	O	O	O	O
Actizera stellata	X	X	O	X	X	O	O	O	X	O	O	O	O	O	O	O	O	O	O	O	O	O	O	O	O	X	O

Species/subspecies														Area number												
	1	2	3	4	5	6	7	8	9	10	11	12	13	14	15	16	17	18	19	20	21	22	23	24	25	26	27
Eicochrysops nandianus	O	O	O	O	O	X	X	O	X	O	O	O	O	O	O	O	O	O	O	O	X	O	O	O	O	O	O
Eicochrysops sebagadis	X	O	O	O	O	O	O	O	O	O	O	O	O	O	O	O	O	O	O	O	O	O	O	O	O	O	O
Euchrysops mauensis mauensis	O	O	O	O	X	O	O	O	O	O	O	O	O	O	O	O	O	O	O	O	O	O	X	O	X	X	O
Euchrysops mauensis abyssiniae	X	O	O	X	X	X	X	O	X	O	O	O	O	O	O	O	O	O	O	O	X	X	X	O	O	X	O
Euchrysops nandensis	O	O	O	X	X	O	X	O	X	O	O	O	O	O	O	O	O	O	O	O	O	O	O	O	O	O	O
Euchrysops crawshayi crawshayi	O	O	O	X	X	X	O	O	X	O	O	O	O	O	O	O	O	O	O	O	O	O	O	O	O	O	O
Euchrysops crawshayi fontainei	O	O	O	O	X	O	O	O	O	O	O	O	O	O	O	O	O	O	O	O	X	O	O	O	X	X	O
Lepidochrysops mpanda	O	O	O	O	O	X	O	O	O	O	O	O	O	O	O	O	O	O	O	O	O	X	O	O	O	O	O
Lepidochrysops elgonae elgonae	O	X	X	O	X	O	O	O	O	O	O	O	O	O	O	O	O	O	O	O	O	O	O	O	O	O	O
Lepidochrysops elgonae moyo	O	X	O	X	O	O	O	O	O	O	O	O	O	O	O	O	O	O	O	O	O	O	O	O	X	O	O
Lepidochrysops jansei	O	O	O	X	O	O	X	O	O	O	O	O	O	O	O	O	O	O	O	O	O	O	O	O	O	O	O
Lepidochrysops kilimanjarensis	O	O	O	O	O	O	O	O	O	O	O	O	O	O	O	O	O	X	O	O	X	X	O	O	O	O	O
Lepidochrysops cupreus	O	O	O	O	O	O	O	O	O	O	O	O	O	O	O	O	O	O	O	O	O	X	O	O	X	O	O
Lepidochrysops species	O	O	O	O	O	O	O	O	O	O	O	O	O	O	O	O	O	X	O	O	X	O	O	O	O	X	O
Lycaena phlaeas aethiopica	X	O	O	X	X	X	X	O	X	O	O	O	O	O	O	O	O	O	X	X	O	O	O	O	X	O	O
Lycaena pseudophlaeas pseudophlaeas	O	X	X	X	O	O	O	O	O	O	O	O	O	O	O	X	O	X	X	O	X	X	O	O	O	O	O
Lycaena abbotti	O	O	O	O	O	O	O	O	O	O	O	O	O	O	O	X	X	O	O	O	O	O	O	O	O	O	O
Metisella carsoni	O	O	O	O	O	O	O	O	O	O	O	O	O	O	O	O	O	X	O	O	O	O	O	O	O	O	O
Kedestes barberae	O	O	O	O	O	O	O	O	O	O	O	O	O	O	O	O	O	O	O	O	O	O	O	O	O	O	O

Appendix 8.2 *Distribution of species and subspecies of butterflies restricted to montane forests in the Eastern Afrotropics. X = presence; O = absence. For explanation of the areas (1–27), see Table 8.1.*

Species/subspecies	Area number																										
	1	2	3	4	5	6	7	8	9	10	11	12	13	14	15	16	17	18	19	20	21	22	23	24	25	26	27
Celaenorrhinus zanqua	O	O	O	O	O	O	O	O	O	O	O	O	X	O	O	O	O	X	O	O	O	O	O	O	O	O	O
Celaenorrhinus kimbozae	O	O	O	O	O	O	O	O	O	O	O	O	X	O	O	O	O	O	O	O	O	O	O	O	O	X	O
Celaenorrhinus kivuensis	O	O	O	O	O	O	O	O	O	O	O	O	O	O	O	O	O	O	O	O	O	O	O	O	O	O	O
Celaenorrhinus species 1	X	X	X	O	O	O	O	O	O	O	O	O	X	O	O	X	O	O	O	O	O	O	X	X	X	O	X
Celaenorrhinus species 2	X	O	O	O	O	O	O	O	X	O	O	O	O	O	X	O	O	O	O	O	O	O	X	X	X	O	X
Abantis meru	X	O	O	O	O	O	O	O	O	O	O	O	O	O	O	O	O	X	O	O	O	O	O	O	O	O	O
Metisella alticola	X	O	X	X	O	X	X	O	X	O	O	O	O	O	O	X	O	X	O	O	O	O	O	X	X	X	X
Metisella nanda	O	X	X	X	X	X	X	X	X	O	O	X	X	O	O	X	O	O	O	X	O	O	X	X	X	X	X
Metisella orientalis	O	O	O	O	O	X	O	O	X	X	O	O	X	O	O	X	O	X	O	X	O	O	X	X	O	O	O
Metisella elgona	O	O	O	O	O	O	O	O	O	O	O	O	O	O	O	O	O	O	O	O	O	O	O	O	O	O	O
Metisella zeta	O	X	O	X	X	O	X	O	X	O	O	X	X	O	X	X	O	X	O	O	O	O	O	O	O	X	X
Metisella quadrisignata	X	O	O	O	O	O	O	O	O	O	O	O	O	O	O	X	X	X	X	X	O	O	X	O	O	O	O
Metisella kambove	X	O	O	O	X	O	O	O	X	O	O	O	X	O	O	X	X	X	X	X	O	X	X	O	O	O	O
Metisella congdoni	O	O	O	O	O	O	O	O	O	O	O	O	O	O	O	X	X	X	O	O	O	O	O	O	O	O	O
Metisella decipiens	O	X	X	O	X	X	O	O	O	O	O	O	O	O	O	X	X	X	O	O	O	O	O	O	O	O	O
Metisella medea	O	O	O	O	O	O	O	O	O	O	O	O	O	O	O	X	O	X	O	O	O	O	O	O	O	O	O
Metisella nyika	O	O	O	O	O	O	O	O	O	O	O	X	X	O	O	X	O	O	O	O	O	X	O	O	O	O	O
Metisella species 1	O	O	O	O	O	O	O	O	O	O	O	X	X	O	O	X	O	O	O	O	O	O	O	O	O	O	O
Metisella species 2	O	O	O	O	O	X	X	O	O	O	O	X	X	O	O	O	O	O	O	X	O	O	O	O	O	O	O
Astictopterus tura	O	O	O	O	O	O	O	O	O	O	O	O	O	O	X	O	O	O	O	O	O	X	O	O	O	O	O
Ampittia parva	O	X	O	O	O	O	O	O	O	X	O	O	O	O	O	O	O	O	O	O	O	O	O	O	O	O	O
Gorgyra bibulus	O	O	O	O	O	O	O	O	O	X	O	O	O	O	O	O	O	O	O	O	O	X	O	O	O	O	O
Ceratrichia bonga	O	X	X	X	O	O	O	O	O	O	O	O	O	O	O	O	O	X	O	O	O	O	O	O	X	X	X
Chondrolepis telisignata	O	O	O	O	O	O	O	O	O	O	O	O	O	O	O	O	X	X	O	O	O	O	O	O	X	X	X
Chondrolepis cynthia	O	O	O	O	X	O	O	O	X	X	O	O	O	O	O	O	O	X	O	O	O	O	O	O	X	X	X
Chondrolepis leggei	O	O	O	O	O	O	O	O	X	O	O	O	O	O	O	O	O	O	O	O	O	O	O	O	O	O	O
Chondrolepis obscurior	O	O	O	O	O	O	O	O	X	O	O	O	O	O	O	X	X	X	X	X	O	O	O	O	O	O	O
Chondrolepis similis	O	O	O	O	O	O	O	O	O	O	O	O	O	O	O	X	O	O	O	O	O	O	O	O	O	O	O
Zenonia crasta	O	O	O	O	O	O	O	O	O	O	O	O	O	O	O	O	O	O	O	O	O	O	O	O	O	O	O
Papilio leucotaenia	O	O	O	O	O	O	O	O	O	O	O	O	O	O	O	O	O	O	O	O	O	O	O	O	X	X	X
Papilio nobilis nobilis	O	X	O	O	O	O	O	O	O	O	O	O	O	O	O	O	O	O	X	O	O	O	O	O	X	O	O
Papilio nobilis leroyi	O	O	O	O	O	O	O	O	O	O	O	O	O	O	O	O	O	X	O	O	O	O	O	O	X	O	X
Papilio pelodorus pelodorus	O	O	O	O	O	O	O	O	O	O	O	O	O	O	O	O	O	O	O	O	O	O	O	O	O	O	O
Papilio pelodorus vesper	O	X	X	X	X	O	O	O	X	X	O	O	O	O	O	X	X	X	O	O	O	O	O	O	X	X	X
Papilio mackinnoni mackinnoni	O	O	O	O	X	X	O	O	X	O	O	O	O	O	O	O	O	O	O	O	O	O	O	O	O	O	O
Papilio mackinnoni isokae	O	O	O	O	O	O	O	O	O	O	O	O	O	O	O	O	O	X	O	O	O	O	O	O	X	X	X
Papilio mackinnoni subspecies	O	O	O	O	O	O	O	O	O	O	O	O	O	O	O	O	O	O	O	O	O	X	X	O	O	O	O
Papilio charopus	O	O	O	O	O	O	O	X	X	O	O	O	O	O	O	O	O	O	O	O	O	O	O	O	X	X	X
Papilio hornimani	O	O	O	O	O	X	X	X	X	X	O	O	O	O	O	X	O	O	O	O	O	O	O	O	O	O	O

Appendix 8.2 (cont.)

| | | | | | | | | | | | | | | | | | | | | | | | | | | | Area number | | |
Species/subspecies	1	2	3	4	5	6	7	8	9	10	11	12	13	14	15	16	17	18	19	20	21	22	23	24	25	26	27
Papilio aethiopis	X	O	O	O	O	O	O	O	O	O	O	O	O	O	O	O	O	O	O	O	O	O	O	O	O	O	O
Papilio chrapkowskii	O	X	X	X	X	X	O	O	X	O	O	O	O	O	O	X	O	O	O	O	O	O	O	O	O	O	O
Papilio desmondi usambarae	O	O	O	X	X	X	X	O	X	X	O	O	X	X	O	X	X	X	O	O	O	O	O	O	O	O	O
Papilio desmondi magdae	O	O	O	O	O	O	O	O	O	O	O	O	O	O	O	O	O	O	O	O	O	O	O	O	O	O	O
Papilio thuraui thuraui	O	O	O	O	O	O	O	O	O	O	O	O	O	O	O	O	O	O	O	O	O	O	O	O	O	O	O
Papilio thuraui cyclopis	O	O	O	O	O	O	O	O	O	O	O	O	O	O	O	O	O	X	O	O	O	O	O	O	O	O	O
Papilio thuraui occidua	O	O	O	O	O	O	O	O	O	O	O	O	O	O	X	O	X	O	X	X	X	O	O	O	O	X	X
Papilio thuraui ufipa	O	O	O	O	O	O	O	O	X	O	O	O	O	O	O	O	O	O	O	O	O	O	O	O	O	O	O
Papilio thuraui ngorongoro	O	O	O	O	O	O	O	O	X	O	O	O	O	O	O	O	O	O	O	O	O	O	O	O	O	O	O
Papilio rex rex	X	O	O	O	O	O	X	X	O	X	O	O	X	O	O	O	O	O	O	O	O	O	O	O	O	O	O
Papilio rex abyssinicus	X	O	O	O	O	O	O	O	O	O	O	O	O	O	O	O	O	O	O	O	O	O	O	O	O	O	O
Papilio rex commixtus	X	O	O	O	O	O	O	O	O	O	O	O	O	O	O	O	O	O	O	O	O	O	O	O	O	O	O
Papilio rex fransiscae	O	X	X	X	X	X	X	O	O	X	O	X	O	O	O	O	O	O	O	X	O	O	O	O	O	X	O
Papilio rex mimeticus	X	O	O	O	O	O	O	O	O	O	O	O	O	O	O	O	O	O	O	O	O	O	O	O	O	O	O
Papilio rex regulus	X	X	O	X	X	X	O	O	O	O	O	O	O	O	X	O	O	O	O	O	O	O	O	O	O	O	O
Papilio echerioides leucospilus	X	X	X	O	O	O	O	O	O	O	O	O	O	O	O	O	O	O	O	O	O	O	X	O	O	O	O
Papilio echerioides oscari	O	O	O	O	X	O	X	X	O	X	O	O	X	X	X	O	O	O	O	O	O	O	O	O	X	X	X
Papilio echerioides wertheri	O	O	O	O	O	O	O	O	O	O	O	O	O	O	O	O	O	O	O	O	O	O	O	O	O	O	O
Papilio jacksoni jacksoni	O	O	O	O	O	X	X	X	O	X	O	O	X	O	X	O	X	X	X	X	X	X	X	O	O	O	O
Papilio jacksoni hecqui	O	O	O	O	O	O	O	O	O	O	O	O	O	O	O	O	O	O	O	O	O	X	X	O	X	O	O
Papilio jacksoni ruandana	O	O	O	O	O	O	O	O	O	O	O	O	O	O	O	O	O	O	O	O	X	X	O	O	O	O	O
Papilio jacksoni kungwe	O	O	O	O	O	O	O	O	O	O	O	O	O	O	O	O	O	O	O	O	O	X	O	O	O	O	O
Papilio jacksoni nyika	O	O	O	O	O	O	O	O	O	O	O	O	O	O	O	O	O	O	O	O	O	O	O	O	O	O	O
Papilio fuelleborni	O	O	O	O	O	O	O	O	O	O	O	O	O	O	O	O	X	X	O	O	O	O	O	O	O	O	O
Papilio sjoestedti sjoestedti	O	O	O	O	O	X	X	O	O	O	O	O	O	O	O	O	O	O	O	O	X	X	X	O	X	X	O
Papilio sjoestedti atavus	X	O	O	O	O	O	O	O	O	O	O	O	O	O	O	O	O	O	O	O	X	X	O	O	O	X	X
Colotis elgonensis elgonensis	O	O	O	O	O	O	X	O	O	O	O	O	O	O	O	O	O	O	O	O	O	O	O	O	X	O	O
Colotis elgonensis kenia	O	X	O	O	X	O	X	O	X	X	O	O	O	O	X	O	O	O	O	O	O	O	O	O	O	O	O
Colotis elgonensis basilewskyi	O	O	O	O	O	O	O	O	X	O	O	O	O	O	O	O	O	O	O	O	O	O	O	O	O	O	O
Colotis elgonensis nobilis	X	O	O	O	O	O	X	X	O	X	O	O	X	O	X	X	X	O	O	O	X	X	X	O	X	X	X
Belenois raffrayi raffrayi	X	O	O	O	O	O	O	O	X	O	O	O	O	O	O	O	O	O	O	O	X	O	O	O	O	O	X
Belenois raffrayi extendens	O	O	O	O	O	X	X	X	O	X	O	O	O	O	X	O	O	X	O	O	X	X	O	O	X	X	O
Belenois raffrayi similis	O	X	O	O	O	O	O	O	O	O	O	O	O	O	O	O	O	O	O	O	O	O	O	O	O	O	O
Belenois margariacea margariacea	O	O	O	O	O	O	O	O	O	O	O	O	O	O	O	O	O	O	O	O	X	O	O	O	O	O	O
Belenois margariacea plutonica	O	O	O	O	O	O	O	O	O	O	O	O	O	O	O	O	O	O	O	O	O	O	O	O	O	O	O
Mylothris crawshayi crawshayi	O	O	O	O	O	O	X	O	O	O	O	O	O	O	X	O	O	O	O	X	X	X	X	O	O	O	O
Mylothris crawshayi leonora	O	O	O	O	O	O	O	X	X	O	O	O	X	O	O	O	O	O	O	O	O	O	O	O	O	O	O
Mylothris sagala sagala	O	O	O	O	O	O	O	O	O	O	O	O	O	O	X	O	O	O	O	O	O	O	O	O	O	O	O
Mylothris sagala seminigra	O	O	O	O	X	X	O	O	O	O	O	O	X	O	O	O	O	O	O	O	O	O	O	O	O	O	O
Mylothris sagala jacksoni	O	X	X	X	X	X	O	O	O	O	O	O	O	O	O	O	O	O	O	O	O	O	O	O	O	O	X

Mylothris sagala dentata
Mylothris sagala swaynei
Mylothris ruandana
Mylothris crocea
Mylothris marginea
Mylothris erlangeri
Mylothris schoutedeni
Mylothris mortoni mortoni
Mylothris mortoni baltis
Mylothris superbus
Mylothris species 1
Mylothris species 2
Alaena bicolora
Alaena madibirensis
Alaena ochracea
Alaena lamborni
Ornipholidotos species
Mimacraea gelinia
Baliochila species
Aslauga orientalis
Spindasis collinsi
Epamera dubiosa
Epamera arborifera
Epamera pseudopollux
Epamera flavilinea
Epamera bansana yalae
Epamera congdoni
Epamera species
Etesiolaus species
Iolaphilus henryi
Iolaphilus montana
Hypolycaena jacksoni
Pilodeudorix zelomima
Pilodeudorix rodgersi
Pilodeudorix species
Virachola edwardsi
Virachola vansomereni
Virachola montana
Anthene rupenzorica
Anthene hobleyi hobleyi
Anthene hobleyi elgonensis
Anthene hobleyi kigezi
Anthene amanica
Anthene species 1
Anthene species 2
Anthene species 3
Anthene species 4

Appendix 8.2 (cont.)

Species/subspecies	\	\	\	\	\	\	\	\	\	\	\	\	\	Area number	\	\	\	\	\	\	\	\	\	\	\	\	\
	1	2	3	4	5	6	7	8	9	10	11	12	13	14	15	16	17	18	19	20	21	22	23	24	25	26	27
Anthene species 5	O	O	O	O	O	O	O	O	O	O	O	O	O	O	O	X	O	O	O	O	O	O	O	O	O	O	O
Anthene species 6	O	O	O	O	O	O	O	O	O	O	O	O	O	O	O	X	O	O	O	O	O	O	O	O	O	O	O
Anthene species 7	O	O	X	O	O	O	O	O	O	O	O	O	O	O	O	X	O	O	X	X	O	O	O	O	O	X	X
Uranothauma crawshayi	O	O	O	X	X	X	O	O	O	O	O	O	O	O	O	O	X	X	X	X	O	O	O	O	O	O	O
Uranothauma cordata	O	O	O	O	O	O	O	O	O	O	O	O	O	O	O	X	X	O	O	O	O	O	O	O	O	O	O
Uranothauma cuneatum	O	O	O	O	X	O	O	O	O	X	O	X	X	O	X	X	X	O	O	O	X	O	O	X	O	X	X
Uranothauma williamsi	O	O	O	O	O	O	O	O	O	O	O	O	O	O	X	X	O	O	O	O	O	O	X	X	O	O	O
Uranothauma lunifer	O	O	O	O	O	X	O	O	O	O	O	O	O	O	O	X	O	O	O	O	O	O	O	O	O	O	O
Uranothauma delatorum	O	O	O	O	X	X	O	O	O	X	O	X	X	O	X	X	X	O	O	O	X	O	X	O	X	X	X
Uranothauma usambarae	O	O	O	O	O	X	O	O	O	O	O	O	O	O	O	X	O	O	O	O	O	O	O	O	O	O	O
Uranothauma nguru	O	O	O	O	O	O	O	O	O	O	O	O	O	O	O	O	O	O	O	O	O	O	O	O	O	O	O
Uranothauma uganda	O	O	O	O	O	O	O	O	O	O	O	O	O	O	O	X	O	O	O	X	O	O	O	O	O	O	O
Uranothauma lukwangule	O	O	O	O	O	O	O	O	O	X	X	O	O	O	O	X	X	O	O	X	X	O	X	O	X	X	X
Uranothauma pseudocrawshayi	O	O	O	O	O	O	X	O	X	O	X	O	X	O	X	X	O	O	O	O	X	O	O	O	O	O	O
Uranothauma kilimensis	O	O	O	O	O	O	X	O	O	O	O	O	O	O	O	X	X	O	O	O	O	O	O	O	O	O	O
Castalius erlti	O	O	O	O	O	O	O	O	O	O	O	O	O	O	O	O	O	X	O	O	O	O	O	O	O	O	O
Castalius margaritaceus	O	O	O	O	O	O	O	O	X	O	O	O	O	O	O	X	X	O	X	X	O	O	X	O	X	X	X
Castalius species	O	O	O	O	O	O	O	O	O	O	X	O	O	O	O	X	O	O	O	X	O	O	O	O	O	O	O
Abisara delicata	O	O	O	O	O	O	O	O	O	O	O	O	O	O	O	X	O	O	O	O	O	O	O	O	O	O	X
Charaxes ansorgei ansorgei	O	O	O	O	O	O	O	O	O	O	O	O	O	O	O	O	O	O	O	O	O	O	O	O	O	O	O
Charaxes ansorgei jacksoni	O	X	X	O	X	O	O	O	O	O	O	O	O	O	O	O	X	X	X	O	O	O	O	O	O	O	O
Charaxes ansorgei ruandana	O	O	O	X	O	O	O	O	O	X	O	O	O	O	O	O	O	O	O	O	X	O	O	O	X	X	X
Charaxes ansorgei ufipa	O	O	O	O	O	O	X	O	O	O	O	O	O	O	O	O	O	O	O	X	O	O	X	O	O	O	O
Charaxes ansorgei levicki	O	O	O	O	O	O	O	O	O	O	O	O	O	O	O	O	X	O	O	O	O	O	O	O	O	O	O
Charaxes ansorgei kilimanjarica	O	O	O	O	O	O	O	O	O	O	O	O	O	O	O	O	O	O	O	O	O	O	O	O	O	O	O
Charaxes ansorgei rydoni	O	O	O	O	O	O	O	O	O	O	O	O	O	O	O	O	O	O	X	O	O	O	O	O	O	O	O
Charaxes lucyae	O	O	O	O	O	O	O	O	O	O	O	O	O	O	O	X	O	O	O	O	O	O	O	O	X	X	O
Charaxes opinatus	O	O	X	O	O	O	O	O	O	X	O	O	O	O	O	O	X	X	X	O	O	O	X	O	O	X	O
Charaxes mafuga	O	O	O	O	X	O	O	O	O	X	X	O	O	O	O	O	O	O	O	O	O	O	O	O	O	X	X
Charaxes usambarae	O	O	O	O	O	X	O	O	O	O	O	O	O	O	O	O	O	O	O	O	O	O	O	O	O	O	O
Charaxes berkeleyi berkeleyi	O	O	O	O	O	O	O	O	O	O	O	O	O	O	O	O	O	O	O	O	O	O	O	O	O	O	O
Charaxes berkeleyi masaba	O	O	O	O	O	O	O	O	O	X	O	O	X	O	O	O	X	X	X	O	O	O	O	O	O	O	O
Charaxes aubyni aubyni	O	X	X	X	X	X	X	O	O	X	O	O	O	O	O	O	X	O	O	O	O	O	O	O	X	X	O
Charaxes aubyni ecketti	O	O	O	O	X	X	X	O	O	X	O	O	O	O	O	O	X	O	X	X	O	O	X	O	X	X	X
Charaxes aubyni australis	O	O	O	O	O	O	O	O	O	O	O	O	O	O	O	O	O	O	O	X	O	O	O	O	O	O	O
Charaxes aubyni subspecies	O	O	X	O	O	X	O	O	O	O	O	O	O	O	O	O	O	O	O	O	O	O	X	O	O	O	O
Charaxes baileyi	O	O	O	O	O	O	O	O	O	O	O	O	O	O	O	O	O	O	O	O	O	O	O	O	O	O	O
Charaxes mccleeryi	O	O	O	O	O	O	O	O	O	O	O	O	X	O	O	O	O	O	O	O	O	O	O	O	O	O	O
Charaxes xiphares maudei	O	O	O	O	O	O	O	O	X	X	O	O	O	O	O	O	O	O	O	O	O	O	O	O	O	O	O
Charaxes xiphares kiellandi	O	O	O	O	O	O	O	O	O	X	O	O	O	O	O	O	O	O	O	O	O	O	O	O	O	O	O
Charaxes xiphares nguru	O	O	O	O	O	O	O	O	O	O	O	X	O	O	O	O	O	O	O	O	O	O	O	O	O	O	O

Charaxes xiphares sichi
Charaxes xiphares kilimensis
Charaxes xiphares burgessi
Charaxes xiphares brevicaudatus
Charaxes xiphares ludovici
Charaxes xiphares woodi
Charaxes nandina
Charaxes acuminatus acuminatus
Charaxes acuminatus mlanji
Charaxes acuminatus nyika
Charaxes acuminatus usambarensis
Charaxes acuminatus teitensis
Charaxes acuminatus oreas
Charaxes acuminatus stonehami
Charaxes acuminatus kigezia
Charaxes species
Cymothoe aurivilli
Cymothoe zombana
Cymothoe melanjae
Cymothoe amaniensis
Cymothoe magambae
Cymothoe collinsi
Cymothoe cottrelli
Kumothales inexpectata
Pseudathyma uluguru
Pseudacraea deludens deludens
Pseudacraea deludens amaurina
Pseudacraea deludens echerioides
Pseudacraea deludens terrena
Neptis lugubris
Neptis smynnertoni
Neptis incongrua
Neptis aurivillii
Neptis occidentalis occidentalis
Neptis kikuyensis
Neptis woodwardi
Salya pseudotrimeni
Hypolimnas anteorta
Precis milonia witgensi
Antanartia schaeneia dubia
Antanartia schaeneia diluta
Antanartia dimorphica dimorphica
Antanartia dimorphica aethiopica
Bematistes scalivittata
Bematistes quadricolor quadricolor
Bematistes quadricolor leptis
Bematistes quadricolor latefascia
Bematistes quadricolor morogoro

Species/subspecies	1	2	3	4	5	6	7	8	9	10	11	12	13	14	15	16	17	18	19	20	21	22	23	24	25	26	27
Bematistes itumbana	O	O	O	O	O	O	O	O	O	O	O	X	O	O	O	O	O	O	O	O	O	O	X	O	O	O	O
Bematistes obliqua kiruensis	O	O	O	O	O	O	O	O	O	O	O	O	O	O	O	O	O	O	O	O	O	O	X	O	O	X	O
Bematistes obliqua elgonensis	O	O	O	O	O	X	O	O	O	O	O	O	O	O	O	O	O	O	O	O	O	O	O	O	O	X	O
Acraea hamata	O	O	O	O	O	O	O	O	O	O	O	O	O	O	O	O	O	O	O	O	O	O	O	O	O	X	X
Acraea amicitiae amicitiae	O	O	O	O	O	X	O	O	O	O	O	O	O	O	O	O	O	O	O	O	O	O	O	O	X	X	O
Acraea amicitiae polychroma	O	O	O	O	O	O	O	O	O	O	O	O	O	O	O	O	O	O	O	X	O	O	O	O	O	X	X
Acraea kraka	O	O	O	O	O	O	O	O	O	O	O	O	O	O	O	O	O	O	O	O	O	O	O	O	O	X	X
Acraea kalinzu	X	O	O	O	O	X	O	O	O	O	O	O	O	O	O	O	O	O	O	O	O	O	O	O	O	X	X
Acraea safie	O	O	O	O	O	O	O	O	O	O	O	O	O	O	O	O	O	O	O	O	O	O	O	O	O	O	O
Acraea melanoxantha	O	O	O	O	X	X	O	O	O	O	O	O	O	O	O	O	O	O	O	O	O	O	O	O	O	O	O
Acraea vuilloti	O	O	O	O	X	O	O	O	O	O	O	X	O	O	O	O	O	O	O	O	O	O	O	O	O	X	X
Acraea burgessi	O	O	X	X	X	X	O	O	O	O	O	O	O	O	X	O	O	O	O	X	O	O	O	O	O	O	O
Acraea ansorgei ansorgei	X	X	O	O	X	O	O	O	X	O	O	O	O	O	O	O	O	O	O	O	O	O	O	O	O	O	X
Acraea ansorgei acuta	O	O	O	O	O	X	O	O	O	O	O	O	O	O	O	O	O	O	X	X	O	O	O	O	O	X	O
Acraea disjuncta disjuncta	O	O	O	O	X	X	O	O	O	O	O	O	O	O	O	O	O	O	O	O	O	O	O	O	O	X	X
Acraea disjuncta kigeziensis	O	O	X	X	O	X	O	O	O	O	O	O	O	O	O	O	O	O	O	O	O	O	O	O	X	X	X
Acraea asboloplintha asboloplintha	O	O	O	O	O	X	X	X	O	X	X	O	O	O	O	X	O	O	O	O	O	O	O	O	O	X	O
Acraea asboloplintha rubescens	O	O	O	O	O	O	O	O	O	O	O	O	O	O	O	O	O	O	O	O	O	O	O	O	O	X	X
Acraea species	O	O	O	O	O	X	O	X	O	X	X	X	O	O	O	X	X	O	O	X	O	O	O	O	X	O	O
Gnophodes grogani	O	O	X	X	X	X	O	O	O	X	X	X	O	O	X	X	X	O	O	X	O	O	O	O	O	X	X
Aphysoneura pigmentaria pigmentaria	O	O	O	O	O	O	O	O	O	X	X	X	O	X	X	X	O	O	O	X	O	O	O	O	O	X	O
Aphysoneura pigmentaria pringlei	O	O	X	X	X	X	O	O	O	O	O	O	O	O	O	O	O	O	O	O	O	O	O	O	O	X	O
Aphysoneura pigmentaria latilimba	O	O	X	O	X	X	O	O	O	O	O	O	O	O	O	O	O	O	O	X	O	O	O	O	X	X	X
Aphysoneura pigmentaria scapulifasc	O	O	O	O	X	O	X	X	O	O	O	O	O	O	O	O	X	X	O	X	O	O	O	O	O	O	O
Bicyclus kenia	O	O	O	O	O	O	O	O	O	X	X	O	O	O	O	O	X	O	O	X	O	O	O	O	O	X	X
Bicyclus jefferyi	O	O	O	O	O	O	O	O	O	X	X	X	O	X	X	X	X	X	O	X	O	O	O	O	X	X	O
Bicyclus danckelmani	O	O	O	O	O	X	O	X	O	X	X	X	O	X	X	X	X	O	O	O	O	O	O	O	X	O	X
Bicyclus aurivillii aurivillii	O	O	O	O	O	O	O	O	O	O	O	O	O	O	O	X	O	O	O	O	O	O	X	X	O	X	X
Bicyclus aurivillii kiruensis	O	O	O	O	O	O	O	O	O	O	O	O	O	O	O	O	O	O	O	O	O	O	X	O	X	O	O
Bicyclus albocinctus	O	O	O	O	O	X	O	O	O	O	O	O	O	O	O	O	O	X	O	O	O	O	O	O	O	O	O
Bicyclus neustetteri	O	O	O	O	O	O	O	O	O	O	O	O	O	X	X	X	X	O	O	O	O	O	O	X	X	X	X
Bicyclus matuta	O	O	O	O	O	O	O	O	O	O	O	O	O	X	X	X	X	O	O	O	O	O	O	X	O	X	O
Bicyclus persimilis	O	O	O	O	O	O	O	O	O	O	O	O	O	X	X	X	O	O	O	O	O	O	X	O	X	O	X
Bicyclus simulis	O	O	O	O	O	O	O	O	O	O	O	O	O	X	X	X	X	O	O	O	O	O	O	O	X	X	X
Bicyclus dentatus	O	X	O	X	X	X	O	O	X	O	O	O	O	X	X	X	O	O	O	O	O	O	X	O	X	X	O
Bicyclus species 1	O	O	O	O	O	O	O	O	O	O	O	O	O	O	O	X	O	O	O	O	O	O	X	X	O	X	O
Bicyclus species 2	O	O	O	O	O	O	O	O	O	O	O	O	O	O	O	O	O	O	O	O	O	O	O	X	O	X	X
Ypthima albida albida	O	O	O	O	O	X	O	O	X	O	O	O	O	O	X	X	O	O	O	O	O	O	O	X	X	X	X
Ypthima albida argentata	O	O	O	O	X	X	O	X	X	O	O	O	O	O	X	O	O	X	O	O	O	O	O	X	X	X	X
Amauris ellioti ellioti	O	O	O	O	O	O	O	X	X	O	O	O	O	O	O	O	O	O	O	O	O	O	O	X	X	X	O
Amauris ellioti ansorgei	O	X	X	O	X	X	O	X	X	O	O	X	X	O	O	O	O	X	O	O	O	O	X	X	X	O	O
Amauris ellioti altumi	O	X	X	O	X	O	O	X	O	O	O	O	X	O	O	O	O	O	O	X	O	O	O	O	O	O	O
Amauris ellioti junia	O	O	O	O	O	O	O	O	O	O	O	X	X	O	X	X	X	X	O	O	O	O	O	O	O	O	O

9 Herpetofauna of the eastern African forests

K. M. HOWELL

Introduction

Most visitors as well as residents of eastern Africa are unappreciative of the variety of amphibians and reptiles found there. With the exception of some people who value snakes for mystical and curative properties, the reptiles are generally great feared and killed whenever encountered, and even amphibians such as frogs and toads are regarded with some suspicion.

Although the layman may be familiar with some of the larger, more conspicuous amphibians and reptiles in the open habitats such as woodlands, and those species which can survive in and around man's dwellings, very few people spend enough time inside natural forest to identify the various amphibians and reptiles found there. These are often cryptically coloured, and some may be found only high in the forest canopy, or hiding inside rotting logs and leaf litter of the forest floor. Thus, all but a very few of those people who live near the forest and work in it are unfamiliar with some of its most interesting inhabitants. Among the amphibians, there are brightly coloured treefrogs which are able to change their colours: a frog, *Leptopelis uluguruensis* with a colour pattern which resembles a small patch of fungus; and a toad, *Bufo brauni* which closely resembles a dead leaf on the forest floor. Although in the breeding season mixed species groups of anurans may congregate in the hundreds, the behaviour of amphibians is seldom observed because they are most active at night. Unlike most related non-forest forms, some of the species have become less dependent on perma-

nent water for reproduction and the caecilians and some toads have internal fertilisation.

Many of the reptiles are also cryptically coloured. Chameleons are able to change their colours and patterns, and one species, *Chamaeleo laterispinis*, closely resembles a lichen in pattern and texture. They also have a bizarre variety of head ornamentation: some have a simple protuberance on the rostrum, but others have as many as three horns extending from the front of the head. While the chameleons are relatively slow, other lizards are much more active. *Holaspis guentheri* has an extremely flattened body which acts as an aerofoil to assist it in making controlled glides from tree to tree. The colour and behaviour patterns of snakes make them very difficult to see unless they are moving, but they too exhibit a variety of forms, from the slender *Thelotornis capensis* to the massive *Bitis gabonica*.

Of all the vertebrates, the amphibians and reptiles of the forests of eastern Africa are the poorest known and receive the least attention from layman and biologist alike. Had it not been for the devoted efforts of Arthur Loveridge (1932a,b, 1933, 1935, 1937, 1942a,b, 1947, 1952, 1953a,b, 1956, 1957), who almost single-handedly tackled the taxonomy and biology of the eastern Africa herpetofauna and provided regional syntheses of his findings, it is unlikely that workers today would have a firm foundation on which to base their studies.

Loveridge was also among the first to realise the uniqueness of the amphibians and reptiles of the eastern African forests, especially of the Usambaras and Ulugurus. In terms of species diversity and endemism the herpetofauna of the other Eastern Arc forests is extremely rich, yet it is only within the last five years that we have gained a further appreciation of this from studies in the Uzungwa and other mountain forests, and are now in a position to reassess Loveridge's earlier work.

I wish, however, to emphasise that there are vast areas of forest in Tanzania unvisited by biologists, and range extensions and species new to science have recently been found even in those areas collected by Loveridge, so it is clear that we have a long way to go before our understanding extends to all of the forest amphibians and reptiles of eastern Africa.

The study of the evolutionary relationships in amphibians and reptiles has changed greatly from the days of simply naming new taxa, and whenever possible I have tried to incorporate studies of ecology and physiology when these are of comparative interest or significance to eastern Africa. For practical reasons many of the available modern techniques of faunal analysis have not yet been used in eastern African forests, and the application of improvements in both laboratory and field techniques will still depend on interested individuals willing to live and work inside the forest. It is my hope that the information presented here will encourage both field workers and those with a more experimental approach to study the fascinating herpetofauna of the eastern forests of Africa and that the renewed interest in these forests will stimulate the conservation of their unique flora and fauna.

Previous studies

Early explorers, travellers and biologists collected amphibians and reptiles from widely scattered localities in eastern Africa. While many early workers specialised in the description of new taxa, Barbour & Loveridge (1928) first monographed the results of intensive collecting and called attention to the high levels of endemism in the Usambara and Uluguru mountains in Tanzania. Loveridge (1932a, 1933, 1935, 1937, 1942a, b) later extended this work, providing basic taxonomic and zoogeographic studies for both

Tanzania and elsewhere in eastern Africa. Uthmoller considered reptiles of mountain forests, but his studies focused on the northern volcanic mountains of Tanzania (Uthmoller, 1937, 1941a, b, 1942). Loveridge (1957) produced a regional checklist in which he summarised taxonomic and distributional findings.

Despite the establishment of a strong taxonomic foundation and the popularisation of his expeditions (Loveridge, 1932b, 1947, 1953a, 1956), few workers followed Loveridge's examples of fieldwork and taxonomic studies. Some of the species endemic to eastern Africa are still known only from a single type specimen or from the type series. With the exception of the work by Schiøtz (1975) on treefrogs, no comprehensive study has been published on any group of amphibians of the eastern forests since that of Loveridge. The situation regarding reptiles is similar. Spawls (1978) provided a distributional list for snakes in Kenya which includes some forest localities and Rasmussen (1981) and Wederkinch (1982) records of snakes and lizards from the Usambara mountains. In a series of taxonomic revisions dealing mainly with southern African forms, Broadley (1966, 1968, 1971, 1980) also included relevant eastern African species. Brygoo & Roux-Estève (1981) revised the forest-dwelling fossorial skinks of the genus *Melanoseps*. Except for these studies, very little recent information exists in summary form on the forest reptiles of eastern Africa. Thus, at a time when herpetological communities in forested areas in West Africa, Central America, Asia and elsewhere are undergoing intensive, detailed ecological investigation (see Toft, 1985 for review), those of the forests of eastern Africa are still largely undocumented and unstudied.

Anyone attempting to examine the distributions of amphibians and reptiles in eastern Africa immediately faces problems regarding both the collection and the identification of individual specimens and the localities where the material of others was collected. Even as expert a collector as the resourceful Arthur Loveridge found it was not an easy matter to collect long series of cryptic, forest-dwelling amphibians and reptiles and the vagaries of the weather seem often to work against the biologist with only a limited amount of time available. The identification of specimens is often fraught with difficulties; as noted above, little revisionary work has been done on some of the groups since the time of Loveridge, and even Schiøtz (1975) was unable to devise a workable key to some forms of treefrogs. Three new species of frogs from the eastern forests have been named since 1981 (Schiøtz, 1982; Grandison, 1983; Grandison & Howell, 1983) and a new toad has been described (Clarke, 1988a). Several new species of reptiles have also been described recently, including two geckos (Perret, 1986) and a chameleon (Klaver & Böhme, 1988) from the Ulugurus and Uzungwas, a new skink from the Uzungwas (Broadley, 1989) and a new gecko from forest on Zanzibar (Pasteur & Broadley, 1988). References in the literature are scattered and there is no field guide to either the amphibians or the reptiles of eastern Africa. This contrasts sharply with the situation in southern Africa, where recent formal studies (Poynton & Broadley, 1985a, b, 1987, 1988) as well as field guides to the amphibians (Stewart, 1967; Passmore & Carruthers, 1979), snakes (Broadley & Cock, 1975) and reptiles in general (Branch, 1988) are available.

Early collectors often simply labelled material 'Africa' and terms used then carried a meaning different from that of today. 'Zanzibar' was sometimes used to mean the mainland coast opposite that island and 'Tanga' could mean any locality within 80 km of the town of that name, including the Usambara mountains. Names of villages change over time, and village sites are abandoned. Several localities separated by hundreds of kilometres may have identical names.

With the limitations above kept in mind, data from older records as well as from recent collecting are incorporated with taxonomic changes made since the time of Loveridge's checklist (Broadley & Howell, 1991; Frost, 1984; Schiøtz, 1975) in an examination of the distribution patterns of the herpetofauna of the eastern forests of Africa.

Methods

As already noted, the collection and identification of amphibians and reptiles of the eastern African forests pose numerous problems. In this chapter I have drawn together published records from a variety of sources, relying heavily, of course, on the classic studies by Loveridge (summarised in Loveridge, 1957). The work of Schiøtz (1975) on treefrogs and Nussbaum (1985) on caecilians has also been included. The sequence of amphibian species names is that in Frost (1984) and of reptiles that of Broadley & Howell (1991).

My own collecting was done as other commitments allowed. The following areas were collected in varying intensities: coastal forests, Pugu Forest Reserve, Kiono Forest Reserve, Jozani Forest Reserve; forests in the Eastern Arc mountains, North Pare, East and West Usambaras, Uluguru and Uzungwa; forests on mountains of volcanic origin, Mount Meru, Ngorongoro Forest Reserve. Numerous friends and colleagues also assisted in collecting at many of the above localities and those it was impossible for me to visit, including the Rondo Plateau and the Nguru, Ukaguru and Rubeho mountains.

Any amphibians and reptiles encountered during walks in the forest were collected and/or noted, and likely hiding places were also investigated. Most of the collecting for anurans was done at night by listening for vocalisations and visiting breeding sites.

Standard preservation techniques were used. Identifications of specimens were confirmed by various authorities as follows: anurans, Ms A. G. C. Grandison (retired) and B. Clarke, British Museum (Natural History), London; caecilians, R. Nussbaum, Museum of Zoology, University of Michigan, Ann Arbor; chameleons, A. F. Stimson, British Museum (Natural History), London; other reptiles, D. G. Broadley, Natural History Museum of Zimbabwe, Bulawayo. Duplicate specimens have been deposited in these institutions and in the collection of the Department of Zoology and Marine Biology, University of Dar es Salaam. Since 1986, material has also been deposited at the California Academy of Sciences, San Francisco.

The forests included in this study: past and present

Any understanding of the forest-dependent fauna present today depends on a knowledge of the geological, climatological and vegetational changes which have taken place over millions of years on the African continent. Although details of such changes, especially during and after the Pleistocene, and the human effects on vegetation remain matters of controversy (see Chapters 4 and 5) the general effects of such changes may be broadly described as follows. It is clear that perhaps during several periods the continent was much more extensively forested than at present. Forest expanded and contracted over time, depending on both geological and climatological vicissitudes; at present, its cover is only a small portion of what it has been at other periods in the past (see Chapter 3). Such expansions and contractions of forest had profound effects on both forest and non-forest organisms. Forests, especially moist tropical forests, are complex ecosystems; it appears that once organisms have specialised and evolved to live within such a complex system, very few are able to exist far outside its bounds. As Kingdon (1971) has pointed out '... forest animals ... are generally fairly rigidly tied to their habitat.' Thus when the distribution of moist forest finally contracted, only some species were able to respond by adapting to a non-forest habitat, such as the drier, more open woodland which characterises much of eastern and southern Africa today. Depending to some extent on the time over which the moister forest was reduced, many forest-dependent species would probably have become extinct as the habitat on which they depended gradually receded. Those which survived probably did so because they were able to adapt to life at the forest edge, or in thicket or other types of degraded forest, or because the area of forest on which they depended remained relatively intact. Such areas have been termed forest 'refugia' (Hamilton, 1976; Diamond & Hamilton, 1980). Once the forests which linked West and East Africa were separated for the final time, perhaps about 20 million years BP (J. Lovett, unpublished data), the

main areas of forest in eastern Africa were fragmented even more. Today, the most extensive and most important of these forests are to be found along the Eastern Arc mountains of Tanzania (see Lovett (Chapter 4, this volume) for a detailed description of the Eastern Arc mountains and forests). Less humid forests have also persisted elsewhere, such as coastal forest.

In a study of the birds of Africa, Moreau (1966) referred to a 'Tanganyika–Nyasa' forest fauna, running from the forests of the Taita Hills in Kenya south to parts of Malawi and northern Mozambique, and I have used his concept to help define the limits of the eastern forests in Africa. Those forests found on the crystalline block mountains of eastern Africa include the Taita Hills, North and South Pare, East and West Usambara, Nguru, Uluguru, Malundwe, Rubeho, Uzungwa and the Mahenge mountains. Barbour & Loveridge (1928) collected extensively in the Usambara and Uluguru forests but other forested areas on crystalline block mountains have received little attention. Biologists have only recently begun to investigate the large forested areas of the Uzungwa scarp.

Earlier collecting by Loveridge (1933) in the Uzungwas was mainly in the plateau forests rather than those of the scarp. He also collected in a large area of southwest Tanzania known as the Southern Highlands. Loveridge used general terms indicating tribal areas to describe the various sections in which he collected, such as Ubena and Ukinga, and botanists have also used these. While no longer in general use, their meaning is widely understood both within and outside Tanzania, and they are retained here for consistency. The geography, geology and terminology of the Southern Highlands area are discussed in detail by Cribb & Leedal (1982).

Forests to the south in Malawi mentioned include the Misuku Hills, the Nyika Plateau, Nchisi Mountain, Mount Mulanje and Mount Thyolo. The vegetational details of these are discussed by Chapman & White (1970).

Evidence from bird distributions (see Moreau, 1966) in northern Mozambique as well as the presence of a species of forest microhylid frog in

Zimbabwe indicates that there may be as yet undocumented connections between the Eastern Arc and more southern mountain forests.

In contrast to the block-faulted mountains and their very ancient forests are the higher mountains of volcanic origin such as Kilimanjaro and Meru in the north and Rungwe in the south (Sampson, 1965). After erupting and killing off whatever vegetation and animals were in the vicinity, these cooled and then were eventually colonised by plants and animals from surrounding areas. Nyamweru (1980) has described these volcanoes and recent eruptions. Coe (1967) has discussed the flora and fauna of the alpine zone of Mount Kenya, and Hedberg (1951) the vegetational belts of the East African mountains. Various aspects of the ecology of Mount Kilimanjaro are included in a special publication (Tanganyika Society, 1965), but there has been no recent herpetological work done on the forests of the mountains of volcanic origin.

Of the once great expanse of coastal forest of eastern Africa, only small fragments remain today. In Kenya, the forests of the Tana River were collected by Loveridge, while Schiøtz (1975) listed treefrogs from some coastal forests. No other work appears to have published on the herpetofauna of the Kenya coastal forests, including those of the Shimba and Mrima hills, Gedi and Arabuko–Sokoke. In Tanzania, the amphibians and reptiles of only some of the forests on Pemba, Zanzibar and Mafia islands (Moreau & Pakenham, 1941; Pakenham, 1983) and at Pugu (Howell, 1981) have been documented. The natural vegetation of several relic patches of coastal forests near Tanga and much of Vikundu near Dar es Salaam has been destroyed since the 1960s, and illegal felling of trees in Pugu Forest Reserve as well as removal of younger trees for building poles has seriously reduced this already small reserve.

Although no recent intensive collecting has been carried out in the Rondo Plateau area, the forests there are of great interest. These were, before cutting in the 1940s and 1950s, the largest *Chlorophora* forests in East Africa (Polhill, 1968) and were collected by Loveridge (1942a, b). The Rondo Plateau is the southernmost end of a chain of coastal forest fragments in Tanzania from the Msumbugwa Forest near Pangani and including Zaraninge Plateau (Kiono), Pugu, and Pande forests.

No information is available to me on the coastal forest herpetofauna of Mozambique, but the finding by Broadley (1990) of new reptile taxa in patches of dune forest and dense evergreen thicket on the islands off the coast of southern Mozambique indicates the potential importance of these habitats.

The forest-dependent amphibians and reptiles of the eastern Africa forests

Table 9.1 lists the amphibian and reptile species regarded as dependent on forest and occurring in the eastern forests of Africa. Taxa which occur in forest but which are also widely found in other habitats (i.e. are not forest-dependent) have been excluded as not useful in an examination of the distributional patterns of strictly forest-dependent animals (see Kingdon, 1971). Species are considered to be forest-dependent if they live in the interior of the forest, or if they live on the forest edge but appear to be dependent on the preservation of the forest habitat for their future survival (see discussion in Stuart, 1983). The difficulties of determining which species of animals are truly forest-dependent have been indicated by Loveridge (1933, 1937) for amphibians and reptiles, Schiøtz (1967, 1975) for treefrogs, Hughes (1983) for snakes and by Moreau (1966) and Stuart (1983) for birds. Kingdon (1971) and Rodgers, Owen & Homewood (1982) discuss mammals of the eastern African forests. Whenever possible, I have relied on personal experience in assessing the forest dependence of taxa as defined by Stuart (1983) above, and despite the limitations inherent in such a listing, feel it provides a reasonable representation of forest-dependent forms in eastern Africa, which for convenience are often referred to simply as 'forest' species.

While there is a general assumption among biologists of the existence of forest-dependent and

Table 9.1. *Forest-dependent amphibians and reptiles of the eastern forests of Africa*

Species	WU	EU	NGU	UKA	ULU	UZU	UKI	UBE	POR	RUN	SH	Others
Amphibia												
ARTHROLEPTIDAE												
+ *Arthroleptis adolfifriderici* (Neiden, 1910)	×	×			×							Zaïre, Rwanda, Uganda, Kenya.
+ *Arthroleptis francei* (Loveridge, 1953)		×										Mt Mulanje, Malawi.
+*Arthroleptis affinis* (Ahl, 1939)	×	×										
+*Arthroleptis reichei* (Neiden, 1910)					×	×			×	×		Misuku Hills, Malawi.
+*Arthroleptis tanneri* (Grandison, 1983)	×											
+*Schoutedenella xenodactyla* (Boulenger, 1909)	×	×			×							
BUFONIDAE												
+*Bufo brauni* (Neiden, 1910)	×	×			×	×						
+ *Bufo taitanus* (Peters, 1878)												Taita Hills, Kenya.
+*Bufo uzunguensis* (Loveridge, 1932)					×							
+ *Mertensophryne micranotis* (Loveridge, 1925)					×							Kilosa, Rondo Plateau, Pugu Forest, Zanzibar; coastal forest, Kenya.
+*Nectophrynoides cryptus* (Perret, 1971)					×							
+*Nectophrynoides minutus* (Perret, 1972)					×							
+*Nectophrynoides tornieri* (Roux, 1906)	×	×			×	×	×					
+*Nectophrynoides viviparus* (Tornier, 1905)					×	×	×		×	×		
+*Nectophrynoides wendyae* (Clarke, 1989, 1988)								×				Rondo Plateau, Mahenge, Lunkandi.
Stephopaedes sp. a										×		Mafia Island.
Stephopaedes sp. b												
HYPEROLIIDAE												
Hyperolius mitchelli (Loveridge, 1953)	×	×		×		×				×		Morogoro, Mahenge, Zanzibar, etc., Malawi, Mozambique, etc.
Hyperolius puncticulatus (Bocage, 1895)	×	×		×		×				×		Malawi, Zimbabwe, South Africa.
Hyperolius rubrovermiculatus (Schiøtz, 1975)												Kwale and Shimba Hills, Kenya.

Hyperoliidae (continued)

Species	Locality
+ *Hyperolius spinigularis* (Stevens, 1971)	Mt Mulanje, Malawi.
+*Hyperolius tanneri* (Schiøtz, 1982)	Kwale, Kenya.
Afrixalus sylvaticus (Schiøtz, 1974)	
+*Afrixalus uluguruensis* (Barbour & Loveridge, 1928)	
+*Phylictimantis keithae* (Schiøtz, 1975)	
+*Leptopelis barbouri* (Ahl, 1929)	
+ *Leptopelis flavomaculatus* (Günther, 1864)	Zanzibar, Kwale, Arabuko–Sokoke, Kenya; Malawi, Zimbabwe, Mozambique.
+*Leptopelis parkeri* (Barbour & Loveridge, 1928)	
+*Leptopelis uluguruensis* (Barbour & Loveridge, 1928)	
+*Leptopelis vermiculatus* (Boulenger, 1909)	

MICROHYLIDAE

Species	Locality
+ *Callulina kreffti* (Nieden, 1910)	Taita Hills, Kenya.
+*Probreviceps macrodactylus* (Nieden, 1926)	
+*Probreviceps uluguruensis* (Loveridge, 1925)	
+*Spelaeophryne methneri* (Ahl, 1924)	near Kilwa; Matengo Hills.
+*Hoplophryne rogersi* (Barbour & Loveridge, 1928)	
+*Hoplophryne uluguruensis* (Barbour & Loveridge, 1928)	
+*Parahoplophryne usambariensis* (Barbour & Loveridge, 1928)	

RANIDAE

Species	Locality
+*Arthroleptides martiensseni* (Nieden, 1910)	
+*Phrynobatrachus kreffti* (Boulenger, 1909)	
Phrynobatrachus ukingensis (Loveridge, 1932)	Misuku Hills, Malawi.
+*Phrynobatrachus uzungwensis* (Grandison & Howell, 1983)	

CAECILIIDAE

Species	Locality
+ *Afrocaecilia taitana* (Loveridge, 1935)	Taita Hills, Kenya
+*Afrocaecilia uluguruensis* (Barbour & Loveridge, 1928)	
+*Boulengerula boulengeri* (Tornier, 1896)	

Table 9.1. (cont.)

Species	WU	EU	NGU	UKA	ULU	UZU	UKI	UBE	POR	RUN	SH	Others
SCOLECOMORPHIDAE												
+ *Scolecomorphus kirkii* (Boulenger, 1822)					x	x		x				Rubeho Mts; Mt Thyolo, Shire highlands, Zomba plateau, Malawi.
+*Scolecomorphus ulunguruensis* (Barbour & Loveridge, 1928)					x							N Pare Mts.
+ *Scolecomorphus vittatus* (Boulenger, 1895)	x	x			x							
Reptilia												
GEKKONIDAE												
+*Urocotyledon wolterstorffi* (Tornier, 1900)		x			x							
+*Lygodactylus conradti* (Matschie, 1892)	x	x			x							
+*Lygodactylus gravis* (Pasteur, 1964)	x	x										
Lygodactylus howelli (Pasteur & Broadley, 1988)												Jozani Forest, Zanzibar; Mafia Island.
+*Lygodactylus williamsi* (Loveridge, 1952)						x						Known only from Kimboza Forest.
Cnemaspis dickersoni (Schmidt, 1919)						x						Southern Ethiopia and Sudan south through Uganda, Kenya to central Tanzania, west to Rwanda and eastern Zaïre.
Cnemaspis africana (Werner, 1895)	x	x										Mt Kilimanjaro, Mt Meru; Taita Hills and 'Athi Plain'.
+*Cnemaspis barbouri* (Perret, 1986)					x							
+*Cnemaspis uzungwae* (Perret, 1986)						x						
CHAMAELEONIDAE												
Bradypodion fischeri (Reichenow, 1887)	x	x	x		x							
Bradypodion tavetanum (Steindachner, 1891)												Mt Kilimanjaro, N Pare Mts, Taita Hills.
Bradypodion uthmoelleri (Müller, 1938)												Mt Hanang, Oldeani, Ngorongoro Crater forest.

Species	1	2	3	4	5	6	7	8	9	Distribution
+*Bradypodion spinosum (Matschie, 1892)	×	×								Shimba Hills, Kenya.
+*Bradypodion tenue (Matschie, 1892)	×	×								
+*Bradypodion oxyrhinum (Klaver & Böhme, 1988)			×							
+*Chamaeleo goetze (Tornier, 1899)			×	×	×	×	×	×	×	Nyika Plateau, Malawi.
+*Chamaeleo tempeli (Tornier, 1899)			×	×	×	×				
+*Chamaeleo werneri (Tornier, 1899)			×	×		×				
*Chamaeleo fuelleborni (Tornier, 1900)					×					
+*Chamaeleo deremensis (Matschie, 1892)	×	×		×						
+*Chamaeleo incornutus (Loveridge, 1932)				×						
+*Chamaeleo laterispinis (Loveridge, 1932)		×								
+*Rhampholeon sp. nov.										
+*Rhampholeon nchisiensis (Loveridge, 1953)	×	×	×	×	×					Nchisi Mt, Malawi. Rondo Plateau; Shire and Mt Thyolo, Zomba, Malawi.
+*Rhampholeon brachyurus (Günther, 1893)			×							
+*Rhampholeon brevicaudatus (Matschie, 1892)	×	×	×							Coastal forest.
+*Rhampholeon temporalis (Matschie, 1892)	×	×								
SCINCIDAE										
+*Proscelotes eggeli (Tornier, 1902)	×									
+*Scelotes uluguruensis (Barbour & Loveridge, 1928)	×		×		×					
Melanoseps ater (Gunther, 1873)	×		×							Southwestern and southeastern Tanzania; Shire Plateau, Misiku Mts, Malawi; northwestern Zambia.
Melanoseps rondoensis (Loveridge, 1942)										Known only from Rondo Plateau.
+*Leptosiaphos rhomboidalis (Broadley, 1989)			×							
LACERTIDAE										
Gastropholis vittata (Fischer, 1886)										Coastal mainland Tanzania and northern Mozambique.
Gastropholis prasina (Werner, 1904)	×									Tanga; Kiono Forest; southeastern Kenya.
Adolfus jacksoni (Boulenger, 1899)	×									Mt Kilimanjaro, Uganda, Kenya and Tanzania to Rwanda, Burundi and Zaire.

Table 9.1. (*cont.*)

Species	WU	EU	NGU	UKA	ULU	UZU	UKI	UBE	POR	RUN	SH	Others
Holaspis guentheri (Gray, 1863)	×	×										Sierra Leone east to Uganda; Angola and northern Mozambique, west to southern Malawi.
TYPHLOPIDAE												
+*Typhlops uluguruensis* (Barbour & Loveridge, 1928)	×	×										
+*Typhlops gierrai* (Mocquard, 1897)	×	×			×	×						
Typhlops rondoensis (Loveridge, 1942)						×	×					Rondo Plateau.
COLUBRIDAE												
Lycophidion meleagris (Boulenger, 1893)	×	×			×							West through Zaïre to Angola.
+*Prosymna ornatissima* (Barbour & Loveridge, 1928)												
+*Geodipsas vauerocegae* (Tornier, 1902)	×	×			×							
+*Geodipsas procterae* (Loveridge, 1922)					×							
Aparallactus guentheri (Boulenger, 1895)		×			×							Tana River south through coastal Kenya, Tanzania, Mozambique, and Malawi to eastern Zimbabwe.
+*Aparallactus werneri* (Boulenger, 1895)	×	×			×							Coastal forest.
+*Philothamnus macrops* (Boulenger, 1895)		×						×				Zanzibar; Rondo Plateau.
Natriciteres variegata (Peters, 1854)										×		South through Mozambique, Malawi, eastern Zimbabwe to Zululand.
+ *Crotaphopeltis tornieri* (Werner, 1897)	×	×			×	×	×			×		Misuku Mts, Malawi.
+*Dipsadoboa werneri* (Boulenger, 1897)	(×)	×			×	×	×					
Daspeltis atra (Sternfeld, 1912)												Southeast Sudan and Ethiopia south to Burundi and northern Tanzania (Mt Kilimanjaro).

Dasypeltis medici (Bianconi, 1859) — Southern Kenya south through Tanzania, Mozambique, Malawi, and eastern Zimbabwe to Zululand.

ATRACTASPIDAE

Atractaspis aterrima (Günther, 1863) — West through Uganda to Guinea.

ELAPIDAE

+*Elapsoidea nigra (Günther, 1888) — Forested and formerly forested areas of subsaharan Africa, Pugu, other coastal forests.

Naja melanoleuca (Hallowell, 1857)

VIPERIDAE

Bitis gabonica (Duméril & Bibron, 1854) — Pugu forest; Rondo Plateau; forested areas from Zululand and Angola north to Sudan and Guinea.

Bitis nasicornis (Shaw, 1802) — Known in Tanzania from only 2 specimens. Guinea east to southern Sudan, Uganda, western Kenya southwest through Rwanda, Burundi, Zaïre to Angola.

Atheris nitschei (Tornier, 1902) — Western Tanzania, Rwanda, Burundi and adjacent Zaïre, northeastern Zambia, and northern Malawi.

+**Atheris ceratophorus* (Werner, 1895)

+**Adenorhinos barbouri* (Loveridge, 1930)

Note: Key: +, species present in Tanganyika–Nyasa forests; ×, found in forest group indicated; *, Tanzanian endemic. WU, West Usambara; EU, East Usambara; NGU, Nguru; UKA, Ukaguru; ULU, Uluguru; UZU, Uzungwa; UKI, Ukinga; UBE, Ubena; POR, Poroto; RUN, Rungwe; SH, Southern Highlands, no further details given; (×), some doubt as to exact locality.

non-forest-dependent species, and of distinct lowland and montane forest faunas in Africa (see Stuart, 1983 for a review), there is less agreement upon the exact altitudinal limits of the latter. The difficulties of determining the position of a particular species with regard to forest dependence as well as to type of forest are compounded by lack of definite data on precise collecting locality and altitude, as noted earlier, and on the exact vegetation type in which a specimen was collected. In some highland and mountain localities in subsaharan Africa the vegetation above 1500 m a.s.l. may be grassland, and above 3500 m may give way to an Afroalpine type (Hughes, 1983). However, in at least some of the Eastern Arc mountains, forest is to be found well above 1500 m. Amphibians and reptiles of the Afroalpine and moorland zones have been specifically excluded from this study; their origins are uncertain (Hughes, 1983).

There also is a dearth of information available on the ecology, behaviour and physiology of forest species. This contrasts with studies on non-forest species. Drewes *et al.* (1977) discussed in some detail the special adaptations which help prevent dehydration and allow basking in full sunlight in a desert frog, *Chiromantis petersi*. Withers *et al.* (1982) studied the physiology of *Hyperolius nasutus*. Physiological studies on montane reptiles (Hebrard, Reilly & Guppy, 1982; Reilly, 1982) have dealt with lizards which live in Afroalpine rather than forest habitat. Western (1974) studied distribution, density and biomass density of lizards in semi-arid northern Kenya, and Kreulen (1979) discussed factors affecting reptile biomass in African grasslands. Only Broadley & Blake (1979) have conducted field studies on a forest chameleon, *Rhampholeon marshalli*, in Zimbabwe, and these were mainly concerned with numbers and movements of the animals rather than with physiological adaptations.

Patterns of distribution of the amphibians and reptiles of the eastern forests of Africa

The distributions of the forest amphibians and reptiles of the eastern forests of Africa are presen-

ted in Table 9.1. These fall into several broad patterns.

Species found in both the eastern and Guineo–Congolian forests

One of the striking features of the distributions is that very few of the species found in the eastern forests of Africa are also found in those to the west. Only a single species of frog, *Arthroleptis adolfifriderici*, falls into this category. Not a single forest-dependent species of the anuran families Ranidae, Hyperoliidae, Bufonidae or Microhylidae is found in both the western and eastern forests, and no caecilian species in shared by the two.

Among the reptiles, the lizard *Cnemaspis dickersoni*, and the snakes *Lycophidion meleagris*, *Atractaspis aterrima* and *Naja melanoleuca* also are found in western forests. Subspecific distinction has been recognised for four of the reptile species found in the western and eastern forests: the lizard *Holaspis guentheri*, the colubrid snake *Natriciteres variegata* and the vipers *Bitis gabonica* and *Atheris nitschei* (see Table 9.2).

Species found in the Tanganyika–Nyasa forests and southern Africa

These species are found in several forests in the Tanganyika–Nyasa system (and in some cases, other isolated forests) but also extend their range into southern Africa. Included here are the frogs *Hyperolius mitchelli* and *Leptopelis flavomaculatus*, the lizards *Melanoseps ater* and *Melanoseps loveridgei* and the snakes *Aparallactus guentheri* and *Dasypeltis medici*. The taxonomy of some of the species in this group is still not without problems: *Hyperolius mitchelli* and *Hyperolius puncticulatus* have only recently been shown to be separable in the field (Schiøtz, 1975) and there may be some confusion of past records. Poynton & Broadley (1987) discuss the two species in southern Africa. Previously many subspecies of *Melanoseps ater* were described; despite a recent revision of the genus by Brygoo & Roux-Estève (1981), the status of some of these remains unclear. Several of the species with this distribution pattern are

Table 9.2. *Subspecies of reptiles found in both the eastern and western forests of Africa*

Subspecies	Location
Holaspis guentheri laevis (Werner, 1895)	northeast Tanzania south to Amatongas and Dondo, Mozambique
Holaspis guentheri guentheri (Gray, 1863)	Uganda, southwest to Angola, northwest to Sierra Leone
Natriciteres variegata sylvatica (Broadley, 1966)	forested areas of southern Tanzania, south through Mozambique and eastern Zimbabwe to Zululand
Natriciteres variegata variegata (Peters, 1861)	Cameroun west to Guinea
Bitis gabonica gabonica (Duméril & Bibron, 1854)	forested areas from Nigeria east to southern Sudan, Uganda and western Kenya south to Angola, Zambia, eastern Zimbabwe and Zululand
Bitis gabonica rhinoceros (Schlegel, 1855)	Guinea to Ghana
Atheris nitschei nitschei (Tornier, 1902)	western Tanzania, Rwanda, Burundi, Uganda and adjacent Zaïre
Atheris nitschei rungweensis (Bogert, 1940)	north from Rungwe to Kigoma; northeastern Zambia and northern Malawi (see Rasmussen & Howell, 1982)

able to survive relatively dry conditions, such as those found at the forest edge, and may even persist in some areas where forest has recently been removed.

Species with a Tanganyika–Nyasa system distribution pattern

This category may be further broken down into those species which occur in one block of or along the length of the Tanganyika–Nyasa system, and those which are restricted to the forests of the Eastern Arc.

Species which extend into the southern portion of the Tanganyika–Nyasa system to the Malawi forests are the frogs *Arthroleptis reichei* and *Hyperolius spinigularis*, the caecilian *Scolecomorphus kirkii*, the chameleons *Chamaeleo goetze* and *Rhampholeon nchisiensis* and the snake *Crotaphopeltis tornieri*.

A single species of anuran, *Arthroleptis francei*, appears to be isolated in forest on Mulanje Mountain (Poynton & Broadley, 1985a).

Species limited to the Eastern Arc mountains of Tanzania fall into several groups: species which occur in several of the forests, those which occur in only the northern or southern portions, and those which are restricted to single mountain blocks.

Those species found along much of the length of the Eastern Arc mountains include the anurans *Arthroleptis affinis*, *Bufo brauni*, *Nectophrynoides tornieri*, *Afrixalus uluguruensis*, *Hyperolius puncticulatus*, *Leptopelis barbouri*, *Leptopelis parkeri*, *Leptopelis uluguruensis*, *Arthroleptides martiensseni* and *Callulina kreffti*, the caecilian *Scolecomorphus vittatus* and the snakes *Typhlops gierrai* and *Atheris ceratophorus*. As already noted at the time of Loveridge, many of these were considered to be Usambara/Uluguru endemics, but recent collecting has revealed their presence in the Uzungwa Mountains.

Species known from only the Usambara and Uluguru and outlying mountains include the frog *Schoutedenella xenodactyla*, the geckos *Urocotyledon wolterstorffi* and *Lygodactylus conradti*, and the snakes *Geodipsas vauerocegae* and *Elapsoidea nigra*. The latter is also recorded from the Nguru Mountains, a poorly collected area which may well yield more species found in both the Usambaras and Ulugurus as its intermediate position might suggest.

Three species are known to occur only in the Uluguru and Uzungwa mountains: the frog *Phrynobatrachus uzungwensis* is at present known only from a single locality, Lupanga peak in the Ulugurus, and from Mwanihana Forest in the Uzungwas. The chameleons *Bradypodion oxyrhinum* and *Chamaeleo werneri* are known from several localities in both ranges.

Moving further south, fewer species appear to have been isolated in both the Uzungwas and the Southern Highlands. The toad *Bufo uzunguensis* is a small species which is extremely difficult to identify and at least one of the paratypes listed by Loveridge in his original description does not belong to this species (Grandison, 1972). The viper *Adenorhinos barbouri* is known from only a few specimens and nothing is known about its biology.

Little is known about the status of the chameleons *Chamaeleo fulleborni* and *Chamaeleo incornutus*, which are not found in the Uzungwas but which occur in the Poroto Mountains in the case of the former and in the Ukinga and Southern Highlands area in the case of the latter. The forest marsh snake *Natriciteres variegata* has been recorded from Rungwe Mountain and the Matengo highlands.

The following species are shared only by the East and West Usambaras: the frogs *Hoplophryne rogersi* and *Phrynobatrachus kreffti*, the caecilian *Boulengerula boulengeri*, the chameleons *Bradypodion spinosum*, *Chamaeleo deremensis* and *Rhampholeon temporalis* and the snake *Dipsadoboa werneri*.

The frogs *Arthroleptis tanneri*, *Hyperolius tanneri* and *Parahoplophryne usambarensis* and the burrowing skink *Proscelotes eggeli* are endemic to the West Usambaras and the gecko *Lygodactylus gravis* is endemic to the East Usambaras.

Some species have been isolated on mountain forests of the Eastern Arc and in smaller, often lower and drier, forests nearer the coast. These include the anurans *Mertensophryne micranotis*, *Leptopelis flavomaculatus* and *Spelaeophryne methneri*. The affinities of a toad in the genus *Stephopaedes* collected in 1989 at Rondo and Kiwengoma Forest are still being evaluated; members of the genus are known from Mahenge in Tanzania, in western Mozambique and in Zimbabwe.

Reptiles with a similar distribution pattern

include the chameleon *Bradypodion tenue*, the pygmy chameleon *Rhampholeon brachyurus*, the lacertid *Gastropholis prasina*, and the snakes *Aparallactus werneri*, *Philothamnus macrops* and *Dasypeltis medici*.

Only one of the Usambara endemics, *Phrynobatrachus kreffti*, can be considered a species with a potentially good dispersing ability, as it can readily move along streams, yet even it is found only in the Usambaras. All the others would appear to be poor dispersers. This is certainly true of the two burrowing forms, the caecilian and skink. The other frogs are known from only very few specimens from restricted localities, and *Arthroleptis tanneri* (Grandison, 1983) and *Hyperolius tanneri* (Schiøtz, 1982) were described only recently. Forest chameleons in general appear to be poor dispersers and this might also be true for forest species of *Lygodactylus*, as other forms are isolated on different mountain forests. The snake genus *Dipsadoboa* contains both forest and non-forest species; only *Dipsadoboa werneri* appears to have been isolated in the Usambara forests.

Currently the Uluguru forests appear to contain the highest number (12) of endemic species of any block in the entire Tanganyika–Nyasa system. These include two toads, *Nectophrynoides cryptus* and *Nectophrynoides minutus*, both only recently recognised as full species distinct from *Nectophrynoides viviparus* (Perret, 1971). Two microhylid frogs, *Probreviceps uluguruensis* and *Hoplophryne uluguruensis* and two caecilians, *Afrocaecilia uluguruensis* and *Scolecomorphus uluguruensis* complete the list of Uluguru endemic amphibians. The endemic reptiles are the lizards *Lygodactylus williamsi* and *Cnemaspis barbouri*. The former is known only from Kimboza Forest and the latter has only been recently described (Perret, 1986). A skink, *Scelotes uluguruensis* and three snakes, *Typhlops uluguruensis*, *Prosymna ornatissima* and *Geodipsas procterae* are all Uluguru endemics; the first two of these are modified for burrowing; little is known about the ecology of the third, which is known only from a few specimens.

The Uzungwa forests harbour five endemic species. A new species of *Nectophrynoides* toad has recently been discovered in forest near Chita (Clarke, 1988a). The treefrog *Phylictimantis keithae* is known only from three localities in the Uzungwa Mountains. *Cnemaspis uzungwae*, a gecko, has recently been described (Perret, 1986). The chameleon *Chamaeleo laterispinis*, which closely resembles the lichen *Parmelia*, is unique in the structure and position of spines on the side of its body and in its pattern of coloration. A newly described skink in the genus *Leptosiaphos* is known only from leaf litter of the Mwanihana Forest Reserve in the Uzungwas (Broadley, 1989).

Taita Hills endemics

Only one forest form is known to be endemic to the forests of the Taita Hills, the caecilian *Afrocaecilia taitana*. Until recently the evidence suggested that the Taita area was or had been completely isolated from the Tanzanian Eastern Arc fauna, but the discovery of the microhylid frog *Callulina kreffti* indicates an affinity with the larger, more southern forests (Beentje, Ndiangui & Mutangah, 1987; Beentje, 1988).

Coastal endemics

Two species of hyperoliid treefrogs, *Hyperolius rubrovermiculatus* and *Afrixalus sylvaticus*, are endemic to Kenyan coastal forest. A new species of *Stephopaedes* toad has been found on Mafia Island in remnant forest (K.M.H., unpublished data). *Melanoseps rondoensis* is a limbless skink known only from the Rondo Plateau. Recent studies by Clarke (1988b) indicate that the dwarf toad *Bufo lindneri*, formerly regarded as endemic to the Dar es Salaam area and coastal forest in the Pugu Hills, is in fact much more widespread and not endemic to coastal forest. A diurnal gecko, *Lygodactylus howelli*, has been found only in groundwater forest on Zanzibar (Pasteur & Broadley, 1988), and on Mafia Island (K.M.H., unpublished data), and the lacertid lizard *Gastropholis vittata* is known only from coastal forest in Tanzania and Mozambique.

Patterns of speciation in the amphibians and reptiles of the eastern forests of Africa

Species found in both the eastern and Guineo–Congolian forests of Africa

Three species of reptiles have recognised subspecies as listed in Table 9.2. In the case of the lizard *Holaspis guentheri*, the eastern populations seem to follow a roughly Eastern Arc distribution pattern south to central Mozambique with the western populations centred in the western forest block. For the viper *Bitis gabonica*, a much more widespread species, the subspecies found in eastern Africa also extends west to Nigeria and to the south, through to southern Africa, and the second subspecies occurs far to the west, from Ghana west to Guinea. Although in Tanzania *B. gabonica* appears to be very much a forest-dependent species, in southern Africa under certain conditions it appears to be sympatric or nearly so with *B. arietans*, a species of much more open country, and in Tanzania, one which is not known to occur in forest. Broadley & Parker (1976) reported a case of *B. gabonica* × *B. arietans* hybridisation following destruction of forest, and Hughes (1983) considers *B. gabonica* to be a forest form which occurs only rarely in savanna.

The situation with regard to *Atheris nitschei* is somewhat different. Members of both subspecies occur within Tanzania; those in the western forests are members of one subspecies which extends west into eastern Zaïre, while those south of Kigoma and in southwestern Tanzania, northern Malawi and northeastern Zambia belong to a second.

Patterns of speciation in species limited largely to the Tanganyika–Nyasa forests

A number of populations of amphibians and reptiles isolated on the various mountain block forests have been designated as subspecies, but not all of these are recognised today (see Table 9.3). Subspecific status has been assigned to fewer species of amphibians than reptiles. For the

Table 9.3. *Subspecies of amphibians and reptiles in Tanganyika–Nyasa forests*

Subspecies	Location
Probreviceps macrodactylus macrodactylus (Nieden, 1926)	Usambaras
Probreviceps macrodactylus loveridgei (Parker, 1931)	Ulugurus, Uzungwa
Probreviceps macrodactylus rungwensis (Loveridge, 1932)	Rungwe
Chamaeleo fischeri fischeri (Reichenow, 1887)	East Usambara, Nguru
Chamaeleo fischeri multituberculatus (Nieden, 1913)	West Usambara
Chamaeleo fischeri uluguruensis (Loveridge, 1957)	Ulugurus
Chamaeleo goetzei goetzei (Tornier, 1899)	Uzungwa, Ubena, Ukinga, Rungwe and Poroto mts
Chamaeleo goetzei nyikae (Loveridge, 1953)	Nyika Plateau
Melanoseps ater ater (Gunther, 1873)	Malawi and Mt Rungwe
Melanoseps ater matengoensis (Loveridge, 1942)	Matengo Highlands
Melanoseps ater uzungwensis (Loveridge, 1942)	Uzungwa
Melanoseps ater longicauda (Tornier, 1900)	N Pare mts, Usambara mts?

former, Loveridge (1957) recognised subspecific epithets only for *Probreviceps macrodactylus*.

More subspecies have been described for reptile populations. In the *Chamaeleo fischeri* complex, much remains to be learned about the status of the various subspecific forms described (see Wederkinch, 1982) in the Usambaras. Loveridge (1957) recognised many subspecies of *Chamaeleo fischeri* on mountains as far away as Mount Kenya, Kilimanjaro and Hanang; I follow Broadley & Howell (1991) in assigning many of these isolated populations specific rank.

Böhme (1982) described a subspecies, *Chamaeleo laterispinis brookesieformis*, from fairly dry woodland in the Uzungwas at Kibao Iyayi, 85 km southwest from the type locality of the species near Mufindi, which is moist evergreen forest. He has since (Böhme, 1987) synonomised it with *C. laterispinis*. Brygoo & Roux-Estève (1981) recognised a number of subspecies of *Melanoseps ater* but have admitted that the validity of at least some of these is questionable. The same authors raised what was regarded as a population that was only subspecifically differentiated to full species level as *Melanoseps loveridgei*.

Diversity

A total of 106 species of forest-dependent amphibians and reptiles are found in the eastern forests of Africa (Table 9.1) and 83.9% (89/106)

occur in the Tanganyika–Nyasa forests. Of those, 84% (75/89) are restricted to those (and some coastal forests); 77.3% (82/106) occur in the Eastern Arc forests and of these, 65.8% (54/82) are strictly endemic to Tanzania's Eastern Arc. If one includes those species restricted to the Eastern Arc forests and coastal forests, the endemics rise to 71.9% (59/82).

Comparing diversities in those forests for which data are available (Table 9.4), greatest diversity is found in the Usambaras and the Ulugurus, and numbers of species decrease as one moves south towards the Uzungwas. Mount Rungwe, volcanic in origin, has fewer forms, and forests in Malawi have even fewer forest-dependent species. The Taita Hills, at the extreme north of the Tanganyika–Nyasa forests, also have very few forest species. The volcanic Mount Kilimanjaro and Mount Meru have no known forest-dependent amphibians and only two forest reptiles. Diversity in the coastal forests is also low.

Are there real differences between the diversities of the two Usambara blocks or between the Usambaras and the Ulugurus? The West Usambaras have been much less thoroughly collected and studied than the East, but even some of the amphibians and reptiles described from the East Usambaras are still known only from one or two specimens. Much more major, long-term field work is needed in the Usambaras and the Ulu-

Table 9.4. *Diversity and endemism in the eastern Africa forest herpetofauna*

Forest(s)	Number of forest species	T/N endemics	T/N endemics to this forest
Usambara (W & E combined)	a: 23	20/23 = 86.9%	5/20 = 25%
	r: 29	16/29 = 55.1%	6/16 = 37.5%
	t: 52	36/52 = 69.2%	11/36 = 30.5%
West Usambara	a: 22	19/22 = 86.39%	2/19 = 20.5%
	r: 23	14/23 = 60.8%	1/23 = 4.3%
	t: 45	33/45 = 73.3%	3/42 = 7.1%
East Usambara	a: 20	17/20 = 85%	0
	r: 26	15/26 = 57.6%	1/26 = 3.8%
	t: 46	32/46 = 69.5%	1/26 = 2.1%
Uluguru	a: 26	23/26 = 88.5%	6/23 = 26%
	r: 24	18/24 = 75%	6/18 = 33.3%
	t: 50	41/50 = 82%	12/41 = 29.2%
Uzungwa	a: 19	18/19 = 94.7%	3/18 = 16.6%
	r: 16	14/16 = 87.5%	3/14 = 21.4%
	t: 35	32/35 = 91.4%	6/32 = 18.7%
Rungwe	a: 3	3/3 = 100%	0
	r: 6	4/6 = 75.1%	0
	t: 9	7/9 = 77.7%	0
Malawi	a: 8	4/8 = 50%	1
	r: 8	4/8 = 50%	0
	t: 16	8/16 = 50%	0
Taita Hills	a: 2	2/2 = 100%	1/2 = 50%
	r: 2	0	0
	t: 4	2/4 = 50%	1/4 = 25%

Note: T/N, Tanganyika–Nyasa forest; a, amphibians; r, reptiles; t, total.

gurus as well as other forests before one can speak of 'significant differences' in species diversities.

Nowhere is this clearer than when examining what at first glance would seem to be an obvious difference in numbers between the Ulugurus and the Uzungwas. At the time of Loveridge (1933, 1952) it was felt that, based on the evidence then available, the major affinities of the Uzungwa herpetofauna were with forests to the west and southwest. But recent collecting at only two localities in the Uzungwa Mountains, Mwanihana and Mufindi, has shown that many species that were at the time of Loveridge believed to be endemic to the Usambara/Uluguru massifs are

present in the Uzungwas: *Nectophrynoides tornieri, Leptopelis uluguruensis, L. barbouri, Arthroleptides martiensseni, Arthroleptis affinis, Callulina kreffti, Bradypodion oxyrhinum, Brookesia brevicaudata, Typhlops gierrai* and *Atheris ceratophorus* are all in this category.

On present evidence there is what appears to be a tapering off of numbers of species as one moves away from the Usambaras and Ulugurus into the Uzungwas, but I prefer to emphasise the similarities between the Usambara/Uluguru herpetofauna and that of the Uzungwas. Bearing in mind that the Uzungwa Mountains represent the largest mountainous forested area in Tanzania and indeed perhaps eastern Africa, and that they are virtually unknown biologically, there is every reason to believe that further work will reveal more species in common with the more northerly mountains of the Eastern Arc as well as others endemic to the Uzungwas, as would be predicted by current island biogeographic theory (MacArthur & Wilson, 1967).

It has not been possible to obtain entirely comparable figures for comparison from other parts of Africa: apparently no complete species lists with sufficiently detailed habitat data have been published. Schiøtz (1967) gave a partial list of frog species (excluding Arthroleptidae) from the forests of West Africa. If the West African forest block is considered in its entirety, there are five genera of West African forest Hyperoliidae not found in the Eastern Arc forests: *Acanthixalus*, *Callixalus*, *Chrysobatrachus*, *Cryptothylax* and *Opisthothylax* (Drewes, 1984). Seven ranid genera present in the west are not found in the forests of the Eastern Arc according to Schiøtz (1967), who noted that for amphibians, richness of the Usambara and Uluguru forests was comparable to that of the forests of Ivory Coast and Ghana (Schiøtz, 1981).

Endemism

The trends seen for diversity also apply when endemism is considered (Table 9.4). Levels of endemism are comparable for the Usambaras and the Ulugurus. In the Usambaras there is a slightly higher level of endemism among reptiles (37.5%) when compared with the Ulugurus (33.3%).

The situation is reversed for the amphibians: the Uluguru forests have 26% of the species endemic, while those of the Usambaras have 21.7%. The herpetofauna of the Uzungwas has only recently been recognised to contain many species formerly believed to be Usambara/Uluguru endemics (see the section above on diversity) and appears to have levels of endemism lower than those of either the Usambaras or Ulugurus. Further south, Mount Rungwe has no endemics, and the forests of Malawi have only a single endemic species of *Arthroleptis*. To the north, the Taita Hills forests have a single endemic amphibian burrowing form.

The coastal forests have only very few forest-dependent species (Table 9.1). Some of these are limited generally to this type of vegetation. Others, in contrast, are limited to very specific sites. Two anurans are known only from Kenya's coastal forests. A skink is endemic to the Rondo Plateau, and a gecko to Jozani Forest on Zanzibar.

While no precise data are available with which to compare endemism in forest reptiles with other African localities, D. Stubbs (*in litt.*) has compiled a preliminary list of endemic amphibians for various African countries. As he notes, the vast majority of locally endemic amphibian species are concentrated in the moist tropical forest belt of Central Africa and the mountain forest areas. He also emphasises that among the small West African states, endemics are extremely localised, and many are known only from type localities. Although Stubbs's list is not restricted to forest species, it does provide at least some relative idea of the differences between various localities in Africa. The continental African countries with the highest levels of endemism are: Cameroun, 62/90; Zaïre, 51/74; Tanzania, 40/51; Ethiopia, 30/31; Angola, 23/24. On Madagascar, all 144 forms are endemic (D. Stubbs, *in litt.*).

The radiation of genera and species in the West African forests is reflected in a number of species with unusual adaptations related to their ecology and reproduction. For example, the treefrog *Acanthixalus spinosus* is apparently mute and breeds in small amounts of water trapped in trees. Several ranid genera not found in the eastern forests, such as *Conraua*, *Petropedetes*, *Astylosternus*, *Scotobleps* and *Trichobatrachus* have been regarded as specialised stream dwellers (but see Wasserburg & Heyer (1983), who argue that tadpoles thought to be adapted to living in swift-flowing streams may actually be specialised for living in or near flowing water but not in a mainstream current).

Laurent (1973), in a comparative survey of the amphibians and reptiles of Africa and South America, describes various adaptations and convergence. I am, however, unable to account for his statement 'In Africa, microhylids are absent from the forest' (p. 261), as *Callulina* and *Probreviceps* are entirely restricted to the eastern forest. He also notes that there is evidence for movement of at least three species of geckos, all non-forest forms, from the coast of West Africa to South America by rafting on floating vegetation and debris. The rafting of reptiles, especially snakes, is known to occur during floods of the Rufiji River in southern Tanzania (Moreau & Pakenham, 1941), but its relevance to the distribution of forest fauna is still unclear.

Origins of the Eastern Arc forest herpetofauna

Given our lack of knowledge about many basic fundamental factors which influenced the present distribution of forest amphibians and reptiles, such as exact timing and nature of the last connections among various forests, climatic variations and the rates at which speciation occurred, any attempted 'explanation' of the origin of the Eastern Arc fauna remains largely speculative. This is especially true considering the lack of fossil evidence for the area. Other than the dinosaurs excavated by the Germans from Tendaguru 60 km northeast of Lindi early in this century (Parkinson, 1930), the only other fossil reptiles recorded from Tanzania are those from Olduvai Gorge (Rage, 1973, 1979; Meylan, 1983). Fossil anurans have only recently been found in South Africa (Van Djik, 1985) but none appear to be known from eastern Africa (Tandy & Keith, 1972; Tihen, 1972).

Species with strong connections to the Guineo–Congolian forests (see Table 9.1) indicate the complexities of interpreting and evaluating the effects of forest separation between the main western and the outlying eastern blocks. Some forms are recognised as subspecies, while others appear to show little variation across their entire range. That many of the Eastern Arc forms may have their origins in the Guineo–Congolian forests is indicated by the fact that almost all the genera present in the eastern forests are also to be found in the forests to the west. This is true even for widely separated forms such as *Nectophrynoides*, with five species in Tanzania, two montane grassland species on Mount Nimba, and two species in Ethiopia (Grandison, 1978). In an examination of forest snakes, Hughes (1983) noted that with the exception of the single montane Kenya endemic viper, *Atheris desaixi*, all the species found in Kenya are also present in Zaïre. He notes further that, in general, there are few endemic forest species of snakes.

In contrast, at least some genera and species appear to have evolved *in situ* in the Eastern Arc forests or to have survived there in isolation and become extinct in all other African forests. Examples of such amphibian genera include the caecilian *Afrocaecilia* and the microhylid frogs *Callulina*, *Probreviceps*, *Hoplophryne* and *Parahoplophryne*. *Callulina kreffti* was until its discovery in the Taita Hills regarded as a Tanzanian endemic. Poynton (1964) regarded *Probreviceps* as primitive to *Breviceps* and noted that while both were originally probably sylvicolous, the burrowing ability of the latter allowed it to leave the forest and occupy drier habitats, and Poynton & Broadley (1967) remarked upon the affinity of *P. rhodesianus*, the only member of the genus found outside Tanzania, and *Breviceps*.

Aside from *Arthroleptides martiensseni* of the Eastern Arc only a single other species in the genus, *A. dutoiti*, is described from Mount Elgon in Kenya, but it may now be extinct or nearly so as it has not been found by biologists during recent searches (R. C. Drewes, personal communication). The only reptile genus that appears to have arisen in the Eastern Arc forests is *Adenorhinos*.

Many of the other species of the Eastern Arc forests appear to have evolved *in situ*. Schiøtz (1967, 1981) noted that while the separation between the eastern and Guineo–Congolian forests is complete at the species level among treefrogs in the family Hyperoliidae, it is not possible to identify species pairs or any other relationship below the genus level linking the eastern and western forests. Those genera with notably large numbers of species include the toads *Nectophrynoides*, the treefrogs *Leptopelis* and the chameleons *Chamaeleo* and *Rhampholeon*. Fewer endemics are found in the coastal forests. The anurans *Hyperolius rubrovermiculatus* and *Afrixalus sylvaticus* are endemic to Kenyan coastal forests; the skink *Melanoseps rondoensis* and the lacertid *Gastropholis vittata* are the only reptiles endemic to the mainland coastal forests of eastern Africa.

Some species are able to exist in both the lower forests on the coast and those of the mountains. These include the anurans *Mertensophryne micranotis*, *Leptopelis flavomaculatus* and *Spelaeophryne methneri*, the lacertid *Gastropholis prasina*, a chameleon *Rhampholeon brevicaudatus* and the snakes *Aparallactus werneri* and *Philothamnus macrops*. The origins of these are difficult to determine. At least two of the anurans, *M. micranotis* and *S. methneri*, escape the hotter, drier conditions of coastal forest by burrowing into crevices in soil and inside rotten logs.

Were these forest species which were able to adapt to drier conditions at the edge of former forest? This would seem likely in the case of *S. methneri* and *R. brevicaudatus* because they have a generally typical Eastern Arc distribution pattern, and the coastal locality at which they were collected, Kilwa, once was much more forested than at present.

A few of the lowland species from drier areas may slowly be penetrating the forests. A new species of gecko in the *Lygodactylus somalicus* species group, *L. howelli*, has been discovered in groundwater forest on Zanzibar (Pasteur & Broadley, 1988). *L. scheffleri uluguruensis* is the name given to an isolated population in the Ulugurus, but the species is generally one of hot, dry areas, with other subspecies from woodland in Kenya. It may now be isolated in the Ulugurus, but have moved in from the drier portions of that range in which woodland is found below moist forest. A parallel situation may exist in the lizards *Agama montana* and *A. agama usambarae*. Moody (1980) notes that the family Agamidae is absent from the tropical forest belt. While these two agamas are found on rocky slopes at the forest edge, their origins would therefore appear to be outside the forest.

The colonising ability of various forest species can be examined by studying the fauna of the forests on the volcanic mountains of eastern Africa. To the northwest of the Eastern Arc on Mount Kilimanjaro, Mount Meru and the Ngorongoro Crater Highlands, very few forest forms are found. The snake fauna contains no forest species from the Eastern Arc. What was formerly regarded as a Kilimanjaro endemic (but a non-forest species) and given specific status as *Crotaphopeltis kageleri* is now treated as a subspecies of a more widely occurring species, *Dipsadoboa shrevei* (Rasmussen, 1986). Only the chameleon *Chamaeleo rudis* (Ngorongoro and Meru), and the skink *Leptosiaphos kilimensis*, with a wide mountain forest distribution, have been able to invade these forests and no species of amphibian or reptile seems to have evolved *in situ* on these mountains. With the exception of *Leptosiaphos kilimensis* and the snake *Dasypeltis atra*, there was little colonisation of Mount Kilimanjaro and Meru forests. Species from the forests of the Eastern Arc mountains were unable to cross the relatively dry, open habitat which separates them from the volcanic mountains.

There is now much evidence to suggest that climatic conditions were once moist enough to have permitted at least contiguous, if not completely continuous, forest belts from much of west to eastern portions of the African continent. During the Pleistocene, this eastern forest was evidently much further fragmented, and as conditions changed, suitable conditions for moist tropical forest largely remained only on the crystalline block and associated mountains. These forests have been termed 'refugia', and they first received intensive study when Moreau (1966) coined the term 'Tanganyika–Nyasa' system when considering their avifauna. Much earlier, Loveridge (1933, 1952) had emphasised the importance of the isolation of the Usambara/Uluguru forests, though evidence available to him at that time suggested that there was less connection between those mountains and the Uzungwas than we now know to be the case. From evidence based on the distribution of amphibians and reptiles as well as birds (Stuart, 1983) it now seems clear that the Uzungwa Mountains as well as others in the south must be included in any postulated refugia. Certain plateau forests, such as Rondo and Kiono, may also have retained sufficient moist forests to permit the survival of some forest-dependent species or to have been colonised by these later.

Laurent (1973) notes that recurrent phases of contraction and expansion of forest provided opportunities for geographic isolation which may explain the multiplicity of forest species in both South America and Africa. Poynton (1982) emphasised this with regard to hybridisation in southern Africa. Laurent indicates that in West Africa six regions were isolated: a Guinea block from Sierra Leone to the Ivory Coast; a Gold Coast block from the Ivory Coast to Ghana; a Nigeria block from southern Dahomey to the Cross River; a Cameroun block from southeastern Nigeria to the Congo estuary, with eastern limits poorly defined; a northern and eastern Congo block; and a southwestern Congo block between the left bank of the Congo River and the gallery forest of the left affluents of the Kasai and Sankuru rivers. In South America he notes that similar 'core areas' have been proposed: a Guiana block, a coastal Venezuelan block, an Andean block, with Colombian Peruvian and Equadorian subdivisions, and a Mato Grosso block.

Endler (1982) and Livingstone (1982) have, however, expressed doubt over the existence of postulated Pleistocene forest refuges, and discuss in detail the problems inherent in defining past refuges using modern distribution patterns. The former author gives specific suggestions for future work and notes the need to examine one or two pairs of postulated refuges in great detail to test the predictions of various models attempting to explain species distributions.

A number of fascinating biological problems related to the origin of the herpetofauna of eastern Africa are highlighted by the distributions of amphibians and reptiles and suggest areas where more research is needed. For example, it is not at all clear exactly what features of the habitat and/or microclimate determine both the broad and local distribution of species. As already noted, there appear to be differences between the forests of West and eastern Africa which make generalisations about species' preference difficult. What in West Africa may be considered a montane species, such as *Typhlops gierrai*, may be found in much lower forests in eastern Africa. Is this related to characteristics of the forest – or to differences of various populations of species in their tolerance to different ecological conditions? The use of habitat ordination techniques such as those employed by Stuart (1983) on birds and Strijbosch (1980) on anurans may help determine answers to these questions.

Kingdon (1971) has emphasised the importance of local environmental conditions and their effects on local populations of mammals, especially with regard to reproduction. Brosset (1968), for example, found that in the forests of Gabon, two populations of the insectivorous bat *Hipposideros caffer* only 6 km apart had completely different reproductive cycles. Broadley (1979) has demonstrated differences in the timing of reproduction in sympatric lizards in dry woodland, but I know of no similar studies of forest species of amphibians or reptiles in eastern Africa. The possible adaptive significance of the extremes of polymorphism seen in treefrogs of the genus *Hyperolius* remains unknown (Muze, 1976). Elsewhere, polymorphism in some anurans has been shown to be related to predation (Caldwell, 1982). Dodd (1981) notes that differences in infrared light reflectance in amphibians and reptiles are of potential importance in thermoregulation mechanisms and/or crypsis, and reported differences among five species of chameleons, but admits his suggestions are speculative. Detailed ecophysiological studies may prove especially valuable in determining the evolutionary history of forest species. For example, Lin & Nelson (1981), in a detailed study of *Chamaeleo jacksoni* and *C. hoehnelii*, concluded on the basis of differences in their reproductive strategies, desiccation tolerance, critical thermal maxima and close proximity of the study site to two different environmental zones, that *C. hoehnelii* was originally restricted to the montane forest belt on Mount Kenya and the Aberdares where low temperature and high humidity are conditions common throughout much of the year. *C. jacksoni*, on the other hand, is adapted to the hotter, drier, more seasonal climate found at lower altitudes.

It is difficult to assess the role of non-forest species which are able to live along paths, roads, and in open areas, both man-made and natural in forests, on forest-dependent faunas. For example, I have regularly found the toad *Bufo gutturalis*, a non-forest form, on roads along forest on Mount

Bondwa in the Ulugurus at altitudes up to 1500 m. In natural clearings in coastal forest, other non-forest amphibians also occur. The role of these species in colonisation and possible competition with forest forms remains unassessed.

Schiøtz (1967, 1975) has called attention to what he termed a 'farm-bush' anuran fauna. He notes that in many parts of Africa, the forest cover has been removed or partially removed, and natural vegetation has been replaced with a mosaic of cultivation and isolated trees and/or patches of forest. Sanderson (1936), in a discussion of habitat preference of amphibians in Nigeria and Cameroun, also noted that non-forest species may be found in clearings around villages surrounded by forest. Schiøtz (1967) suggested that at least for the treefrogs, a possible explanation may be that post-breeding dispersal in the dry season would account for the occupation of new areas. He notes that this process might take some years, and that Sanderson's observations were made 20 years after the initial clearing of a forest for a village. Schiøtz's own observations were made much more recently after clearing had taken place. He did not give precise figures and noted that the non-forest species found were still relatively few.

Likewise, when forest cover is reduced by large mammals (see Kortlandt, 1984 for a review), by natural die-off or by old trees which fall, open up areas within the forest and change the character of the vegetation, how do various forest-dependent species persist? At least some species are able to survive in the forest edge in recently cleared areas, and some even appear to remain as long as adequate cover of some dense vegetation persists, such as *Bufo brauni* (personal observations), *Naja melanoleuca* and *Bitis gabonica* (Ionides & Pitman, 1965). But Stewart & Pough (1983) have shown that for a tropical forest frog the number of nest and retreat sites may be important factors in the regulation of populations, and it is unlikely that such species would be able to retain their population levels once forest had been cleared or the character of the vegetation drastically altered.

The effect of man on the eastern forests and associated herpetofauna

When considering the past and present distribution of animal species in eastern Africa, it is necessary to understand geological, climatic and evolutionary changes which have occurred on a vast scale. The effect of man on the present distribution of plants and animals is often overlooked except in the most obvious sense that he has been responsible for much recent forest reduction that has often completely destroyed forest communities. In eastern Africa there are no exact data available on forest amphibian and reptile species which have become extinct in historical times; mention has already been made of the frog *Arthroleptides dutoiti* on Mount Elgon, populations of which appear to have been greatly reduced, probably owing to logging operations. The species has not been collected since its description in 1935, despite several attempts to obtain more specimens of it. In southern Africa, Poynton (1986) has discussed man-made ecological changes which appear to have virtually eliminated a non-forest frog.

Evidence exists, however, that man may have been present in at least some of the forests of eastern Africa thousands of years ago (Hamilton, Taylor & Vogel, 1986; Hamilton & Bensted-Smith, 1989) and that his activities such as clearing and burning the forest may have had long-term effects on the distribution of vegetation. I have suggested above that clearings in forest may be colonised by non-forest species, possibly by means of post-breeding dispersal of juveniles from outside the forest.

Another direct effect that man may have had in the past and has at present is the creation of conditions which favour hybridisation of populations that under natural conditions would remain ecologically isolated. Chapin (1948), for example, documented hybridisation in populations of the Paradise flycatchers *Terpsiphone rufiventer* and *T. viridis* (Aves: Passeriformes: Muscicapidae) resulting from forest clearing in eastern Africa. Hybridisation in anurans after environmental disturbance is well documented in North America (Blair, 1941; Jones, 1973; Feder, 1979; Gerhardt,

Guttman & Karlin, 1980). Given the rapid defor-estation occurring in eastern Africa, it would not be surprising to find hybrid swarms occurring, although as yet none has been reported.

Concern over forest destruction in eastern Africa is not new; both German and British col-onial administrations established forest reserves. Champion (1933) and Loveridge (1933) expres-sed concern over deforestation and soil erosion in eastern Africa's forests. Loss of forest, soil ero-sion, reduction in stream flow and poor agri-cultural productivity are still major problems in the region today and have increased with the burgeoning human population. In the Usambaras, the only forests for which detailed information is available, some natural forest cover has been reduced by 50% from 1968 up to now (Rodgers & Homewood, 1982). Hamilton & Bensted-Smith (1989) have reviewed the history of govern-ment forest policy in Tanzania and have des-cribed the ecological effects of both small and large-scale logging on the Usambara forests.

Little heavy industry is present in the vicinity of the eastern African forests, but an exception is a large paper pulp processing plant in the Uzungwa mountains near Mufindi. The polluting potential of the paper industry is well known. Amphibians which rely on streams and rivers for their breed-ing sites are potentially at risk if water quality is altered, and there has also been concern expres-sed that airborne pollutants may have severe negative effects on scarp forest above the factory (Christiansson & Åshvuud, 1985).

While the main threat to the herpetofauna of the Eastern Arc forests is forest destruction, the past few years have seen the development of com-mercial collecting of live amphibians and reptiles for the pet trade in Europe. Endemics such as certain species of chameleons from the Uzungwas and Usambaras have been exported and reports indicate that many of these animals die during transport. Some of the species collected are notoriously difficult to maintain in captivity and it is likely that almost all die after only a short time. The Tanzania Wildlife Division has allowed such uncontrolled exploitation to take place. In no way can this be considered wise use of a resource, and I suggest that no commercial exploitation for the

pet trade in live animals be permitted from the species of the Eastern Arc forests.

Further research

A sound knowledge of the inter- and intraspecific relationships of these faunas is, of course, necess-ary in order to assess the significance of much needed physiological and ecological studies on the forest amphibians and reptiles of eastern Africa. As previously noted, the herpetofaunas of many of the eastern forests have yet to be studied in any detail, and herpetologists have only just begun to study mountain ranges such as the Rubeho and the Uzungwas. But in addition to the general surveys utilising new techniques of capture (Vogt, 1987) needed in all the forests, there is also a need to apply biochemical techniques to study evolutionary relationships. Such methods as microcomplement fixation and the study of mitochondrial and ribosomal DNA sequences (Guttman, 1985) would allow refined insight into the ages and relationships of each herpetofaunal assemblage relative to the others within the Eastern Arc system.

Short (1984) commented on the research needed for studies on birds in tropical forests. I believe his observations apply generally to all animal groups and have modified them only slightly. Research priorities in tropical forests should include: (i) identifying and characterising threatened habitats and their faunas, (ii) success-ful establishment of sites where long-term studies can be carried out, (iii) immediate identification of areas that are doomed and facilitation of observa-tions and salvage collecting at such sites, (iv) the mapping of species' ranges, (v) determination of densities of animals studied in diverse habitats and (vi) monitoring of changes in populations.

Very much associated with the suggested studies is the need to train and provide conditions favourable to research for local, indigenous researchers. While some short-term survey work can be done by visiting biologists, the need for long-term, continuing work is also great. The financial realities of eastern Africa are such that without international cooperation and financial and academic assistance, one cannot expect the much-needed research to be carried out. Short (1984) has clearly stated the problems involved in ornithology, and a parallel situation exists for other disciplines. While there are large numbers of both professional and amateur herpetologists in Europe and North America with the most sophisticated equipment at their disposal, these areas have relatively low diversities of amphibians and reptiles. The forests of eastern Africa contain an incredibly rich herpetofauna, yet there are meagre resources and almost no trained indigenous specialists able to devote themselves to its study and conservation.

Indigenous herpetologists are needed not only to carry out long-term scientific studies, but also to use their knowledge of diverse local languages and customs to initiate educational and conservation efforts effective within the prevailing sociological, political, and cultural environments. For example, in Tanzania, certain tribal groups have strong cultural beliefs and practices concerning reptiles. It is unlikely that someone from outside such a group would have access to its ethnoherpetological data, and even less likely that he would be able to apply them to conservation efforts.

Until steps are taken to ensure the preservation of the natural forests of eastern Africa and until international academic and financial cooperation is forthcoming to assist in the research and training of indigenous African biologists, the future of the unique Eastern Arc herpetofauna must be regarded as insecure.

Acknowledgements
The following kindly provided information on specimens in their care: D. G. Broadley, B. Clarke, R. C. Drewes, A. G. C. Grandison, R. Nussbaum, and A. F. Stimson.

Many people have helped me in the field and by collecting in areas I was unable to visit personally. I would especially like to thank Charles A. Msuya who accompanied me to many forests, often under difficult conditions; without his technical skill as well as support, fieldwork would not have been possible. W. A. Rodgers arranged my first visit to the Uzungwas and S. N. Stuart offered field support and scientific skills in the Usambaras. Mr and Mrs J. Tanner of Mazumbai welcomed me, as they did all visitors, with unfailing hospitality. J. C. and J. Lovett not only provided transport for many trips but made me welcome in their camps and house. Ms E. Boswell and T. C. E. Congdon as well as many others at Mufindi provided support, expertise and specimens. Special thanks go to Mrs Helga Voigt for obtaining the rare vipers of the Uzungwas. A. Braunlich, D. Emmrich, C. Fox, Ms J. Beakbane, the late T. Grant, H. Grossman, P. Hardison, Ms V. Bound, S. Telford and L. Kisoma all provided specimens. Jan Kielland, lepidopterist extraordinaire of Tanzania, provided numerous specimens from remote forests obtained under the most arduous of conditions. N. Baker not only provided specimens but offered unflagging logistical and other forms of support. I gratefully thank all others who contributed specimens, especially my collegues and students of the Department of Zoology and Marine Biology, University of Dar es Salaam.

D. G. Broadley, J. Cadle, B. Clarke, and R. C. Drewes, and J. C. Poynton all provided useful comments and criticism on earlier drafts of this manuscript, and I especially thank D. G. Broadley for sharing his knowledge and experience of the reptiles.

References

BARBOUR, T. & LOVERIDGE, A. (1928). A comparative study of the herpetofauna of the Uluguru and Usambara Mountains, Tanganyika Territory with description of new species. *Memoirs of the Museum of Comparative Zoology at Harvard College* 50, 87–265.

BEENTJE, H. J. (ed.) (1988). An ecological and floristic study of the forests of the Taita Hills, Kenya. *Utafiti (Occasional Papers of the National Museums of Kenya)* 1, 23–66.

BEENTJE, H. J., NDIANGUI, N. & MUTANGAH, J. (1987). Forest islands in the mist. *Swara* 10, 20–1.

BLAIR, A. (1941). Variation, isolating mechanisms and hybridization in certain toads. *Genetics* 26, 398–417.

BOHME, W. (1982). Ein neues Chamaleon aus Tanzania, mit Bemerkungen über Mimese bei

Echsen (Reptilia: Sauria). *Bonner Zoologische Beiträge* **33**, 349–61.

BOHME, W. (1987). Selten Reptilien aus dem Uzungwe-Gebirge, Sud Tanzania. *Herpetofauna* **9**(48), 27–34.

BRANCH, W. R. (1988). *A Field Guide to the Snakes and Other Reptiles of Southern Africa.* Capetown: Struik.

BROADLEY, D. G. (1966). A review of the genus *Natriciteres* Loveridge (Serpentes: Colubridae). *Arnoldia Rhodesia* **9**, I–II.

BROADLEY, D. G. (1968). A review of the African cobras of the genus *Naja*, (Serpentes: Elapidae). *Arnoldia Rhodesia* **3**(29), 1–44.

BROADLEY, D. G. (1971). A revision of the African snake genus *Elapsoidea* Bocage (Elapidae). *Occasional Papers of the National Museums of Rhodesia, Ser. B* **4**(32), 577–626.

BROADLEY, D. G. (1979). A field study of two sympatric 'annual' lizards (genus *Ichnotropis*) in Rhodesia. *South African Journal of Zoology* **14**, 133–8.

BROADLEY, D. G. (1980). A revision of the African snake genus *Prosymna* Gray (Colubridae). *Occasional Papers of the National Museums of Rhodesia, Ser. B* **6**(7), 481–556.

BROADLEY, D. G. (1989). A reappraisal of the genus *Panaspis* Cope, with the description of a new species of *Leptosiaphos* (Reptilia: Scincidae) from Tanzania. *Arnoldia Zimbabwe* **9**(32), 439–49.

BROADLEY, D. G. (1990). The herpetofaunas of the islands off the coast of south Mozambique. *Arnoldia Zimbabwe* **9**(35), 469–93.

BROADLEY, D. G. & BLAKE, D. K. (1979). A field study of *Rhampholeon marshalli marshalli* on Vumba mountain, Rhodesia (Sauria: Chamaeleonidae). *Arnoldia Rhodesia* **8**(34), 1–6.

BROADLEY, D. G. & COCK, E. V. (1975). *Snakes of Zimbabwe.* Harare: Longman Zimbabwe.

BROADLEY, D. G. & HOWELL, K. M. (1991). A checklist of the reptiles of Tanzania, with synoptic keys. *Syntarsus* **1**, 1–70.

BROADLEY, D. G. & PARKER, R. H. (1976). Natural hybridization between the Puffadder and Gaboon Viper in Zululand. *Durban Museum Novitates* **11**, 77–83.

BROSSET, A. (1968). La permutation du cycle sexuel saisonnier chez le chiroptère *Hipposideros caffer*, au voisinage de l'équateur. *Biologia Gabonica* **4**, 325–41.

BRYGOO, E. R. & ROUX-ESTEVE, R. I. (1981). Un genre de lézards scincines d'Afrique: *Melanoseps.*

Bulletin du Musee National d'Histoire Naturelle (Paris), Sér. 4 **3**, 911–30.

CALDWELL, J. P. (1982). Disruptive selection: a tail color polymorphism in *Acris* tadpoles in response to differential predation. *Canadian Journal of Zoology* **60**, 2818–27.

CHAMPION, A. M. (1933). Soil erosion in Africa. *Geographical Journal* **82**, 130–9.

CHAPIN, J. (1948). Variation and hybridization among the paradise flycatchers of Africa. *Evolution* **2**, 111–26.

CHAPMAN, J. D. & WHITE, F. (1970). *The Evergreen Forests of Malawi.* Oxford: Commonwealth Forestry Institute.

CHRISTIANSSON, C. & ÅSHVUUD, J. (1985). Heavy industry in a rural tropical ecosystem. *Ambio* **14**, 123–33.

CLARKE, B. T. (1988a). The amphibian fauna of the East African rain forests including the description of a new species of a toad, genus *Nectophrynoides* Noble (Anura: Bufonidae). *Tropical Zoology* **1**, 169–77.

CLARKE, B. T. (1988b). Real vs apparent distributions of dwarf amphibians: *Bufo lindneri* Mertens 1955 (Anura: Bufonidae) – a case in point. *Amphibia–Reptilia* **10**, 297–306.

COE, M. J. (1967). *The Ecology of the Alpine Zone of Mount Kenya.* The Hague: Junk.

CRIBB, P. J. & LEEDAL, G. P. (1982). *A Field Guide to the Common Mountain Flowers of Southern Tanzania.* Rotterdam: Balkema.

DIAMOND, A. W. & HAMILTON, A. C. (1980). The distribution of forest passerine birds and Quaternary climatic change in Africa. *Journal of Zoology (London)* **191**, 379–402.

DODD, C. K., JR (1981). Infrared reflectance in chameleons (Chamaeleonidae) from Kenya. *Biotropica* **13**, 161–4.

DREWES R. C. (1984). A phylogenetic analysis of the Hyperoliidae (Anvra): treefrogs of Africa, Madagascar and the Seychelles Islands. *Occasional Papers of the California Academy of Sciences* **139**, 1–70.

DREWES, R. C., HILLMAN, S. S., PUTHAM, R. W. & SOKOL, O. M. (1977). Water, nitrogen and ion balance in the desert treefrog *Chiromantis petersi* Boulenger (Anura: Rhacophoridae) with comments on the structure of the integument. *Journal of Comparative Physiology* **116**, 257–67.

ENDLER, J. A. (1982). Pleistocene forest refuges: fact or fancy? In *Biological Diversification in the Tropics*, ed. G. T. Prance, pp. 641–57. New York: Columbia University Press.

FEDER, J. H. (1979). Natural hybridization and genetic divergence between the toads *Bufo boreas* and *Bufo punctatus*. *Evolution* **33**, 1089–97.

FROST, D. (ed.) (1984). *Amphibian Species of the World*. Lawrence, KA: Association of Systematics Collections.

GERHARDT, H. C., GUTTMAN, S. & KARLIN, A. A. (1980). Natural hybrid between *Hyla cinerea* and *Hyla gratiosa*: Morphology, vocalization and electrophoretic analysis. *Copeia* 1980 No. 4, pp. 577–89.

GRANDISON, A. G. C. (1972). The status and relationship of some East African earless toads (Anura: Bufonidae) with description of a new species. *Zoologische Mededeelingen* **46**, 30–47.

GRANDISON, A. G. C. (1978). The occurrence of *Nectophrynoides* (Anura: Bufonidae) in Ethiopia. A new concept of the genus with a description of a new species. *Monitore zoologico italiano N.S.* supplemento XI, No. 6, 119–72.

GRANDISON, A. G. C. (1983). A new species of *Arthroleptis* (Anura: Ranidae) from the West Usambara mountains, Tanzania. *Bulletin of the British Museum (Natural History), Zoology* **45**, 77–84.

GRANDISON, A. G. C. & HOWELL, K. M. (1983). A new forest species of *Phrynobatrachus* (Anura: Ranidae) from Morogoro Region, Tanzania. *Amphibia–Reptilia* **4**, 117–24.

GUTTMAN, S. J. (1985). Biochemical studies of anuran evolution. *Copeia* 1985, 292–309.

HAMILTON, A. C. (1976). The significance of patterns of distribution shown by forest plants and animals in tropical Africa for the reconstruction of Upper Pleistocene palaeoenvironments: a review. *Palaeoecology of Africa, the Surrounding Islands and Antarctica* **9**, 63–97.

HAMILTON, A. C. & BENSTED-SMITH, P. (ed.) (1989). *Forest Conservation in the East Usambara Mountains, Tanzania*. Gland, Switzerland: International Union for Conservation of Nature and Natural Resources.

HAMILTON, A. C., TAYLOR, D. & VOGEL, J. C. (1986). Early forest clearance and environmental degradation in south-west Uganda. *Nature* **320**, 164–7.

HEBRARD, J. J., REILLY, S. M. & GUPPY, M. (1982). Thermal ecology of *Chamaeleo hoehnolii* and *Mabuya varia* in the Aberdare Mountains: constraints on heterothermy in an alpine habitat. *Journal of the East Africa Natural History Society and National Museum* **176**, 1–6.

HEDBERG, O. (1951). Vegetation belts of the East Africa Mountains. *Svensk Botanisk Tidskrift* **45**, 140–202.

HOWELL, K. M. (1981). Pugu Forest Reserve: biological values and development. *African Journal of Ecology* **19**, 73–82.

HUGHES, B. (1983). African snake faunas. *Bonner Zoologische Beiträge* **34**, 311–56.

IONIDES, C. J. P. & PITMAN, C. R. S. (1965). Notes on three east African venomous snake populations. *Puku* No. 3, 87–95.

JONES, J. M. (1973). Effects of thirty years of hybridization on the toads *Bufo americanus* and *Bufo woodhousii fowleri* at Bloomington, Indiana. *Evolution* **27**, 435–48.

KINGDON, J. (1971). *East African Mammals: An Atlas of Evolution in Africa*, Vol. I. London: Academic Press.

KLAVER, C. & BOHME, W. (1988). Systematics of *Bradypodion tenue* (Matschie, 1892) (Sauria: Chamaeleonidae) with a description of a new species from the Uluguru and Uzungwa Mountains, Tanzania. *Bonner Zoologische Beiträge* **39**, 381–93.

KORTLANDT, A. (1984). Vegetation research and the 'bulldozer' herbivores of tropical Africa. In *Tropical Rainforest*, ed. A. C. Chadwick and C. L. Sutton, pp. 205–26. Special Publication of Leeds Philosophical and Literary Society.

KREULEN, D. (1979). Factors affecting reptile biomass in African grasslands. *American Naturalist* **114**, 157–65.

LAURENT, R. F. (1973). A parallel survey of equatorial amphibians in Africa and South America. In *Tropical Forest Ecosystems in Africa and South America: a Comparative Review*, ed. B. J. Meggers, E. S. Ayensu and W. D. Duckworth, pp. 259–66. Washington, DC: Smithsonian Institution Press.

LIN, J.-Y. & NELSON, C. E. (1981). Comparative reproductive biology of two sympatric tropical lizards *Chamaeleo jacksoni* Boulenger and *Chamaeleo hoehnelii* Steindachner (Sauria: Chamaeleonidae). *Amphibia–Reptilia* **3/4**, 287–311.

LIVINGSTONE, D. A. (1982). Quaternary geography of Africa and the Refuge Theory. In *Biological Diversification in the Tropics*, ed. G. T. Prance, pp. 523–36. New York: Colombia University Press.

LOVERIDGE, A. (1932a). New reptiles and amphibians from Tanganyika Territory and Kenya Colony. *Bulletin of the Museum of Comparative Zoology at Harvard* **72**, 375–87.

LOVERIDGE, A. (1932b). *Many Happy Days I've Squandered*. New York: Harper & Brothers.

LOVERIDGE, A. (1933). Reports on the scientific results of an expedition to the southwestern highlands of Tanganyika Territory. I. Introduction and zoogeography. *Bulletin of the Museum of Comparative Zoology at Harvard* 75, 1–43.

LOVERIDGE, A. (1935). Scientific results of an expedition to rain forest regions in eastern Africa. I. New reptiles and amphibians from East Africa. *Bulletin of the Museum of Comparative Zoology at Harvard* 79, 3–19.

LOVERIDGE, A. (1937). Scientific results of an expedition to rain forest regions in eastern Africa. IX. Zoogeography and itinerary. *Bulletin of the Museum of Comparative Zoology at Harvard* 79, 481–541.

LOVERIDGE, A. (1942a). Scientific results of a fourth expedition to forested areas in East and Central Africa. IV. Reptiles. *Bulletin of the Museum of Comparative Zoology at Harvard* 91, 237–373.

LOVERIDGE, A. (1942b). Scientific results of a fourth expedition to forested areas in East and Central Africa. V. Amphibians. *Bulletin of the Museum of Comparative Zoology at Harvard* 91, 377–436.

LOVERIDGE, A. (1947). *Tomorrow's a Holiday*. New York: Harper & Brothers.

LOVERIDGE, A. (1952). Grant No. 914 (1946). Ecological studies on the vanishing vertebrate fauna of rain forest remnants in tropical East Africa. *Year-book of the American Philosophical Society*, pp. 164–5.

LOVERIDGE, A. (1953a). *I Drank the Zambezi*. New York: Harper & Brothers.

LOVERIDGE, A. (1953b). Zoological results of a fifth expedition to East Africa. III. Reptiles from Nyasaland and Tete. *Bulletin of the Museum of Comparative Zoology at Harvard* 110, 143–322.

LOVERIDGE, A. (1956). *Forest Safari*. London: Lutterworth Press.

LOVERIDGE, A. (1957). Checklist of the reptiles and amphibians of East Africa (Uganda, Kenya, Tanzania, Zanzibar). *Bulletin of the Museum of Comparative Zoology at Harvard* 117, 153–362.

MACARTHUR, R. H. & WILSON, E. O. (1967). *The Theory of Island Biogeography*. Monographs of Population Biology. Princeton, NJ: Princeton University Press.

MEYLAN, P. (1983). Lizards and snakes of Olduvai Gorge, Kenya [*sic*]. *Anthroquest* 26, 12.

MOODY, S. M. (1980). Phylogenetic and historical biogeographical relationships of the genera in the family Agamidae (Reptilia: Lacertilia). PhD thesis, University of Michigan.

MOREAU, R. E. (1966). *The Bird Faunas of Africa and its Islands*. London: Academic Press.

MOREAU, R. E. & PAKENHAM, R. W. (1941). The land vertebrates of Pemba, Zanzibar and Mafia: a zoogeographical study. *Proceedings of the Zoological Society of London, Ser. A* 110, 97–128.

MUZE, E. S. (1976). Studies on the sex ratio and polymorphism in *Hyperolius puncticulatus* (Rhacophoridae) at Amani, Tanzania. MSc thesis, University of Dar es Salaam.

NUSSBAUM, R. (1985). Systematics of caecilians (Amphibia: Gymnophiona) of the family Scolecomorphidae. *Occasional Papers of the Museum of Zoology, University of Michigan* 713, 1–49.

NYAMWERU, C. (1980). *Rifts and Volcanoes*. Nairobi: Nelson Africa.

PAKENHAM, R. H. W. (1983). The reptiles and amphibians of Zanzibar and Pemba islands, with a note on the freshwater fishes. *Journal of the East Africa Natural History Society and National Museum* 171, 1–40.

PARKINSON, J. (1930). *The Dinosaur in East Africa*. London: H. F. & G. Witherby.

PASSMORE, N. I. & CARRUTHERS, V. C. (1979). *South African Frogs*. Johannesburg: Witswatersrand University Press.

PASTEUR, G. & BROADLEY, D. G. (1988). A remote, insular species of the *Lygodactylus somalicus* superspecies (Sauria: Gekkonidae). *Amphibia–Reptilia* 9, 237–44.

PERRET, J.-L. (1971). Les espèces du genre *Nectophrynoides* d'Afrique (Batriciens, Bufonides) *Annales de la Faculté des Sciences du Cameroun* 6, 99–109.

PERRET, J.-L. (1986). Révision des espèces africaine du genre *Cnemaspis* Strauch, sous-genre *Ancylodactylus* Muller (Lacertilia, Gekkonidae) avec la description de quatre espèces nouvelle. *Revue suisse de Zoologie* 93, 457–505.

POLHILL, R. (1968). Tanzania. In *Conservation of Vegetation in Africa South of the Sahara*, ed. I. Hedberg and O. Hedberg, pp. 166–78. Acta Phytogeographica Suecica, Vol. 54.

POYNTON, J. C. (1964). The Amphibia of Southern Africa. *Annals of the Natal Museum* 17, 1–334.

POYNTON, J. C. (1982). On species pairs among southern African amphibians. *South African Journal of Zoology* 17, 67–74.

POYNTON, J. C. (1986). *Hyperolius argus* (Anura) in

Natal: taxonomy, biogeography and conservation. *South African Journal of Zoology* 18, 149–52.

POYNTON, J. C. & BRADLEY, D. G. (1967). A new species of *Probreviceps* (Amphibia) from Rhodesia. *Arnoldia* 3(14), 1–3.

POYNTON, J. C. & BRADLEY, D. G. (1985a). Amphibia Zambesiaca 1. Scolecomorphidae, Pipidae, Microhylidae, Hemisidae, Arthroleptidae. *Annals of the Natal Museum* 26, 503–53.

POYNTON, J. C. & BRADLEY, D. G. (1985b). Amphibia Zambesiaca 2. Ranidae. *Annals of the Natal Museum* 27, 115–81.

POYNTON, J. C. & BRADLEY, D. G. (1987). Amphibia Zambesiaca 3. Rhacophoridae and Hyperoliidae. *Annals of the Natal Museum* 28, 161–229.

POYNTON, J. C. & BRADLEY, D. G. (1988). Amphibia Zambesiaca 4. Bufonidae. *Annals of the Natal Museum* 29, 447–90.

RAGE, J. C. (1973). Fossil snakes from Olduvai, Tanzania. *Fossil Vertebrates of Africa* 3, 1–7.

RAGE, J. C. (1979). Les serpents de la Rift Valley – un aperçu général. *Bulletin de la Société Géologique de France* 21, 329–30.

RASMUSSEN, J. B. (1981). The snakes from the rain forest of the Usambara Mountains, Tanzania: a checklist and key. *Salamandra* 17, 173–88.

RASMUSSEN, J. B. (1986). On the taxonomic status of *Dipsadoboa werneri* (Boulenger), *D. shrevi* (Loveridge), and *Crotaphopeltis hotomboeia kageleri* Uthmoller (Boiginae, Serpentes). *Amphibia–Reptilia* 7, 51–73.

RASMUSSEN, J. B. & HOWELL, K. M. (1982). The current status of the rare Usambara Mountain Forest Viper, *Atheris ceratophorus* Werner, 1895, including a probable new record of *A. nitschei rungweensis* Bogert, 1940, and a discussion of its validity (Reptilia, Serpentes, Viperidae). *Amphibia–Reptilia* 3, 269–77.

REILLY, S. M. (1982). Ecological notes on *Chamaeleo schubotzi* from Mount Kenya. *Journal of the Herpetological Association* 28, 1–3.

RODGERS, W. A. & HOMEWOOD, K. M. (1982). Species richness and endemism in the Usambara mountain forests, Tanzania. *Biological Journal of the Linnean Society* 18, 197–242.

RODGERS, W. A., OWEN, C. F. & HOMEWOOD, K. M. (1982). Biogeography of East African forest mammals. *Journal of Biogeography* 9, 41–54.

SAMPSON, D. N. (1965). The geology, volcanology and glaciology of Kilimanjaro. *Tanganyika Notes and Records* 64, 118–24.

SANDERSON, I. T. (1936). The amphibians of the Mamfe Division, Cameroon. Ecology of the frogs. *Proceedings of the Zoological Society of London* 1936, 165–208.

SCHIØTZ, A. (1967). The treefrogs (Rhacophoridae) of West Africa. *Spolia Zoologica Musei Hauniensis* 25, 1–346.

SCHIØTZ, A. (1975). *The Treefrogs of Eastern Africa.* Copenhagen: Steenstrupia.

SCHIØTZ, A. (1981). The Amphibia in the forested basement hills of Tanzania: a biogeographical indicator group. *African Journal of Ecology* 19, 205–7.

SCHIØTZ, A. (1982). Two new *Hyperolius* (Anura) from Tanzania. *Steenstrupia* 8, 269–76.

SHORT, L. (1984). Priorities in ornithology: the urgent need for tropical research and researchers. *Auk* 101, 892–3.

SPAWLS, S. (1978). A checklist of the snakes of Kenya. *Journal of the East Africa Natural History Society and National Museum* 31(67), 1–18.

STEWART, M. (1967). *Amphibians of Malawi.* Albany: State University of New York Press.

STEWART, M. & POUGH, F. H. (1983). Population density of tropical forest frogs: relation to retreat sites. *Science* 221, 570–2.

STRIJBOSCH, H. (1980). Habitat selection by amphibians during their terrestrial phase. *British Journal of Herpetology* 6, 93–8.

STUART, S. N. (1983). Biogeographic and ecological aspects of forest bird communities in eastern Tanzania. PhD thesis, University of Cambridge.

TANDY, M. & KEITH, R. (1972). *Bufo* in Africa. In *Evolution in the genus* Bufo, ed. W. F. Blair, pp. 119–70. Austin: University of Texas Press.

TANGANYIKA SOCIETY (1965). Kilimanjaro. *Tanganyika Notes and Records* 64.

TIHEN, J. A. (1972). The fossil record. In *Evolution in the genus* Bufo, ed. W. F. Blair, pp. 8–13. Austin: University of Texas Press.

TOFT, C. A. (1985). Resource partitioning in amphibians and reptiles. *Copeia* 1985, 1–21.

UTHMOLLER, W. (1937). Beitrag zur Kenntnis der Schlangenfauna des Kilimandscharo (Tanganyika Territory, ehemaliges Deutsch-Ostafrika). *Temminckia* 11, 97–134.

UTHMOLLER, W. (1941a). Beitrag zur Kenntnis der Schlangenfauna Nordost-Ostafrikas (Tanganyika Territory). Ergebnisse der Ostafrika-Expedition Uthmoller–Bohmann. 5A Schlangen I. *Zoologischer Anzeiger* 135, 225–42.

UTHMOLLER, W. (1941b). Beitrag zur Kenntnis der

Schlangenfauna Nordost-Ostafrikas (Tanganyika Territory, ehem. Deutsch-Ostafrika). Ergebnisse der Ostafrika-Expedition Uthmoller–Bohmann. 7A Schlangen II. Teil. *Zoologischer Anzeiger* **136**, 193–206.

UTHMOLLER, W. (1942). Die Schlangen Ostafrikas in ihrem Lebenstraum . . . der Vulkane Kilimandscharo, Meru, Hanang, der Großen Ostafrikanischen Bruchstufe und des Usambaragebirges. *Archiv für Naturgeschichte Leipzig (NF)* **10**, 1–70.

VAN DIJK, D. E. (1985). An addition to the fossil Anura of Southern Africa. *South African Journal of Science* **81**, 207–8.

VOGT, R. C. (1987). Techniques: you can set drift fences in the canopy. *Herpetological Review* **18**, 13–14.

WASSERBURG, R. J. & HEYER, R. (1983). Morphological correlates of subaerial existence in leptodactylid tadpoles associated with flowing water. *Canadian Journal of Zoology* **61**, 761–9.

WEDERKINCH, E. (1982). The lizard fauna of the Usambara Mountains. pp. 237–8. In Rodgers, W. A. & Homewood, K. M., Species richness and endemism in the Usambara mountain forests, Tanzania. *Biological Journal of the Linnean Society* **18**, 197–242.

WESTERN, D. (1974). The distribution, density and biomass density of lizards in a semi-arid environment of northern Kenya. *East African Wildlife Journal* **12**, 49–62.

WITHERS, P. C., HILLMAN, S. S., DREWES, R. C. & SOKOL, O. M. (1982). Water loss and nitrogen excretion in sharp-nosed reed frogs (*Hyperolius nasutus*: Anura, Hyperoliidae). *Journal of Experimental Biology* **97**, 335–43.

10 The zoogeography of the montane forest avifauna of eastern Tanzania

S. N. STUART, F. P. JENSEN, S. BRØGGER-JENSEN AND R. I. MILLER

Introduction

The evolution and zoogeography of the montane forest avifaunas of Africa have long been a source of fascination to ornithologists (e.g. Chapin, 1923, 1932; Moreau, 1933, 1952, 1954, 1963, 1966; Dowsett, 1971, 1980a,b; Hamilton, 1976; Diamond & Hamilton, 1980; Stuart, 1981a, 1983, 1986; Jensen & Stuart, 1985). The montane forests are characterised by many altitudinally restricted species which usually have disjunct distributions reflecting the patchiness of the available habitat. Many lowland species occur in Africa's montane forests, but do not necessarily do so as isolated populations; these species are not considered here. For the purposes of this chapter we have defined montane forest species as those which do not normally occur below a certain altitude (in eastern Tanzania, few such species occur below 700–900 m a.s.l. during the breeding season), and which seem to be dependent upon forest for their survival (usually for nesting sites and food). Thus species restricted to forest at intermediate elevations (usually between 700 and 1500 m), such as the Banded Green Sunbird *Anthreptes rubritorques*, are considered here to be montane.

It is not always easy to decide what is a forest species and what is a montane species, since the distinctions are not necessarily hard and fast ones (see Stuart, 1983 for a fuller discussion of this problem). One complication is that many forest-dependent species are able to survive in cleared areas within a few kilometres of forest (Stuart, 1983). This 'overspill' effect is still very poorly understood from an ecological perspective. Seasonal altitudinal movements can make it difficult to decide whether or not a species is 'montane'. Certain species are montane only through part of their ranges. Examples include the Barred Long-tailed Cuckoo *Cercococcyx montanus*, Olive Woodpecker *Mesopicos griseocephalus*, Grey Cuckoo-shrike *Coracina caesia*, Pink-footed Puffback *Dryoscopus angolensis*, Lagden's Bush-shrike *Malaconotus lagdeni*, Olive Thrush *Turdus olivaceus*, Rufous-cheeked Ground-robin *Sheppardia bocagei*, Black-throated Apalis *Apalis jacksoni*, White-winged Apalis *A. chariessa*, Brown-headed Apalis *A. alticola* and Dusky Flycatcher *Muscicapa adusta*. In these cases we have included the montane populations in the analysis and discussions in this chapter. Other species can be considered basically lowland but have isolated montane subspecies which are included in this chapter, these being Green Barbet *Stactolaema olivacea* (subsp. *belcheri* included), Elliot's Woodpecker *Mesopicos elliotti* (subspp. *johnstoni* and *kupeensis* included), Tiny Greenbul *Phyllastrephus debilis* (subsp. *albigula* included), Lühder's Bush-shrike *Laniarius luehderi* (subspp. *brauni* and *amboimensis* included), Many-coloured Bush-shrike *Malaconotus multicolor* (subspp. *nigrifrons* and *graueri* included), Masked Apalis *Apalis binotata* (subsp. *personata* included) and Margaret's Batis *Batis margaritae* (nominate subsp. included).

On the basis of the distribution of montane

forest birds, Moreau (1966) recognised seven Montane Forest Groups in Africa.

1. Cameroun. This includes the mountains of western Cameroun, principally those from Mount Cameroun to the Bamenda Highlands, extending into Nigeria at the Obudu and Mambilla plateaux. The mountains on the island of Bioko (formerly Fernando Po), Equatorial Guinea, are also part of the Cameroun Group as defined by Moreau.
2. Angola. This is the smallest of Moreau's Groups, restricted to the highlands around Mount Moco and parts of the narrow escarpment which runs down the western side of the country.
3. East Zaïre. This comprises the montane forests on both sides of the rift valley. This area extends from the Lendu Plateau in Zaïre and Mount Rwenzori on the Uganda/Zaïre border, south through Rwanda and Burundi to Mahale Mountain in western Tanzania and the Marungu Highlands in southeastern Zaïre.
4. Kenya. This Group is centred on the Kenyan Highlands, extending north to the isolated mountains of northern Kenya (i.e. Kulal and Marsabit), northeastern Uganda, and southern Sudan (the Imatong Mountains), and south to the mountains of northern Tanzania (as far as the Mbulu Mountains, the Crater Highlands, Mount Meru and Mount Kilimanjaro).
5. Tanzania/Malawi. This Group is centred on the mountains of eastern Tanzania, extending north to the Pare Mountains, and also to the Taita Hills and Kasigau in southeastern Kenya. The southern part of the Group includes all the mountains of Malawi (including those just across the border in Zambia), and the mountains in Mozambique north of the Zambezi River.
6. South of the Zambezi. This includes the mountains of eastern Zimbabwe and those in Mozambique south of the Zambezi River. It also includes the forests of South Africa, though here, in the more

temperate climate, the isolation of the such areas is much less distinct.
7. Ethiopia. The entire highland region of Ethiopia.

Within each of these Groups, the montane forest avifauna is said to have a similar species composition. However, Moreau does not provide any species lists for these forests, or Groups, and hence a detailed appraisal of the data on which he based his conclusions is not possible. Nevertheless, Moreau's Montane Forest Groups have found general acceptance amongst ornithologists and naturalists who consider the Groups to be generally correct, if not self-evident. Minor adjustments to the Group boundaries have been made by Dowsett (1971), but these are also not substantiated by any species lists or analysis.

This chapter starts by reviewing the montane forest avifauna of Africa as a whole, and then focuses in particular on the forests of eastern Tanzania. Ornithologists have concentrated on the forests of eastern Tanzania because of their large number of rare and endemic species. In particular, attention has been paid to the Usambara Mountains, where R. E. Moreau lived between 1928 and 1946. Moreau made major contributions to African zoogeography, with a particular focus on montane forest avifaunas, culminating in his widely cited book (Moreau, 1966). More recently, the eastern Tanzanian forests have received attention on account of the critical conservation problems which they pose (Stuart, 1983, 1985; Collar & Stuart, 1985, 1988). In this chapter, an overview is presented of the zoogeography of the eastern Tanzanian montane forest avifauna, paying particular attention to its distinctiveness, diversity, levels of endemism and likely origins.

The African montane forest avifauna

Purpose and methods of analysis

In recent years, many more species lists for individual montane forests have been published, thus making it possible to study regional variations in species composition more accurately than

Moreau (1966) was able to do in his analysis. In this chapter we undertake a numerical and statistical analysis of montane forest bird distribution in Africa, based upon newly available data as well as other unpublished records. We then compare the results of our analysis with those obtained by previous workers, notably Moreau (1966) and Diamond & Hamilton (1980).

Fifty forests, or general forest areas, were considered in the analysis. Information on the species composition of the montane avifaunas was compiled from published records (principally Prigogine, 1953, 1960, 1971, 1978, 1984, 1985; Urban & Brown, 1968; Hall & Moreau, 1970; Clancey,

1971; Beesley, 1972; Snow, 1978; Britton, 1980; Dowsett, 1981; Irwin, 1981; Pinto, 1983; Stuart, 1983, 1986; Mann, 1985; Stuart & Jensen, 1985; Hillman, 1986; Stuart et al., 1987). We are also particularly grateful to the late Dr A. Prigogine for supplying data on montane forest bird distribution in the Albertine Rift forests, and to D. C. Moyer for providing his unpublished data from the Ufipa Plateau. For Cameroun and Tanzania, we have also made use of data collected during our own fieldwork. The species lists are probably complete for some forests, and near complete for the remainder. Details of the forests are given in Table 10.1.

Table 10.1. *Montane Forest Groups and bird species richness in 50 African forests*

Forest	Country	Montane Forest Group (following Moreau, 1966)[a]	(following this chapter)[b]	Number of species
Aberdare Mountains	Kenya	KE	CEA	37
Angolan Escarpment	Angola	ANG	None	12
Bale Mountains	Ethiopia	AB	ETH	18
Bioko	Equatorial Guinea	CAM	CAM	32
Crater Highlands	Tanzania	KE	CEA	26
East Uzungwa	Tanzania	TN	EC	42
Imatong	Sudan	KE	CEA	25
Impenetrable Forest	Uganda	ECO	AR	67
Inyanga Highlands	Zimbabwe/Mozambique	SOZ	SOZ	16
Itombwe	Zaïre	ECO	AR	83
Kahuzi	Zaïre	ECO	AR	67
Kakamega/Nandi	Kenya	KE	CEA	35
Kilimanjaro	Tanzania	KE	CEA	31
Lendu Plateau	Zaïre	ECO	AR	37
Mahale Mountain	Tanzania	ECO	None	26
Mau	Kenya	KE	CEA	38
Mount Cameroun	Cameroun	CAM	CAM	42
Mount Elgon	Kenya/Uganda	KE	CEA	42
Mount Gorongosa	Mozambique	SOZ	SOZ	17
Mount Kabobo	Zaïre	ECO	AR	51
Mount Kenya	Kenya	KE	CEA	43
Mount Kupe	Cameroun	CAM	CAM	35
Mount Manenguba	Cameroun	CAM	CAM	29
Mount Meru	Tanzania	KE	CEA	30
Mount Moco	Angola	ANG	None	12
Mount Mulanje	Malawi	TN	SM	21
Mount Oku	Cameroun	CAM	CAM	34
Mount Rungwe	Tanzania	TN	EC	35
Mount Selinda	Zimbabwe	SOZ	SOZ	13

Table 10.1. (*cont.*)

Forest	Country	Montane Forest Group (following Moreau, 1966)[a]	(following this chapter)[b]	Number of species
Mount Thyolo	Malawi	TN	SM	17
Mountains west of L. Edward	Zaïre	ECO	AR	70
Mufindi	Tanzania	TN	EC	34
Nguru	Tanzania	TN	EC	31
Nguruman	Kenya	KE	CEA	22
North Pare	Tanzania	TN	CEA	27
Nyika Plateau	Malawi/Zambia	TN	EC	31
Nyungwe Forest	Rwanda	ECO	AR	60
Obudu Plateau	Nigeria	CAM	CAM	35
Rumpi Hills	Cameroun	CAM	CAM	35
Rwenzori	Uganda/Zaïre	ECO	AR	64
Simien Mountains	Ethiopia	AB	ETH	10
South Pare	Tanzania	TN	CEA	23
Southwest Ethiopia	Ethiopia	AB	ETH	20
Taita Hills	Kenya	TN	None	14
Ufipa Plateau	Tanzania	None	None	16
Uluguru	Tanzania	TN	EC	41
Ukaguru	Tanzania	TN	EC	28
Usambara	Tanzania	TN	EC	41
Virunga Volcanoes	Rwanda/Zaïre/Uganda	ECO	AR	56
Vumba	Zimbabwe	SOZ	SOZ	16

Note: [a]Montane Forest Groups following Moreau (1966): AB, Abyssinia; ANG, Angola; CAM, Cameroun; ECO, East Congo; KE, Kenya; SOZ, South of Zambezi; TN, Tanzania.
[b]Montane Forest Groups following this chapter: AR, Albertine Rift; CAM, Cameroun; CEA, Central East Africa; EC, East Coast Escarpment; ETH, Ethiopia; SM, South Malawi; SOZ, South of Zambezi.

The 50 forests considered in this analysis have a combined total of 163 montane forest species (following the definition of a montane forest species outlined in the Introduction to this chapter). In order to assess the biogeographic affinities of these forests in terms of the species composition of their montane bird communities, a cluster analysis was performed on these data, based on Sorensen's Index (Q_s), which is given by:

$$Q_s = \frac{2j}{a+b}$$

where a = the number of species in forest 1, b = the number of species in forest 2, j = the number of species common to both forests.

The index ranges from zero for two forests with no species in common, to one for two forests with all species in common. In Figure 10.1 a cluster analysis dendrogram is presented. In terms of their montane bird species composition these forests fall into seven major Groups (indicated in capital letters on Figure 10.1). In addition five areas, the Taita Hills, Mahale Mountain, Ufipa Plateau, Angolan Highlands and Angolan Escarpment, have an unusual assemblage of species and do not fit into any of the seven Groups. Of these major Groups, it can be seen from Figure 10.1 that two of them, Cameroun and Ethiopia, are particularly distinct. The remaining five Groups all cluster between Q_s 0.6 and 0.7. Within each Group the Q_s value between montane forest avifaunas is almost always between 0.8 and 1.0.

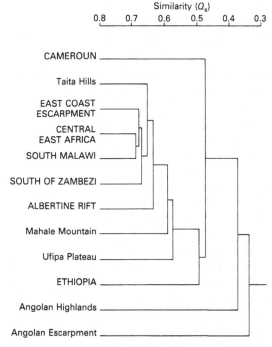

Figure 10.1. Cluster Analysis of the major avifaunal Montane Forest Groups in Africa, in terms of their species composition (see text for details). Major Montane Forest Groups are indicated in capitals.

These five Groups can be given the following names:

1. East Coast Escarpment (east Tanzania to north Malawi)
2. Central East Africa (north Tanzania, Kenya, east Uganda)
3. South Malawi (south Malawi and north Mozambique)
4. South of Zambezi (mainly east Zimbabwe and south Mozambique)
5. Albertine Rift (east Zaïre, west Uganda, Rwanda, Burundi)

These results are discussed in greater detail in the following section.

Discussion

Montane forest Groups

The 50 forests (listed in Table 10.1) chosen for this analysis are widely spread throughout Moreau's Montane Forest Groups. The results of the cluster analysis shown in Figure 10.1 show a number of important groupings in terms of the species composition of the avifaunas. The groupings are discussed in turn.

1. The Angolan forests. These are easily the most distinct of all the montane forests in Africa in terms of their species composition. Surprisingly, they are also very distinct from each other. The Angolan escarpment forests (with only 12 montane species, six of which are endemic), have a Q_s of only 0.333 with Mount Moco, and they are even more dissimilar from every other forest. The highest level of similarity between Mount Moco and any other forest is only 0.368, with the Crater Highlands in northern Tanzania. Clearly the Angolan forests do not constitute a Montane Forest Group in their own right, but rather an anomaly, outside the main variation in the African montane forest avifauna. Other examples of such anomalies, not quite as distinct but nevertheless not fitting into any Montane Forest Group, are presented below. The Angolan forests are very species-poor. The escarpment forests are characterised by high endemism, whereas those in the highlands around Mount Moco have a much lower level of endemism, but include a combination of species quite unlike that found anywhere else.

2. The Cameroun highlands. The isolated montane forests of Cameroun are, as stated by Moreau (1966), a distinct Montane Forest Group in their own right (see Figure 10.1). The Group includes the Obudu Plateau in Nigeria and the mountains on Bioko (Fernando Po). The Group is very dissimilar from any others, and the highest Q_s value between any of the Cameroun forests and any other forest is only 0.472 (between Mount Oku and Mau in Kenya). The Cameroun Group is relatively homogeneous in its species composition (see Jensen & Stuart, 1985), and the lowest Q_s within the Group is 0.727 between Mount Oku and Fernando Po (at opposite ends of the mountain chain). Mount Oku is the most distinct member of the Group, and continuing forest clearance on its slopes is likely to increase this distinctiveness as a result of local extinctions of

species restricted to the lower parts of the montane forest (Stuart, 1986).

3. The Ethiopian mountains. The forests of Ethiopia also form a very distinct Montane Forest Group, in accordance with Moreau's (1966) hypothesis. The Simien Mountains in northern Ethiopia are fairly dissimilar from the forests further south, principally on account of their paucity in species (only 10 montane forest species). The Q_s between the Simien and Bale Mountains is 0.666. However, only one species occurring in the Simien Mountains is absent from both the Bale Mountains and southwestern Ethiopia, Erckel's Francolin *Francolinus erckelii*. The montane forest avifauna in the Simien Mountains is, therefore, best considered to be a depauperate version of the Ethiopian Montane Forest Group, which is better represented further south in the country. Not surprisingly, the highest Q_s value between any Ethiopian forest and any other is 0.489 between the mountains of southwestern Ethiopia and the Imatong Mountains in southern Sudan. This suggests that movement of non-endemic montane forest species between Ethiopia and the other montane regions of Africa probably took place along this route.

The remaining Montane Forest Groups are each less distinct than those in Cameroun and Ethiopia. Nevertheless, it can be seen from Figure 10.1 that between Q_s values of 0.6 and 0.7 most of the forests fall into five main groupings, excluding a few oddities, namely the Taita Hills, Mahale Mountain and Ufipa Plateau, discussed later. These five Groups are discussed below.

4. The Albertine Rift. The forests along the Albertine Rift form a Montane Forest Group, much as suggested by Moreau (1966). However, the Mahale Mountains in western Tanzania are not part of this Group. Also the most northerly member of the Group, the Lendu Plateau in eastern Zaïre, is a relatively species-poor, aberrant member, analogous to the relationship between the Simien Mountains and the rest of the Ethiopian Montane Forest Group (see above). Otherwise, variation within the Albertine Rift Group is relatively slight, with no particular difference between the eastern and western sides of

the rift valley. The highest Q_s value between any Albertine Rift forest and any other is 0.633 between the Itombwe Mountains and Mount Elgon on the Kenya/Uganda border.

5. The East Coast Escarpment. This Montane Forest Group corresponds to Moreau's Tanzania/Malawi Group, except that its area is more limited. The Pare Mountains, the Taita Hills, the mountains of southern Malawi and most of those in northern Mozambique are not part of this Group. The Group is therefore centred on the mountains of eastern Tanzania and northern Malawi, from the Usambaras, south to the Nyika Plateau (and probably at least as far as the Viphya Mountains in central Malawi and the Njesi Plateau in northern Mozambique). Dowsett (1971) considered the mountains of southern Malawi and northern Mozambique to be in a separate Group from those of eastern Tanzania and northern Malawi; his contention is borne out by this analysis. Dowsett also considered the Tanzania/Malawi Group to have its northern limit in the Nguru Mountains. This analysis shows the northern limit to be the Usambaras, which fall clearly into the East Coast Escarpment Montane Forest Group, while the Pare Mountains fall clearly into Central East African Montane Forest Group (see below) *contra* Moreau (1966). The relatively low Q_s value between the Usambaras and the South Pares (0.656) is remarkable in view of the very short distance between these mountains. Variation within the East Coast Escarpment Montane Forest Group exhibits some interesting patterns. The three southernmost forests, Mufindi, Mount Rungwe and Nyika Plateau, form one subgroup. The three wet, east-facing escarpments, Nguru, Uluguru and Mwanihana, form another subgroup. The Usambaras are slightly more distinct, this probably being a reflection of the small Kenyan influence in their avifauna referred to by Dowsett (1971). The dry Ukaguru Mountains are an aberrant and depauperate member of the East Coast Escarpment Group, and are analogous, in this respect, to the Lendu Plateau and Simien Mountains in their respective Groups. The highest Q_s between any East Coast Escarpment forest and

any other is 0.676 between the Usambaras and the North Pares. These results are discussed in greater detail later in this chapter.

6. The Central East African Mountains. This Montane Forest Group is broadly equivalent to Moreau's (1966) Kenya Group. All the Kenyan highlands fall into this Group, excepting the Taita Hills (and presumably Kasigau), but including Mount Elgon and the montane elements in the Kakamenga/Nandi area. The Imatong Mountains in southern Sudan form the northwestern limit of the Group, and so the intervening mountains in northeastern Uganda (not considered in this analysis) probably belong here also. The southern limits of the Group are in northern Tanzania, south to the Crater Highlands (presumably also including the Mbulu Highlands and Mount Hanang to the south) and the North and South Pare Mountains. The subdivisions within the Central East African Montane Forest Group are interesting. Rather than corresponding to the eastern and western sides of the rift valley, there are two different groupings. One is a northwest subgroup, consisting of Mount Kenya, the Aberdare Mountains, Mau, Mount Elgon and Kakamenga/Nandi. The other is a southeast subgroup, consisting of the North and South Pare Mountains, Mount Kilimanjaro, Mount Meru, the Crater Highlands and the Nguruman Hills. A closer examination of distribution patterns reveals the reason for this pattern. The northwest subgroup has a preponderence of species derived from the Albertine Rift Montane Forest Group, whereas the southeast subgroup is dominated by species characteristic of the Eastern Coast Escarpment Montane Forest Group. The Central East African Montane Forest Group has very few unique endemics. This Group is best considered a zone of overlap between the Albertine Rift and Eastern Coast Escarpment centres of endemism. The Group shows increasing similarities to the Albertine Rift on the Mount Elgon side, and to the East Coast Escarpment on the Pare Mountains side. The Imatong Mountains form a third subgroup within the Central East African mountains. This must be attributable, in part, to their relative isolation, and possibly also to their position as an 'access point' to the Ethiopian Highlands (see above).

7. Southern Malawi. These mountains (Mulanje and Thyolo) are included by Moreau (1966) in his Tanzania/Malawi Montane Forest Group, and by Dowsett (1971) in a Southeastern Montane Forest Group, which also includes the mountains south of the Zambezi River. This analysis shows the mountains of southern Malawi to form a small Montane Forest Group of their own (see Figure 10.1), which presumably also includes most of the poorly studied mountains of northern Mozambique, notably Chiperone and Namuli. The highest Q_s between any forest in the Southern Malawi Montane Forest Group, and any other forest is 0.682, between Mount Mulanje and, curiously, the South Pare Mountains.

8. South of the Zambezi. This corresponds exactly to Moreau's Group of the same name. The forest avifaunas considered in this analysis are all very similar to each other, though greater differences might be expected if the more depauperate South African forests are included. The highest Q_s between any forest in this Group and any other is 0.667 between the Inyanga Highlands on the Zimbabwe/Mozambique border, and Mount Thyolo.

9. Some anomalies. In addition to the Angolan forests discussed above, there are three other forests included in this analysis which cannot be placed in any Montane Forest Group. These are also isolated, species-poor forests with unusual species combinations (analogous to the Angolan forests). Doubtless other such anomalies must exist, for instance in northern Somalia, and the Marungu Highlands in southeastern Zaïre, but these are not considered in this chapter. The three anomalies identified in this analysis are:

(a) Taita Hills. Treated by Moreau (1966) as part of the Tanzania/Malawi Group, but in fact dissimilar to all other forests (its highest Q_s is 0.649 with South Pare).

(b) Mahale Mountain. Though considered by Moreau (1966) and Prigogine (1985) as part of the Albertine Rift Group, with which it undoubtedly has some links, it is in fact closest in species composition to the northwest subgroup of the Central East African Group (but its highest Q_s is only 0.588 with Mount Elgon).

(c) Ufipa Plateau. Recognised by Moreau (1966) as a very difficult forest to categorise, its highest Q_s is only 0.571 with the Crater Highlands.

Patterns of species richness

The montane bird species totals for each of the 50 forests used in this analysis are given in Table 10.1. One of the most noticeable aspects of these figures is the extraordinary species richness of the Albertine Rift Montane Forest Group. The richest single forest area is the Itombwe Mountains in eastern Zaïre, which, with 83 montane species, has approximately half the total number of montane forest species in all of Africa. The Lendu Plateau (37 species) is considered species-poor compared to the other Albertine Rift forests (see above), but this area is not impoverished when compared with the rest of Africa.

Most of the forests in the Cameroun, Eastern Coast Escarpment and Central East African Montane Forest Groups have moderate species richness, usually between 30 and 43 species, with only a few exceptions. The Ethiopia, Southern Malawi and South of the Zambezi Montane Forest Groups are all species-poor.

Table 10.2 presents combined species totals for each of the Montane Forest Groups identified by the cluster analysis dendrogram in Figure 10.1. This table also draws on additional data, because there are a few montane forest species which

occur in the region, but not in the forests used for this analysis. Examples include the Prince Ruspoli's Turaco *Tauraco ruspolii* in the Ethiopian Group, and the Dappled Mountain-robin *Modulatrix orostruthus* in the Southern Malawi Group. Once again, the same patterns of species richness can be seen, with the highest diversity in the Albertine Rift Group.

Endemic and characteristic species

Table 10.2 also presents data on the number of endemic species (i.e. those occurring nowhere else) and the percentage of endemism within each of the Montane Forest Groups. These results are discussed below.

1. Cameroun. This Group has the highest level of endemism, hence the remarkable distinctiveness of its avifauna remarked upon earlier. Characteristic species (i.e. those endemics which occur through virtually all forests within the Group) are the Cameroun Mountain Greenbul *Andropadus montanus*, Cameroun Olive Greenbul *Phyllastrephus poensis*, Cameroun Mountain Robin-chat *Cossypha isabellae*, Green Longtail *Urolais epichlora*, Cameroun Blue-headed Sunbird *Nectarinia oritis* and Little Olive-back *Nesocharis shelleyi*.

2. Ethiopia. Despite its poverty in species, the Ethiopian Group is relatively rich in endemics, characteristic species being the Black-winged Lovebird *Agapornis taranta*, White-cheeked Turaco *Tauraco leucotis*

Table 10.2. *Avian species richness and endemism in the African Montane Forest Groups*

Montane Forest Group	Total number of species	Number of endemics to the Group	Percentage endemism
Cameroun	50	20	40.0
Ethiopia	23	8	34.8
Albertine Rift	87	32	36.8
East Coast Escarpment	57	16	28.1
Central East Africa	59	2	3.4
South Malawi	23	1	4.3
South of Zambezi	18	3	16.7

Table 10.3. *The number and percentage of species shared between certain Montane Forest Groups*

Montane Forest Group	Number of species shared with AR[a]	Percentage of species shared with AR	Number of species shared with EC[a]	Percentage of species shared with EC
Cameroun	28	56	14	28
Ethiopia	13	57	9	39
Albertine Rift	—	—	27	31
East Coast Escarpment	27	47	—	—
Central East Africa	46	78	38	64
South Malawi	15	65	21	91
South of Zambezi	10	56	14	78

Note: [a]AR, Albertine Rift; EC, East Coast Escarpment.

and Abyssinian Catbird *Parophasma galinieri*.

3. Albertine Rift. This Group has by far the largest number of endemics, a reflection of the species richness in this area. The characteristic species are the Handsome Francolin *Francolinus nobilis*, Red-faced Woodland Warbler *Phylloscopus laetus*, Masked Apalis, Blue-headed Sunbird *Nectarinia alinae* and Dusky Crimson-wing *Cryptospiza jacksoni*.

4. East Coast Escarpment. This is another Group with a high level of endemism. However, most of the endemics are very rare, occurring in only a few forests, so there are no completely characteristic species, the nearest ones being Sharpe's Akalat *Sheppardia sharpei*, Spot-throat *Modulatrix stictigula* and Chapin's Apalis *Apalis chapini*.

5. Central East Africa. Being a transition zone between the previous two Groups, this Group has only two endemics, Jackson's Francolin *Francolinus jacksoni* and Abbott's Starling *Cinnyricinclus femoralis*, and no truly characteristic species.

6. Southern Malawi. There is only one endemic and characteristic species, the Thyolo Alethe *Alethe choloensis*.

7. South of the Zambezi. Of the three endemic species, only one, Barratt's Scrub Warbler *Bradypterus barratti*, can be considered characeric of this Group (including the forests in South Africa).

Relationships between the montane forest groups
The seven Montane Forest Groups that can be identified in Figure 10.1 are all valid in terms of the species composition of the forests. However, this cluster analysis does not reveal much about how these Groups originated. Table 10.2 reveals that four of the Groups (Cameroun, Ethiopia, Albertine Rift and the East Coast Escarpment) have high endemism. It therefore seems that the characteristic elements of these Groups are the result of *in situ* evolution. Insight into the origins of the other three Groups can be obtained from Table 10.3, which demonstrates that the Albertine Rift and East Coast Escarpment have acted as the colonising sources for the Groups with low levels of endemism.

1. Central East Africa Group. With only two endemics (see Table 10.2), and with 78% of its species occurring in the Albertine Rift and 64% in the East Coast Escarpment, the Central East Africa Group is clearly a merging zone between its two neighbours, as suggested above. There is also a small Ethiopian influence in the Central East African avifauna. Fourteen species are shared between these two Groups, and two species, Rueppell's Robin-chat *Cossypha semirufa* and the Montane White-eye *Zosterops abyssinica*, occur in no other Groups.

2. Southern Malawi Group. This Group is clearly a distinctive derivative from the East Coast Escarpment Group, 91% of its species occurring there.
3. South of the Zambezi Group. This is another distinctive derivative from the East Coast Escarpment Group, 78% of its species occurring there.

The other Groups are all more distinct, although a comparison of Tables 10.2 and 10.3 shows that all but two of the Cameroun Group's non-endemic species occur in the Albertine Rift Group. In terms of the distinctiveness of the species composition of individual forests, seven Montane Forest Groups can be recognised. However, in terms of biogeographic origins, there are only four significant Groups: Cameroun, Ethiopia, Albertine Rift and the East Coast Escarpment.

Refugia

The results of this analysis concur with those of Hamilton (1976) and Diamond & Hamilton (1980). Three major centres of endemism for montane forest birds, these corresponding to the Cameroun, Albertine Rift and East Coast Escarpment Montane Forest Groups, can be identified. Hamilton (1976) points out that these centres of endemism are situated very close to similar concentrations of lowland forest birds, and suggests that they represent refugia where forest survived during the driest periods of the Pleistocene. Smaller refugia may have existed in areas where other endemics occur, notably Ethiopia and the Angolan Escarpment.

Table 10.2 emphasises the particularly high number of species and endemics around the Albertine Rift, and this area may be the main centre of evolution of the African montane forest avifauna. Many of the more widespread montane species, occurring in Cameroun and also in East Africa, could well have originated from the Albertine Rift area. The Cameroun and East Coast Escarpment forests have also been important sites for the evolution of montane species. However, because of their less central geographi-

cal positions, they have probably been less important as sources from which colonising species have spread. It could be argued that one reason for the high species richness in the Albertine Rift forests is that they have 'received' colonising species from both the East Coast Escarpment and Cameroun, in addition to their being an important centre of evolution. Also, both the southern Malawi and South of the Zambezi Group are derived to a considerable extent from the East Coast Escarpment Group. The Central East Africa Group is clearly derived from both the Albertine Rift and East Coast Escarpment Groups (see above).

Conservation

It is important that adequate conservation measures be taken in each of the Montane Forest Groups in order to ensure the preservation of the maximum possible biological diversity. At present, forest conservation is reasonably effective for the South of the Zambezi Group, but elsewhere, the situation is far from satisfactory (see Mackinnon & MacKinnon, 1986). Several of the Central East African forests are included in National Parks, though, from a species conservation point of view, these forests are not the highest priorities. However, the International Union for the Conservation of Nature (IUCN) is now making progress in forest conservation in the Usambara Mountains, and other conservation projects are being carried out in the Uzungwa Mountains, Mount Rwenzori, the Impenetrable Forest and the Virunga Volcanoes. Much more work is needed on the montane forests of eastern Zaïre, most notably the Itombwe Mountains. In Ethiopia the government has made impressive progress in its efforts to conserve forest on the Bale Mountains, and in Cameroun the International Council for Bird Preservation (ICBP) is involved in conservation work, especially on Mount Oku. Two areas where more effective conservation is very urgently needed are the mountains of southern Malawi, where the area of forest remaining is tiny, and the Escarpment of Angola. More details on the priorities for forest bird conservation in Africa are given by Collar & Stuart (1985, 1988).

An overview of the Eastern Tanzanian montane forest avifaunas

In terms of species composition, therefore, the East Coast Escarpment forests constitute one of the seven distinct Groups of African montane avifaunas. This distinctive East Coast Escarpment montane forest avifauna, which we discuss in the remainder of this chapter, can be considered to range from the Usambaras in the north to the Nyika and Viphya Plateaux in the south (see Figure 10.2). It is likely that the Matengo Highlands in southern Tanzania and the Njesi Plateau in northern Mozambique also belong here, though their avifaunas are not well known.

In this section, we carry out an analysis of the East Coast Escarpment avifauna, concentrating on eight particular sites. For half these sites, the montane forest avifaunas are very well known, and it is unlikely that many additional species will be found, for instance in the Usambaras, Ulugurus, at Mufindi, or on the Nyika Plateau. The possibility remains of one or two additional species still being found in the East Uzungwas and Mount Rungwe. The Nguru and Ukaguru avifaunas have been slightly less well sampled (as indicated by the discovery of the elusive Iringa Ground-robin *Sheppardia lowei* in the Ukagurus in 1989: N. E. Baker and E. M. Boswell, *in litt.*), but even in these sites it is unlikely that more than a handful of montane forest species remain unrecorded. Despite the incompleteness of the data, we doubt that any new discoveries in the Ngurus or Ukagurus will result in significant changes to the conclusions we make in this chapter. There are, of course, several other mountain ranges in eastern Tanzania that are too poorly known to be included in this analysis. These include the Pare Mountains, Rubeho Mountains, Mahenge Highlands, Njombe Highlands, Kipengere Mountains, Livingstone Mountains, and Matengo Highlands. The picture painted in this chapter is the best possible on available data; further explorations will inevitably result in new surprises.

Diversity and endemism in the East Coast Escarpment forests

Fifty-seven montane forest species are known from the East Coast Escarpment forests, of which 16 are endemics (i.e. occurring nowhere else) (see Table 10.2). This level of diversity is similar to that of the Cameroun and Central East Africa Montane Forest Groups, and much higher than that of the species-poor Ethiopia, South Malawi and south of Zambezi Groups. The East Coast Escarpment Group is, however, impoverished in comparison with the Albertine Rift Group, which has 87 montane forest species, 32 of them endemics (see Table 10.2). In terms of the numbers of endemic species, the East Coast Escarpment is one of the three major centres in Africa, along with Cameroun and the Albertine Rift (Ethiopia representing a smaller centre of endemism).

Table 10.4 presents data on eight forests within the East Coast Escarpment Group for which the avifaunas are well known. Basic inventory work on these avifaunas has been described by Loveridge (1922, 1933), Friedmann (1928), Sclater & Moreau (1932–3), Bangs & Loveridge (1933), Moreau (1935, 1940), Moreau & Moreau (1937), Fuggles-Couchman (1939, 1984, 1986), Williams (1951), Friedmann & Stager (1964), Ripley & Heinrich (1966, 1969), Benson & Benson (1977), Britton (1978, 1981), Dowsett & Dowsett-Lemaire (1979), Stuart (1980, 1981a, 1981b, 1983, 1985), Stuart & Turner (1980), Stuart & van der Willigen (1980), Turner (1980), Stuart *et al.* (1981, 1987), Stuart & Jensen (1981, 1985), van der Willigen (1981), Jensen & Stuart (1982), Jensen (1983), Dowsett-Lemaire (1989) and Jensen & Brøgger-Jensen (1992).

In terms of species diversity, these forests have between 27 and 42 montane species, the highest diversity being in the Usambaras, Ulugurus and Eastern Uzungwas (Mwanihana and Chita). There is a general trend to decreasing diversity with increasing distance from the Equator (from Usambara *c.* 5° S to Nyika Plateau (*c.* 11° S). Another trend is for higher diversity in areas of high annual rainfall, such as Usambara and Ulu-

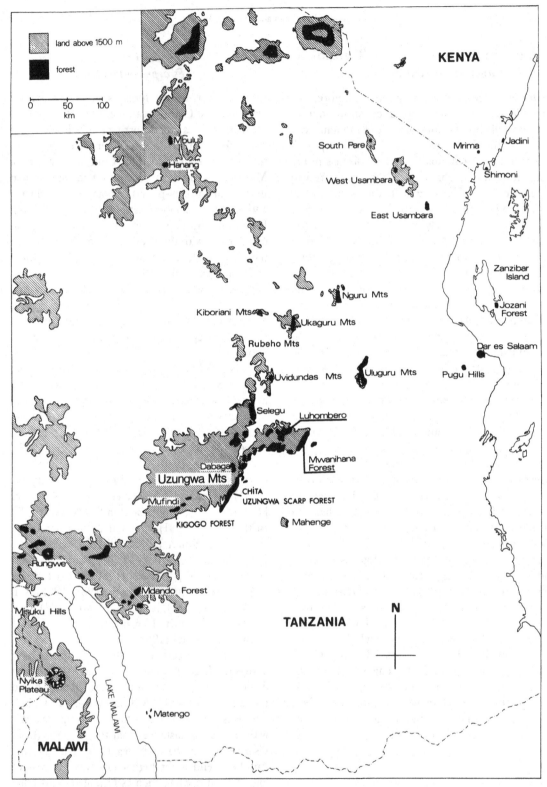

Figure 10.2. Distribution of east coast escarpment montane forest from northern Tanzania to northern Malawi.

Table 10.4. *Diversity and endemism of montane birds in eight forest areas within the East Coast Escarpment Group*

Forest area	Number of montane species	Number of East Coast Escarpment endemic species	Percentage of East Coast Escarpment endemics	Number of species endemic to each forest
Usambara	41	9	22.0	2
Nguru	31	7	22.6	0
Ukaguru	28	6	21.4	0
Uluguru	41	9	22.0	2
East Uzungwa	42	10	23.8	1
Mufindi	34	4	11.8	0
Mount Rungwe	35	3	8.6	0
Nyika Plateau	31	2	6.5	0

guru (1800–3000 mm annually). Among the more northern forests, the dry Ukagurus (perhaps 1200 mm annually) in the rain shadow of the Ulugurus, have a remarkably small number of montane species. Conversely, Mount Rungwe in the species-poor southern part of the East Coast Escarpment has a larger than expected number of montane species, presumably related to its very high rainfall.

The level of endemism shows a similar trend to species diversity in relation to latitude and annual rainfall (see Table 10.4). Only a few of the 16 endemic species range further south than Mufindi. The wet forests of Usambara, Uluguru and East Uzungwa are particularly important for endemics, and interestingly, only one of the

endemic species, the Iringa Ground-robin, is a dry montane forest dweller (and even this occurs in a few wet forests).

Three of the forests have endemics that occur nowhere else. These are the Usambaras, the only locality for the Usambara Eagle Owl *Bubo vosseleri* and the Usambara Ground-robin *Sheppardia montana*, the Ulugurus with the Uluguru Bush-shrike *Malaconotus alius* and Loveridge's Sunbird *Nectarinia loveridgei*, and the Eastern Uzungwas with the Rufous-winged Sunbird *N. rufipennis* (this last species being restricted to the Mwani-hana Forest). The distributions of all the 57 montane forest birds in these eight forests are given in Table 10.5, together with the identities of the 16 endemic and five near-endemic species.

Table 10.5. *The distribution of montane forest birds in the East Coast Escarpment forests*

Species	USA	NGU	UKA	ULU	EUZ	MUF	RUN	NYP
Mountain Buzzard *Buteo oreophilus*	×	×	×	×	×	×	×	×
Olive Pigeon *Columba arquatrix*	×	×	×	×	×	×	×	×
Lemon Dove *Aplopelia larvata*	×	×	×	×	×	×	×	×
Hartlaub's Turaco *Tauraco hartlaubi*	×							
Barred Long-tailed Cuckoo *Cercococcyx montanus*	×	×	×	×	×	×	×	

Table 10.5. (*cont.*)

Species	USA	NGU	UKA	ULU	EUZ	MUF	RUN	NYP
Usambara Eagle Owl *Bubo vosseleri*[a]	×							
Bar-tailed Trogon *Apaloderma vittatum*	×	×	×	×	×		×	×
Moustached Green Tinkerbird *Pogoniulus leucomystax*	×		×	×	×	×	×	×
Olive Woodpecker *Mesopicos griseocephalus*	×	×		×	×	×	×	×
Grey Cuckoo-shrike *Coracina caesia*	×	×		×	×		×	
Mountain Greenbul *Andropadus tephrolaemus*	×	×	×	×	×	×	×	×
Stripe-cheeked Greenbul *A. milanjensis*	×	×	×	×	×	×	×	×
Shelley's Greenbul *A. masukuensis*	×	×	×	×	×	×	×	
Olive Mountain Greenbul *Phyllastrephus placidus*	×	×	×	×	×	×	×	
Tiny Greenbul *P. debilis albigula*[a]	×	×						
Fuelleborn's Black Boubou *Laniarius fuelleborni*	×	×	×	×	×	×	×	×
Black-fronted Bush-shrike *Malaconotus multicolor nigrifrons*	×	×	×	×	×	×	×	×
Uluguru Bush-shrike *M. alius*[a]				×				
White-starred Forest-robin *Pogonocichla stellata*	×	×	×	×	×	×	×	×
Swynnerton's Forest-robin *Swynnertonia swynnertoni*[b]					×			
White-chested Alethe *Alethe fuelleborni*[b]	×	×	×	×	×	×	×	×
Olive-flanked Ground-robin *Cossypha anomala*[b]			×	×		×	×	×
Usambara Ground-robin *Sheppardia montana*[a]	×							
Iringa Ground-robin *S. lowei*			×		×	×		
Sharpe's Akalate *S. sharpei*[a]	×		×	×		×	×	
Spot-throat *Modulatrix stictigula*[a]	×	×	×	×	×	×	×	

Table 10.5. (*cont.*)

Species	USA	NGU	UKA	ULU	EUZ	MUF	RUN	NYP
Dappled Mountain-robin *Modulatrix orostruthus*[b]	×				×			
Olive Thrush *Turdus olivaceus*	×	×		×	×	×	×	×
Orange Ground-thrush *T. gurneyi*	×	×	×	×	×	×	×	×
Mountain Illadopsis *Trichastoma pyrrhopterum*								×
African Hill Babbler *Alcippe abyssinica*	×		×	×	×	×	×	×
Mountain Yellow Warbler *Chloropeta similis*				×		×		×
Evergreen-forest Warbler *Bradypterus lopesi*	×	×	×	×	×	×	×	×
Yellow-throated Woodland Warbler *Phylloscopus ruficapilla*	×	×		×	×		×	×
Brown Woodland Warbler *P. umbrovirens*				×				
White-winged Apalis *Apalis chariessa*[b]				×	×			
Bar-throated Apalis *A. thoracica*	×	×	×	×	×	×	×	×
Brown-headed Apalis *A. alticola*					×	×	×	×
Chapin's Apalis *A. chapini*[a]		×	×	×	×	×	×	×
Long-billed Apalis *A. moreaui*[a]	×							
Red-capped Forest Warbler *Orthotomus metopias*[a]	×	×	×	×	×	×		
Mrs Moreau's Warbler *Bathmocercus winifredae*[a]			×	×	×			
Dusky Flycatcher *Muscicapa adusta*	×	×	×	×	×	×	×	×
White-tailed Crested Flycatcher *Elminia albonotata*	×	×	×	×	×	×	×	×
Banded Green Sunbird *Anthreptes rubritorques*	×	×		×	×			
Montane Double-collared Sunbird *Nectarinia ludovicensis*								×
Eastern Double-collared Sunbird *N. mediocris*	×					×	×	×

Table 10.5. (*cont.*)

Species	USA	NGU	UKA	ULU	EUZ	MUF	RUN	NYP
Loveridge's Sunbird *N. loveridgei*[a]				×				
Moreau's Sunbird *N. moreaui*[a]		×	×		×			
Rufous-winged Sunbird *N. rufipennis*[a]					×			
Thick-billed Seed-eater *Serinus burtoni*					×	×	×	
Oriole Finch *Linurgus olivaceus*	×			×		×	×	×
Usambara Weaver *Ploceus nicolli*[a]	×			×	×			
Red-faced Crimson-wing *Cryptospiza reichenovii*	×	×	×	×	×	×	×	×
Kenrick's Starling *Poeoptera kenricki*	×		×	×	×			
Waller's Chestnut-winged Starling *Onychognathus walleri*	×	×		×	×	×	×	×
Sharpe's Starling *Cynnyricinclus sharpii*	×						×	

Note: [a]A taxon endemic to the East Coast Escarpment forests (16 species).
[b]A near-endemic species occurring in only a few localities outside the East Coast Escarpment.
Compared with montane forests elsewhere in Africa, the East Coast Escarpment forests have reasonably high endemism (see Table 10.1) but moderate species diversity. Table 10.4 gives some comparisons of species numbers between certain forest areas (data taken from Miller *et al.*, 1989).
Key: USA, Usambara; NGU, Nguru; UKA, Ukaguru; ULU, Uluguru; EUZ, East Uzungwa; MUF, Mufindi; RUN, Mount Rungwe; NYP, Nyika Plateau.

In terms of species diversity, the East Coast Escarpment forests are impoverished compared with those in the Albertine Rift, on a par with Cameroun and Central East Africa, and rich compared with Ethiopia, South Malawi and South of Zambezi (see Table 10.1).

Patterns of distribution within the East Coast Escarpment

Although the montane avifaunas within the East Coast Escarpment are similar in terms of their species composition, they are by no means homogenous. Only 15 of the 57 montane species occur in all the forests for which data are provided in Tables 10.2 and 10.3. These include none of the endemic species, though one near endemic, the White-chested Alethe *Alethe fuelleborni*, is ubiquitous through these areas.

A cluster analysis dendrogram of the eight montane forest bird communities is presented in Figure 10.3. The same methodology is employed as for the analysis shown in Figure 10.1. The impoverished Ukaguru community is one of the most distinct species assemblages. The Usambaras are equally distinct, partly owing to their large number of rare species, and also to a few affinities with the Central East African forests, notably the presence of Hartlaub's Turaco *Tauraco hartlaubi* and the absence of Chapin's

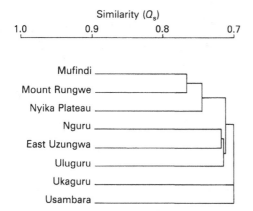

Figure 10.3. Cluster analysis of eight East Coast Escarpment forests in terms of their species composition (see text for details).

Apalis. The remaining bird communities fall into two groupings, a northern one including East Uzungwa, Nguru and Uluguru (characterised by several rare endemics) and a southern one comprising Mufindi, Mount Rungwe and the Nyika Plateau (in which rare endemics are almost absent).

More could be learnt of the distribution patterns within the East Coast Escarpment forests if more inventory work could be carried out in the unknown and poorly explored areas. These areas are mentioned in the Conclusion.

Patterns of subspeciation within the East Coast Escarpment

Jensen & Stuart (1985) and Stuart (1986) have provided a description of the subspeciation patterns of montane forest birds in the Cameroun Group, but no such study has previously been carried out for the East Coast Escarpment. In Table 10.6 we give distributional details for each of the 19 species with more than one subspecies

Table 10.6. *The distributions of subspecies of montane forest birds in the East Coast Escarpment*

Subspecies	USA	NGU	UKA	EUZ	MUF	RUN	NYP	ULU
Mesopicos griseocephalus kilimensis	×	×		×				×
Mesopicos griseocephalus persimilis					×	×	×	
Andropadus tephrolaemus usambarae	×							
Andropadus tephrolaemus chlorigula		×	×	×	×			
Andropadus tephrolaemus fusciceps						×	×	
Andropadus tephrolaemus neumanni								×
Andropadus milanjensis striifacies	×	×	×	×	×			×
Andropadus milanjensis olivaceiceps						×	×	
Laniarius fuelleborni usambaricus	×	×	×	?				×
Laniarius fuelleborni fuelleborni					×	×	×	
Alethe fuelleborni usambarae	×	×	×	×	×			×
Alethe fuelleborni fuelleborni						×	×	
Cossypha anomala grotei			×		×			×
Cossypha anomala macclounii						×	×	
Sheppardia lowei lowei				×	×			
Sheppardia lowei subsp. nov.			×					
Sheppardia sharpei usambarae	×	×						
Sheppardia sharpei sharpei					×		×	×
Sheppardia sharpei bangsi								×
Modulatrix orostruthus amani	×							
Modulatrix orostruthus sanjei				×				
Turdus olivaceus roehli	×							
Turdus olivaceus nyikae		×		×	×	×	×	×

Table 10.6. (*cont.*)

Subspecies	USA	NGU	UKA	EUZ	MUF	RUN	NYP	ULU
Alcippe abyssinica abyssinica	×							
Alcippe abyssinica stierlingi			×	×	×	×		×
Alcippe abyssinica stictigula							×	
Phylloscopus ruficapilla minulla	×	×		×				×
Phylloscopus ruficapilla johnstoni						×	×	
Apalis thoracica murina	×	×	×			×		
Apalis thoracica injectivora				×	×			
Apalis thoracica youngi							×	
Apalis throacica uluguru								×
Apalis chapini chapini		×	×	×	×			×
Apalis chapini strausae						×	×	
Orthotomus metopias metopias	×	×	×					
Orthotomus metopias pallidus				×	×			
Orthotomus metopias altus								×
Elminia albonotata subcaerulea	×	×	×	×	×	×		×
Elminia albonotata albonotata							×	
Muscicapa adusta murina	×							
Muscicapa adusta fuelleborni		×	×	×	×	×		×
Muscicapa adusta subadusta							×	
Nectarinia mediocris usambarica	×							
Nectarinia mediocris fuelleborni					×	×	×	
Ploceus nicolli nicolli	×							
Ploceus nicolli anderseni				×				×

Note: ?, species present but subspecific identity uncertain.
Key to locations as in Table 10.5.

occurring in the East Coast Escarpment. The remaining 38 species all occur through the East Coast Escarpment as only one race. Table 10.6 is arranged with the forests in geographical sequence from the Usambaras (north) to the Nyika Plateau (south); however, the Ulugurus, which form an isolated block to the east of the main East Coast Escarpment with several endemic subspecies, are placed on the far right of the table. This enables the reader to obtain an easier impression of characteristic subspecific distributions along the mountain chain. For the recognition of subspecies we have generally followed Benson & Benson (1977) and Britton (1980), with subsequent revisions from Jensen & Stuart (1982), Franzmann (1983), Jensen, Brøgger-Jensen & Petersen (1985), Stuart & Jensen (1985), Stuart *et al.* (1987) as well as results from our recent research (Jensen & Brøgger-Jensen, 1992) and new information on the Iringa Ground-robin received from N. E. Baker and E. M. Boswell (*in litt.*).

In Table 10.7 we present an analysis of the subspeciation patterns shown in Table 10.6. Ignoring the isolated Uluguru Mountains, Table 10.7 shows the frequency of subspecific distribution boundaries between the various forest blocks. We treat the Ngurus and Ukagurus as one block because they are geographically close together, and many species are absent from the dry Ukagurus. As can be seen, most of the subspecific breaks occur in four locations:

1. between Usambara and Nguru/Ukaguru;

Table 10.7. *The frequency of subspecific boundaries at various locations within the East Coast Escarpment*

Location (i.e. between these mountain blocks)	Frequency of subspecific distribution boundaries
Usambara–Nguru/Ukaguru	4
Nguru/Ukaguru–East Uzungwa	3
East Uzungwa–Mufindi	1
Mufindi–Mount Rungwe	6
Mount Rungwe–Nyika Plateau	4

2. between Nguru/Ukaguru and the Uzungwas;
3. between south Uzungwa (Mufindi) and the Southern Highlands (Mount Rungwe);
4. between the Southern Highlands and North Malawi (Nyika Plateau).

It seems likely, therefore, that geographical and vegetational factors have restricted movements of at least certain species of birds at these points. It can also be seen from Table 10.5 that several species have disjunct distributions, or reach their distributional limits at these same locations.

There is also a considerable degree of subspecific discontinuity between the Ulugurus and the nearest part of the East Coast Escarpment. Table 10.8 gives details of the single forest endemic montane subspecies in the East Coast Escarpment.

From this table the relative isolation of the Ulugurus, with five endemic subspecies, in addition to the two endemic species mentioned earlier, is apparent. From a biogeographic point of view, the study of subspeciation is valuable in that it provides circumstantial evidence of small-scale movements of birds between certain mountain blocks but not others (perhaps during a period when the forest cover in the lowlands between the mountains was more extensive than it is now). Such small-scale occasional movements between mountains would be very difficult to detect by direct observation because there are so few ornithologists in Tanzania. It is possible that movements between mountains could be related to the seasonal altitudinal movements undertaken by certain species, discussed in the following section.

Seasonal altitudinal movements

Most montane forest birds in Africa have usually been considered to be sedentary. However, in the

Table 10.8. *The single forest endemic montane subspecies of the East Coast Escarpment mountains*

Forest	Subspecies
Usambara	*Modulatrix orostruthus amani*
	Apalis moreaui moreaui
	Ploceus nicolli nicolli
Ukaguru	*Sheppardia lowei* subsp. nov.
Uluguru	*Andropadus tephrolaemus neumanni*
	Sheppardia sharpei bangsi
	Phylloscopus umbrovirens fugglescouchmani
	Apalis thoracica uluguru
	Orthotomus metopias altus
Uzungwas (including Mufindi)	*Swynnertonia swynnertoni rodgersi*
	Modulatrix orostruthus sanjei
	Apalis thoracica injectivora
Nyika Plateau	*Trichastoma pyrrhopterum nyasae*
	Nectarinia ludovicensis whytei

southern, more seasonal parts of the continent there is considerable evidence of post-breeding movements of montane forest species to lower altitudes. Evidence of such movements has been provided for several species in Malawi (Benson & Benson, 1977) and Zimbabwe (Irwin, 1981). Nearer the Equator, few such movements have been documented until recent years; the most notable example is that of the Northern Double-collared Sunbird *Nectarinia preussi* in the vicinity of Mount Cameroun, whose marked annual migrations were well described by Serle (1950, 1954). Serle (1964) also noted smaller scale seasonal altitudinal movements on the southeastern slopes of Mount Cameroun among several other montane forest species, and more evidence for these movements has since been provided by Stuart (1986). However, for the species-rich Albertine Rift forests, J.-P. Vandeweghe (personal communication) has found no evidence of any seasonal migrations; nor has any evidence of this phenomenon been forthcoming from the relatively well studied Central East African forests.

In the East Coast Escarpment forests seasonal movements have been best studied in the Usambaras by Stuart (1983). In a three-year study, Stuart found evidence of at least 11 species undertaking such movements. The general pattern among these species seems to be one of a sedentary, territorial adult population, sometimes occurring at a low density, and a larger, at least partially migratory surplus of immature and non-territorial (presumably young) adults. Among the more migratory species, such as the White-starred Forest-robin *Pogonocichla stellata*, probably the bulk of the population migrates to lower elevations. Some individuals move as far as the East African coast from where there are recent records from Mrima Hill in Kenya (Britton, Britton & Coverdale, 1980) and the Pugu Hills in Tanzania (Baker, 1984), all outside the breeding season. Another very migratory montane species is the Barred Long-tailed Cuckoo, which has also been found in coastal Kenya and Tanzania (Britton, 1977; Stuart & Jensen, 1985), although this is a difficult species to study and there is considerable evidence that its movements are more complex

than just a simple altitudinal shift (see Stuart & Jensen, 1985). Other montane forest species which are known to undertake small vertical seasonal migrations in the Usambaras are the African Hill Babbler *Alcippe abyssinica*, Shelley's Greenbul *Andropadus masukuensis*, Mountain Greenbul *A. tephrolaemus*, Stripe-cheeked Greenbul *A. milanjensis*, White-chested Alethe, Sharpe's Akalat, Orange Ground-thrush *Turdus gurneyi*, Black-fronted Bush-shrike and Kenrick's Starling *Poeoptera kenricki*. With all these species, censuses have shown that part of the population moves to lower elevations, often between 300 and 600 m, during the cold non-breeding season (Stuart, 1983). The general downwards movement takes place in March and April, and the birds return to their breeding areas in September and October. There is one record of a Shelley's Greenbul, netted at Amani in the East Usambaras at 900 m on 17 April 1980, which was retrapped on the escarpment below at 600 m on 20 August 1981 (Stuart, 1983).

There is also evidence of seasonal vertical movements of lowland forest birds in the Usambaras. At least eight such species make annual seasonal movements, and with most of them their populations move into montane and submontane forest to breed (Stuart, 1983).

Fieldwork carried out in the foothill forests of the Uluguru Mountains (especially in Kimboza Forest), and in the Magombera Forest in the plain at the foot of the Uzungwa escarpment near Mwanihana Forest, has shown that several montane species move into these areas (both at only 300 m) in the months of July and August (probably longer) (Stuart & Jensen, 1981, 1985; Stuart *et al.*, 1987). Species which are known to undertake altitudinal movements in these areas are as follows (Ul, Uluguru; Uz, Uzungwa): Olive Pigeon *Columba arquatrix* (Uz), Lemon Dove *Aplopelia larvata* (Ul, Uz), Barred Long-tailed Cuckoo (Ul, Uz), Bar-tailed Trogon *Apaloderma vittatum* (Ul), Mountain Greenbul (Ul), Stripe-cheeked Greenbul (Ul), White-chested Alethe (Ul, Uz), White-starred Forest-robin (Ul, Uz), Spot-throat (Ul), Orange Ground-thrush (Uz), Evergreen-forest Warbler *Bradypterus lopesi* (Ul), White-tailed Crested Flycatcher *Elminia albonotata* (Ul), Black-fronted Bush-shrike (Ul, Uz) and Kenrick's Starling (Ul).

More research, especially detailed population studies, is very likely to show that seasonal altitudinal movements are of greater significance than has hitherto been realised. This means that effective conservation of a forest bird community requires adequate protection of the entire altitudinal spectrum of forest upon which the component bird species depend.

The origin of the East Coast Escarpment montane avifauna

Jensen & Stuart (1985) and Stuart (1986) discussed the origin of the Cameroun montane forest avifauna and found it to be derived from a wide variety of sources. The same can be said of the East Coast Escarpment montane forest avifauna.

The 21 endemic and near endemic forms, listed in Table 10.5, can be considered to have had four basic origins. These are as follows.

1. Endemics which evolved *in situ*. These are species which seem to have originated in the East Coast Escarpment itself. Nine species fall into this category, seven of which are 'ancient' species without close relatives. These 'ancient' forms comprise *Sheppardia montana*, *S. lowei* and *Cossypha anomala*, and the genus *Modulatrix* (a genus of uncertain affinities endemic to the East Coast Escarpment forests). Swynnerton's Forest-robin *Swynnertonia swynnertoni* and the Rufous-winged Sunbird are also 'ancient' forms which probably originated in eastern Tanzania. By contrast, Loveridge's Sunbird and Moreau's Sunbird *Nectarinia moreaui* are examples of recently evolved species in the East Coast Escarpment. Stuart & van der Willigen (1980) have suggested that Moreau's Sunbird is a hybrid form between Loveridge's Sunbird (the Uluguru endemic) and the widespread Eastern Double-collared Sunbird *N. mediocris*. More recent studies being

carried out by A. J. Beakbane have revealed a number of sunbirds of this superspecies which are difficult to identify to species; it must now be doubted whether Moreau's Sunbird constitutes a valid taxon.

2. West African derived species. Seven of the East Coast Escarpment endemics have close relatives in the montane and lowland forests of West and Central Africa. These seven are best considered isolated forms of species with more westerly distributions, which have since diverged to the specific level. They are the Usambara Eagle Owl, White-chested Alethe, Sharpe's Akalat, Chapin's Apalis, White-winged Apalis, Mrs Moreau's Warbler *Bathmocercus winifredae* and Banded Green Sunbird. These all form superspecies with West African species (see Hall & Moreau (1970) and Snow (1978) for details).

3. Species of Asian origin. The Red-capped Forest Warbler *Orthotomus metopias* is a tailor-bird of a genus otherwise consisting of Asian species. The Long-billed Apalis *Apalis moreaui* should probably also be placed in *Orthotomus* and is likely to be closely related to the Red-capped Forest Warbler (Fry, 1976; Stuart, 1981b). Interestingly, there are also a few species of probable Asian origin in the Albertine Rift forests, most notably the Itombwe Owl *Phodilus prigoginei* and the African Green Broadbill *Pseudocalyptomena graueri*. The presence of these species in African montane forests suggests a much wetter climate in Southwest Asia in the past than exists today.

4. Species of lowland origin. Three East Coast Escarpment endemics have probably originated from lowland species which occur nearby. The Tiny Greenbul is merely the distinctive montane race (*albigula*) of the East African lowland form (*rabai*). The Uluguru Bush-shrike is probably a montane forest derivative of the widespread Grey-headed Bush-shrike *Malaconotus blanchoti*. And the Usambara

Weaver *Ploceus nicolli* is most closely related to the Olive-headed Golden Weaver *P. olivaceiceps* (Franzmann, 1983).

The remaining 36 species, which are neither endemic nor near-endemic, fall into two groups:

5. Six species are confined to montane forest in the East Coast Escarpment but range widely through lower altitude woodlands and forests in some other parts of Africa. These are the Barred Long-tailed Cuckoo, Olive Woodpecker, Black-fronted Bush-shrike, Olive Thrush, Brown-headed Apalis and Dusky Flycatcher. The origins of all these widespread and adaptable species cannot be readily determined.

6. The remaining 30 species are all widespread (more or less) montane forest species. Evidence from present-day distribution patterns suggests that at least some of these probably speciated in the East Coast Escarpment but have since spread well beyond its limits. The likely such species are the Moustached Green Tinkerbird *Pogoniulus leucomystax*, Stripe-cheeked Greenbul, Yellow-throated Woodland Warbler *Phylloscopus ruficapilla*, Bar-throated Apalis *Apalis thoracica*, Eastern Double-collared Sunbird and Kenrick's Starling. Many of the other species probably originated elsewhere in Africa, most likely in the species-rich Albertine Rift forests, though possibly also in Cameroun.

The East Coast Escarpment Montane Forest Group, therefore, has diverse origins. The montane species communities occurring in Africa today are clearly composed of species which have evolved, dispersed and contracted in large part independently of each other. Stuart (1983) carried out several analyses of bird distribution from biogeographical, ecological and morphometric angles. In each case the evidence for community-wide interspecific competition having a major influence on bird distributions was equivocal. It seems likely that many species which occur together today did not evolve sympatrically and so it is not surprising that the montane communities are now composed of species with diverse origins.

Table 10.3 presents data on the relationships between certain Montane Forest Groups. It can be seen that 91% of the South Malawi species and 78% of the South of Zambezi species occur in the East Coast Escarpment. These two southern Montane Forest Groups are best considered to be composed of distinctive species assemblages, almost all of which are probably derived from the eastern Tanzanian forests. The Central East African Montane Forest Group is best considered a merging zone between the East Coast Escarpment and the Albertine Rift, with 64% of its species occurring in the former and 78% in the latter. Species that originated in the East Coast Escarpment seem to have spread both to the north and to the south. It should also be noted from Table 10.3 that nearly half (47%) of the East Coast Escarpment species occur in the Albertine Rift. It is therefore probable that the forests of eastern Tanzania have 'received' species from this latter source (which is the major centre of montane species diversity and endemism in Africa) in addition to 'contributing' species to other parts of the continent.

Conclusion

From the overview presented in this chapter, it is clear that enough is now known of the zoogeography of the eastern Tanzanian montane forest avifauna to give a description of its relationships to avifaunas elsewhere in the continent, and of its diversity, endemism, distribution patterns, subspeciation and seasonal movements as well as its likely origins. However, this overview requires one major qualification. Many parts of the East Coast Escarpment are yet to be explored. For instance, almost nothing is known of the northern extension of the Nguru Mountains west of Handeni, nor of the Rubeho Mountains, the Uvidunda Mountains, most parts of the Uzungwas (especially the northern extension at Image), nor most parts of the Southern Highlands. Mountains lying outside the main escarpment, notably Malundwe, Mahenge, the Mbarika

Mountains, the Matengo Highlands and the Njesi Plateau (in Mozambique), are equally little known. The ornithological discoveries in the Mwanihana Forest and Chita Forest (East Uzungwas) since 1980 (see Stuart *et al.*, 1981, 1987; Stuart & Jensen, 1981; Jensen & Stuart, 1982; Jensen, 1983; Jensen & Brøgger-Jensen, 1992) have greatly changed the overall picture of the ornithology of the East Coast Escarpment forests. It is probable that future discoveries will necessitate a major revision of this chapter. The first task for future researchers is, therefore, basic inventory work in the unexplored areas. This is important, not least, in adequately determining conservation priorities.

The second important research task would be ecological research on the requirements of individual species. Very little is known of the ecology of nearly all these species, and only three major ecological studies have been carried out, those by Moreau (1935), Stuart (1983), and Dowsett-Lemaire (1989). At a time when pressing conservation problems demand increased action, a clearer knowledge of the ecological requirements of species of conservation concern is urgently required. Specific research needs include the following:

What are the minimum area requirements for individual species? It is known that several species can survive in tiny forest patches, probably because they can cross between them, provided the distances are not too great. But how far can forest birds travel? What are the long-term effects of habitat fragmentation and increased isolation of individual forest patches?

Much more work needs to be done on the phenomenon of seasonal altitudinal movements, since these have a considerable bearing on the sorts of conservation strategies that should be employed. Are the foothill forests at the bottom of the escarpments essential as wintering grounds for many montane species? Or do these species also migrate to other wintering sites further afield that remain to be discovered? What proportion and age/sex class of the population of each species migrates?

As elsewhere in the tropics, the effects of selective logging of forests on bird species need to be evaluated in greater detail. Are the current forest management practices sufficient to preserve the full range of species (including birds)? If not, what changes are needed?

The most important need is, of course, the conservation and wise management of the eastern Tanzanian forests. They represent a unique biological heritage, as well as the water catchment for much of Tanzania. It would indeed be a tragedy if many of these fascinating forests disappear before the species within them are discovered; the accompanying environmental deterioration that would result from wholesale forest loss would be an equal tragedy for the people of Tanzania.

Summary

In this chapter a description of our current knowledge of the montane forest avifauna of eastern Tanzania is given in relation to the biogeography of the African montane forest avifauna as a whole. The forests of eastern Tanzania are shown to be one of the main centres of diversity and endemism for montane forest birds on the African continent (the others being the Albertine Rift and Cameroun). The analysis focuses on the seven best known areas in Tanzania (the Usambaras, Ngurus, Ukagurus, Ulugurus, East Uzungwas, Mufindi and Mount Rungwe) and one in northern Malawi (the Nyika Plateau). The avifaunas of these areas are compared with those of similar habitats elsewhere on the African continent, and are analysed in terms of diversity, endemism, distribution patterns, subspeciation and likely origins. The Usambaras, Ulugurus and East Uzungwas emerge as the areas of highest diversity and endemism and, hence, conservation importance. Priorities for future research include surveys in unexplored areas, and studies on minimum area requirements for individual species of concern, seasonal altitudinal movements and their implications for conservation, and the effects of various management regimes (including selective logging) on the forest avifauna.

References

BAKER, N. E. (1984). Recent coastal records of the White-starred Forest Robin *Pogonocichla stellata* in Tanzania. *Scopus* **8**, 51–2.

BANGS, O. & LOVERIDGE, A. (1933). Reports on the scientific results of an expedition to the southwest highlands of Tanganyika Territory. 3. Birds. *Bulletin of the Museum of Comparative Zoology at Harvard* **75**, 143–221.

BEESLEY, J. S. S. (1972). Birds of the Arusha National Park, Tanzania. *Journal of the East Africa Natural History Society and National Museum* **132**, 1–32.

BENSON, C. W. & BENSON, F. M. (1977). *The Birds of Malawi*. Limbe: Montfort Press.

BRITTON, P. L. (1977). A Kenya record of *Cercococcyx montanus* at sea-level. *Scopus* **1**, 23–4.

BRITTON, P. L. (1978). The Andersen Collection from Tanzania. *Scopus* **2**, 77–85.

BRITTON, P. L. (ed.) (1980). *Birds of East Africa*. Nairobi: East Africa Natural History Society.

BRITTON, P. L. (1981). Notes on the Andersen Collection and other specimens from Tanzania housed in some West German museums. *Scopus* **5**, 14–22.

BRITTON, P. L., BRITTON, H. A. & COVERDALE, M. A. C. (1980). The avifauna of Mrima Hill, south Kenya coast. *Scopus* **4**, 73–8.

CHAPIN, J. P. (1923). Ecological aspects of bird distribution in tropical Africa. *American Naturalist* **57**, 106–25.

CHAPIN, J. P. (1932). The birds of the Belgian Congo. Part 1. *Bulletin of the American Museum of Natural History* **65**.

CLANCEY, P. A. (1971). *A Handlist of the Birds of Southern Mozambique*. Lourenço Marques: Instituto de Investigação Cientifica de Moçambique.

COLLAR, N. J. & STUART, S. N. (1985). *Threatened Birds of Africa and Related Islands. The ICBP/IUCN Red Data Book*. Cambridge: International Council for Bird Preservation and International Union for Conservation of Nature and Natural Resources.

COLLAR, N. J. & STUART, S. N. (1988). *Key Forests for Threatened Birds in Africa*. Cambridge: ICBP/IUCN.

DIAMOND, A. W. & HAMILTON, A. C. (1980). The distribution of forest passerine birds and Quaternary climatic change in Africa. *Journal of Zoology (London)* **191**, 379–402.

DOWSETT, R. J. (1971). The avifauna of the Makutu Plateau, Zambia. *Revue de zoologie et de botanique africaines* **84**, 312–33.

DOWSETT, R. J. (1980a). Extinctions and colonisations in the African montane island avifaunas. *Proceedings, 4th Pan-African Ornithological Congress*, pp. 185–97.

DOWSETT, R. J. (1980b). Post-Pleistocene changes in the distributions of African montane forest birds. *Acta XVII Congress of International Ornithology*, pp. 787–92.

DOWSETT, R. J. (1981). The past and present distribution of montane birds in Malawi. *Nyala* **7**, 25–45.

DOWSETT, R. J. & DOWSETT-LEMAIRE, F. (1979). The Mountain Buzzard *Buteo tachardus* in central Africa. *Scopus* **3**, 14–18.

DOWSETT-LEMAIRE, F. (1989). Ecological and biogeographical aspects of forest bird communities in Malawi. *Scopus* **13**, 1–80.

FRANZMANN, N.-E. (1983). A new subspecies of the Usambara Weaver *Ploceus nicolli*. *Bulletin of the British Ornithological Club* **103**, 49–51.

FRIEDMANN, H. (1928). A collection of birds from the Uluguru and Usambara Mountains, Tanganyika Territory. *Ibis* (12)**4**, 74–99.

FRIEDMANN, H. & STAGER, K. E. (1964). Results of the 1964 Cheney Tanganyikan Expedition. Ornithology. *Contributions in Science, Los Angeles* **84**, 3–50.

FRY, C. H. (1976). On the systematics of African and Asian tailorbirds (Sylviidae). *Arnoldia (Rhodesia)* **8**(6), 1–15.

FUGGLES-COUCHMAN, N. R. (1939). Notes on some birds of the Eastern Province of Tanganyika Territory. *Ibis* (14)**3**, 76–106.

FUGGLES-COUCHMAN, N. R. (1984). The distribution of, and notes on, some birds of Tanzania. *Scopus* **8**, 1–17, 73–8, 81–92.

FUGGLES-COUCHMAN, N. R. (1986). Breeding records of some Tanzanian birds. *Scopus* **10**, 20–6.

HALL, B. P. & MOREAU (1970). *An Atlas of Speciation in African Passerine Birds*. London: Trustees of the British Museum (Natural History).

HAMILTON, A. C. (1976). The significance of patterns of distribution shown by forest plants and animals in tropical Africa for the reconstruction of Upper Pleistocene palaeoenvironments: a review. *Palaeoecology of Africa, the Surrounding Islands and Antarctica* **9**, 63–97.

HILLMAN, J. C. (1986). *Bale Mountains National Park Management Plan*. Addis Ababa: Wildlife Conservation Organisation.

IRWIN, M. P. S. (1981). *The Birds of Zimbabwe*. Salisbury (Harare): Quest Publishing.

JENSEN, F. P. (1983). A new species of sunbird from Tanzania. *Ibis* **125**, 447–9.

JENSEN, F. P. & BRØGGER-JENSEN, S. (1992). The forest avifauna of the Uzungwa Mountains, Tanzania. *Scopus* **15**, 65–83.

JENSEN, F. P., BRØGGER-JENSEN, S. & PETERSEN, G. (1985). The White-chested Alethe *Alethe fuelleborni* in Tanzania. *Scopus* **9**, 127–32.

JENSEN, F. P. & STUART, S. N. (1982). New subspecies of forest birds from Tanzania. *Bulletin of the British Ornithological Club* **102**, 95–9.

JENSEN, F. P. & STUART, S. N. (1985). The origin and evolution of the Cameroon montane forest avifauna. In *Proceedings of the International Symposium on African Vertebrates: Systematics, Phylogeny and Evolutionay Ecology*, ed. K.-L Schuchmann, pp. 181–91. Bonn: Zoologisches Forschungsinstitut und Museum Alexander Koenig.

LOVERIDGE, A. (1922). Notes on East African birds (chiefly nesting habits and stomach contents) collected 1915–1919. *Proceedings of the Zoological Society of London* (1922), 837–62.

LOVERIDGE, A. (1933). Reports on the scientific results of an expedition to the south-western highlands of Tanganyika Territory. I. Introduction and zoogeography. *Bulletin of the Museum of Comparative Zoology of Harvard* **75**, 5–43.

MACKINNON, J. & MACKINNON, K. (1986). *Review of the Protected Areas System in the Afrotropical Realm*. Gland, Switzerland: International Union for Conservation of Nature and Natural Resources, and United Nations Environment Programme.

MANN, C. F. (1985). An avifaunal study in Kakamega Forest, Kenya, with particular reference to species diversity, weight and moult. *Ostrich* **56**, 236–62.

MILLER, R. I. STUART, S. N. & HOWELL, K. M. (1989). A methodology for analyzing rare species distribution patterns utilizing GIS technology: the rare birds of Tanzania. *Landscape Ecology* **2**(3), 173–89.

MOREAU, R. E. (1933). Pleistocene climatic changes and the distribution of life in East Africa. *Journal Ecology* **21**, 415–35.

MOREAU, R. E. (1935). A synecological study of Usambara, Tanganyika Territory, with particular reference to birds. *Journal Ecology* **23**, 1–43.

MOREAU, R. E. (1940). Distributional notes on some East African birds. *Ibis* (14)**4**, 454–63.

MOREAU, R. E. (1952). Africa since the Mesozoic, with particular reference to certain biological problems. *Proceedings of the Zoological Society of London* **121**, 869–913.

MOREAU, R. E. (1954). The distribution of African evergreen forest birds. *Proceedings of the Linnean Soceity of London* **165**, 35–46.

MOREAU, R. E. (1963). Vicissitudes of the African biomes in the late Pleistocene. *Proceedings of the Zoological Society of London* **141**, 395–421.

MOREAU, R. E. (1966). *The Bird Faunas of Africa and its Islands*. London: Academic Press.

MOREAU, R. E. & MOREAU, W. M. (1937). Biological and other notes on some East African birds. *Ibis* (14)**1**, 152–75, 321–45.

PINTO, A. A. DA R. (1983). *Ornitologia de Angola*, Vol. 1. Lisbon: Instituto de Investigacão Cientifica Tropical.

PRIGOGINE, A. (1953). Contribution à l'étude de la faune ornithologique de la région à l'ouest du Lac Edouard. *Annales, Musée Royal du Congo Belge, Sciences zoologiques* **24**, 1–117.

PRIGOGINE, A. (1960). La faune ornithologique du massif du Mont Kabobo. *Annales, Musée Royal du Congo Belge, Sciences zoologiques* **85**, 1–46.

PRIGOGINE, A. (1971). Les oiseaux de l'Itombwe et de son hinterland, 1. *Annales, Musée Royal de l'Afrique Centrale, Sciences zoologiques* **185**, 1–298.

PRIGOGINE, A. (1978). Les oiseaux de l'Itombwe et de son hinterland, 2. *Annales, Musée Royal de l'Afrique Centrale, Sciences zoologiques* **223**, 1–134.

PRIGOGINE, A. (1984). Les oiseaux de l'Itombwe et de son hinterland, 3. *Annales, Musée Royal de l'Afrique Centrale, Sciences zoologiques* **243**, 1–146.

PRIGOGINE, A. (1985). Conservation of the avifauna of the forests of the Albertine Rift. *ICBP Technical Publication* **4**, 277–95.

RIPLEY, S. D. & HEINRICH, G. H. (1966). Comments on the avifauna of Tanzania, 1. *Postilla* **96**, 1–45.

RIPLEY, S. D. & HEINRICH, G. H. (1969). Comments on the avifauna of Tanzania, 2. *Postilla* **134**, 1–21.

SCLATER, W. L. & MOREAU, R. E. (1932-3). Taxonomic and field notes on some birds of north-eastern Tanganyika Territory. *Ibis* (13)**2**, 487–522, 656–83; **3**, 1–33, 187–219, 399–440.

SERLE, W. (1950). A contribution to the ornithology of the British Cameroons. *Ibis* **92**, 343–76, 602–38.

SERLE, W. (1954). A second contribution to the ornithology of the British Cameroons. *Ibis* **96**, 47–80.

SERLE, W. (1964). The lower altitudinal limit of the montane forest birds of the Cameroon Mountain, West Africa. *Bulletin of the British Ornithological Club* **84**, 87–91.

SNOW, D. W. (ed.) (1978). *An Atlas of Speciation in African Non-passerine Birds.* London: Trustees of the British Museum (Natural History).

STUART, S. N. (1980). Birds that vanish while we turn our backs. *Oryx* **15**, 367–70.

STUART, S. N. (1981a). A comparison of the avifaunas of seven East African forest islands. *African Journal of Ecology* **19**, 133–51.

STUART, S. N. (1981b). An explanation of the disjunct distributions of *Modulatrix orostruthus* and *Apalis* (or *Orthotomus*) *moreaui*. *Scopus* **5**, 1–4.

STUART, S. N. (1983). Biogeographical and ecological aspects of forest bird communities in Eastern Tanzania. PhD dissertation, University of Cambridge.

STUART, S. N. (1985). Rare forest birds and their conservation in Eastern Africa. *ICBP Technical Publication* **4**, 187–96.

STUART, S. N. (ed.) (1986). *Conservation of Cameroon Montane Forests* Cambridge: International Council for Bird Preservation.

STUART, S. N., HOWELL, K. M., VAN DER WILLIGEN, T. A. & GEERTSEMA, A. A. (1981). Some additions to the avifauna of the Uzungwa Mountains, Tanzania. *Scopus* **5**, 46–50.

STUART, S. N. & JENSEN, F. P. (1981). Further range extensions and other notable records of forest birds from Tanzania. *Scopus* **5**, 106–15.

STUART, S. N. & JENSEN, F. P. (1985). The avifauna of the Uluguru Mountains, Tanzania. *Gerfaut* **75**, 155–97.

STUART, S. N., JENSEN, F. P. & BRØGGER-JENSEN, S. (1987). Altitudinal zonation of the avifauna in Mwanihana and Magombera Forests, eastern Tanzania. *Gerfaut* **77**, 165–86.

STUART, S. N. & TURNER, D. A. (1980). Some range extensions and other notable records of forest birds from eastern and northeastern Tanzania. *Scopus* **4**, 36–41.

STUART, S. N. & VAN DER WILLIGEN, T. A. (1980). Is Moreau's sunbird *Nectarinia moreaui* a hybrid species? *Scopus* **4**, 56–8.

URBAN, E. K. & BROWN, L. H. (1971). *A Checklist of the Birds of Ethiopia.* Addis Ababa: Haile Selassie I University Press.

WHITE, F. (1978). The Afromontane Region. In *Biogeography and Ecology of Southern Africa*, ed. M. J. A. Werger, pp. 463–513. The Hague: Junk.

WHITE, F. (1981). The history of the Afromontane archipelago and the scientific need for its conservation. *African Jounral of Ecology* **19**, 33–54.

WILLIAMS, J. G. (1951). Notes on *Scepomycter winifredae* and *Cinnyris loveridgei*. *Ibis* **93**, 469–70.

11   Mammals in the forests of eastern africa

J. KINGDON AND K. M. HOWELL

Introduction

Aims

It is the purpose of this chapter to list and identify the mammalian forest fauna of eastern Africa, updating the list with new information. We provide some simple measures of affinity and distinctness in the forest mammal communities of far eastern Africa. We consider to what extent the affinities might reflect eastward dispersal from a westerly focus or a residual forest fauna impoverished by climatic degradation. Two principal levels of endemism are identified: those species that are broadly distributed in the forests of eastern Africa and those that have a very restricted distribution within this region. Further categories of endemics are also identified, restricting species to (i) montane, (ii) lowland and (iii) coastal forest types, including those found on Zanzibar, Pemba and Mafia Islands. The possible importance of colonisation, competition and isolation in the evolution of distinct species or subspecies is discussed.

A list of forest-dependent mammals from eastern and southwestern Tanzania has been assembled and is compared with a list of comparable mammals from western Uganda (Budongo, Kibale, Kalinzu, Kayonza and Bwindi). Those species held in common are listed in Table 11.1, with local subspecies shown separately under their regional heading. From these lists some simple measures of endemism have been estimated.

Table 11.1. *Forest mammals from two regions of eastern Africa*

Western Uganda	Species in common	East and southwest Tanzania
Insectivora		
TENRECIDAE		
Potamogale velox	–	–
Micropotamogale ruwenzorii	–	–
SORICIDAE		
Myosorex blarina	–	*Myosorex geata*
–	–	*Sylvisorex howelli*
Sylvisorex lunaris	–	–
	Sylvisorex megalura	
Scutisorex somereni	–	–
Crocidura maurisca	–	–
–	–	*Crocidura nigricans*
	Crocidura hildegardae	

Table 11.1. (*cont.*)

Western Uganda	Species in common	East and southwest Tanzania
	Crocidura bicolor	
–	–	*Crocidura tansaniana*
–	–	*Crocidura telfordi*
–	–	*Crocidura usambarae*
CHRYSOCHLORIDAE		
Chrysochloris stuhlmanni stuhlmanni	*Chrysochloris stuhlmanni*	*Chrysochloris stuhlmanni tropicalis*
Chrysochloris stuhlmanni fosteri		
MACROSCELIDIDAE		
Rhynchocyon cirnei stuhlmanni	*Rhynchocyon cirnei*	*Rhynchocyon cirnei petersi*
		Rhynchocyon cirnei chrysopygus

Chiroptera

PTEROPODIDAE		
Myonycteris torquata	–	*Myonycteris relicta*
–	–	*Pteropus comoroensis*
–	–	*Pteropus voeltzkowi*
	Stenonycteris lanosus	
	Lissonycteris angolensis	
Micropteropus pusillus	–	–
Epomops franqueti	–	–
	Eidolon helvum	
Hypsignathus monstrosus	–	–
Megaloglossus woermanni	–	–
–	–	*Epomophorus wahlbergi*
EMBALLONURIDAE		
Taphozous peli	–	–
–	–	*Taphozous hildegardae*
NYCTERIDAE		
Nycteris nana	–	–
	Nycteris grandis	
RHINOLOPHIDAE		
Rhinolophus ruwenzorii	–	–
–	–	*Rhinolophus swinnyi*
Rhinolophus alcyone	–	–
Hipposideros cyclops	–	–
Hipposideros camerunensis	–	–
	Hipposideros commersoni	
VESPERTILIONIDAE		
Pipistrellus nanulus	–	–
	Eptesicus tenuipennis	
	Mimetillus maloneyi	
Glauconycteris gleni	–	–
Glauconycteris superba	–	–
Glauconycteris humeralis	–	–
MOLOSSIDAE		
Myopterus whitleyi	–	–
Tadarida nanula	(synonymous?)	*Tadarida brachypterus*
Tadarida congica	–	–

Table 11.1. (*cont.*)

Western Uganda	Species in common	East and southwest Tanzania
Tadarida thersites	–	–
Tadarida leonis	–	–
Tadarida aloysiibaudiae	–	–

Primates

LORISIDAE
Perodicticus potto	–	–
	Galago demidovii	
Galago thomasi	–	*Galago garnetti*
–	–	*Galago zanzibaricus*
Galago inustus	–	–

CERCOPITHECIDAE
Colobus b. tephrosceles	*Colobus badius*	*Colobus b. kirkii, C. b. gordonorum*
	Colobus guereza	
Colobus p. ruwenzorii, C. p. adolfi-friederici	*Colobus polykomos*	*Colobus p. sharpei*
Cercocebus albigena	–	*Cercocebus galeritus* subsp. nov.
Cercopithecus l'hoesti	–	–
Cercopithecus m. stuhlmanni, C. m dogetti	*Cercopithecus mitis*	*Cercopithecus m. moloneyi, C. m. albogularis*
Cercopithecus neglectus	–	–

PONGIDAE
| *Gorilla gorilla* | – | – |
| *Pan troglodytes* | – | – |

Rodentia

ANOMALURIDAE
| | *Anomalurus derbianus* | |

MURIDAE
Deomys ferrugineus	–	–
–	–	*Beamys hindei*
Cricetomys emini	–	–
–	–	*Cricetomys cosensi*
	Dendromys mystacalis	
Praomys jacksoni	–	–
–	–	*Praomys delectorum*
Praomys aeta	–	–
Praomys stella	–	–
Malacomys longipes	–	–
Colomys goslingi	–	–
Oenomys hypoxanthus	–	–
	Lophuromys flavopunctatus	
	Lophuromys sikapusi	
Lophuromys woosnami	–	–
Hybomys univittatus	–	–
Thamnomys venustus	–	–
Thamnomys rutilans	–	–
–		*Otomys anchietae*
	Otomys denti	

Table 11.1. (*cont.*)

Western Uganda	Species in common	East and southwest Tanzania
Otomys irroratus	–	–
–	–	*Otomys angoniensis*
Otomys typus	–	–
HYSTRICIDAE		
Atherurus africanus	–	–
SCIURIDAE		
Funisciurus carruthersi	–	–
Funisciurus pyrrhopus	–	*Funisciurus (Paraxerus) lucifer*
Funisciurus alexandri	–	*Funisciurus palliatus*
Funisciurus boehmi	–	*Funisciurus vexillarius*
	Heliosciurus rufobrachium	
Protoxerus strangeri	–	–
Heliosciurus ruwenzorii	–	–

Hyracoidea

PROCAVIIDAE		
	Dendrohyrax arboreus	
Dendrohyrax dorsalis	–	*Dendrohyrax validus*

Artiodactyla

SUIDAE		
Hylochoerus meinertzhageni	–	–
BOVIDAE		
Tragelaphus euryceros	–	–
Neotragus batesi	–	*Neotragus moschatus*
Cephalophus nigrifrons	–	–
Cephalophus callipygus	–	*Cephalophus harveyi*
Cephalophus rubidus	–	*Cephalophus adersi*
–	–	*Cephalophus spadix*
Cephalophus sylvicultor	–	–
	Cephalophus monticola	

Pholidota

MANIDAE		
Manis tricuspis	–	–

Carnivora

MUSTELIDAE		
Aonyx congica	–	–
VIVERRIDAE		
Genetta victoriae	*Genetta tigrina*	–
Genetta servalina intensa	*Genetta servalina*	*Genetta servalina lowei*
	Nandina binotata	
Bdeogale nigripes	–	*Bdeogale crassicauda*
Crossarchus alexandri	–	–
FELIDAE		
Felis aurata	–	–

Table 11.2. *Comparison of major taxa of forest and non-forest mammals*

	West Uganda forests	East and southwest Tanzania forests	East and southwest Tanzania non-forest
Primates	13	8	2
Pangolin	2	0	1
Hyraxes	2	2	2
Orycteropus	0	0	1
Insectivores	11	11	12
Fruit bats	8	7	4
Insect bats	20	7	43
Squirrels	7	4	4
Other rodents	21	9	37
Hares	0	0	4
Carnivores	8	4	24
Large herbivores	2	0	24
Duikers and pygmy antelope	6	5	3
Totals	100	57	161

The interest of mammals

There are indisputably far fewer species in the East African forests than in Central and West Africa. Furthermore, the shared occurrence of various rodent, insectivore and primate species, blue duiker *Cephalophus monticola*, palm civet *Nandinia binotata*, and the anomalure *Anomalorus derbianus*, has encouraged the view that these forests are merely impoverished outliers of the Guineo–Congolian block. The small number of species in common (see Table 11.2) implies that tenuous bridges between these regions have been traversed by only a limited number of versatile or mobile species or that exceptional genetic stability has been maintained in disjunct populations over very long periods.

The distinctness of each endemic species can be examined for clues to its origins. Do its affinities with western species suggest a modified descendant from an earlier specifically forest connection? Are there indications of derivation from a non-forest lineage? Could they be relicts of former dominant or widespread populations that once embraced a wider spectrum of habitats? Are its origins opaque? Every endemic can, to a greater or lesser extent, be described as a localised specialist but the likelihood of varied adaptive backgrounds lends every eastern endemic a peculiar interest. There must have been interchanges between the major biotopes of forest and various grades of wooded grasslands and the climatic oscillations of the Plio–Pleistocene (see Chapters 2–5, this volume) would have created opportunities for animals to adapt towards moister or drier regimes even if the core areas of forest were relatively stable.

The eastern forest mammal fauna is not only less numerous in species but the counterparts from the two regions differ both at specific level and at the more subtle level of ecological niche. Those species that are common to east and west are likely to occupy different niches because the eastern community has a different, often simpler composition. Fewer or different food plants may imply greater specialisation, while fewer or different competitors might broaden the niche.

Previous studies

Matschie (1895) compiled the first checklist for what was then Deutsch Ost Afrika. Loveridge (1923, 1928) and Allen & Loveridge (1927, 1933) reported on mammals collected for the Harvard Museum of Comparative Zoology and the Smithsonian Institution. Moreau & Pakenham (1941) analysed the island faunas of Pemba, Zanzibar and Mafia. C. J. P. Ionides made casual observations and collections in southeastern Tanzania during the 1950s. In 1951 Swynnerton & Hayman published an inventory of Tanganyika's mammal fauna and between 1971 and 1982 one of us brought out a study of East African mammals that was comprehensive at the time of its publication (Kingdon, 1971–82). Rodgers, Owen & Homewood (1982) and Grubb (1983) published papers on the biogeography of East African forest mammals and Struhsaker (1981) discussed how East African primate species may be related to western populations. Pakenham (1984), Swai (1984) and Mturi (1984) have dealt mainly with the relationship of Zanzibar forest mammals to those further west. Howell & Jenkins (1984), Jenkins (1984) and Hutterer (1986) have expanded our knowledge of shrew taxonomy and distribution and Kock & Howell (1988) that of forest bats.

There is still much basic collecting and documentation to be done on the smaller mammals of Tanzania's eastern forests. For example, in 1981 a new subspecies of mangabey monkey was found in the Uzungwa Mountains (Homewood & Rodgers, 1981), and in the years 1984–6 three new species of shrews were described from the Uluguru and Usambara mountains. Even older collections, such as those for the British Museum by Willoughby P. Lowe in the 1930s have, on re-examination, turned up undescribed forms, such as the extremely isolated and distinct servaline genet, *Genetta servalina lowei*, from Dabaga in the Uzungwa Mountains.

There is also uncertainty about the taxonomic status of some species. For example, it is still not known whether the smaller galagos belong to two or three distinct species. The Zanzibar galago, *Galago zanzibaricus* (which was once regarded as a subspecies of *G. senegalensis*) is widely distributed through the forests of the eastern coasts and mountains. The Senegal bushbaby, *G. senegalensis*, has also been recorded from forest as well as its more usual habitat in drier thickets and woodlands. The overall distribution of *G. zanzibaricus* is currently in some doubt, because it may have been confused in some areas with the minuscule

G. demidovii. Very small animals collected from the Rondo Plateau in 1952 have been allocated to this species (Swynnerton & Hayman, 1951; S. Bearder, *in litt.*). It is now thought possible that *G. demidovii* occurs in a discontinuous scatter from Zaïre across to the Southern Highlands in Malawi and to the Indian Ocean littoral.

One of us (J. K.) has worked mainly in the Southern Highlands, while K. H. has visited and trapped in several forest areas from the Usambaras through the Uzungwa as well as coastal forest and thicket (Howell, 1981). Primatologists K. Homewood, S. Wasser and F. Mturi have also been active in making observations in forest areas. Ecologist W. A. Rodgers has participated in and stimulated studies of mammals, and it was during a brief expedition by Homewood and Rogers that the Sanje mangabey was discovered. There can be little doubt that systematic study will augment the species listed here and add new species or subspecies to the inventory. The ratios and figures quoted in subsequent pages should therefore take into account our incomplete knowledge.

Measures of endemism in Eastern African forest fauna

African mammalian fauna from forests tends to be radically different from that of the savannas that surround it. Considering what a minuscule proportion of Tanzania is covered in moist forest, only about 2%, this fauna is astonishingly rich: 24% of the mammal fauna for the eastern and southwestern half of Tanzania is forest-adapted or forest-dependent. A few species, mainly the larger ones and some bats, inhabit both forest and non-forest but this is not the rule in most groups. Typical species found in both forest and non-forest habitats are elephant, buffalo, bushbuck, blotched genet and civet. Of bats, the giant leaf-nosed bat *Hipposideros commersoni*, the hairy bat *Myotis bocagei* and the pipistrelle *Pipistrellus nanus* are examples. The proportion of forest mammal species restricted to montane areas is 11%. About 9% are known only from lowland forest, while the rest range through both highland and lowland forests. Nonetheless, very distinctive local populations such as montane hyrax and squirrels are sometimes restricted to a particular altitude. When the eastern forest fauna is compared with that of western Uganda, overall numbers are much lower, 57–100 species. However, this mammalian impoverishment is taxonomically uneven and mainly concerns primates, bats and rodents.

When the two forest faunas are compared order by order with the numbers of non-forest mammals in east and southwest Tanzania, much greater disproportions appear. As could be expected, large herbivores, carnivores and arid-adapted rodents and bats all greatly outnumber related forest types in both Tanzania and Uganda. Not surprisingly, forest primates, fruit bats and duikers outnumber savanna forms.

There are nine species that are narrowly endemic to eastern forests (and islands) and another 11 species that are widespread through the forests of the eastern mountains and littoral (including degraded thicket-like forests); this more broadly defined endemism accounts for 35% of the eastern fauna. Most of these 20 species may have equivalent species in the main forest block but they are, nonetheless, a very distinctive array with well-established adaptations to local conditions. This equivalence will be discussed below.

When the forest species in eastern and southwestern Tanzania are compared with western Uganda, 29 are shared, which represents 50% of the total. Most seem to be successful and widely distributed animals with relatively broad ecological tolerance, and they presumably combine this with an aptitude for colonisation across degraded habitat or along riverine strips. A number of these are bats and carnivores with broad spectrum diets, but the golden mole *Chrysochloris stuhlmanni*, and red colobus *Colobus badius*, are less adaptable in this respect and separation of eastern and western populations may be of very long standing.

When closely related species in the two regions are matched, 60% of the eastern forests mammals broadly resemble their counterparts in the main Guineo–Congolian forest block. However, gross comparison at the species level conceals the extent of endemism in the mountains and islands, which number over 40 local and sometimes very distinct subspecies.

Impoverishment

A major difference between the two regions is that many mammals from the Guineo–Congolian forest block tend to be sympatric specialists belonging to generic radiations, while their eastern congeners are likely to be single, locally adapted allopatric isolates. This applies especially to monkeys, squirrels and duikers. Larger forest specialists are absent (apes, giant pangolin, giant hog and bongo) as are a range of larger frugivorous bats and slow-flying insectivorous bats. These mostly find the eastern boundaries of their range along the western Rift Valley in Uganda or, in the case of the bongo and giant hog, Mount Kenya.

The small extent of eastern forests may be sufficient to explain the absence of large forest mammals which need extensive ranges. Likewise, the climatic vicissitudes of the Plio–Pleistocene anthropogenic degradation (Hamilton, 1981) could have so diminished the area or degraded the quality of these forests that they became uninhabitable for some species, especially those that live at low densities beneath a closed canopy in lowland rain forest. Narrow corridors could be equally inhibiting for colonisation or recolonisation by low density specialists that are dependent upon a measure of climatic stability, continuity of habitat and species diversity in foods or prey.

The absentees from eastern Africa are in themselves significant. The potto is effectively wholly arboreal, making it a poor disperser. The forest pangolins are likely to depend upon year-long density and a long-term reliability of ants and termites that might not be met in the east, especially at higher altitudes. For aquatic and semi-aquatic insectivores or rodents such as *Potamogale*, *Scutisorex*, *Deomys* and brush-tailed porcupines the drying out of rivers would have posed impossible vicissitudes, especially for species with invertebrate diets. Large herbivores with wide ranges that included non-forest areas would be exposed to higher levels of competition and predation; for bongo and giant hog such exposure could have been critical to survival.

The number of mammal species within any one forest block is likely to have a relationship to (i) the overall area of forest, (ii) its ecological diversity, (iii) its climatic history, especially its relative stability in the past and (iv) its human history, especially past population density and hunting pressure. Thus the Uzungwa forests, which cover the largest area and a broad spectrum of altitudes and edaphic conditions, have a more diverse mammal community. This is most evident in primates, where the Sanje mangabey and red colobus occur together with the more widely distributed guenons and pied colobus and in the occurrence of exclusively forest species such as the servaline genet, *Genetta servalina lowei*.

The diverse origins of eastern endemics

The Guineo–Congolian rain forest block may not be the only source of mammal fauna in eastern Tanzania. It is likely that many forest organisms derive from more arid habitats and it is possible that the eastern forests retain populations in which adaptation to forest-dwelling is less advanced than in closely related types that live further to the west. If this is so, it invests the Tanzanian forest communities with some potential for an understanding of how non-forest organisms adapt to moister conditions as well as forest forms coping with drier regimes.

During very moist periods, engulfment of large areas of formerly dry country by forest would have selected for many arid country species to adapt to the new conditions. This would affect local populations in specific localities as well as larger populations across very broad fronts running all the way between the Atlantic and Indian Ocean shores. Cycles of afforestation in central Africa and especially over the Zaïre basin would have been comprehensive and their effect on mammal populations would have been correspondingly extensive.

Similar developments along the eastern seaboard would have been synchronised but less extensive in scope, and only a few populations would have connected up with their western counterparts during each climatic pulse. Would the frequent recurrence of comprehensive reafforestation over central Africa have exerted more pressure for adaptive change? Could eastern populations have remained genetically less advanced because they were detached from such developments? There certainly appear to be several eastern species that are more conservative in their adaptations than their western equivalents.

Consider some of the pairs listed in Table 11.3. *Dendrohyrax dorsalis* is a long-limbed, entirely arboreal lowland forest animal with a long skull and muzzle. By contrast, *D. validus* shows proportions that are closer to those of the more terrestrial *Heterohyrax* and *Procavia*. The adoption of arboreal habits appears to be a late and advanced condition in hyraxes. It is therefore striking that

Table 11.3. *Western forest equivalents of Eastern forest endemic mammal species*

Western equivalents of related forms	Eastern endemics
Perodicticus potto (possible equivalent)	*Galago garnetti*
Galago inustus	*Galago zanzibaricus*
Dendrohyrax dorsalis	*Dendrohyrax validus*
Numerous specialist forms of squirrels	*Funisciurus palliatus/lucifer* (medium-sized generalized complex)
Cricetomys emini	*Cricetomys cosensi* and *Beamys hindei*
more wholly arboreal *Praomys* spp.	*Praomys delectorum*
other *Thamnomys* spp.	*Thamnomys cometes*
Bdeogale nigripes	*Bdeogale crassicauda*
Neotragus batesi	*Neotragus moschatus*
Cephalophus callipygus and *C. nigrifrons*	*Cephalophus harveyi* and *C. natalensis*

the more conservative form of tree hyrax should be restricted to the Tanzanian mountains and islands, having presumably been displaced from the surrounding lowland forests, which are now inhabited by the very widespread *D. arboreus*, which is intermediate in form between *D. validus* and *D. dorsalis* (Kingdon, 1971). *D. arboreus* is in turn a montane species in western Uganda, while the lowland forests are in the process of being colonised by *D. dorsalis* (which is replacing *arboreus*). It is as though three layers of advancing arborealists are superimposed on each other. However, the hyrax story may have still further subtleties, because the rocky alpine habitats on Mount Rungwe are occupied by the semi-arboreal *Heterohyrax brucei lademanni* although it is entirely ringed by forest where *Dendrohyrax arboreus* is dominant. This suggests that a *Heterohyrax* which had already adapted to subalpine Mount Rungwe might have pre-empted later colonisation by *Dendrohyrax*.

The suni, *Neotragus moschatus*, has an extensive range all down eastern Africa. It represents a stable and successful but broadly very conservative bovid type that is almost certainly of non-forest origin. Its closest relatives, *N. batesi* in the Congolian block and *N. pygmaeus* in upper Guinea, are wholly forest dwellers but appear to be more derived species than suni (see Kingdon, 1982a).

The greater galagos have recently been sub-divided into two distinct species (Olsen, 1979). One of these is the eastern endemic *Galago garnetti*, which is more terrestrial and can tolerate very much drier thickets and more open forests than its ecological equivalent the potto, *Perodicticus potto*, which feeds on similar foods and shares several aspects of its ecological niche with the greater galago. *G. crassicaudatus* (until recently the species to which *G. garnetti* was assigned) occupies drier country including arid woodland, and the absence of any large galagos anywhere in the main forest block suggests that galagos, especially the larger species, were never true forest species. If this also applies to the smaller species this would put the eastern endemic *G. zanzibaricus* into a crucial intermediate position because it resembles both *G. demidovii* in the main

forest block and the Senegal galago *G. senegalensis* in the savannas. A thick-tailed dwarf galago that builds leaf nests and may be semi-social has recently been discovered to coexist in forest south of the Rufiji River with the much commoner and more widespread *G. zanzibaricus*.

The dog mongooses also conform to a pattern of disjunction between easterly and westerly forest forms. *Bdeogale crassicauda* is a more conservative species that is widely distributed in eastern African woodlands and forests. The Zanzibar Island race, *B. c. tenuis*, has been suggested to be the stranded fragment of a still earlier population. Further up the coast another race, *B. c. omnivora*, may represent an incipient species. This mongoose is also stranded in the northernmost coastal enclave of Sokoke Forest and is intermediate in colouring and tooth structure between *B. crassicauda* and *B. nigripes*, which is a specialist forest mongoose that has a patchy distribution between Cameroun and Mount Kenya. Stranding is likely to be an appropriate way of describing the origins of several Zanzibar endemics but especially a giant rat of peculiar affinities, *Cricetomys cosensi*. This pouched rat is related not only to the widespread *C. gambianus* but also to the eastern endemic *Beamys hindei* which, perhaps because it is extremely trap-shy, has been recorded in only a few montane and coastal localities between the mountains around Lake Nyasa and the East African coast. The interesting point about *C. cosensi* is that its closest affinities are with *C. emini*, a western forest species that is unknown from Tanzania. *C. gambianus* occupies a wide variety of relatively moist habitats on the mainland, where it is sympatric with *Beamys* in the few places where the latter has been found.

If *C. gambianus* is the most recently evolved type of pouched rat, which seems likely, the relict status of *Beamys*, a smaller but demonstrably less evolved member of the same lineage, is more comprehensible. What is uncertain is the degree to which *Beamys* is physiologically or adaptively restricted to montane forest, riverine and lowland well-wooded habitats, or is limited by competition with other rodents, notably *Cricetomys*. Its forest or near-forest adaptation could therefore be a by-product of its evolutionary and competitive history. Similar considerations could explain the stranding of *C. cosensi* on Zanzibar while *C. emini* is now tied to the main forest block. Both *emini* and *cosensi* are capable of behaving like *gambianus* in that they will invade villages, towns and cultivation within their habitat. The ecological and adaptive success of *C. gambianus* may therefore be the single most important factor in limiting the distribution of *C. emini* and possibly *Beamys*. Both may be intrinsically less exclusive forest mammals than they appear and this could apply to other eastern endemics.

Red duikers are common in most African forest areas. In Tanzania, the larger *Cephalophus harveyi* has a predominantly northern and central distribution, while *C. natalensis* occurs in the south. Intermediate forms, which may be hybrids, are known from what could be an overlap zone running inland from the general vicinity of Dar es Salaam. An equivalent niche in the rich lowland forests of Uganda is occupied by the substantially larger *C. callipygus*. In Kenya presumed hybrids between *C. callipygus* and *C. harveyi* have been collected along the zone of contact on the eastern face of the Mau.

Equatorial montane habitats from Cameroun to Mount Kenya are inhabited by the long-legged, long-hooved *C. nigrifrons*. This species prefers boggy, wet areas within the high rainfall belts. Equivalent habitats in Tanzania are occupied by the less specialised *C. harveyi*. It is a typical pattern that there are more specialisations and more species in the Guineo–Congolian forests while the east possesses fewer species of greater adaptability and a broader ecological range.

A group that typifies endemism along the littoral and in the mountains of eastern Africa is the *Funisciurus palliatus/lucifer* complex of squirrels. A succession of discretely differing populations of *F. palliatus* is strung down the coast from Somalia to Zululand. On the mountain massifs inland there is a variety of mountain squirrels, which one of us has united into a single species, *F. lucifer* (Kingdon, 1974b). A more advanced and successful species of squirrel, *Heliosciurus rufobrachium*, is the only medium-sized tree squirrel in the drier surrounding woodlands but is sympatric with *Funisciurus* types in the richer and more diversi-

fied forests further west (except at higher altitudes). *Funisciurus* is presumably competitively inferior in dry woodland today, but was it always unable to cross dry woodland? Alternatively, could the arrival of *Heliosciurus* have inhibited it?

Subsequent isolation of the mountain squirrel complex has been sufficiently lengthy to have led to differentiation into at least five distinct forms living on some 12 mountain massifs. In a single small locality on the West Usambaras there is a very small population of a unique species, *F. vexillarius*. One of us (Kingdon, 1974b) has suggested that this squirrel could be a hybrid swarm derived from mixing between *F. lucifer byatti* and *F. palliatus* but the population appears to be monomorphic and so may represent an independent offshoot from the *F. lucifer* complex. Competition is but one factor restricting such endemics to montane areas. Conditions have undoubtedly been more benign but also more severe in the past. Today's forest patches and galleries may be vestiges of larger areas that existed in the past and the core of their habitats may or may not have remained relatively stable, but their proximity to much drier environments would always have been closer than in most western forests.

Mammal groups probably vary greatly in their responses to geographic isolation, fragmentation of populations, ecological constriction or community impoverishment. A smaller choice of food plants may force one species to become more specialised in its diet while fewer potential competitors allow another to broaden its niche. A species prone to colour variation (such as the squirrels and the giant elephant shrews) will more readily generate regional varieties or subspecies through isolation than a less variable one.

Where eastern and western populations of the same species can be compared it may be worth testing these aspects of their environment to see if the survival of more versatile types has been favoured in the east. If some forest species have been recruited from non-forest populations it will be interesting to determine to what extent they could have lost their tolerance for relatively dry conditions, whether their restriction is primarily attributable to competitive exclusion in more difficult habitats, or both. It is possible that western populations may have progressed further towards specialised forest-living and moisture dependence than some eastern populations.

Island or land bridge?

There has been much theoretical discussion as to whether Tanzanian montane biota belong to oceanic island analogues or to broken land bridge islands (Diamond & Hamilton, 1980; Diamond, 1981; Stuart, 1981). Recruitment from non-forest sources has had scant discussion. As far as mammals are concerned, we have tried to show that all three elements seem to be present in the mammalian fauna of eastern Tanzania. Furthermore, intrinsically 'land bridge' type species (such as *Bdeogale* and *Cricetomys*) can be identified in the fauna of actual islands such as Zanzibar (Pakenham, 1979).

It could be argued that phyletically ancient groups such as the golden moles and elephant shrews are more likely to belong to land bridge fauna, while recently evolved, versatile species are more likely to be oceanic island type colonists. This, however, takes little account of widely differing rates of speciation within mammalian taxa. Some isolated species might have been able to maintain their basic genotype over millions of years while others differentiate very readily in response to local peculiarities of climate. Sometimes both elements may be present in the same species. For example, *Rhyncocyon* is known to be a very conservative genus. Its anatomical peculiarities and ecological niche date back to the Oligocene. Its colouring, on the other hand, is highly labile and variable, so the criteria for distinguishing endemic subspecies can ultimately turn on our proclivity to seize upon the more obvious and superficial aspects in an array of museum skins.

Forest endemics of restricted distribution in eastern Africa

Several important mammalian endemics have a much more restricted distribution than the species listed above, but discussion of subspeciation in local populations involves rather different considerations. When isolate subspecies of forest mammals from far eastern Africa are examined they fall into three very distinct categories: montane, lowland and island.

Seventeen subspecies of mammals have been described from the islands of Zanzibar, Pemba and Mafia. An equal number have been described from the mountain massifs of eastern and southwestern Tanzania. The only clearly defined discrete focus for lowland forest mammals on the mainland is found in the narrow enclaves of forest that run up from Mozambique to the Kenya coast, from which some eight subspecies have been named. All share the common characteristics of physical and ecological isolation. Major rivers along the eastern seaboard are known to define the boundaries of some savanna mammal distributions, notably the Rufiji. It remains to be seen whether forest mammals have also been limited by rivers or by Pleistocene oceanic incursions up the Rufiji estuary. When the detailed composition of subspeciating mammals is examined it is clear that broadly the same groups are involved, namely cercopithecid primates, insectivores, squirrels, viverrids, carnivores and duikers. In the impoverished communities to which they belong, each of these groups seems to fill a single, rather broader niche than their western congeners and, in most instances, they are relatively numerous (or have been until recently), even if only in very restricted localities.

Several species from the Kenya coastal enclave have markedly different forms in Tanzania, notably *Rhyncocyon cirnei chrysopygus*, *Colobus badius rufomitratus* and *Bdeogale crassicauda omnivora*. It is possible that this community once had some direct westerly connection across what is now arid and once volcanically active land.

It has been postulated elsewhere (Kingdon, 1982a) that the overall evolutionary trend in the main line 'red' duiker lineage has been a phylogenetic increase in body size. On this criterion Ader's duiker, at 6.5–12 kg, is the most conservative species in the entire red duiker radiation. It may therefore be significant that we find it 'stranded' on two marginal forest outliers, Zanzibar and the Kenya coastal enclave.

Cephalophus spadix is another duiker restricted

to a few eastern enclaves but with a larger, very closely related form, *C. sylvicultor*, ranging right through the main forest block. It too appears to owe its existence as a unique endemic to the inability or delay of *C. sylvicultor* in colonising these mountain forests.

Another mammal that may be stranded in the eastern mountains is *Myosorex geata*, a very isolated representative of the most conservative genus of African shrews. Other species occur on Uganda mountains, and *Myosorex* is relatively common in the temperate cul-de-sac of South Africa, which has served as a last refuge for numerous ancient groups. The golden moles, *Chrysochloris*, have a similar distribution pattern and it would seem that both these highly specialised insectivores may belong to very stable 'older' types which have enjoyed a long head-start in developing their specialisations.

None of these specialised insectivores are exclusively forest animals but they may be typical of eastern isolates as a whole in tending to combine phyletic conservatism and a refined specialisation to (relatively moist) local conditions.

The levels of phyletic age must vary enormously from taxon to taxon, and it seems very unlikely that the mammal fauna has been acquired over one or two recruitments. It seems much more probable that the fauna has developed piecemeal over a very great and unknown period of time and that by no means all its members derive as immigrants from the lowland forests of the Zaïre basin.

References

ALLEN, G. M. & LOVERIDGE, A. (1927). Mammals from the Uluguru and Usambara Mountains, Tanganyika Territory. *Proceedings of the Boston Society of Natural History* 38, 413–41.

ALLEN, G. M. & LOVERIDGE, A. (1933). Reports on the scientific results of an expedition to the southwestern highlands of Tanganyika Territory. *Bulletin of the Museum of Comparative Zoology, Harvard* 75, 47–140.

DIAMOND, A. W. (1981). Reserves as oceanic islands: lessons for conserving some East African montane forests. *African Journal of Ecology* 19, 21–6.

DIAMOND, A. W. & HAMILTON, A. (1980). The distribution of forest passerine birds and Quaternary climatic change in tropical Africa. *Journal of Zoology (London)* 191, 379–402.

GRUBB, P. (1983). The biogeographic significance of forest mammals in eastern Africa. In *Proceedings of the 3rd international colloquium on the ecology and taxonomy of African small mammals*, Antwerp, 20–24 July 1981, ed. E. van der Straeten, W. N. Verheyen and F. de Vree, pp. 75–85. *Annales, Musée Royal de l'Afrique Centrale, Sciences Zoologiques* 237 (i–iv), 1–227.

HAMILTON, A. (1981). Quaternary history of African forests: relevance to conservation. *African Journal of Ecology* 19, 1–6.

HOMEWOOD, K. M. & RODGERS, W. A. (1981). A previously undescribed mangabey from southern Tanzania. *International Journal of Primatology* 2, 47–55.

HOWELL, K. M. (1981). Pugu Forest Reserve: biological values and development. *African Journal of Ecology* 19, 73–81.

HOWELL, K. M. & JENKINS, P. D. (1984). Records of shrews (Insectivora, Soricidae) from Tanzania. *African Journal of Ecology* 22, 67–8.

HUTTERER, R. (1986). Diagnosen neuer Spitzmause aus Tansania (Mammalia: Soricidae). *Bonner Zoologische Beiträge* 37, 23–33.

JENKINS, P. (1984). Description of a new species of *Sylvisorex* (Insectivora: Soricidae) from Tanzania. *Bulletin of the British Museum (Natural History), Zoology* 47, 65–76.

KINGDON, J. (1971). *East African Mammals: An Atlas of Evolution in Africa*, Vol. I. London: Academic Press.

KINGDON, J. (1974a). *East African Mammals*, Vol. IIA. London: Academic Press.

KINGDON, J. (1974b). *East African Mammals*, Vol. IIB. London: Academic Press.

KINGDON, J. (1977). *East African Mammals*, Vol. IIIA. London: Academic Press.

KINGDON, J. (1979). *East African Mammals*, Vol. IIIB. London: Academic Press.

KINGDON, J. (1982a). *East African Mammals*, Vol. IIIC. London: Academic Press.

KINGDON, J. (1982b). *East African Mammals*, Vol. IIID. London: Academic Press.

KOCK, D. & HOWELL, K. M. (1988). Three bats new for mainland Tanzania. *Senckenbergiana Biologica* 68, 223–39.

LOVERIDGE, A. (1923). Notes on East African mammals collected 1920–1923. *Proceedings of the Zoological Society of London* (1923) 685–739.

LOVERIDGE, A. (1928). Field notes on vertebrates collected by the Smithsonian Chrysler East African expedition of 1926. *Proceedings of the US National Museum* 73(7), 1–69.

MATSCHIE, P. (1895). Die Saugethiere Deutsch-Ost-Afrikas. In *Deutsch-Ost-Afrika, 3: Die Thierwelt Ost Afrikas. Wirbelthiere*, Teil 1: i–xvviii, ed. K. Mobius, pp. 1–167. Berlin: D. Reimer.

MOREAU, R. W. & PAKENHAM, R. H. M. (1941). The land vertebrates of Pemba, Zanzibar and Mafia: a zoo-geographical study. *Proceedings of the Zoological Society of London, Ser. A* **110**, 97–128.

MTURI, F. (1984). Man's utilization of natural resources and the endangered Zanzibar red colobus monkey (*Colobus badius kirkii*). In *Proceedings of a Symposium on the Role of Biology in Development*, University of Dar es Salaam, Sept. 1983, ed. J. Middleton, A. Nikundiwe, F. Banyikwa and J. R. Mainoya, p. 113.

OLSEN, T. R. (1979). Studies on aspects of the morphology and systematics of the genus *Otolemur* Coquerel, 1859 (Primates: Galagidae). PhD thesis, University of London.

PAKENHAM, R. H. W. (1979). *The Birds of Zanzibar and Pemba: An annotated Check-list*. No. 2, pp. 1–134. London: British Ornithologists' Union.

PAKENHAM, R. H. W. (1984). *The Mammals of Zanzibar and Pemba Islands*. Harpenden, 84 pp.

RODGERS, W. A., OWEN, C. F. & HOMEWOOD, K. M. (1982). Biogeography of East African forest mammals. *Journal of Biogeography* **9**, 41–54.

STRUHSAKER, T. (1981). Forest and primate conservation in East Africa. *African Journal of Ecology* **19**, 99–114.

STUART, S. N. (1981). A comparison of the avifaunas of seven east African forest islands. *African Journal of Ecology* **19**, 133–52.

SWAI, I. (1984). Problems associated with wildlife conservation in Zanzibar. In *Proceedings of a Symposium on the Role of Biology in Development*, University of Dar es Salaam, Sept. 1983, ed. J. Middleton, A. Nikundiwe, F. Banyikwa and J. R. Mainoya, pp. 113–14.

SWYNNERTON, G. H. & HAYMAN, R. W. (1951). A checklist of the land mammals of the Tanganyika Territory and Zanzibar protectorate. *Journal of the East African Natural History Society* **20**, 274–392.

12   Ecology of the Zanzibar red colobus monkey, *Colobus badius kirkii* (Gray, 1968), in comparison with other red colobines

FATINA A. MTURI

Introduction

Colobus monkeys belong to the family Cercopithecidae, subfamily Colobinae. The actual classification of red colobus at specific and subspecific level is equivocal (Verheyen, 1962; Dandelot, 1968; Rahm, 1970; Kingdon, 1971; Struhsaker, 1975). For instance, the Zanzibar red colobus is considered a distinct species by Verheyen (1962) and Dandelot (1968), while others classify it at subspecific level (Rahm, 1970; Kingdon, 1971; Struhsaker, 1975). This problem arises as a result of the allopatric distribution and high affinities of the red colobus. The classification adopted in this chapter follows that of Rahm (1970), Struhsaker (1975, 1980), Kingdon (1971, 1981) and Rodgers (1981), in which all the 14 red colobus types are maintained as subspecies of *Colobus badius* Kerr.

The 14 subspecies of red colobus occur in a patch-like allopatric distribution across tropical Africa from Senegal to Zanzibar (Rahm, 1970; Kingdon, 1971; Struhsaker, 1975). Most populations are primarily adapted to mature low and medium altitude rain, riverine and groundwater forests (Struhsaker, 1975; Rodgers, 1981). However, some, such as *Colobus badius temminckii* (Kuhl), also occur in savanna woodland (Struhsaker, 1975; Starin, 1981). The Zanzibar red colobus are also found in mangrove forest, thickets of secondary growth and sometimes in cultivated

areas: habitats that are also reported for the Gambia colobus *C. b. temminckii* (Starin, 1981).

Five allopatric subspecies of red colobus occur in East Africa. Three of these are found in eastern African forests (Figure 12.1). These are *Colobus badius rufomitratus* (Peters) in Tana River Forest Reserve in Kenya; *C. b. gordonorum* (Matschie) in south central Tanzania (Uzungwa Mountains – Kilombero scarp, Mwanihana Forest Reserve, Magombero Forest Reserve); *C. b. kirkii* in Zanzibar and Uzi islands (Struhsaker, 1975, 1981; Struhsaker & Leland, 1980; Rodgers, 1981). However, a colony of *C. b. kirkii* was also found in Pemba Island at Ngezi Forest Reserve in 1978 by the Zanzibar Government Forestry Department (see Figure 12.1).

All three red colobus subspecies of eastern Africa are rare and endangered (International Union for the Conservation of Nature, 1966, 1986; Kingdon, 1971). *C. b. kirkii* is estimated to number 1469 animals (Silkiluwasha, 1981). *C. b. rufomitratus* has decreased to a very critical level from an estimate of 1246–1796 animals to 300 animals (Marsh, 1978a, 1985). Thus it qualifies as the rarest and most endangered of the red colobus subspecies.

Of the other Eastern Africa red colobus subspecies, *C. b. rufomitratus* has been the subject of several surveys and detailed ecological studies (Groves, Andrews & Horne, 1974; Andrews, Groves & Horne, 1975; Marsh, 1978a, b, c,

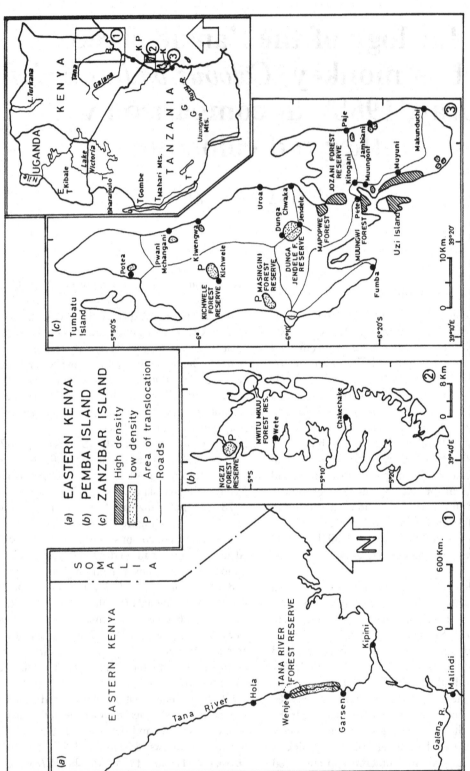

Figure 12.1. The distribution of red colobus in East Africa. R, *Colobus badius rufomitratus* in eastern Kenya (*a*); K, *C. b. kirkii* in Pemba (*b*) and Zanzibar (*c*); G, *C. b. gordonorum* in south central Tanzania; T, *C. b. tephrosceles* in western Tanzania and Uganda; E, *C. b. Elliot* in western Uganda.

1979a, b, 1981a, b, c, 1985), and there are a number of brief ecological studies on *C. b. kirkii* (Harper *et al.*, 1974; Robins, 1976; Struhsaker & Leland, 1980; Silkiluwasha, 1981; Mturi, 1983; Swai, 1983) as well as a detailed one (Mturi, 1991). Reports on *C. b. gordonorum* are more scanty (Struhsaker & Leland, 1980; Rodgers, 1981; Wasser, Chapter 13).

This chapter examines the feeding ecology and some aspects of ranging behaviour and social organisation of the Zanzibar red colobus monkey in comparison with other subspecies of the red colobus. Comparative evidence described in this chapter demonstrates many ways in which habitat differences and biogeographical isolation can generate variability at the subspecific level.

History

The isolation of some mammals endemic to montane and forest 'islands' of eastern Africa has been suggested to be attributable partly to a long series of climatic changes (especially in the late Quaternary period) (Hamilton, 1981), and partly to competitive exclusion and replacement in the intervening countryside (Homewood, 1978; Kingdon, 1971, 1981; Rodgers, Owen & Homewood, 1982). These factors also explain much of the distribution of the endemic subspecies of red colobus in eastern Africa.

The common ancestor of the three subspecies of red colobus is generally believed to have spread from the western refuge forest of Central Africa along the southern route (Kingdon, 1971, 1981; Rodgers *et al.*, 1982). The Tana River red colobus is, however, supposed to have been isolated from the others much earlier during the very arid period 25 000–12 000 years Before Present (BP) when the forest disappeared (Rodgers *et al.*, 1982). *C. b. kirkii* and *C. b. gordonorum* are believed to have become isolated from one another only recently, when rising sea levels at the end of the last glacial period, around 10 000 BP, isolated Zanzibar from the mainland (Rodgers *et al.*, 1982). These two subspecies are therefore neoendemic. The Tana *C. b. rufomitratus* is much closer to the Uganda and western Tanzania *C. b. tephrosceles* (Elliot) in pelage and vocalisation than

it is to the geographically closer Zanzibar *C. b. kirkii* and Iringa *C. b. gordonorum* (Struhsaker, 1975, 1981; Struhsaker & Leland, 1980). *C. b. rufomitratus* and *C. b. kirkii* skulls have the common feature of persistent metopic suture (Kingdon, 1971; Brand-Jones in Rodgers *et al.*, 1982). However, this feature alone does not justify linking the two subspecies closely together because it has been suggested that it is among features characteristic of drift in small relict populations (Brand-Jones in Rodgers *et al.*, 1982). On the other hand, pelage, facial pattern and vocalisations of *C. b. kirkii* and *C. b. gordonorum* have higher affinities with each other than with any other *C. badius* subspecies (Kingdon, 1971; Struhsaker, 1975; Struhsaker & Leland, 1980). Thus these subspecies of red colobus, which have only recently been isolated from each other, show highest affinities.

Today's distribution of red colobus in Eastern Africa appears to have resulted from both past climatic changes and human activities. The extent to which human disturbance has helped to mould tropical African forest ecosystems varies and is mostly poorly understood. Humans, however, could have influenced the distribution of red colobus in eastern Africa. For example, Rodgers (1981) reported three sites of past and one of present existence of red colobus in areas well outside the published range but which are in line between the ranges of *C. b. gordonorum* and *C. b. kirkii*. Kingdon (1971) and Williams (1966, cited in Marsh, 1978a) reported *C. b. rufomitratus* in the Tana River and Arabuko forests, but the monkeys were later absent in Arabuko Forest (Marsh, 1978a). The absence of red colobus in such areas where they were previously sighted most likely results from human disturbances. This may be elimination by direct killing, since they are considered a delicacy by local people (Nishida, 1972; Harper *et al.*, 1974; Marsh, 1978a; Pakenham, 1979; Rodgers *et al.*, 1982). Destruction and reduction of their habitat may have reduced and eliminated the monkey populations as well.

The range of all three eastern Africa subspecies has been fragmented mostly as the result of human activities, particularly agriculture and selective felling of trees (Struhsaker, 1975;

Marsh, 1978a, 1979a, 1981b; Silkiluwasha, 1981; Mturi, 1983, 1991). Man has also influenced the distribution of the Zanzibar red colobus through translocation as reported above (see Figure 12.1).

Methods

The study site

The study of the Zanzibar red colobus was conducted mainly at Jozani Forest Reserve, and Kichanga, a small area just 100 m south of Jozani Forest Reserve; in addition, there were three opportunistic study sites: Pete, Muungwi and Muungoni (Figure 12.2).

Preparation of the study site

The entire Jozani Forest Reserve was divided into nine east–west grid paths at 500 m intervals and one north–south path. The southern part of the Reserve was divided into 100 m² quadrats. This was the intensive study site (Figure 12.2).

Animal study

The field research on the Zanzibar red colobus covered the period January 1980–April 1981. Three groups (I, II and III) were habituated prior to the systematic study. To obtain data on behaviour, groups I and II, which ranged in the southern part of Jozani Forest Reserve (Figure 12.2), were each subject to a period of 5 days per month of systematic observations, between April 1980 and March 1981. Systematic observations were also conducted three days monthly on group III, which ranged in Kichanga (Figure 12.2) between August 1980 and March 1981.

The monkeys were followed from dawn (0615–0700 h) until dusk (1830–1900 h) during the observation periods. The behaviour of all visible individuals was recorded for a 5-minute period every 15 minutes. Data collection procedures ensured that no less than three-quarters of all group members were sampled once every hour, providing adequate representation of the group during each independent hourly record for statistical analyses. The plant species and plant

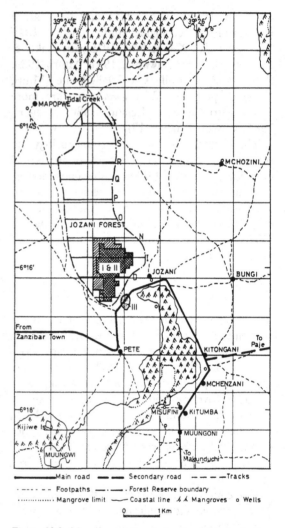

Figure 12.2. Map showing the study sites. Main study sites (I), Jozani Forest Reserve and Kichanga. Opportunistic study sites (II), Pete, Muungwi and Muungoni. The shaded area (III) is the intensive study site. East–west transects DINOPQRST are shown at 500 m intervals, and one north–south transect.

part, here referred to as item being fed upon by each focal animal, was recorded whenever visible. Food item categories were similar to those used by Struhsaker (1975) and Marsh (1978a, 1981a). The map location of the group was noted every 15 min.

Total systematic study times per group were 660, 648 and 264 h for groups I, II and III, respectively. The remaining data came from *ad libitum* observation of the study groups, as well as of other groups in Jozani Forest, and in Muungoni, Pete and Muungwi.

Vegetation study

Woody plants of 2.5 m or above were sampled along a 2.5 m wide strip on either sides of grid paths to determine their density, basal area and canopy abundance. Density was determined from the number of trees sampled. Canopy cover was determined by computing an index of the crown of each sampled plant from the product of its crown depth and breadth. Basal area was calculated from measurements of tree girths at 1.5 m height above the ground.

Phenological condition of the top 10 food species of the study groups (groups I and II) was examined at monthly intervals for 12 months. The items assessed were leaf buds, young, mature and old leaves, unripe and ripe fruits and flowers together with flower buds. Each item was given an abundance score based on visual assessment using field-glasses where appropriate. The score varied from 0 to 10, 0 representing absence of the item and 10 the maximum crop that a tree was estimated to be capable of supporting. Each month an abundance index (A_n) was computed, which took into account the canopy cover of the trees using the formula from Marsh (1978a):

$$A_n = \frac{\sum_{i=1}^{n} (A_i \times P_i)}{P_n}$$

where A_n is the weighted mean abundance index for the item on n species in the sample; A_i is the mean abundance score of the item on all specimens of species i in the sample; P_i is the proportion of the sum of canopy indices of all trees in the group's range belonging to species i; and P_n is the proportion of the sum of canopy indices of all trees within the groups range made up of n species in the phenology sample.

Results

Diet composition

Examination of the diets of the three study groups revealed that a large proportion of their food came from a small number of plant species. Table 12.1 shows the percentage contributions of 38 plant species in the annual diet of the three study groups. Similar trends were observed on both a monthly and a daily basis, illustrated by the monthly variation in contribution to diet of the ten top food species for the three study groups (Table 12.2) and the daily contribution of five food species to the diet of individuals in group I (Table 12.3). This is consistent with reports for other red colobus species (Clutton-Brock, 1972, 1975a; Struhsaker, 1975; Gatnot, 1978; Marsh, 1978a, 1981a). Table 12.4 compares the contribution to the diet of Nos 1, 5, 10, and 11 of the top food species of this study with those of the above studies.

The monkeys fed on a variety of food items: leaf buds, young leaf blades, young leaf petioles, mature leaf blades, mature leaf petioles, leaf galls, flower buds, open flowers, unripe fruits, ripe fruits, seeds, young bark, young stems, earth, dead wood and charcoal. All but the last two items were also reported as food items for red colobus monkeys studied elsewhere. Table 12.4 shows the item contribution to the diet of *Colobus badius* in five studies (Struhsaker, 1975, 1978a; Clutton-Brock, 1972, 1975a; Gatnot, 1978; Marsh, 1978a, 1981a; and this study). Consumption of dead wood was also observed in the Kibale *C. b. tephrosceles* (Struhsaker, 1975), but charcoal eating has not been recorded in any other red colobus nor in other primates.

There was variation between days and between months in the proportions of the several items that comprised the diet, as reported for other red colobus subspecies (Clutton-Brock, 1972, 1975a; Struhsaker, 1975; Marsh, 1978a, 1981a; Gatnot, 1978). The Zanzibar red colobus, like the other subspecies, preferred to eat young growth. Figure 12.3 shows the monthly contribution of different food items to the diet of Zanzibar red colobus.

Diversity in the diet

Despite a large contribution to the diet by only a few species, the Zanzibar red colobus monkeys were observed to have a diverse diet at species and item level. Seventy-seven different species were

Table 12.1. *Relative abundance and use of 38 plant species by* Colobus badius kirkii

Plant food species	Group I AD (%)	BA (%)	SR	R	Group II AD (%)	BA (%)	SR	R	Group III AD (%)	BA (%)	SR	R
Bridelia micrantha (Hochst.) Baill.	22.3	5.2	4.6	7	16.9	4.7	3.8	7.5	1.5	1.0	1.6	7
Ficus mucuso Ficalho	15.7	4.5	3.7	8	15.5	4.2	4	6	0.6	0.7	0.9	9
Calophyllum inophyllum L.	12.8	23.6	0.6	17	8.6	18.9	0.5	12	*	*		
Vitex doniana Sweet	9.0	12.1	0.8	16	12.5	17.5	0.8	11	0.4	0.3	1.5	8
Syzygium cuminii (L.) Skeels	8.1	3.8	2.3	12	19.3	4.8	4.3	4	*	0.2		
Cocos nucifera L.	6.7	1.2	5.8	5	0.3	1.1	0.3	13	13.4	37.6	0.4	12
Mangifera indica L.	3.3	0.6	5.6	6	1.4	0.5	2.8	9	17.3	5.3	3.4	4
Casuarina equisetifolia L.	2.8	0.4	8.1	3	2.0	0.2	12.4	2	0	0		
Albizia gummifera (J.F. Gmel) C.A. Sm	3.0	0.2	16.8	1	0.8	0.2	4.1	5	*	*		
Syzygium cordatum (Hochst.) Krauss	1.7	1.6	1.1	15	6.1	1.7	3.8	7.5	*	0.1		
Macaranga capensis (Baill.) Sim	1.4	0.1	12.5	2	*	*			0	0		
Eugenia capensis (Eckl. & Zeyh.) Sond.	1.0	0.9	1.3	14	1.6	1.1	1.5	10	0	*		
Ficus natalensis Hochst.	0.6	0.2	2.7	10	4.1	0.3	14.3	1	*	*		
Ficus sur Forssk.	0.7	*			0.3	*			0.3	0.6	0.5	11
Albizia glaberrima (Schum. & Thonn.) Benth.	0.6	0.2	3.5	9	1.5	0.2	7.6	3	*	*		
Apodytes dimidiata Arn.	0.3	0.2	1.9	13	*	0.1			0	0		
Polyspheria multiflora Hiern	0.2	1.2	0.2	18	0.3	1.1	0.3	12	0	*		
Maytenus heterophylla (Eckl. & Zeyh.) N. Robson	0.8	0.1	6.9	4	*	*			0	0		
Myxostylon aethiopicum (Thunb.) Loes.	0.5	0.2	2.4	1	*	0.2			0	*		
Artocarpus altilis (Parkinson) Forsbert	0.1	*			0.1	*			0.8	0.5	1.7	6
Terminalia catapa L.	0.1	*			0.1	*			47.2	4.7	10.5	1
Sorindeia madagascariensis DC.	*	*			0.1	*			1.2	0.2	5.2	3
Syzygium malaccensis (L.) Merr. & Perr.	*	*			*	*			4.0	0.7	5.7	2
Ceiba petandra (L.) Gaertn.	0	0			0	0			3.9	1.5	2.7	4
Syzygium aromaticum (L.) Merr. & Perr.	0	0			0	0			2.6	0.8	3.3	5
Sideroxylon enerme L.	0	0			0	0			0.2	0.3	0.6	10
Ficus lutea Vahl	0.5	*			0.1	*			0	0		
Ficus scassellatii Pamp.	0.2	*			0.3	*			1.9	*		
Macphersonia hildebrandtii O. Hoffm.	0.3	*			*	*			0	*		
Paidium guajava L.	0.6	*			*	*			0	0		
Annona senegalensis Pera.	0.2	*			*	*			0	0.3		
Antidesma venosum Tul.	*	*			0.7	*			*	*		
Aporrhiza nitida Gilg	0	0			0.5	*			0	0		
Citrus lemon (L.) Burm. f.	*	*			*	*			0.1	*		
Tamarindus indica L.	0	0			0	0			0.2	*		
Allophyllus pervillei Blume	0.1	*			*	*			0	0		
Suregada zanzibariensis Baill.	0.1	*			*	*			0	*		
Ficus ingens (Miq.) Miq.	0.1	*			*	*			0	0		

Note: *, values below 0.1; 0, no record; R indicates selection rank.

$$\text{Selection ratio (SR)} = \frac{\text{\% Feeding record in annual diet (AD)}}{\text{\% Basal Area (BA)}}$$

Ratios were not calculated for species with Basal Area (BA) below 0.1%.

Table 12.2. *Monthly contribution to the diet of 10 species for groups I, II and III*

| Plant species | Group | Monthly percentage of feeding records | | | | | | | | | | | |
		1980 Apr.	May	June	July	Aug.	Sept.	Oct.	Nov.	Dec.	1981 Jan.	Feb.	Mar.
Bridelia micrantha	I	39.9	14.3	1.1	1.8	0.7	0.2	26.7	26.5	50.2	17.2	26.9	38.6
	II	26.8	19.1	0.6	0	0	0	20.3	19.4	33.8	30.5	14.4	25.4
Calophyllum inophyllum	I	14.8	20.8	16.4	14.3	13.4	10.7	8.9	6.6	8.1	24.2	10.7	7.6
	II	12.1	11.6	16.8	24.4	6.2	3.7	2.4	0	9.2	15.6	3.8	8.6
Ficus mucuso	I	11.0	29.8	46.0	6.9	12.5	15.3	15.5	8.0	1.1	17.8	18.2	6.3
	II	10.4	22.1	41.0	32.2	12.9	10.2	5.4	9.0	14.4	11.9	23.0	9.4
Syzygium cuminii	I	2.8	3.9	5.8	14.5	7.7	9.3	22.3	7.4	5.4	5.3	7.6	3.9
	II	16.9	12.7	12.2	0	15.0	32.7	40.0	24.1	8.7	20.8	18.6	20.1
Syzygium cordatum	I	0	2.2	4.5	5.2	2.0	0.4	0	1.0	0.4	1.8	1.5	2.5
	II	0	3.4	0	3.3	0.6	1.3	7.2	25.4	6.6	3.7	5.6	10.5
Vitex doniana	I	19.1	13.5	3.1	2.8	16.8	28.2	10.4	7.6	2.9	1.9	7.7	8.0
	II	25.5	12.4	1.5	1.6	28.2	20.6	15.1	6.5	4.5	4.2	17.6	10.4
Casuarina equisetifolia	I	0	0	1.1	22.6	4.1	0.7	0	0	1.6	0.7	0.6	4.2
	II	1.7	1.1	0	0	4.5	11.7	0.6	0	2.1	0.3	0.3	1.3
Cocos nucifera	I	6.7	5.1	4.3	5.6	8.2	4.2	3.2	7.2	7.5	6.7	10.2	8.6
Mangifera indica	I		0.3	6.5	10.9	0.7	3.1	0	7.4	1.3	3.4	3.6	2.8
Abizia gummifera	I		0	0.2	3.6	5.9	12.8	1.2	2.6	1.8	0.8	2.7	3.5
Ficus natalensis	II	1.3	0	14.4	15.0	5.6	7.6	0.8	1.5	0.3	0	5.6	2.9
Eugenia capensis	II	0.4	0.4	0.3	1.3	5.8	0.9	1.8	2.6	0.6	0	1.8	1.0
Albizia glaberrima	II	0	0	0.6	12.7	7.3	0	0	0.2	0	0	0	0
Terminalia catapa	III					57.2	51.2	43.8	48.0	37	56.9	40	43.3
Syzygium mallacensis	III					0	0	6.7	0	0.7	8.2	16.2	0
Mangifera indica	III					23.7	20.4	19.6	21.8	6.6	16.4	18.5	16.2
Ceiba petandra	III					0.8	2.5	12.4	9.9	0.7	1.9	1.9	4.4
Cocos nucifera	III					11.9	16.4	9.3	9.1	30.8	7.9	11.2	10.8
Syzygium aromaticum	III					0	0	0	0	14.4	0	1.5	2.6
Sorindocia madagascariensis	III					0	3.5	1.0	0.8	0.3	0	1.2	2.8
Bridelia micrantha	III					0	0	0	8.7	1.3	0.5	0.4	1.0
Ficus scassellatii	III					0	0	0	0	0	6.1	2.7	3.4
Paulinia pinnata	III					0	2.5	1.0	0	1.6	0	0.8	2.3

Table 12.3. *Daily contribution to group I diet of five top ranking species in November 1980*

| Species | Daily percentage of feeding records | | | | |
	Day 1	2	3	4	5
Vitex doniana	11.7	22.8	20.6	45.8	38.9
Calophyllum inophyllum	16.0	20.2	7.1	3.3	8.4
Syzygium cuminii	16.0	24.6	1.6	4.2	1.1
Albizia gummifera	22.3	10.5	17.5	10.0	3.2
Ficus mucuso	0	12.3	6.3	22.5	36.8
Records made by 5 top ranking species	75.6	90.4	71.4	86.7	91.5
Total number of food species recorded	15	8	16	10	11

Table 12.4. *Red colobus* (C. badius) *diet in five studies*

Study site	Z			T	F	G	K
Study group	I	II	III	M			CW
Length of study period (months)	12	12	8	12	12	9	12
Mean % of feeding records							
Leaf buds	14.6	21.6	8.3	16.4	17.5	2.8	14.5
Young leaf (incl. petioles)	32.1	31.8	55.8	36	24	32	27.2
Mature leaf (incl. petioles)	7.3	6.3	2.9	11.5	5.4	44.1	23.7
Leaf stalk	7.3	5.6	3.1	1.5	1.9[b]		19.0
Flowers and floral buds	10.6	5.4	8.1	6.2	8.7	6.8	
Fruit and seeds	31.7	31.2	23.5	25	35.9	11.4	5.6
Bark	1.9	1.1	0	0.2	3.7		
Earth	*	*	*	0.2	0.2		
Dead wood	0.4	0.2	0.2				*
Total number of food species	63	62	42	22	39	59	56
Records made by top ranking species %	22.3	19.3	47.2	29.4	43.6	15.4	15.4[a]
Records made by 5 top rank species %	67.9	72.8	85.8	78	76.6[b]	56	55.9[a]
Records made by 10 top rank species %	85.4	88	94	92.3	–	–	–
Records made by 11 top rank species %	86.8	85.4	94.8	93.6	91	–	–
Mean index of species diversity	2.123	1.971	1.559	1.925	–	–	2.61[a]
Mean index of item diversity	1.466	1.402	1.186				2.21[a]

Note: Z, Zanzibar (this study); T, Tana, Kenya (Marsh, 1978a); F, Fathala (Gatnot, 1978); G, Gombe (Clutton–Brock, 1975a); K, Kibale, Uganda (Struhsaker, 1978a).
[a]Value from Struhsaker, 1975.
[b]Value from Gatnot, 1975 in Marsh, 1981a.
*Feeding records below 0.1%.

observed to be eaten by the monkeys. Groups I, II, and III were observed to use as food 63, 62 and 42 food species, respectively. The species diversity of the food of the Zanzibar red colobus compares with 59 food species for Gombe *C. b. tephrosceles* (Clutton-Brock, 1972, 1975a) and 56 for Kibale *C. b. tephrosceles* (Struhsaker, 1975), but only 22 food species for Tana River *C. b. rufomitratus* (Marsh, 1978a, 1981a) and 39 for Senegal *C. b. temminckii* (Gatnot, 1978) (see Table 12.4).

The number of plant species used on any one day by the Zanzibar red colobus was also diverse. The mean number of species used each day by group I was 12 (SD ±4.07, range 5–21); for group II was 9 (SD ±3.02, range 5–18) and for group III 7.5 (SD ±3.19, range 4–12). These values are

similar to those reported for the Tana River *C. b. rufomitratus*, which had a mean number of species of 15.2 (range 11–18) (Marsh, 1981a).

Diet diversity was further examined by calculating diversity indices. The Shannon–Weiner diversity index H (Wilson & Bossert, 1971) was calculated for food species and item, using the formula

$$H = - \sum_{i=1}^{n} P_i \log_e P_i$$

Figure 12.3. Relationship between abundance in the phenology sample and contribution to the monthly diet of six specific items: young leaves, mature leaves, leaf buds, flower and floral buds, unripe fruit, and ripe fruits. O–O, Abundance index, Group I; ×--×, abundance index, Group II; ●–●, percentage of monthly diet, Group I; △--△, percentage of monthly diet, Group II.

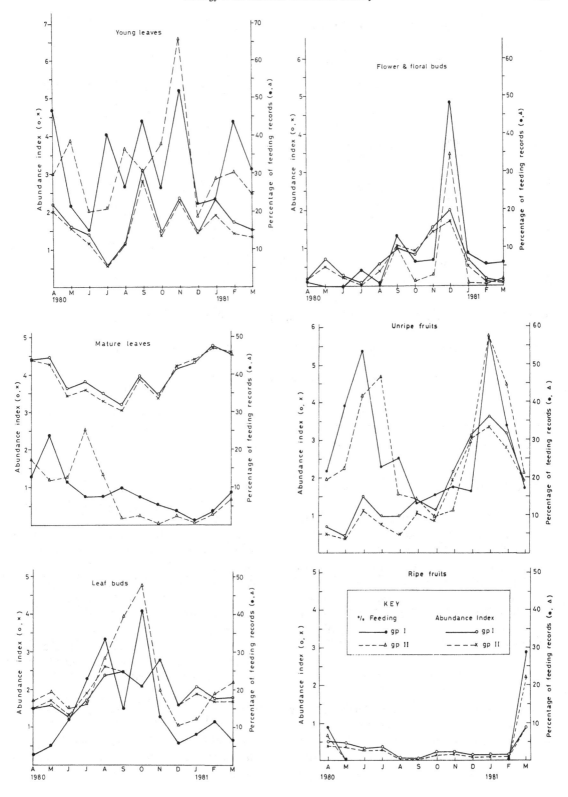

Table 12.5. *Food species diversity in three studies of red colobus (number of food species used shown in parentheses)*

| | Zanzibar (this study) | | | Tana (Marsh, 1981a) | Kibale (Struhsaker, 1975) |
	Group I	Group II	Group III	Group M	Group CW
Apr.[e]	1.664	1.811		1.753[d]	1.95[b]
May	2.002(15)	2.072(15)		1.949	2.98[b]
June	1.719(17)	1.508(11)		1.874	2.22[b]
July	2.106(16)	1.705(9)		1.745	2.65[b]
Aug.	2.341(25)	2.205(17)	1.112(5)	1.501	2.65[b]
Sept.	2.274(26)	1.955(12)	1.415(9)	1.981	2.55[b]
					2.73[a]
Oct.	2.021(16)	1.795(14)	1.698(12)	2.012[c]	2.71[b]
				2.131[d]	2.46[a]
Nov.	2.506(20)	2.033(17)	1.427(8)	2.132[c]	2.84[b]
				2.208[d]	3.03[a]
Dec.	1.700(27)	2.220(30)	1.675(15)	1.923[c]	2.75[b]
				1.992[d]	2.40[a]
Jan.[f]	2.497(37)	2.053(27)	1.396(10)	1.838[d]	2.83[b]
					2.41[a]
Feb.	2.318(29)	2.106(18)	1.765(16)	2.104	2.15[b]
					2.77[a]
Mar.	2.326(30)	2.192(19)	1.990(22)	2.026	2.62[b]
					2.62[a]
Average	2.12	1.97	1.56	1.93	2.60

Note: [a]1970; [b]1971; [c]1973; [d]1974; [e]1980; [f]1981.

where P_i represents the proportion of observations for each food species or item.

Table 12.5 shows food species diversity indices for Zanzibar, Kibale and Tana red colobus monkeys. Table 12.6 shows food item diversity indices for *C. b. kirkii* and Kibale red colobus monkeys. Both tables show that Kibale red colobus consistently had the highest indices followed by groups I and II of the Zanzibar red colobus, then Tana red colobus and finally group III of the Zanzibar red colobus.

Food preference

An important influence on the extent of feeding on different species and items may have been their abundance within the groups' range. There was significant correlation between contribution to the diet of the top ten species and their abundance measured in percentage of Basal Area (group I: $r_5 = 0.770$, $P < 0.01$; group II: $r_5 = 0.745$, $P < 0.05$; group III: $r_5 = 0.758$, $P < 0.01$). Feeding on these ten most important food species, which contributed 85.4, 88 and 94% of the three groups' annual diet respectively, was significantly related to their abundance. The important plants in the area also were the important food species for the monkeys.

To examine food preferences further, a selection ratio for each species was calculated from the formula selection ratio SR:

$$SR = \frac{\%\ of\ feeding\ records\ accounted\ for\ by\ species\ A}{\%\ of\ total\ basal\ area\ accounted\ for\ by\ species\ A}$$

Basal area is closely related to canopy size (Bennett, 1983; Mturi, 1991). Because it is measured

Table 12.6. *Food item diversity indices in two studies of red colobus*

Month	Zanzibar (this study)			Kibale (Struhsaker, 1975)
	Group I	Group II	Group III	Group CW
April	1.55[d]	1.89[d]		1.93[b]
May	1.49	1.40		2.32[b]
June	1.23	1.43		2.25[b]
July	1.48	1.37		2.35
August	1.35	1.64	1.09[d]	2.32
September	1.81	1.61	1.32	2.31
October	1.54	1.17	1.31	1.95[a]
				2.32[b]
November	1.58	0.98	0.91	2.50[a]
				2.18[b]
December	1.38	1.41	1.31	2.20[a]
				2.35[b]
January	1.17[e]	1.01[e]	1.12[e]	2.18[b]
				2.31[c]
February	1.35	1.23	1.26	2.48[b]
				1.88[c]
March	1.66	1.68	1.18	2.03[b]
				2.13[c]
Average	1.47	1.40	1.19	2.21

Note: [a]1970; [b]1971; [c]1972; [d]1980; [e]1981.

accurately it is a better guide to the relative size of canopy than use of canopy index, which is estimated. Several studies have used canopy index in calculating selection ratio (Struhsaker, 1975; Marsh, 1978a, 1981a); however, the use of basal area is becoming more popular (Waterman & Choo, 1981; McKey *et al.*, 1981; Bennett, 1983). Selection ratio serves as an approximate indicator of the differential use of species in relation to their abundance and thus helps to explain the differences between species in their contribution to the diet.

The Zanzibar red colobus preferred some plant species more than would be expected from their abundance: these had high selection ratios and ranks. Other species were eaten less than expected from their abundance: these had low selection ratios and ranks (Table 12.1). Similar observations have been made for Tana, Kibale and Gombe red colobus (respectively, Marsh, 1978a,

1981a; Struhsaker, 1975; Clutton-Brock, 1975a). However, the three studies used different formulae in calculating selection ratios, which limits direct comparison.

Selection for plant part was investigated relating its contribution to diet and its abundance index (A_n) from phenological measures. Spearman rank correlation coefficients for six items over 12 months were calculated. Variation of the proportion in the diet for young leaves, unripe fruits and mature leaves did not correlate significantly with the abundance of these items in the canopy. A significant correlation was found for leaf buds ($r_s = 0.608$, $P < 0.05$; $r_s = 0.566$, $P < 0.05$), ripe fruits ($r_s = 0.618$, $P < 0.05$; $r_s = 0.651$, $P < 0.05$) and flower and floral buds ($r_s = 0.574$, $P < 0.05$; $r_s = 0.695$, $P < 0.05$), for groups I and II, respectively. Marsh (1981a) observed that variation in the proportion of fruit, flowers and leaf buds in the monthly diet did not correlate significantly with the overall abundance of these items, but a correlation was found for young leaves ($r_s = 0.618$, $P < 0.05$). A significant correlation might be expected when the use of an item is consistently limited by its abundance; failure to show such a correlation resulted from the lack of selectivity of the monkeys in the choice of these food items relative to others.

Selectivity was examined further by plotting the monthly contribution to the diet of six items as a function of their abundance indices (see Figure 12.3). Preferred food items that had marked (seasonal) variation in their abundance were heavily consumed when they were abundant (e.g. leaf buds, flowers and flower buds).

Monthly variation in food species and item contribution to the diet was reported in other red colobus studies (Clutton-Brock, 1972, 1975a; Struhsaker, 1975; Marsh 1978a, 1981; Gatnot, 1978). As in this study, they showed that the variation was influenced by the abundance of specific food items and their preferences by the monkeys.

Other food habits

Idiosyncrasies of diet

Differences in the acceptability of food to three subspecies of the red colobus were examined by considering Kibale (Struhsaker, 1975), Gombe (Clutton-Brock, 1975a), Tana (Marsh, 1981a) and Zanzibar (this study), all of which showed some diet overlap at both species and genus level. For instance, 11 plant food species were used by two or more subspecies of red colobus. Of these, *Albizia gummifera* was eaten by the red colobus in all the above studies; *A. glaberrima* by all except Kibale red colobus; *Tamarindus indica* and *Antidesma venosum* by Tana and Zanzibar; *Vitex doniana, Bridelia micrantha* and *Syzygium cumini* by Zanzibar and Gombe; *Blighia unijugata* Bak. and *Harrisonia abyssinica* Oliv. by Tana and Gombe; and *Ficus mucuso* by Kibale and Zanzibar red colobus. Table 12.7 summarises dietary overlap among these subspecies of red colobus. There were 12 cases of overlap at the generic level: one genus was used by all red colobus, four genera by the Zanzibar and Gombe red colobus, one genus

by Tana and Zanzibar, one by Kibale and Zanzibar, and five by Tana and Kibale red colobus. Examination of beta diversity, i.e. diversity between communities, revealed that tree species found in the habitats of the monkeys are generally different. Thus the low diet overlap may be attributable to high beta diversity among the habitats of the monkeys.

There were, however, cases where a plant species was present yet was ignored by the red colobus, although it was consumed by red colobus elsewhere. For instance, *Phoenix reclinata* Jacq., found in Tana and Zanzibar, was not eaten by red colobus in Tana while the Zanzibar red colobus consumed its fruits with relish. *Pluchea discorides* (DC.), found in Tana and Zanzibar, was used only by the Tana red colobus. There were similar cases at generic level. *Anthocleista schweinfurthii* Gilg is found in Gombe and was eaten by the Gombe red colobus, constituting 6.3% of the diet. *A. grandiflora* Gilg is abundant in Jozani Forest Reserve, Zanzibar (2.33% relative density and 2.86% canopy index), but was not fed upon by the Zanzibar red colobus. Two other genera were found in both Gombe and Zanzibar but were observed to be consumed only by the Gombe red colobus. There are three *Diospyros* species in Zanzibar, none of which was observed to be eaten; a fourth species, *D. mespiliformis* (Hochst.) A. DC., found in Tana but not in Zanzibar, was the least preferred food (0.8% of the annual diet) by the Tana red colobus, although it was the most abundant tree in Tana (canopy index 23.7%). Another *Diospyros* species was fed on by the Kibale red colobus (0.14% of diet), however, *D. natalensis* (Harv.) Brenan, found in both Tana and Zanzibar, was not consumed by the red colobus monkeys in either location.

Animal food

Foraging for animal food has not been reported with certainty in any red colobus. There was, however, a suspicion that the Zanzibar red colobus diet included invertebrates: the monkeys were occasionally observed to pick up and eat unidentified items from both dead and living tree trunks, which did not appear to have fungi or

Table 12.7. *Diet overlap in four subspecies of red colobus*

Species	% V	% D	F	FL	LB	YL	ML	LP	S
			Food item						
Albizia gummifera (J.F. Gmel.) C.A. Sm.	G 1.3	G 3.6	G	G	G	G	G	G	G
	T 9.5	T 9.6	T	T	T	T	T	T	T
	Z 0.4	K 0.88	K	K	K	K	K	K	K
		Z 3.0	Z	Z	Z	Z	Z	Z	Z
Albizia glaberrima (Schumach & Thonn.) Benth.	G 3.7	G 13.2	G	T	G	G	G	G	G
	T 1.9	T 2.9	T	T	T	T	T	T	T
	K 0.3	K 4			K	K	K		
	Z 0.1	Z 0.6	Z	Z	Z	Z	Z	Z	Z
Ficus natalensis Hochst	T 1.9	T 4.2	T		T	T	T		
	K –	K 1.23	K			K			
	Z 0.3	Z 0.57	Z		Z	Z	Z		Z
Tamarindus indica L.	T 1.2	T 2.3	Z		T	T	T		
	Z *	Z 0.2			Z	Z			
Antidesma venosum Tul.	T 1.2	T 0.1	Z		Z	T	T		
	Z 0.1	Z 0.1				Z			
Vitex doniana Sweet	G 2.1	G 0.2			G	G			
	Z 4.2	Z 9.0	Z	Z	Z	Z	Z	Z	Z
Bridelia micrantha (Hochst.) Baill.	G 0.8	G 0.2			G	G			G
	Z 3.6	Z 22.27	Z	Z	Z	Z			Z
Syzygium cuminii (L.) Skeels	Z 0.5	Z 8.15	Z	Z	Z	Z	Z	Z	Z
	G 0.3	G 0.1	G						
Blighia unijugata Bak.	G 1.8	G 0.4				G			
	T 4.0	T 1.9	T	T		T	T		
		Z 0.4							
Harrisonia abyssinica Oliv.	G 0.3	G 0.1					G		
		T 0.1				T			
Ficus mucuso Ficalho	K 0.6	K 3.8	K			K	K		K
	Z 0.7	Z 15.72	Z			Z	Z		Z

Note: G, Gombe *C. b. tephrosceles*, source Clutton–Brock, 1975a; K, Kibale *C. b. tephrosceles*, source Struhsaker, 1975; T, Tana *C. b. rufomitratus*, source Marsh, 1981a; Z, Zanzibar *C. b. kirkii*, source this study.
% V, percentage in vegetation; % D, percentage in diet; *, one tree.
% V for Zanzibar is from group I range.
F, fruit; FL, flowers; LB, leaf bud; YL, young leaf; ML, mature leaf; LP, leaf petiole; S, stem.

lichens on them. Indirect evidence of invertebrate, particularly arthropod, feeding in red colobus has also been reported by Struhsaker (1975, 1978a, b), who found larvae, pupae and adult insects present on dead tree trunks that the Kibale red colobus were feeding on. He also observed *C. b. temminckii* and *C. b. badius* picking something (either dirt or termite) from a termite tunnel, and saw *C. b. preussi* eat something out of a piece of dead wood that was then dropped. It is likely that feeding on animal matter is an infrequent behaviour involving only small-sized and immobile or slow-moving animals, hence the lack of firm observations of such behaviour.

Drinking

Drinking of free water is not very common among red colobus. It has not been observed in the Zanzibar red colobus or Iringa red colobus (S. K.

Wasser, personal communication). Throughout his study, lasting more than four years, Struhsaker (1975) observed the Kibale *C. b. tephrosceles* drinking on only two occasions. In Mahale a male *C. b. tephrosceles* was observed to drink water once during casual observation (Nishida, 1972). There have been no reports of drinking among the other red colobus.

Group composition

The Zanzibar red colobus live in heterosexual social groups (multi-male and female groups), like most other red colobus studied to date (Clutton-Brock 1972, 1975a; Nishida, 1972; Struhsaker, 1975; Struhsaker & Leland, 1980; Gatnot, 1978; Starin, 1981; Wasser, this volume). The exception is the Tana red colobus (*C. b. rufomitratus*), which generally live in single-male groups, with a maximum of two adult males per group (Marsh, 1978a, 1979a, 1981b). The range in number of adult males per group varied: *C. b. kirkii* had 2–5; Kibale *C. b. tephrosceles*, 2–10 (Struhsaker, 1975); Senegal *C. b. temminckii*, 1–13 (Gatnot, 1975, cited in Marsh, 1979a); and Tana *C. b. rufomitratus*, 1 or 2 (Marsh, 1978a, 1979a, 1981b). The number of adult females per adult male in the bisexual groups of these subspecies was 5.9, 2.6, 2.0 and 6.6, respectively. Solitary animals were not uncommon; they were on all occasions adult and subadult males among, *C. b. kirkii*, *C. b. tephrosceles* and *C. b. rufomitratus*; in *C. b. temminckii* solitary adult or subadult females were occasionally observed. There were uneven sex ratios in the red colobus which were most marked in the Tana red colobus followed by the Zanzibar red colobus.

Like the other red colobus, *C. b. kirkii* group composition was observed to vary, as a result of births, immigrations, emigrations and disappearance of animals in the groups. A birth peak occurred between November and January, when three-quarters or more of the adult females were observed to have infants. Infant mortality was very high: more than half of the infants born did not survive the first 6 months. This was also reported by Silkiluwasha (1981).

Group membership dynamics of the Zanzibar

C. b. kirkii were found to be generally similar to those reported for the Kibale *C. b. tephrosceles* by Struhsaker (1975) and Struhsaker & Leland (1979). The adult male group membership at these sites was very stable and their social relationships were close, having strict linear dominance ranks. Young animals, juveniles and subadults transferred between groups, mostly during intergroup encounters. However, groups I and II were each joined by two solitary *C. b. kirkii* subadult males. Starin (1981) reported that in the Gambia *C. b. temminckii* all females voluntarily leave their natal group, before or just after maturity, and most males are forced out of the group by resident adult males before adulthood. The Tana *C. b. rufomitratus* migrations were by adult and subadult females and adult males but never by independently moving juveniles (Marsh, 1978a, 1979a,b). Clutton-Brock (1975b) reported that in Gombe *C. b. tephrosceles* males transferred from one group to another. In all the red colobus where females transferred (in Tana, Gambia and Kibale) it has been observed that they did so very easily, while adult male transfer was very difficult.

Intraspecific relations

In the Zanzibar *C. b. kirkii* relations between members of the same group were mostly peaceful. The same was observed for *C. b. tephrosceles* at Gombe (Clutton-Brock, 1975b), and Kibale (Struhsaker, 1975; Struhsaker & Leland, 1979). However, relationships between adult and subadult males were generally hostile in Tana *C. b. rufomitratus* (Marsh, 1978a, 1979a,b) and Gambia *C. b. temminckii* (Starin, 1981).

Intergroup relations varied from tolerant to aggressive among Zanzibar red colobus, as was the case for Kibale red colobus (Struhsaker, 1975) and Gambia red colobus (Starin, 1981). Groups at these localities frequently tolerated the proximity of another group. There were extreme cases where two groups mixed and stayed together (e.g. for as long as 30 min in Zanzibar) before they separated. In contrast, intergroup relations in Tana red colobus were generally aggressive (Marsh, 1978a, 1979a, b). In all of

Table 12.8. *Range size, range overlap, group size and group density of the red colobus in three sites: Zanzibar, Tana and Kibale*

	C. b. kirkii Jozani, Zanzibar (this study)	C. b. rufomitratus Tana, Kenya (Marsh, 1979a)	C. b. tephrosceles Kibale, Uganda (Struhsaker, 1975)
Habitat type	Semi-evergreen forest	Semi-evergreen forest	Rain forest
Number of groups encountered	20	13	15
Range in group size	18–56	12–30	19–70+
Range in study group size	38–56[a], 28–44[b]	16–33	19–22
Study group range area (ha)	59[a], 61[b]	9	35.3
Study group range overlap (%)	71[a], 69[b]	32.8	98
Group density (gp/km^2)	2.1	11.8	5.9

Note: [a]Group I; [b]Group II.

these subspecies the majority of intergroup interactions involved groups avoiding proximity through vocal communication. In a few cases one group displaced another group.

Variation in intergroup relations in red colobus corresponded with variation in their ranging patterns (Table 12.8). Zanzibar *C. b. kirkii* were observed to have extensive range overlap with no tendency towards exclusive use of particular localities. The same was reported for Kibale *C. b. tephrosceles* by Struhsaker (1975) and Gambia *C. b. temminckii* (Starin, 1981). The Gombe *C. b. tephrosceles* had little range overlap between groups, and this was confined to the peripheral parts of the range (Clutton-Brock, 1975b). The Senegal *C. b. temminckii* were observed to have largely or wholly exclusive use of their ranges (Gatnot, cited in Marsh, 1979a); the Tana *C. b. rufomitratus* also had little range overlap, and the shared range was rarely used by one of the groups (Marsh, 1978a, 1979a).

Group size, density, social relations and ranging patterns

The Tana *C. b. rufomitratus* had the smallest groups and highest group density. The Zanzibar *C. b. kirkii* and Kibale *C. b. tephrosceles* had larger range size and greater range overlap, although the former had the lowest group density and smaller group size (Table 12.8). Thus, in these three

subspecies of red colobus, group density, range size and intergroup relations are consistent with the reported association between high density, small ranges and more exclusive range use in other colobines (Yoshiba, 1968; Rudran, 1973; Oates, 1977).

Polyspecific associations

Species were considered to be in association when members of the different species were spatially intermingled. Polyspecific associations among monkeys are common and have been reported in many studies (Haddow, 1952; Jones & Sabater Pi, 1968; Aldrich-Blake, 1968; Bernstein, 1967; Gartlan & Struhsaker, 1972; Struhsaker, 1975, 1978a; Starin, 1981; Davies, 1984; Wasser, Chapter 13).

The Sykes monkey (*Cercopithecus mitis albogularis*), which is endemic to Zanzibar at subspecific level, was found to associate with *C. b. kirkii* in a number of activities. It is the only large, diurnal monkey sympatric with the Zanzibar red colobus. They were observed to associate in resting, playing, feeding, allogrooming, and even moving together from one place to another. A systematic study of the three groups showed 482 occasions of association. The red colobus groups associated more with solitary *C. mitis*: there were 155, 157 and 108 observations for groups I, II and III associating with solitary *C. mitis*, compared with 24, 17, and 21 observed associations with groups of *C. mitis*, respectively. In Kibale, *C. b. tephrosceles* (Struhsaker, 1975) and in Mwanihana, *C. b. gordonorum* (Wasser, Chapter 13) also associated with groups of *C. mitis*. On no occasion in the Zanzibar study was a solitary red colobus observed to associate with *C. mitis*. On the other hand, solitary red colobus in Kibale (Struhsaker, 1975), Gambia (Starin, 1981) and Mwanihana (Wasser, Chapter 13) associated with other primate species. In Kibale and Mwanihana more than two primate species were found together simultaneously on some occasions.

In both Kibale and Zanzibar red colobus associations were mostly initiated and disbanded by the other primates. However, in Gambia, red colobus approached vervet monkeys (*Cercopithecus aethiops*) in most cases (Starin, 1981). The duration of the associations varied but it was very short among the Zanzibar red colobus and *C. mitis*, 1–15 min, compared with 130.4 and 121.8 min between the Kibale red colobus and *C. ascanius* and *C. mitis*, respectively (Struhsaker, 1975). Associations involving Gambia red colobus generally lasted for several days up to a year (Starin, 1981). In the Iringa red colobus some associations are of long duration; S. K. Wasser (personal communications) reports repeated observations of the same polyspecific grouping over a period of several days.

Resting was the most common activity recorded during the association of the Zanzibar red colobus and *C. mitis*, comprising 56.6% of the records. Feeding accounted for 21.9%, allogrooming 5.5%, and playing 3.8% of the records. In the Kibale red colobus the most common social activity was grooming followed by playing (Struhsaker, 1975). In both Zanzibar and Kibale red colobus, play mostly involved young animals. Interspecific aggression between the red colobus and the other primate species was not observed in Zanzibar, while there was occasional aggression in Kibale (Struhsaker, 1975) and Gambia (Starin, 1981).

Discussion

The Zanzibar red colobus food habits, group composition, intraspecific relations and primate polyspecific associations show great similarities to those of other red colobus that have been studied. There are very few behavioural aspects in which any particular subspecies of red colobus stands unique. In the majority of cases a behaviour is common in two or more subspecies irrespective of their geographical proximity. This section discusses the similarities and differences among the red colobus in the above respects, and attempts to show why the differences and similarities exist.

The red colobus populations so far studied are mainly folivores. This is likely to be related to feeding adaptations they have for utilising leaves, namely, by possession of a large chambered stomach analogous to that of ruminants (Moir, 1965; Bauchop & Martucci, 1968; Kay, Hoppe & Maloiy, 1976). They prefer young leaves, with high nutrient and low antifeedant content. (An antifeedant is defined as a plant chemical compound potentially harmful to herbivores.) Young leaves commonly contain more protein and less fibre (Dougall & Drysdale, 1964; Struhsaker, 1975; Hladik, 1977; Waterman & Choo, 1981; Mturi, 1991), and also have less toxin than mature leaves (Waterman & Choo, 1981; Mturi, 1991). In ruminants, foods containing a high proportion of fibre take longer to ferment (Janis, 1976). Thus it may be important for red colobus to utilise young leaves, which are more digestible, especially as colobines are near the bottom of the size range for which fore-stomach fermentation is viable (Parra, 1978).

All the red colobus monkeys studied are very selective in their feeding. This may be aimed at maximising the intake of nutrients while minimising the intake of deleterious compounds (Freeland & Janzen, 1974; Westoby, 1974; Pulliam, 1975; Glander, 1982; Waterman, 1984). Selectivity for particular plant parts is observed in red colobus; for example, mature leaf petioles, which contain less fibre, are preferred to mature leaf blades (Waterman & Choo, 1981; Baranga, 1982; Mturi, 1991). The petioles also have higher digestibility and fermentable carbohydrates, and less total phenolics and condensed tannins, than blades (Mturi, 1991). This chemistry of leaf petioles may have facilitated the high specialisation for leaf petioles seen in *C. b. gordonorum* (see Wasser, Chapter 13).

There is a variation in the importance of different plant parts to the diet among red colobus subspecies. Red colobus living in habitats with low tree species diversity – Fathala (Senegal), Tana (Kenya) and Jozani (Zanzibar) – ate more fruits and seeds and fewer mature leaves (including petioles) than those in habitats with high tree diversity – Kibale (Uganda), Gombe (Tanzania) and Mwanihana (Tanzania). The switch from folivory to frugivory may be an adaptation to avoid or reduce the intake of leaves that contain toxic secondary products and are generally indigestible, especially when mature. If the leaves of trees in the Tana, Jozani and Fathala habitats contain generally high concentrations of toxic secondary compounds, the monkeys would be expected to feed preferentially on acceptable nutrient-rich items when present; even if these nutrient-rich items also contain toxic secondary compounds, the additional nutrients may offset the costs of processing high toxin concentrations (Moir & Harris, 1962; Balch & Campling, 1965; Freeland & Janzen, 1974; McKey, 1978). It may also be that in habitats of lower tree species diversity, mature leaves of only a few species were worth exploiting when considering the presence of toxic secondary compounds, nutrients and digestibility. An alternative explanation may be that the red colobus populations of Gombe and Kibale (*C. b. tephrosceles*) and the population at Mwanihana (*C. b. gordonorum*) were able to become better adapted to exploit a diet containing a higher proportion of mature leaves. This may be related to the types of stomach microbes with which they have coexisted through their evolutionary history. They may have some physiological and/or anatomical adaptations for exploiting such food items. For instance, the grey langur (*Presbytis entellus*) and purple-faced langur (*P. senex*) in Sri Lanka tend to select ripe and unripe fruits, respectively. This has been correlated with differences in their intestinal mucosae (Hladik, 1977; Clutton-Brock, 1977). An additional and perhaps complementary

explanation may be that the use of more mature leaves by the red colobus of Kibale, Gombe and Mwanihana is an adaptation towards reducing interspecific competition, as suggested for *G. b. gordonorum* by Wasser (Chapter 13). Moreover, the observation of less mature leaf feeding in the Zanzibar red colobus, serves to support Marsh's (1981a) hypothesis that mature leaves will be eaten more often in forests of high tree and primate species diversity.

Feeding on animal matter is uncommon in red colobus and other colobines: it has been observed in *Colobus guereza* (Oates, 1977), *C. satanas* (McKey, 1978), *Presbytis entellus* and *P. senex* (Hladik, 1977) and *P. rubicunda* (Davies, 1984). This may be because colobines obtain some animal protein through digestion of fore-stomach microbes (Parra, 1978).

Drinking of water is uncommon among red colobus and other colobines (as already mentioned), possibly because water produced physiologically and during fermentation of food may meet the monkeys' water requirements.

Charcoal eating is unique to the Zanzibar red colobus. One of the benefits of charcoal eating may be absorption of gases. Charcoal tablets are used in human medicine as an absorbent for the purpose of reducing flatulence (Cook & Martin, 1951; Hall, Thompson & Strother, 1981); however, Potter, Ellis & Levitt (1985) showed that activated charcoal does not influence gas formation *in vitro* or *in vivo*. Charcoal may also absorb toxins from the gut: because of its absorbent properties activated charcoal is used in treatment of human poisoning (Cook & Martin, 1951; Potter *et al.*, 1985). Thus, although the actual cause of charcoal eating behaviour is unknown, its effects may be beneficial to the monkeys. The behaviour may be unique to the Zanzibar red colobus because it forages more frequently at ground level and also because of greater availability in Zanzibar from charcoal kilns and bush fires.

Red colobus may also show preferences (selectivity) between plant species for high nutrients and against antifeedants (fibre and toxins). For example, *Albizia gummifera* has high concentrations of protein relative to other species in Jozani (Zanzibar) (Mturi, 1991). More than 85% of mature leaf blade feeding in *C. b. kirkii* was on *A. gummifera* and *A. glaberrima*. Moreover, Marsh (1981a) reported that almost 50% of all mature leaf feeding records for his *C. b. rufomitratus* study group were from *Acacia robusta* (Burch.). In East Africa *Acacia* spp. typically contain higher concentrations of protein in their leaves than most other genera (Dougall & Drysdale, 1964). The genus *Diospyros* generally contains naphthoquinone derivatives which are toxic to many groups of animals (Hegenaur, in McKey, 1978). Species of this genus are generally ignored by red colobus in Zanzibar; they are weakly selected in Tana (Marsh, 1981a) and Kibale (Struhsaker, 1975), despite their high abundance there. This genus was also rarely eaten by black-and-white colobus in Kibale (Oates, 1977), probably for the same reasons.

The observation that plant species common in several field sites were eaten by red colobus in some areas and ignored in others partly may be attributable to local variation in their nutrient and toxins content. It has been demonstrated that the amount of chemical defence (toxins) in plants against herbivores is influenced by soil type (Gartlan, McKey & Waterman, 1978; McKey, 1978). In coastal Cameroun, leached nutrient-deficient soils are associated with flora showing extreme chemical defence against herbivores (Gartlan *et al.*, 1978). An alternative reason for the observation is that differences in abundance of potential food sources across field sites may affect food selectivity, causing species to be eaten in one area but ignored in others.

Monthly and seasonal variation in species and item selection by red colobus could have similar explanations. Concentrations of secondary compounds (chemical defence) may vary in the same area and plant at different times of day and year (Huges & Genset, 1973). Other factors such as abundance of food species and item must also play a major role in monthly and seasonal variation in feeding.

All red colobus maintained a diverse diet at species and item level; the red colobus of Mwanihana, which specialises on leaf petioles, is an exception (Wasser, this volume).

Plants contain a wide variety of chemical compounds, both nutrients and antifeedants, in innumerable combinations of the various plant parts and species (Janzen, 1978). It may be advantageous, indeed necessary, to diversify the diet in order to obtain an optimal mix of nutrients (Chivers, 1977; Hladik, 1977; Oates, 1978), and to take small amounts of a range of antifeedants that may be more tolerated than larger quantities of a few (Freeland & Janzen, 1974). Thus, there may be upper limits at which particular toxins could be taken without diminishing the net value of the food eaten. High food species and item diversity amongst red colobus may enable them to avoid exceeding the upper limits of particular toxins that some species and item contain if eaten in large quantities. However, high diversity may also be necessary for a balanced diet. For example, the red colobus of Mwanihana specialised on leaf petioles (from several plant species); this plant part may be phytochemically highly acceptable, as discussed above, and thus leaf petioles and the other plant parts may satisfy their nutritive requirements.

The red colobus populations studied show variation in foraging levels. The red colobus of Zanzibar and Senegal forage at all levels from ground level to canopy tops; this may be an attempt to meet their nutrient requirements. Given the dangers of predation on the ground (e.g. python and monitor lizard in Zanzibar) the nutritional benefits of ground plants must be high. Black-and-white colobus (*Colobus guereza*) also come to the ground, to feed on water plants in order to obtain minerals (Oates, 1977, 1978). Other red colobus that forage exclusively above ground layers (mainly above 10 m), probably get all their nutrients by so doing.

The red colobus show great similarities in group composition, consisting of multi-male and female groups except the Tana red colobus, whose groups tend to include one or a maximum of two adult males per group. The presence of more adult and subadult females than males in red colobus groups may be explained by the parallel existence of solitary and/or extra-group males. It also may be due to higher mortality rate in males. In all red colobus, it is the adult and subadult males that are mainly involved in aggressive encounters with other conspecific groups, or with predator or competitor species.

Transfer from one group to another is very difficult for adult male red colobus. It involves a lot of aggression as reported in the Tana (Marsh, 1978a, 1979a) and Gambia red colobus (Starin, 1981). These aggressive encounters sometimes cause injuries which may eventually result in death. Killings are not uncommon in aggressive encounters during male transfer in Gambia (Starin, 1981). The red colobus in Tana tolerate one or occasionally two adult males in a group while the rest are forced to range in unfavourable areas. They eventually become unhealthy and sometimes die as a result (Marsh, 1978a, 1979a). All these factors account for a higher mortality rate amongst males, which may result in the unequal sex ratio seen in these red colobus monkeys.

Marsh (1978a, 1979a) suggests that the extreme sex ratio within groups of the Tana red colobus is attributable to high extra-group male mortality related to the marginal nature of their habitat. He adds that seasonal mating in the Senegal red colobus (in Fathala, a habitat similar to Tana) increases the cost–benefit ratio to dominant males, excluding others to a point at which the behaviour is maladaptive. Thus they have more than one male in a group. This suggestion has support from the Zanzibar red colobus study. The Zanzibar red colobus in Jozani, a habitat similar to Tana (Kenya) and Fathala (Senegal) and with seasonality in birth like the Fathala red colobus, has multi-male and female groups (see also Wasser, Chapter 13). However, the Kibale red colobus exhibits aseasonal births and multi-male groups (Struhsaker, 1975).

The high permeability of groups of red colobus could be explained in terms of avoidance of inbreeding. In those groups with mobile adult males, adult females with infants or those that are pregnant may move to prevent their infants from being killed by immigrant adult males, as in the Tana (Marsh, 1978a, 1979b) and Gambia red colobus (Starin, 1981). Adult female transfer may also be related to mate choice (Marsh, 1979b; Starin, 1981).

There is variation in intergroup relations and ranging patterns both between and within the subspecies of red colobus. Ranging pattern vary from total range overlap to exclusive use of range; intergroup aggression increases in the same direction. Such variation is also reported in other colobines (*Presbytis entellus*: Jay, 1963; Yoshiba, 1968; *P. melalophos*: Bennett, 1983; *Colobus guereza*: Dunbar & Dunbar, 1974; Oates, 1977) and in other primates, including *Cercopithecus aethiops* (Kavanagh, 1981) and *Papio ursinus* (Hamilton, Buskirk & Buskirk, 1976). Various explanations have been proposed for these differences, all relating to the environment, including differences in degree of predation pressure, range size, population density, and distribution and abundance of food. Therefore, it is not surprising that in red colobus variation occurs even within subspecies.

The Tana red colobus exhibit high exclusive range use, male intolerance and uneven sex ratio compared to other red colobus. Such behaviour patterns may eventually lead to morphological changes such as sexual dimorphism. For instance, it has been reported that body size dimorphism in red colobus is slight or lacking (Lentenegger Kelly, in Marsh, 1978a), but this may not apply to the Tana red colobus (Marsh, 1978a).

All red colobus subspecies showed polyspecific associations with primates sympatric to them. The type of activities performed during such associations suggest that they were not always chance affairs but were sometimes deliberate acts. Even if they happened by chance, whenever they occurred they appeared to be encouraged, based on the type of activities which occur when different species come together (e.g. mutual grooming and playing). The monkeys form polyspecific associations as they gain many of the advantages of social life such as grooming (especially in solitary individuals) and efficiency of predator detection and feeding. Moreover, the disadvantage of interspecific competition for resources like food and mates at times may be quite low. Polyspecific associations are further discussed in the following chapter by Wasser (this volume).

In conclusion, the information gathered here illustrates how habitat and phylogeny can interact to yield the comparative differences and similarities found among the red colobus. The different subspecies have either retained their ancestral behaviour patterns, or have adapted convergently or divergently in response to similar or different ecological conditions experienced over evolutionary time. Such specialisations and changes, illustrated here by the red colobus, are among the evolutionary mechanisms through which new species may evolve.

Of the 14 subspecies of red colobus five are classified as rare and endangered (IUCN, 1966, 1986); these include all three subspecies found in eastern Africa. The forest habitats where these monkeys are found are disappearing owing to human destructive activities such as agriculture, tree felling, settlement, industry and fire (Marsh, 1978a, 1985; Rodgers, 1981 and Chapter 14; Struhsaker, 1981; Mturi, 1983, 1991 and personal observations; Wasser & Lovett, Chapter 1; Wasser, Chapter 13). Therefore to protect these red colobus subspecies and other fauna and flora there is a great need to protect the forests in eastern Africa. Hopefully, incentive to protect these areas can be facilitated by increasing public awareness that forests provide vital water catchment to the country's predominantly arid land.

Summary

Food habits, group composition, ranging patterns, intraspecific relations and primate polyspecific associations of the Zanzibar red colobus monkey *Colobus badius kirkii* are described and compared with other subspecies of red colobus monkeys (*C. b. tephrosceles* in Kibale, Uganda and Gombe, Tanzania; *C. b. rufomitratus* in Tana, Kenya; *C. b. gordonorum* in Mwanihana, Tanzania; *C. b. temminckii* in Fathala, Senegal and Abuko, Gambia).

Food habits of the red colobus subspecies were very similar with only few exceptions. They used a variety of plant species and plant parts, and were very selective in their feeding. Feeding on animal matter and drinking of free water were infrequent. However, the Mwanihana red colobus specialised on leaf petioles of various species while charcoal eating was unique to the Zanzibar red colobus. The subspecies in habitats with high tree diversity had more diverse diet and used more mature leaves than those in habitats of low tree diversity.

The red colobus showed similar group composition, with few exceptions, forming multi-male and female groups except the Tana red colobus. Solitary individuals, adult and subadult males (and, in the Gambia red colobus, adult and subadult females) also occurred. Groups of red colobus had more adult and subadult females than respective males. Among the subspecies of red colobus there was variation in age and sex of individuals that transferred between groups.

Variation in intra- and inter-group relations and ranging patterns was observed both between and within subspecies of red colobus. In the Zanzibar, Kibale and Gombe red colobus intragroup relations were mostly peaceful; in the Tana and Gambia red colobus adult and subadult males were generally hostile to each other. Inter-group relations varied from tolerant to aggressive in Zanzibar, Kibale and Gambia red colobus; but were basically aggressive in Tana red colobus. The Zanzibar, Kibale and Gambia red colobus exhibited extensive range overlap; Gombe, Senegal and Tana red colobus had largely or wholly exclusive use of their ranges. All red colobus subspecies showed polyspecific associations with primate species sympatric to them.

The comparison shows how habitat and phylogeny interact to yield the differences and similarities observed among the red colobus. Such specialisations and changes among the red colobus demonstrate the evolutionary mechanisms through which new species may evolve.

All the three subspecies of red colobus in eastern Africa are rare and endangered. There is, therefore, an urgent need to conserve them and other species by protecting the forests of eastern Africa.

Acknowledgements

I thank the Office of Ministry of Natural Resources and Tourism and the Department of Forestry who permitted field work in Zanzibar, and to the University of Dar es Salaam for financial support. Drs K. Homewood, J. I. Pollock, A. M. Nikundiwe, W. A. Rodgers and A. H. Omari gave advice and encouragement throughout the study. I thank Mr L. Mwasumbi for help with plant identification and Dr K. M. M. Lwiza for writing programs for data analysis. Finally, I thank Drs S. Wasser, J. Lovett and Professor P. Jewell who commented on this paper at various stages and Ms Paullete Chiasson who typed the initial draft.

References

ALDRICH-BLAKE, F. P. G. (1968). A fertile hybrid between two *Cercopithecus* species in the Bundongo Forest, Uganda. *Folia Primatologica* 9, 15–21.

ANDREWS, P., GROVES, C. & HORNE, J. (1975). Ecology of the Lower Tana flood plain (Kenya). *Journal of the East African Natural History Society and National Museum* 151, 1–31.

BALCH, C. C. & CAMPLING, R. C. (1965). Rate of passage of digesta through the ruminant digestive tract. In *Physiology of Digestion in Ruminants*, ed. R. W. Dougherty, pp. 108–23. London: Butterworth.

BARANGA, D. (1982). Nutrient composition and food preferences of colobus monkeys in Kibale forest Uganda. *African Journal of Ecology* 20, 113–21.

BAUCHOP, T. & MARTUCCI, R. (1968). Ruminant-like digestion of the langur monkey, *Science* 161, 698–700.

BENNETT, E. L. (1983). The banded langur. Ecology of a colobine in the West Malaysian rain-forest. PhD thesis, University of Cambridge.

BERNSTEIN, I. S. (1967). Intertaxa interactions in a Malayan primate community. *Folia Primatologica* 7, 198–207.

CHIVERS, D. J. (1977). The feeding behaviour of Siamang (*Symphalangus syndactylus*). In *Primate Ecology: studies of feeding and ranging behaviour in lemurs, monkeys and apes*, ed. T. H. Clutton-Brock, pp. 355–382. London: Academic Press.

CLUTTON-BROCK, T. H. (1972). Feeding and ranging behaviour of the red colobus. PhD thesis, University of Cambridge.

CLUTTON-BROCK, T. H. (1975a). Feeding behaviour of red colobus and black-and-white colobus in East Africa. *Folia Primatologica* 23, 165–207.

CLUTTON-BROCK, T. H. (1975b). Ranging behaviour of red colobus (*Colobus badius tephrosceles*) in the Gombe National Park. *Animal Behaviour* 23, 706–22.

CLUTTON-BROCK, T. H. (1977). Some aspects of intraspecific variation in feeding and ranging behaviour in primates. In *Primate Ecology: Studies of Feeding and Ranging Behaviour in Lemurs, Monkeys and Apes*, ed. T. H. Clutton-Brock, pp. 539–56. London: Academic Press.

COOK, F. E. & MARTIN, E. W. (1951). *Remington's Practice of Pharmacy*. Easton: Mack Publishing Co.

DANDELOT, P. (1968). *Preliminary Identification Manual for African Mammals*. Washington, DC: Smithsonian Institution Press.

DAVIES, A. G. (1984). An ecological study of the red leaf monkeys (*Presbytis rubicunda*) in the dipterocarp forest of northern Borneo. PhD thesis, University of Cambridge.

DOUGALL, H. W. & DRYSDALE, V. M. (1964). The chemical composition of Kenya browse and pasture herbage. *East African Wildlife Journal* 2, 86–120.

DUNBAR, R. I. M. & DUNBAR, E. P. (1974). Ecology and population dynamics of *Colobus guereza* in Ethiopia. *Folia Primatologica* 21, 188–208.

FREELAND, W. J. & JANZEN, D. H. (1974). Strategies in herbivory by mammals: the role of plant secondary compounds. *American Naturalist* 108, 269–89.

GARTLAN, J. S., McKEY, D. B. & WATERMAN, P. G. (1978). Soils, forest structure and feeding behaviour of primates in a Cameroon coastal rain-forest. In *Recent Advances in Primatology*, Vol. 1. *Behaviour*, ed. D. J. Chivers and J. Herbert. London: Academic Press.

GARTLAN, J. S. & STRUHSAKER, T. T. (1972). Polyspecific associations and niche separation of rain-forest anthropoids in Cameroon, W. Africa. *Journal of Zoology (London)* 168, 221–65.

GATNOT, B. L. (1975). Ecologie d'un colobe bai (*Colobus badius temminckii* Kuhn 1820) dans un milieu marginal en Sénégal. Doctoral dissertation, Universitaire de Paris.

GATNOT, B. L. (1978). Characteristics of the diet of West African red colobus. In *Recent Advances in Primatology*, Vol. 1. *Behaviour*, ed. D. J. Chivers and J. Herbert, pp. 253–56. London: Academic Press.

GLANDER, K. E. (1982). The impact of plant secondary compounds on primate feeding behaviour. *Yearbook of Physical Anthropology* 25, 1–18.

GROVES, C., ANDREWS, R. & HORNE, J. F. M. (1974). The Tana colobus and mangabey. *Oryx* 12, 565–75.

HADDOW, A. J. (1952). Field and laboratory studies on an African monkey, *Cercopithecus ascanius schimidtii* Matschie. *Proceedings of the Zoological Society of London* 122, 297–394.

HALL, G. H. Jr, THOMPSON, H. & STROTHER, A. (1981). Effects of orally administered activated charcoal on intestinal gas. *American Journal of Gastroenterology* **75**, 192–6.

HAMILTON, A. C. (1981). The Quaternary history of African forests: its relevance to conservation. *African Journal of Ecology* **19** (182), 1–6.

HAMILTON, W. J., BUSKIRK, R. E. & BUSKIRK, W. H. (1976). Defence of space and resources by chacma (*Papio ursinus*) baboon troops in African desert and swamp. *Ecology* **57**, 1263–72.

HARPER, D. M., KINGSTON, T. J., ROBINS, R. J. & TAYLOR, C. B. (1974). The Oxford Expedition to Zanzibar. *Bulletin of the Oxford Exploration Club* **12**(17), 157–88.

HLADIK, C. M. (1977). A comparative study of the feeding strategies of sympatric species of leaf monkeys: *Presbytis senex* and *Presbytis entellus*. In *Primate Ecology: studies of feeding and ranging behaviour in lemurs, monkeys and apes*, ed. T. H. Clutton-Brock, pp. 323–53. London: Academic Press.

HOMEWOOD, K. M. (1978). Feeding strategy of the Tana mangabey *Cerocebus galeritus galeritus* (Mammalia: Primates). *Journal of Zoology (London)* **186**, 375–91.

HUGES, D. W. & GENSET, K. (1973). Alkaloids. *Phytochemica* **2**, 118–70.

IUCN (1966). *Red Data Sheets*, Vol. 1. *Mammalia*, Morges, Switzerland: International Union for the Conservation of Nature and Natural Resources.

IUCN (1986). *Red Data List of Threatened Animals*. Gland, Switzerland: IUCN.

JANIS, C. (1976). The evolutionary strategy of the Equidae and the origins of rumen and caecal digestion. *Evolution* **30**, 757–74.

JANZEN, D. H. (1978). Complications of interpreting the chemical defenses of trees against tropical arboreal plant-eating vertebrates. In *The Ecology of Arboreal Folivores*, ed. G. G. Montgomery, pp. 73–84. Washington, DC: Smithsonian Institution Press.

JAY, P. (1963). The Indian langur monkey. In *Primate Social Behaviour*, ed. C. Southwick, pp. 114–23. Princeton, NJ: Van Nostrand.

JONES, C. & SABATER Pi, J. (1968). Comparative Ecology of *Cercocebus albigena* (Gray) and *Cercocebus torguatus* (Kerr) in Rio Muni, W. Africa. *Folia Primatologica* **9**, 99–113.

KAVANAGH, M. (1981). Variable territoriality among tantalus monkeys in Cameroon. *Folia Primatologica* **36**, 76–98.

KAY, R. N. B., HOPPE, P. & MALOIY, G. M. O. (1976). Fermentative digestion of food in the colobus monkey, *Colobus polykomos. Experientia* **32**, 485–7.

KINGDON, J. (1971). *East African Mammals*, Vol. 1. London: Academic Press.

KINGDON, J. (1981). Where have the colonists come from? A zoogeographical examination of some mammalian isolates in Eastern Africa. *African Journal of Ecology* **19**, 115–24.

MARSH, C. W. (1978a). Ecology and social organization of the Tana River red colobus *Colobus badius rufomitratus*. PhD thesis, University of Bristol.

MARSH, C. W. (1978b). Comparative activity budgets of red colobus. In *Recent Advances in Primatology*, Vol. 1, ed. D. J. Chivers and J. Herbert, pp. 250–1. London: Academic Press.

MARSH, C. W. (1978c). Problems of primate conservation in a patchy environment along the lower Tana River, Kenya. In *Recent Advances in Primatology*, Vol. 2, Conservation, ed. D. J. Chivers and W. Lane-Petter, pp. 85–6. London: Academic Press.

MARSH, C. W. (1979a). Comparative aspects of social organization in Tana River red colobus. *Colobus badius rufomitratus. Zeitschrift für Tierpsychologie* **51**, 337–62.

MARSH, C. W. (1979b). Female transference and mate choice among Tana River red colobus. *Nature* **281**, 568–9.

MARSH, C. W. (1981a). Diet choice among red colobus (*Colobus badius rufomitratus*) on the Tana River, Kenya. *Folia Primatologica* **35**, 147–78.

MARSH, C. W. (1981b). Ranging behaviour and its relation to diet selection in Tana River red colobus (*Colobus badius rufomitratus*). *Journal of Zoology (London)* **195**, 473–92.

MARSH, C. W. (1981c). Time budget of Tana River red colobus. *Folia Primatologica* **35**, 30–50.

MARSH, C. W. (1985). A Resurvey of Tana River Primates and their Forest Habitat. Report for the Institute of Primate Research, National Museum of Kenya and the Department of Wildlife Conservation and Management.

McKEY, D. B. (1978). Soils, vegetation and seed eating by black colobus monkeys. In *The Ecology of Arboreal Folivores*, ed. G. G. Montgomery, pp. 423–37. Washington, DC: Smithsonian Institution Press.

McKEY, D. B. (1979). The distribution of secondary compounds within plants. In *Herbivores, Their*

Interactions with Secondary Plant Metabolites, ed. G. A. Rosenthal and D. H. Janzen, pp. 55–133. London: Academic Press.

McKEY, D. B., GARTLAN, J. S., WATERMAN, P. G. & CHOO, G. M. (1981). Food selection by black colobus monkeys (*Colobus satanas*) in relation to plant chemistry. *Biological Journal of the Linnean Society* 16, 115–46.

MOIR, R. J. (1965). The comparative physiology of ruminant-like animals. In *Physiology of Digestion in the Ruminant*, ed. R. W. Dougherty, pp. 1–44. Washington: Butterworth.

MOIR, R. J. & HARRIS, L. E. (1962). Ruminal flora studies in sheep: influence of nitrogen intake upon ruminal function. *Journal of Nutrition* 77, 285–98.

MTURI, F. A. (1983). Man's utilization of natural resources and the endangered Zanzibar red colobus monkey (*Colobus badius kirkii*). In *Proceedings of the Symposium on the Role of Biology in Development*, ed. J. Middleton, A. Nikundiwe, F. Banyikwa and J. R. Mainoya, p. 113.

MTURI, F. A. (1991). The feeding ecology and behaviour of the red colobus monkey (*Colobus badius kirkii*). PhD thesis, University of Dar es Salaam.

NISHIDA, T. (1972). A note on the ecology of the red colobus monkeys (*Colobus badius tephrosceles*) living in the Mahali Mountains. *Primates* 13, 57–64.

OATES, J. F. (1977). The guereza and its food. In *Primate Ecology: studies of feeding and ranging behaviour in lemurs, monkeys and apes*, ed. T. H. Clutton-Brock, pp. 275–321. London: Academic Press.

OATES, J. F. (1978). Water-plant and soil consumption by guereza monkey. (*Colobus guereza*): a relationship with minerals and toxins in the diet? *Biotropica* 10, 241–53.

PAKENHAM, R. H. W. (1979). *The Birds of Zanzibar and Pemba: An Annotated Check-list*. Surrey: The Gresham Press.

PARRA, R. (1978). Comparison of foregut and hindgut fermentation in herbivores. In *Ecology of Arboreal Folivores*, ed. G. G. Montgomery, pp. 205–29. Washington, DC: Smithsonian Institution Press.

POTTER, T., ELLIS, C. & LEVITT, M. (1985). Activated charcoal: *In vivo* and *in vitro* studies of effect on gas formation. *Gastroenterology* 88, 620–4.

PULLIAM, H. R. (1975). Diet optimization with nutrient constraints. *American Naturalist* 109, 765–8.

RAHM, U. H. (1970). Ecology, zoogeography and systematics of some African forest monkeys. In *The Old World Monkeys: evolution, systematics and behaviour*, ed. J. R. Napier and P. H. Napier, pp. 589–626. London: Academic Press.

ROBINS, R. J. (1976). The composition of Jozani Forest, Zanzibar. *Botanical Journal of the Linnean Society* 72, 223–34.

RODGERS, W. A. (1981). The distribution and conservation status of colobus monkeys in Tanzania. *Primates* 22(1), 33–45.

RODGERS, W. A., OWEN, C. F. & HOMEWOOD, K. M. (1982). Biogeography of East African forest mammals. *Journal of Biogeography* 9, 41–54.

RUDRAN, R. (1973). Adult male replacement in one-male troops of purple-faced langurs (*Presbytis senex senex*) and its effects on population structure. *Folia Primatologica* 19, 166–92.

SANTON, N. (1975). Herbivore pressure on two types of tropical forest. *Biotropica* 7, 8–11.

SILKILUWASHA, F. (1981). The distribution and conservation status of the Zanzibar red colobus. *African Journal of Ecology* 19, 187–94.

STARIN, E. D. (1981). Monkey moves. *Natural History* 90(9), 37–43.

STRUHSAKER, T. T. (1975). *The Red Colobus Monkey*. Chicago: University of Chicago Press.

STRUHSAKER, T. T. (1978a). Food habits of five monkey species in the Kibale Forest, Uganda. In *Recent Advances in Primatology*, Vol. 1. *Behaviour*, ed. D. J. Chivers and J. Herbert, pp. 225–48. London: Academic Press.

STRUHSAKER, T. T. (1978b). Inter-relations of red colobus monkeys and rainforest trees in the Kibale forest, Uganda. In *The Ecology of Arboreal Folivores*, ed. G. G. Montgomery, pp. 397–422. Washington, DC: Smithsonian Institution Press.

STRUHSAKER, T. T. (1980). Comparison of the behaviour and ecology of red colobus and red tail monkeys in the Kibale forest, Uganda. *African Journal of Ecology* 18, 33–51.

STRUHSAKER, T. T. (1981). Forest and primate conservation in East Africa. *African Journal Ecology* 19, 99–114.

STRUHSAKER, T. T. & LELAND, L. (1979). Socioecology of five sympatric monkeys species in the Kibale Forest, Uganda. *Advances in the Study of Behaviour* 9, 159–228.

STRUHSAKER, T. T. & LELAND, L. (1980). Observations on two rare and endangered populations of red colobus monkeys in East Africa. *African Journal of Ecology* 18, 191–216.

SWAI, I. S. (1983). Wildlife Conservation Status in Zanzibar. MSc thesis, University of Dar es Salaam.

VERHEYEN, W. N. (1962). Contribution à la craniologie comparés des primates. *Annales, Musée Royal de l'Afrique Centrale, Sciences zoologiques* 105, 1–256.

WATERMAN, P. G. (1984). Food acquisition and processing as a function of plant chemistry. In *Food Acquisition and Processing in Primates*, ed. D. J. Chivers, B. A. Wood and A. Bilsborough, pp. 117–211. New York: Plenum Press.

WATERMAN, P. G. & CHOO, G. M. (1981). The effects of digestibility-reducing compounds in leaves on food selection by colobinae. *Malays. Appl. Biol.* 10, 147–62.

WESTOBY, M. (1974). The analysis of diet selection by large generalist herbivores. *American Naturalist* 108, 290–304.

WILSON, E. O. AND BOSSERT, W. H. (1971). *A Primer of Population Biology*. Stanford, CT: Sinauer Associates.

YOSHIBA, K. (1968). Local and intergroup variability in ecology and social behaviour of common Indian langurs. In *Primates: Studies in Adaptation and Variability*, ed. P. C. Jay, pp. 217–42. New York: Holt, Rinehart and Winston.

13 The socioecology of interspecific associations among the monkeys of the Mwanihana rain forest, Tanzania: a biogeographic perspective

SAMUEL K. WASSER

Introduction

Although much of East Africa is arid savanna, about 2% of Tanzania contains rich tropical rain forest. The majority of this is montane forest, found on a chain of ancient (80 million years old) block-fault mountains; these mountains, which stem from the Pare Mountains in the north to the Southern Highlands in the south, have been termed the Eastern Arc Mountain Chain (Lovett, 1985). Most mountains in the chain have been isolated from one another since the Pleistocene (Hamilton, 1982; Kingdon, 1989) by a sea of arid woodland savanna. The isolation of the montane forests, in conjunction with a relatively stable climate (Hamilton, 1982), has resulted in significantly fewer overall species of nearly all taxa examined when compared with the more continuously distributed Guineo–Congolian forest (Stuart, 1981; Rodgers, Owen & Homewood, 1982; Lovett, 1985; Lovett, Bridson & Thomas, 1988; Kingdon, 1989 and several chapters in this volume). [The Guineo–Congolian forest is the main forest block extending across Central Africa, from Lake Victoria to Liberia: White, 1981.] The isolation of the Eastern Arc mountains also has produced high rates of endemism in almost every major taxonomic group (Stuart, 1981; Rodgers *et al.*, 1982; Hamilton, 1988; Lovett *et al.*, 1988;

Kingdon, 1989 and several chapters in this volume). This endemism is particularly striking among the plants: over 25% of the 2000 plant species found in Tanzanian forests are endemic (Lovett, 1985).

The majority of chapters in this book illustrate the consequences of biogeographic isolation in the Eastern Arc montane forests on patterns of endemism and species distributions. This chapter takes a somewhat different focus, examining how biogeographic isolation has affected interactions at the community level. In particular, patterns of interspecific association among primates in the Uzungwa montane forest are compared with those between similar primate species in the Guineo–Congolian forest block.

Interspecific associations are defined as groupings in which individuals of more than one species are spatially and socially intermingled during movements, as well as during periods of resting, socialising and feeding (Struhsaker, 1981a). Such associations are, by nature, the outcome of community level interactions, making them particularly interesting from a biogeographic point of view. Isolation-related changes in species distributions at one taxonomic level (e.g. tree and hence plant food species) can lead to a cascade of adjustments throughout the community, the effects of which may be seen on patterns of interspecific associations among forest primates. These associations can, in turn, affect behavior and niche breadth of sympatric species, facilitate hybridisation, and perhaps even enhance rates of speciation (Gautier-Hion, 1988).

The plant and primate communities of the Uzungwa Mountain Forest

The Uzungwa Mountains contain some of the richest forest in the Eastern Arc chain (Rodgers & Homewood, 1982; Rodgers, Chapter 14). They are the only mountains in East Africa that have forest on the escarpment ranging from 300 to 2600 m a.s.l., receiving up to 2500 mm in annual rainfall (Lovett et al., 1988). Plant species in the Uzungwa forests show marked altitudinal variation; however, a number of important plant food species in the Guineo–Congolian forests either are absent or are present at very low densities in the Eastern Arc (Lovett, Chapter 4). Unlike the other montane forests in the Eastern Arc, these southernmost forests also have a distinct dry season (June–September) and a single rainy season (November–May), which also may have contributed to the forest plant biogeography of these communities (Lovett, Chapter 4). The composition and interactions of the primate community are likely to have been affected by these latter conditions as well.

The Uzungwa Mountains contain the only forest in the Eastern Arc that has representatives of *Cercopithecus*, *Cercocebus* and *Colobus* monkeys (the three major genera of Old World monkeys found in forests throughout Central Africa). Moreover, two of the three endemic subspecies of primates found in Tanzania are found only in the Uzungwa Mountains: the recently discovered Sanje crested mangabey (*Cercocebus galeritus sanjei*: Homewood & Rodgers, 1981) and the Uhehe red colobus (*Colobus badius gordonorum*). [Tanzania's third endemic subspecies is the Zanzibar red colobus (*C. b. kirkii*) of Zanzibar Island: Mturi, Chapter 12.] The other two forest primates found in the Uzungwas are the black-and-white or pied col-

obus (*C. angolensis palliatus*) and the Sykes monkey (*Cercopithecus mitis moloneyi*).* As elsewhere, the red colobus and the mangabey typically occur in multi-male and female groups in the Uzungwas, whereas the black-and-white colobus and the Sykes monkey typically occur in single-male, multi-female groups (personal observation).

Forest monkey biogeography

The red colobus and the mangabey are narrowly adapted to rain forest or large stretches of riverine forest in Central Africa, such as in the Tana River in Kenya. The Sykes monkey and the black-and-white colobus also prefer forest but have a much higher tolerance for drier woodland habitat. These patterns are reflected by a more patchy distribution of the former two species compared with the latter two throughout Tanzania. For example, the more common Central African red colobus, *C. b. tephroseles*, is found at several locations along the heavily forested western border of Tanzania, including Gombe Stream, Mahale Mountains, Mbizi and Biharamulo. However, the only red colobus populations found in Tanzania east of the arid corridor are *C. b. gordonorum* in the Uzungwa Mountains/Magombera Forest, and *C. b. kirkii* in a few patches of forest on Zanzibar Island (Rodgers, 1981; see also Mturi, this volume). *C. b. kirkii* appears to be the most evolutionarily conservative of these three subspecies (Kingdon, 1981). However, *C. b. gordonorum* most closely resembles *C. b. kirkii* (Kingdon, 1981; Rodgers, 1981; Struhsaker, 1981b), showing considerable evolutionary conservatism as well. Thus far, the only mangabeys reported in Tanzania (*Cercocebus galeritus sanjei*) occur in the Uzungwa Forest (Homewood & Rodgers, 1981). As a genus, mangabeys also tend to show considerable evolutionary conservatism (Waser, 1984). In contrast to red colobus and mangabeys, black-and-white colobus (*C. guereza*) are found in dry and wet montane forests in northern Tanzania, including Mount Kilimanjaro, Mount Meru, and

---

[1] Judging from coat colour, the Sykes monkey is probably a hybrid between *C. m. moloneyi* and *C. m. monoides* (J. Kingdon, personal communication).

the Serengeti and Arusha National Parks. *C. angolensis* has been reported in at least 42 different localities in Tanzania, including virtually all southern montane forests in the Eastern Arc, and a variety of riverine forest and woodland habitats in southern and eastern Tanzania (Rodgers, 1981). The Sykes monkey, *Cercopithecus mitis*, shows a somewhat similar distribution to that of the black-and-white colobus in Tanzania (Gautier-Hion *et al.*, 1988).

Mammals show evidence of both southern and northern routes of migration into the Eastern Arc, although some species in the arc undoubtedly originated there (Kingdon, 1981, 1989; Rodgers *et al.*, 1982; Kingdon & Howell, Chapter 11). Both Kingdon (1981, 1989) and Rodgers *et al.* (1982) argue convincingly for a predominantly southern migration route for mammals in the Uzungwas. Based in part on the conservatism and patchy distribution of the red colobus and mangabey subspecies in the Eastern Arc, Kingdon (1981, 1989; see also Kingdon & Howell, Chapter 11) further argues that the southern connections between the Tanzania coast and the Guineo–Congolian forest block are 'particularly ancient'. *C. b. gordonorum* and *C. g. sanjei* accordingly appear to have experienced long-term biogeographical isolation in the Uzungwas over evolutionary time (see also Kingdon & Howell, Chapter 11). The assumption of long-term isolation among these primate species is an important component of the hypothesis generated in this chapter for the origin of their interspecific associations in the Uzungwas.

Interspecific associations in a biogeographic context

Struhsaker (1981a) reviewed the interspecific association data in five different forests throughout Central Africa. Interspecific associations were common between the guenons, mangabeys and red colobus, albeit relatively uncommon among the black-and-white colobus. Associations between black-and-white colobus and red colobus were particularly rare, with one exception: the black-and-white and the red colobines in the Magombera/Uzungwa region were

in mixed species groups with one another in over half of their sightings. This chapter examines how rates of interspecific associations among these primates may have been affected by the biogeographic isolation of the Uzungwa forest community.

Methods

Study area

The Mwanihana Forest is located in the northeastern part of the Uzungwa Mountains, due west of Sanje Village (lat. 8° 10' S, long. 35° 50' E). The study area consisted of a series of circuits. Two routes were formed by parallel trails running east–west along each side of the Sanje River. Each of these trails accessed a narrow strip approximately 0.15 km wide and 2.25 km long, from 450 to 1050 m altitude. Two additional loop trails were cut: one to the north of Sanje River for approximately 2 km, and the other to the south for approximately 1 km. A small number of censuses also were conducted at higher altitudes, up to 1700 m, using 'pit sawyer's trails' that connected to the northern transect route.

Census data

Data were collected from August to November 1984, and February, June–October and December 1986. Individual trails were walked on a rotating basis, five mornings per week beginning at 0700 h (an additional afternoon transect beginning at 1530 h was occasionally walked as well). A minimum of 20 censuses per month were conducted during each of these months, with the exception of September 1984. During each census, data were collected on the size, composition, map location and altitude of all monkey groups sighted, as well as on monkey diets, behaviour and predation. Individual monkey groups were not habituated as we chose instead to concentrate on as many different groups as possible over the study period.

Observers tried to maximise observation of feeding and other behaviour whenever a group was sighted. At the end of each observation, observers searched extensively under the feeding trees for additional confirmation of food types and especially plant parts eaten. These latter data were included in the sample only when the specimen was obviously fresh, and the exact plant part/species could not be determined by direct observation. Each plant species and part eaten by a given monkey species during a sighting was counted only once (i.e. received a count of 1.0, regardless of the number of individuals per monkey species observed feeding on it). Diets (plant food species and parts/species) were calculated for each monkey species by summing frequency data from (i) direct observations of both single and mixed species groups, and (ii) collections found under feeding trees for single monkey species sightings only. (This latter restriction to single species sightings served to prevent erroneous assignment of collected plant parts eaten when more than one monkey species was simultaneously feeding in the same tree.) The durations of observation/sighting were on average too brief (mean $\pm$ SE $= 22 \pm 0.4$ min) to allow calculations of feeding rates per item of plant food.

All samples of unknown plant food species were collected and pressed for subsequent identification. Plant identifications were made by Jon Lovett and by the herbarium staff of the Department of Botany, University of Dar es Salaam.

Analyses

Observation time varied in this study across primate species as well as by month, based on their availabilities in the study area. Availability of different plant parts/species also varied monthly. For these reasons, all data were summarised in one-month blocks. Association frequencies were compared across months after normalising association data around the monthly availabilities of each monkey species in the comparison; observed (O) association frequency data were subtracted from expected (E) association frequencies, where E was based on the availabilities of each species during that month. These ($O - E$) differences then were compared using the Wilcoxon matched-pairs signed-rank test (Siegel, 1956) across months.

Calculation of E is illustrated in equation 1 for a comparison of the tendency of red colobus (R) to associate with black-and-white colobus (BW) versus with mangabeys (M). The expected frequency of R associating with BW in this comparison (E_{BW}) was calculated as the proportion of the total BW sightings (BW$_{TOT}$ = freq. single + freq. mixed species sightings of BW) among the total number of BW and M sightings [BW$_{TOT}$ + M$_{TOT}$ − (freq. mixed sightings of BW and M)] over each one-month block, multiplied by the total number of mixed group sightings between R and BW and R and M in that month:

$$E_{BW} = BW_{TOT}/(BW_{TOT} + M_{TOT} - BW\&M_{MIXED})$$
$$\times [R\&BW_{MIXED} + R\&M_{MIXED}] \qquad (1)$$

A comparable E_M was calculated for mangabeys by substituting M$_{TOT}$ for BW$_{TOT}$ in the numerator of equation 1. Each expected frequency (E) was then subtracted from its respective observed frequency (O) in that month to obtain the 'normalised' values ($O - E$) compared in the Wilcoxon matched-pairs signed-rank test.

Similarly, equation 2 shows the expected value for R to associate in single (R$_S$) versus mixed species groups with BW:

$$E_R = R_{TOT}/(BW_{TOT} + R_{TOT} - R\&BW_{MIXED})$$
$$\times (R_S + R\&BW_{MIXED}) \qquad (2)$$

E_{BW} in the second comparison was obtained by substituting BW$_{TOT}$ for R$_{TOT}$ in the numerator, and BW$_S$ for R$_S$ in equation 2.

Dietary overlap was calculated by summing the percentages of each food species (or part/food species) in the diet of primate species i that was shared by primate species j in a given month (Struhsaker, 1978).

Results

The red colobus and black-and-white colobus were most heavily concentrated at elevations between 400 and 1000 m. Sykes monkeys appeared to be evenly distributed at all elevations, whereas mangabeys were more heavily concentrated at and above 1000 m. These altitudinal variations parallel altitudinal variation in habitat type described by Lovett *et al.* (1988). Sykes monkeys and mangabeys also were common in disturbed areas at lower elevations.

We were unable to distinguish between most social groups of a given monkey species. Based on the number of sightings per species per unit area, and several estimates of home range size for these species throughout Central Africa (Struhsaker, 1975, 1978; Oates, 1977; Homewood, 1978; Rudran, 1978; Waser, 1984), we estimated a minimum of 10 red colobus, 10 black-and-white colobus, 8 Sykes monkey, and 6 mangabey groups in the study area.

Community social structure

Figure 13.1 illustrates the percentage of 477 sightings in which each group type was encountered for 10 min or more over the study period (presumed to be long enough to observe reliably whether the group contained individuals from more than one species). Red colobus and black-and-white colobus were observed most frequently (50% and 48%, respectively, of all sightings over the entire study). Sykes monkeys were the next most frequently observed (26% of all sightings) followed by the mangabey (13%). Mangabey sightings were low in part because these monkeys had a very large home range that took them out of the area for several months of the study period.

All species were found in mixed species groupings with one another (Figure 13.1). However, this tendency was most marked between the red colobus and the black-and-white colobus, even after controlling for species availabilities; these two species were found in mixed groups with one another as often as they were found in single species groups ($P = 0.5$, Wilcoxon signed-rank test: see Methods, p. 269). In contrast, both red and

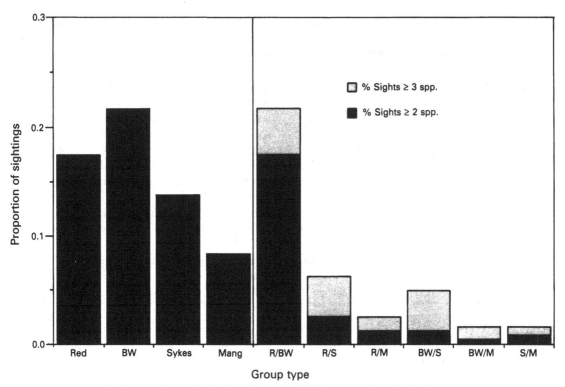

Figure 13.1. Proportion of sightings of each group type over the 11 month study period. Red=red colobus; BW=black-and-white colobus; S=Sykes monkey; M=mangabey. Hatched bars represent mixed species sightings that contained 3 or more species.

black-and-white colobus were found in single species groups significantly more often than in mixed species groups with Sykes monkeys or with mangabeys ($P < 0.01$ in each case).

Both colobines associated significantly more often in mixed groups with one another than they did with mangabeys ($P < 0.002$ for each comparison) or with Sykes monkeys ($P < 0.05$ for each comparison; Figure 13.1). Moreover, half of the mixed groups containing red colobus and Sykes monkeys also contained black-and-white colobus; the reverse was not the case (Figure 13.1). Mangabeys and Sykes monkeys were found more often in single species than in any type of mixed species group ($P < 0.01$ for each comparison); they showed no significant mixed species preference, with one exception: Sykes monkeys were found with red colobus significantly more often than with black-and-white colobus ($P < 0.05$; Figure 13.1).

T-tests revealed no significant differences in the number of individuals per species when sighted in single versus mixed species groups. Mean group sizes were: red colobus 23.3 (range 1–75), black-and-white colobus 4.0 (range 1–20), mangabey 10.2 (range 1–40), and Sykes monkey 3.4 (range 1–15). Mean group sizes probably were underestimated.

Dietary habits and overlap

The dietary diversity of each monkey species was calculated for plant food species, plant part, and plant part/species over the entire 11 month study period, using the Shanon–Wiener Diversity Index (Wilson & Bossert, 1971). A list of the five most preferred plant food species per monkey species is presented in Appendix 13.1. Differences in observation times prohibited statistical analyses of between species comparisons. However, within species comparisons of dietary diversity at the plant species versus plant part and plant part/species levels were possible in Table 13.1; these latter comparisons reflect the degree to which the plant part preferences of each monkey species resulted in increased specialisation or generalisation of its diet.

The red colobus diet was more diverse than that of the black-and-white colobus in terms of plant food species (row 1, Table 13.1). By contrast, plant part preferences of the black-and-white colobus were more diverse than those of the red colobus (row 2, Table 13.1). In fact, the black-and-white colobus is the only species in the table whose plant part diversity exceeded its plant species diversity. This also was reflected in the plant part/species diversity (row 3, Table 13.1); only in the black-and-white colobus was the plant part/species significantly greater than its plant species diversity ($P = 0.025$, Chi-square: Attneave, 1959). These findings are consistent with other studies of these or related monkey species (e.g. Struhsaker & Oates, 1975; Struhsaker, 1978; Homewood, 1978; Rudran, 1978; Waser, 1984). Overall, the plant part/species diversity was greatest for the red colobus,

Table 13.1. *Shanon–Wiener index of plant food species, plant part, and plant part per species diversities among the forest monkeys; number of plant species in diet as well as proportion of study time each monkey species was observed also are listed*

	Red colobus	Black-and-white colobus	Sykes monkey	Mangabey
Plant species[a]	**3.08**	2.14	**2.67**	**2.50**
Plant part[a]	1.75	**2.32**	1.66	1.54
Plant part per species	3.88	3.40[b]	3.03	2.30
Number of plant species in diet	45	21	21	16
Proportion of study time	0.32	0.32	0.22	0.14

Note: [a]The larger value per column in rows 1 and 2 are in bold type.
[b]Plant part/species diversity (row 3) > plant species diversity (row 1) at the $P < 0.05$ level.

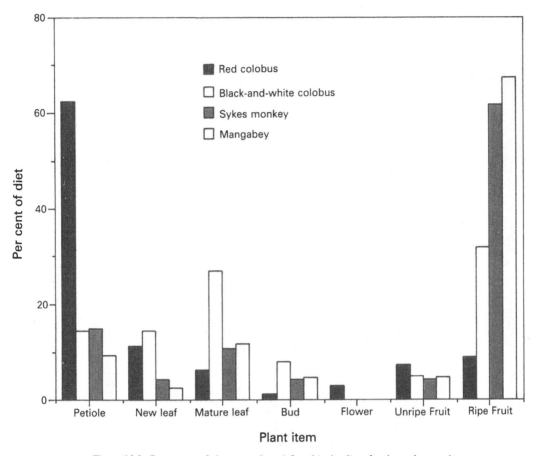

Figure 13.2. Percentage of plant parts (items) found in the diet of each monkey species.

followed by the black-and-white colobus, Sykes monkey and mangabey.

The specific plant part preferences of each of the forest monkey species are shown in Figure 13.2. The red colobus diet contained a high proportion of mature petioles (only 1.0% of the petioles shown in the figure are from new leaves). In fact, the red colobus ate petioles of 35 out of the 45 plant species (78%) that we recorded in their diet. The black-and-white colobus diet was fairly generalised in terms of plant parts eaten, whereas the mangabey and Sykes monkey diets were both quite specialised, on ripe fruits. Both mangabeys and Sykes monkeys were observed to eat insects as well. On a number of occasions, the mangabeys were observed catching and eating crabs from rivers throughout the forest. Their preference for crabs also was apparent from their

faecal composition, which on occasion consisted almost entirely of crab carapaces.

Dietary overlaps were examined between all study species at both plant species and plant part/species levels (Figure 13.3). The highest overlap in plant food species (solid bars) occurred between the Sykes monkeys and the mangabeys, and between the red and the black-and-white colobus. Differences in plant species overlap per month between the red colobus and the black-and-white colobus was significantly greater than those between the red colobus and the Sykes monkey and between the black-and-white colobus and the Sykes monkey ($P < 0.05$, Wilcoxon signed-rank test; Figure 13.3). No other plant species overlap comparisons were significant.

Dietary overlap in part/plant species (Figure 13.3) among the mangabey and the Sykes monkey

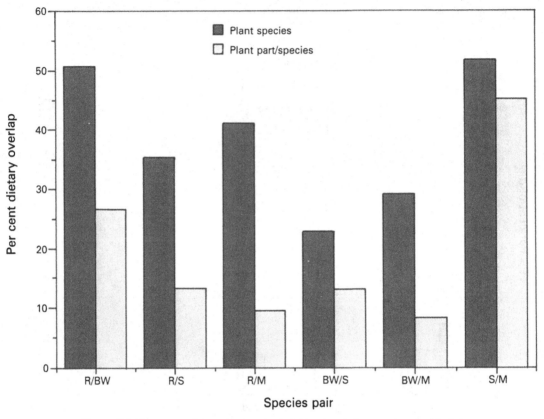

Figure 13.3. Dietary overlap in plant species (solid bars) and plant part/species (shaded bars) among each forest monkey species pair. Abbreviations as in Figure 13.1.

was equivalent to their plant species overlap, presumably because of their common preference for ripe fruit (Figure 13.2). In contrast, the unique preference for leaf petioles among the red colobus resulted in its having a significantly lower part/plant species overlap than plant species overlap with all other primate species (with black-and-white colobus, $P < 0.01$; with Sykes monkey, $P < 0.02$; and with mangabey, $P < 0.04$). The part/plant species overlap also was significantly lower than the plant species overlap among the black-and-white colobus and the mangabey ($P < 0.04$, Wilcoxon signed-rank test; Figure 13.3).

Discussion

Interspecific associations were quite prevalent among the monkeys in the Uzungwa montane for-

est. They were most common between red colobus and black-and-white colobus, and least common between the Sykes monkey and mangabey. Moreover, these tendencies persisted even after adjusting for differences in availabilities of each primate species. Dietary overlap appeared to be an important predictor of these interspecific associations. Thus, the highest rates of interspecific association were found among the two primate species that had high plant species overlap *in conjunction with* relatively low plant part/species overlap. High plant species overlap might be expected to favour mixed-species associations because different primate species would have more similar needs to forage in the same species of tree at any point in time (i.e. conflicts of interest over the tree species in which to forage would be low). However, high plant species overlap also may have associated costs owing to interspecific competition. These costs could be reduced (and associations more readily favoured) between monkey species that fed on different parts of the same plant species, as appeared to be the case among the red and the black-and-white colobines.

The evolution of interspecific associations

Waser (1982) proposed that mixed-species associations could result from chance associations at common food sources. Such associations should be largely dependent upon primate species' availabilities, and their dietary overlap at the plant species level. Chance association at a common food source is unlikely to explain fully the mixed-species association patterns found in the Uzungwas. High dietary overlap in plant food species occurred between red colobus and black-and-white colobus, as well as between mangabeys and Sykes monkeys – the respective species pairs most commonly and least commonly found to associate interspecifically. What distinguished these two species pairs was their respective dietary overlaps at the plant part/species level, being relatively low between the red colobus and the black-and-white colobus and relatively high between the mangabey and the Sykes monkeys.

Dietary overlap at the plant part level is relatively low between the red colobus and the black-and-white colobus owing to specialisation on leaf stems by red colobus and generalisation across several plant parts by the black-and-white colobus. Plant part overlap remains high among the mangabey and Sykes monkey as each species appears to have specialised on ripe fruits (Figures 13.2 and 13.3). Combined, these data support the hypothesis that interspecific competition is reduced by divergent plant part preferences among the red and the black-and-white colobus, despite their tendency to feed on the same plant species. Mangabeys and Sykes tend to share many of the same plant species as well; however, unlike the colobines, interspecific competition between them should have been comparatively high because of their common preference for fruit. Several lines of evidence support the latter assertion. Gautier-Hion (1988: p. 274) states: at all sites, '*Cercopithe-*

cus diets tend to differ the most when fruit production is lowest. 'This may indicate that fruit is a limiting resource and that competition for fruit is the prime factor leading to dietary shifts.' Cords (1987) found that *Cercopithecus mitis* and *Cercocebus ascanius* were more frugivorous at Kakamega than at Kibale, suggesting that this may have resulted from 'competitive release' caused by absence of the mangabey, *Lophocebus albigena*, and other fruit eaters at Kakamega. Finally, Kingdon (1981) suggests that competition with *Cercopithecus* sp. is a primary reason for the patchy distribution of *Cercocebus galeritus* throughout East Africa.

Over 25% of the black-and-white colobus diet consisted of mature leaves in the Uzungwas. It therefore could be argued that the red colobus preference for leaf stems still generated interspecific competition for leaf parts. This is unlikely. As can be seen in Appendix 13.1, the majority of plant species overlap between red colobus and black-and-white colobus occurred on two species, *Albizia gummifera* and *Tabernaemontana pachysiphon*. The black-and-white colobus ate primarily pods and secondarily new leaves, whereas the red colobus diet was largely restricted to mature petioles of the former species. Their overlap on the latter species, which was superabundant in the Mwanihana, was almost entirely on leaf stems; this overlap already was reflected in the data shown in Figures 13.2 and 13.3.

A variety of benefits from mixed species associations have been proposed in the literature, including enhanced predator detection and avoidance, and improved foraging efficiency (Struhsaker, 1981a; Gautier-Hion, Quris & Gautier, 1983; Cords, 1987; Waser, 1987). These benefits, combined with the reduced competitive costs just described, would raise further the net benefit of mixed species associations among the red and the black-and-white colobines (see also Cords, 1990). Based on comparative evidence, Struhsaker (1981a) argued that protection against predation by large eagles was probably the major benefit derived from mixed-species groupings among primates in Africa and Asia. A relatively high crowned-hawk eagle density in the Uzungwas makes this hypothesis relevant to this forest as well. Censuses of the area by S. Stuart (unpublished data) revealed an average of 2–3 crowned-hawk eagle sightings per day in the Uzungwas, compared with a rough estimate of one every 4–5 days reported for Kibale (Struhsaker, 1981a). All monkey species in the Uzungwas also gave alarm calls, as well as responding to the alarms of other primate species, whenever these eagles were flying overhead. Crowned-hawk eagles were observed with freshly predated monkeys on two occasions in the Uzungwas: one was an adult female Sykes monkey and the other an unidentified infant (personal observation). Pythons, leopards and humans probably predate these monkeys as well.

Biogeographic bases of interspecific associations

Why have the red and the black and white colobines failed to form mixed species associations with one another wherever else they have been studied (Struhsaker, 1981a)? Plant food species overlap between the two colobines in the Uzungwas is four times greater than that in the well-studied Kibale Forest (Struhsaker, 1978). Biogeographic isolation may be partly responsible for this high dietary overlap, resulting in relatively lower availability of suitable plant food species from which to choose in the Uzungwas. In fact, the mean dietary overlap across all monkey species in the Uzungwas is 2.7 times higher than that mean in Kibale (39.25 vs 15.4, respectively). Virtually all censuses in the isolated Eastern Arc forests have found lower species diversities than those found in the more vast, continuously distributed Guineo–Congolian forests, which includes the species-rich Kibale (Stuart, 1981; Rodgers et al., 1982; Kingdon, 1989 and several chapters in this volume). In fact, the majority of the main plant food genera in the diets of the red and the black-and-white colobines at Kibale (Struhsaker, 1978) are absent or at very low densities in the Uzungwas (Lovett, Chapter 4); only a few individuals of *Celtis* sp. and no *Markhamia* sp. are found in the Uzungwas (Lovett et al., 1988) – the two most important food species of the red and the black-and-white colobus at Kibale.

Comparisons with other large montane forests

in the Eastern Arc that have high altitudinal ranges comparable to those in the Uzungwas (e.g. the Nguru and Usambara montane forests) further support the above views. Dry season tolerant plants such as *Albizia* are absent or at very low densities in these other Eastern Arc forests (Lovett, Chapter 4). Red colobus (and mangabeys) are absent in these forests as well (Rodgers, 1981), despite the presence of many plant species in common with the Uzungwas. These findings suggest that the red colobus (and mangabey) may have difficulty surviving on much of the available plant material typically found in the Eastern Arc. Such a condition also would have contributed to the high overlap on primary food species such as *Albizia* among the Uzungwa colobines.

Competitive pressures resulting from high plant food species overlap among the Uzungwa colobines probably selected for their divergence at the plant part level; this point is supported by two factors. (i) There is extreme specialisation of the Uzungwa red colobus on petioles. Petioles comprise almost 60% of the red colobus diet in the Uzungwas (Figure 13.2). They comprise <20% of the red colobus (*Colobus badius tephrosceles*) diet in Kibale; the Kibale red colobus diet is more evenly spread over petioles, young leaves, and leaf and flower buds (Struhsaker, 1978). A picture similar to Kibale is seen among the red colobus (*C. b. rufomitratus*) in the riverine forest of Tana River (Marsh, 1981). The diet of the Zanzibar red colobus (*C. b. kirkii*) – the closest relative to *C. b. gordonorum* in the present study (Kingdon, 1981; Rodgers, 1981; Struhsaker, 1981b) – contains about one quarter the amount of leaf stems found in the Uzungwa red colobus diet (Mturi, Chapter 12). Black-and-white colobus are absent from the Tana forest and Zanzibar Island. (ii) Black-and-white colobus in the Uzungwas appeared to be more generalised at the plant part level compared with black-and-white colobus at Kibale (*C. guereza*); 56% of the latter species' diet consists of young leaves compared with less than 20% in the Uzungwas.

In summary, the potential for interspecific competition between the Kibale colobines, coupled with relatively higher availability of suitable plant food species, appear to have selected for dietary divergence at the plant species level among the Kibale colobines. This contrasts with the Uzungwa colobines where plant species overlap appears to have been necessarily high and dietary divergence instead occurred primarily at the plant part level. These patterns may have facilitated interspecific associations among the Uzungwa colonies compared with those in Kibale, uniquely enabling the former to feed in the same species of tree with minimal interspecific competition.

Conclusion

The results of this study require replication because of unhabituated animals and sparse sightings of some species during certain months (e.g. mangabeys). Nevertheless, some conclusions can be made with these precautions in mind.

Low diversity of suitable plant food species initially may have resulted in interspecific competitive pressures in the Uzungwas that forced a form of character displacement: the red colobus became more specialised and perhaps the black-and-white colobus more generalised at the plant part level, maximising feeding differences between them on the same food species. Such evolutionary changes would have decreased the costs of their interspecific associations, regardless of the benefit(s).

It is difficult to explain why character displacement appears to have evolved over competitive exclusion among the Uzungwa colobines (see also Gautier-Hion, 1988). It could be that random processes, accentuated by biogeographic isolation, caused character displacement to win the evolutionary race over competitive exclusion. Similar events were noted by Diamond *et al.* (1988); they reported the evolution of character displacement (vs competitive exclusion) in myzomelid honeyeaters isolated on an island in the Bismarck Archipelago.

A number of factors could have facilitated the evolution of character displacement in the Uzungwas. First, the costs of dietary divergence among the colobines may have been minimal; leaf stems tend to be relatively low in dietary fibre,

total phenolics and condensed tannins while being high in digestibility and fermentable tannins relative to leaf blades (see Mturi, Chapter 12). Specialising on leaf stems accordingly may have enabled the red colobus to eat a wide variety of otherwise unpalatable plant food species. By contrast, the black-and-white colobus may have been able to meet its nutritional needs by exaggerating its well-known tendency to generalise on plant food parts rather than on plant food species (Clutton-Brock, 1974; Oates, 1977).

Pre-existing potential benefits from mixed-species associations (e.g. predation avoidance: Struhsaker, 1981a; Gautier-Hion & Tutin, 1988; Cords, 1990) also may have increased the rate through which character displacement could have evolved. The pre-existence of such benefits is suggested by the comparative prevalence of mixed species associations throughout Central Africa (Struhsaker, 1981a; Waser, 1987). Wasser (1982) described how phenotypic differences between individuals can complement one another to a degree that facilitates the evolution of reciprocal associations between non-kin. This kind of evolutionary event could have occurred among the Uzungwa colobines, for example, if different visual acuities (which could result from species-specific differences in search images for particular food parts) made mixed groups more able than single-species groups to spot predators at safe distances (see also Judd, 1943; Pollitzer, 1972; Terborgh, 1983; Gautier-Hion, 1988).

If correct, the Uzungwa scenario suggests how isolation of a forest community can lead to a cascade of behavioural differences between separated populations of the same or closely related monkey species. Community differences in food availability can lead to differences in dietary specialisation, which can in turn change the whole social structure of organisms – in this case causing distantly related species to move as a single social unit. Such behavioural shifts undoubtedly can lead to other changes as well, as natural selection continues to fine-tune each monkey species to its newly formed niche. It is perhaps appropriate to point out that mechanisms such as these could incidentally lead to the evolution of completely new species, much like those that have now been described for a number of taxonomic groups found in the Eastern Arc Montane Forest Chain.

Summary

Grouping and foraging patterns were studied among the Uhehe red colobus (*Colobus badius gordonorum*), the black-and-white colobus (*C. angolensis palliatus*), the Sykes monkey (*Cercopithecus mitis moloneyi*) and the Sanje crested mangabey (*Cercocebus galeritus sanjei*) in the Mwanihana Forest of the Uzungwa Mountains, Tanzania. The Uzungwa Mountains are part of the Eastern Arc mountain chain – a string of ancient, biogeographically isolated, forested mountains that stretch across eastern Tanzania. All primate species in the Mwanihana Forest were found to associate interspecifically during the 11-month study period. However, the highest rates of interspecific association were found between the two colobines (>50% of their sightings; 25% of all monkey sightings). The red and the black-and-white colobus rarely associate with one another wherever else they co-occur. Overlap in plant food species among the colobines in the Uzungwas was also was relatively high compared with other locations, whereas dietary overlap at the part per plant species was comparatively low. These conditions sharply contrast with those between the Sykes monkey and the mangabey; both of these species show strong preferences for fruit throughout their range. Accordingly, their dietary overlap in the Mwanihana was high at plant species as well as plant part levels; they also rarely formed mixed species associations with one another. Results suggest that dietary divergence at plant part level may have reduced competition between the Uzungwa colobines, while preserving their tendencies (through high plant species overlap) to feed in the same species of tree. Such dietary patterns may have resulted from a comparatively lower availability of potential plant food species in the biogeographically isolated Uzungwa Mountains compared with elsewhere. The possibility that these events represent a form of character displacement is discussed.

Acknowledgements

This project began as part of a combined effort by several botanists and zoologists to characterise the wildlife of the Uzungwa forest. The project was initiated by Alan Rodgers and Jon Lovett, with the support of the (then) World Wildlife Fund and the National Geographic Society. Permission to carry out this study was granted from the Serengeti Wildlife Research Institute, the Tanzanian National Parks, the Tanzanian Department of Forestry, and the Tanzanian National Scientific Research Council. Jon Lovett provided valuable assistance in plant identification, as well as information on plant species distributions across Central Africa. Assistance on plant identifications also was provided by the Botany Department of the University of Dar es Salaam. We are particularly grateful for the help of several field assistants, including Alison Starling, Preston Hardison, Virginia Bound, Ayoub Njalale, Agricola Lihuru, and Langson Kisoma. Special thanks go to Alison Starling, without whom this project never would have begun, and to Mpishi Jon Lemani, who bakes the finest forest bread in all of Tanzania. Valuable comments on the manuscript were provided by Carolyn Crockett, Jon Lovett, Fatina Mturi Omari, Gordon Orians, Rudy Rudran, Tom Struhsaker, Lauren Marra Wasser, and two anonymous reviewers. This research was supported by grants to the author from the National Geographic Society and the World Wildlife Fund.

References

ATTNEAVE, F. (1959). *Application of Information Theory to Psychology. A Summary of Basic Concepts, Methods and Results.* New York: Henry Holt and Co.

CLUTTON-BROCK, T. H. (1974). Feeding behaviour of red colobus and black and white colobus in East Africa. *Folia Primatologica* 22(2–3), 178–207.

CORDS, M. (1987). *Mixed-species associations of Cercopithecus monkeys in the Kakamega Forest, Kenya.* University of California Publications in Zoology, No. 117, pp. 1–109.

CORDS, M. (1990). Vigilance and mixed-species association of some East African forest monkeys. *Behavioral Ecology and Sociobiology* 26, 297–300.

DIAMOND, J., PIMM, S. L., GILPIN, M. E. & LECROY, M. (1988). Rapid evolution of character displacement in myzomelid honeyeaters. *American Naturalist* 134, 675–708.

GAUTIER-HION, A. (1988). Polyspecific associations among forest guenons: ecological, behavioural and evolutionary aspects. In *A Primate Radiation: Evolutionary Biology of the African Guenons*, ed. A. Gautier-Hion, G. Bourlière, J.-P. Gautier and J. Kingdon, pp. 452–76. Cambridge University Press, Cambridge.

GAUTIER-HION, A., BOURLIÈRE, G., GAUTIER, J.-P. & KINGDON, J. (ed.) (1988). *A Primate Radiation: Evolutionary Biology of the African Guenons.* Cambridge: Cambridge University Press.

GAUTIER-HION, A., QURIS, R. & GAUTIER, J. P. (1983). Monospecific vs. polyspecific life: A comparative study of foraging and antipredator tactics in a community of *Cercopithecus* monkeys. *Behavioural Ecology and Sociobiology* 12, 325–35.

GAUTIER-HION, A. & TUTIN, C. E. G. (1988). Simultaneous attack by adult males of a polyspecific troop of monkeys against a crowned hawk eagle. *Folia Primatologica* 51, 149–51.

HAMILTON, A. C. (1982). *Environmental history of Africa: A study of the Quaternary.* London: Academic Press.

HAMILTON, A. C. (1988). Guenon evolution and forest history. In *A Primate Radiation: Evolutionary Biology of the African Guenons*, ed. A. Gautier-Hion, G. Bourlière, J.-P. Gautier, and J. Kingdon, pp. 13–34. Cambridge: Cambridge University Press.

HOMEWOOD, K. M. (1978). Feeding strategy of the Tana mangabey *Cercocebus galeritus galeritus* (Mammalia: Primates). *Journal of Zoology (London)* 186, 375–91.

HOMEWOOD, K. M. & RODGERS, W. A. (1981). A previously undescribed mangabey from Southern Tanzania. *International Journal of Primatology* 2, 47–55.

JUDD, D. (1943). Color blindness and the detection of camouflage. *Science* 97, 544–6.

KINGDON, J. (1981). Where have the colonists come from? A zoogeographical examination of some mammalian isolates in eastern Africa. *African Journal of Ecology* 19, 115–24.

KINGDON, J. (1989). *Island Africa.* Princeton, NJ: Princeton University Press.

LOVETT, J. (1985). Moist forests of eastern Tanzania. *Swara* 8(5), 8–10.

LOVETT, J., BRIDSON, D. M. & THOMAS, D. (1988). A preliminary list of the moist angiosperm flora of Mwanihana Forest Reserve, Tanzania *Annals of the Missouri Botanical Garden* 75, 874–85.

MARSH, C. W. (1981). Diet choice among red colobus (*Colobus badius rufomitratus*) on the Tana River, Kenya. *Folia Primatologica* 35, 147–78.

OATES, J. F. (1977). The social life of a black-and-white colobus monkey, *Colobus guereza. Zeitschrift für Tierpsychologie* 45, 1–60.

POLLITZER, W. S. (1972). Discussion. In *Genetics, Enviornment, and Behavior: Implications for Educational Policy*, ed. L. Ehrman, G. S. Omen and E. Caspari. New York: Academic Press.

RODGERS, W. A. (1981). The distribution and conservation status of colobus monkeys in Tanzania. *Primates* 22, 33–45.

RODGERS, W. A. & HOMEWOOD, K. M. (1982). Biological values and conservation prospects for the forests and primate populations of the Uzungwa Mountains, Tanzania. *Biological Conservation* 24, 285–304.

RODGERS, W. A., OWEN, C. F. & HOMEWOOD, K. M. (1982). Biogeography of East African Forest Mammals. *Journal of Biogeography* 9, 41–54.

RUDRAN, R. (1978). Socioecology of the blue monkeys (*Cercopithecus mitis stuhlmanni*) of the Kibale Forest, Uganda. *Smithsonian Contributions to Zoology* No. 249.

SIEGEL, S. (1956). *Nonparametric Statistics for the Behavioural Sciences.* New York: McGraw-Hill.

STRUHSAKER, T. T. (1975). *The Red Colobus Monkey.* Chicago: University of Chicago Press.

STRUHSAKER, T. T. (1978). Food habits of five monkey species in the Kibale Forest, Uganda. In *Recent Advances in Primatology*, Vol. 1 *Behaviour*, ed. D. J. Chivers, pp. 225–48. London: Academic Press.

STRUHSAKER, T. T. (1981a). Polyspecific associations among tropical rain-forest primates. *Zeitschrift für Tierpsychologie* 57, 268–304.

STRUHSAKER, T. T. (1981b). Vocalizations, phylogeny and paleogeography of red colobus (*Colobus badius*). *African Journal of Ecology* 19, 265–83.

STRUHSAKER, T. T. & OATES, J. F. (1975). Comparison of the behavior and ecology of red colobus and black and white colobus monkeys in Uganda: a summary. In *Socioecology and Psychology of Primates*, ed. R. H. Tuttle, pp. 103–24. The Hague: Mouton Publishers.

STUART, S. (1981). A comparison of avifaunas of seven East African forest islands. *African Journal of Ecology* 19, 133–52.

TERBORGH, J. (1983). *Five New World Primates.* Princeton, NJ: Princeton University Press.

WASER, P. M. (1982). Primate polyspecific associations: do they occur by chance? *Animal Behaviour* 30, 1–8.

WASER, P. (1984). Ecological differences and behavioral contrasts between two mangabey species. In *Adaptations for Foraging in Nonhuman Primates*, ed. P. S. Rodman and J. G. H. Cant, pp. 195–216. New York: Columbia University Press.

WASER, P. M. (1987). Interactions among primate species. In *Primate Societies*, ed. B. B. Smuts, D. L.

Cheney, R. M. Seyfarth, R. W. Wrangham and T. Seyfarth, R. W. Wrangham and T. T. Struhsaker. Chicago: University of Chicago.

WASSER, S. K. (1982). Reciprocity and the trade-off between associate quality and relatedness. *American Naturalist* 119, 720–31.

WHITE, F. (1981). The history of the Afromontane archipelago and the scientific need for its conservation. *African Journal of Ecology* 19, 33–54.

WILSON, E. O. & BOSSERT, W. H. (1971). *A Primer of Population Biology.* Stamford, CT: Sinauer.

Appendix 13.1.   *Composite list of the five most preferred plant food species in the diets of each monkey species found in the Mwanihana Forest, Tanzania. Percentage of these plant species in the diets of each monkey species also is given. [Data included in this table are from direct observations of both single and mixed species groups, and collections found under feeding trees for single monkey species sightings only (see methods); n = 188 for red colobus (RC); n = 72 for black and white colobus (BW); n = 48 each for mangabey (M) and Sykes monkey (S).]*

Scientific name	Vernacular name (Swahili/Kihehe)	RC	BW	M	S
Afrosersalisia cerasifera	mkumbulu	12.7		18.8	6.3
Albizia gummifera	mtanga mwekundu	15.9	12.5	12.5	
Landolfia kirkii	lingombe		4.2	6.3	18.8
Tabernaemontana pachysiphon	mlowo lowo	12.7	20.8	25.1	18.8
Uapaca paludosa	mtuwu	4.8		6.3	6.3
Parkia filicoidea	mlenda	3.2	8.3	6.3	6.3
Erythrophleum suaveolens	mwahe	7.9	4.2		
Bequaertiodendron natalense	mkeregeti				12.5
Lagynias pallidiflora	mhomangabaku	3.2		12.5	
Entada pursaetha	lifute	3.2	16.7		

PART IV
Conservation

14 The conservation of the forest resources of eastern Africa: past influences, present practices and future needs

W. A. RODGERS

Introduction

Merely understanding the biology of eastern Africa's closed forests is not sufficient if we wish to maintain the resource as a functional natural community for posterity. Conservation needs action: it needs management inputs into both the resource itself and the human populations who depend on the resources for their livelihood and in so doing often degrade the resource. There is still a need for biology, however, especially for a resource as complex as the tropical forests, where even the identification of the component species remains problematical.

We are witnessing the loss of forests and loss of forest species all over the tropics; eastern Africa is no exception, but here we have had only a small and fragmented resource base of forest to start with. These forests are important for water catchment and timber, and their area is coveted for agricultural development. Pressures on the forest land are growing and are often incompatible. It is a sad paradox that now, in the 1990s, we often *know how* we could achieve conservation and how we can share the 'cake', but we often *lack the political and financial will* to do so.

This chapter traces the history of land use in the forest areas in eastern Africa, and examines the causes of incompatibility between land usages, and some theoretical and practical implications for conservation. The chapter concludes with a rather lengthy discussion of conservation needs, in what are intended as practical terms. I make little apology for length: there is a dearth of information on inputs for forest conservation; East Africa's conservation movement has been rather strongly focused on the larger mammals in the past. The irony is that whilst the survival of the mammals is a luxury, the forests are essential for our survival! This is increasingly understood by many administrators and planners; in Tanzania this has led to the recent preparation of new forest policies and action plans for long-term sustainable use and conservation of the forest resources. Unfortunately, this realisation has coincided with a period of extreme economic difficulty. There are no funds to implement the provisions of new and enlightened policies. Forest conservation inputs will therefore depend largely on foreign aid assistance in the coming decade. This chapter may help stress to donor organisations the necessity for such assistance.

The historical pattern of land use in forested areas

The precolonial period

Palaeoecologists and prehistorians have interpreted the changing climate and vegetation cover of the late Pleistocene and Holocene periods in eastern Africa (see Hamilton, 1982 for a full

review and summaries by Griffiths, Chapter 2 and
Lovett, Chapter 3). They argue that forest cover
was at maximum spread some 6000–8000 years
BP (before present) and has decreased since then
owing partly to a gradually drying climate but
especially to the increasing effects of clearing and
burning by hunter-gatherer and agricultural man.
This chapter discusses the effect of man in
greater detail.

The end of the late stone age and the start of an
iron age culture in many parts of East Africa some
2000 years ago would have led to intensified forest clearing for wood for iron-smelting (Davidson,
1961; Soper, 1967; Clarke, 1969; Hamilton,
1974). Morrison & Hamilton (1974), in a detailed
study of a forest system in Uganda, suggested that
such clearing began in the valleys, then the lower
slopes, and eventually spread to the upper slopes.
The beginning of the Azanian culture of eastern
and central Africa, traces of which have been seen
on the Uluguru, Usagara and Southern Highland
mountains of Tanzania, date from this period.
Until recently there was little informed discussion
on the development of forest-living people over
the ensuing millennium. Lundgren & Lundgren
(1972) mention mountain forest areas cleared
during minor dry periods (e.g. AD 1000) and then
kept open by fire and grazing; Wood (1965) and
Soper (1967) further discuss the possibility of
man-induced clearings in the forests of Mounts
Meru and Kilimanjaro. Dale (1954) quotes Fosbrooke as saying that a period of aridity from AD
1600 to 1750 would have caused a spread of
people from the dry plains to the moister forests.
Fosbrooke & Sassoon (1965) have described a
stone-bowl culture from the present-day forest–
grassland boundary on Kilimanjaro. Odner
(1971) in a more detailed discussion of Kilimanjaro mentions digging sticks and stone bowls from
an estimated date of 1500 BP. Sutton (1966) says
that dwarf hunting people of the forest, the
Wakonyingo, had a pottery and defensive pit
culture, and that they pre-dated the present-day
Bantu cultivators. Odner (1971) gives a date of at
least 400 years for the Wakonyingo pottery.

Hamilton (1990) provides a more exhaustive
discussion of probable major impacts of early Iron
Age peoples on what is usually considered to be
primary or climax natural forest, with especial
reference to the Usambara Mountains. He, and a
detailed paper by Schmidt (1990), present
evidence of archaeological finds, mainly pottery,
in the surface soil layers of tall natural forest sites
in the East Usambaras, e.g. Kwanguni. This they
ascribe to early Iron Age settlements some 2000
years ago. Schmidt (1990) argues that this was
part of a settlement pattern invading forests all
over East Africa between 500 BC and AD 500. He
describes a later wave of village settlements and
clearings in the later iron age, at about AD 900–
1000, e.g. sites at Mtai in the East Usambaras.
This pattern of clearing may have allowed the
growth of the present large old *Ocotea* trees so
prominent in the Usambara forests. However, this
settlement pattern then vanished as contemporary
travellers (e.g. Moreau, 1935) showed extensive
forest cover in the East Usambaras. Farler (1879),
in a report from Mlinga Mission, discusses large-
scale depopulation in the 19th century. Lovett
(Chapter 3) suggests that the high incidence of
secondary forest on the Western Uzungwas indicates a past higher agricultural population.

Kjekshus (1977), in a detailed debate on East
African demography, concluded that human
populations were growing slowly prior to
European conquest in the late 19th century. He
suggested that the elaborate development of
irrigation, manuring and erosion control systems
on many mountain areas indicated a long and
stable occupation of those areas, allowing such
agricultural practices to evolve (e.g. the Chagga of
Kilimanjaro and the Matengo of the Southern
Highlands). Marealle (1949) estimated the
Chagga people to have been on Kilimanjaro for at
least 400 years.

Long association with forest may have led to
the evolution of land use practices allowing stable
ecological conditions and traditions of soil and
water conservation. These traditions are based on
individual or family ownership of land which
allows the use of permanent tree crops as well as
irrigation and anti-erosion measures. This is
evident in the Chagga (Allan, 1965) and Matengo
(Kjekshus, 1977), as well as the Wasambaa
systems of the Usambara (Farler, 1879; Eichhorn,
1911–1923; Dobson, 1955; Feierman, 1970).

Such patterns contrast with more recent colonisers of mountain land, such as the Waluguru people who moved into the hill forests less than 200 years ago to escape land arguments on the plains (Young & Fosbrooke, 1960; Temple, 1972). Here land ownership is vested in the clan and title to individual plots changes frequently. Thus there is little or no investment in permanent crops, irrigation or anti-erosion works.

Patterns were different in the coastal lowlands where communities may have been more mobile, the Wangoni and Wasegeju being immediate examples of people who have undertaken major movements in the 1800s. This would have led to cultivation practices with longer fallow periods and involved considerable forest/thicket clearing (Kjekshus, 1977). Gum copal commerce and the development of coconut plantations grew rapidly in the early 1800s and much forest would have been cut as a consequence (Rodgers, 1992).

Forest land did have significance in people's lives, exemplified by the many sacred forest patches or *kayas* of South Kenya (Oxford University, 1981; Hawthorne, Chapter 5), and smaller groves still visible near Dar es Salaam and elsewhere. Much has been written about the major role of forest wildlife in people's diets in West Africa (e.g. Martin, 1983); far less is written about such relationships in East Africa. Forests are sources of bush meat (e.g. duiker and other small antelope, pig, cane rat, hyrax, birds, etc.), and hunting is still widespread (e.g. Kelsey & Langton (1984) for present-day practices in Arabuko–Sokoke forest in north coastal Kenya, and Alpers (1975) for past practice in South Tanzania). The Wahehe people of the Uzungwa Mountains are less conservative than most East African people in their diet, and eat a wide variety of forest animals: crabs, snakes, hornbills, hyrax, duiker, black-and-white colobus and blue monkey (W.A.R., personal observations). Tribal and local social factors are of great importance in determining dietary habits, and hence hunting pressures. Fleuret (1979, 1980) discusses the importance of wild plants in people's lives in West Usambaras. Forest trees may provide significant famine foods, such as *Parinari excelsa* in the Usambaras (J. C. Lovett, personal communication).

Forest land use in colonial times: the German period

With the arrival of German colonialism at the end of the 19th century a relatively stable pattern of land use changed. Human population numbers probably decreased, and shifting cultivation in denser thicket increased at the expense of semi-permanent settlement (Kjekshus, 1977). Rodgers (1975) described the Wangindo people in south coastal Tanzania as leaving large valley settlements and moving deeper into less fertile forested land on upper slopes and plateaux to escape from German pressures. Much montane arable land within the natural forest was alienated by the new settlers and local people would have changed their settlement patterns accordingly, e.g. the Chagga moving higher up Mount Kilimanjaro.

Many forest products were used by the Germans in commerce, e.g. wild rubber (*Landolphia* and *Saba* species), and indigenous timber was exported. Communications, and the siting of towns and missions in climatically preferable, highland forest areas (e.g. Lushoto and Amani in the Usambaras; Morningside, Bwakira and Bunduki in the Ulugurus; Kisarawe in the Pugu Hills) led to greater pressure on these forest lands.

On the other hand, the early colonists were well aware of the importance of forested catchments for water supplies. Schnabel (1990) details the early history of German forestry in Tanzania. The first forester appointed in 1892 led to a full Department of Forestry and Wildlife by 1912. Volkens (1897) and Siebenlist (1914) stressed the need for forest reservation and even reafforestation in German East Africa. Hitchins (1907) did the same for Kenya. The Germans did instigate policies of forest reservation for water and for the control of timber cutting (Anon., 1902). Whilst several accounts describe the early history of wildlife conservation in Tanzania (e.g. Matzke, 1976; Rodgers & Lobo, 1981), less has been written about the beginnings of forestry. Hamilton & Mwasha (1990) give a detailed account of forestry development in the eastern Usambara Mountains. Schabel (1990) summarises the major German achievements. By 1914, 231 separate forest reserves were gazetted, covering more than 7500 km²; these were mainly mountain and coastal forests. The first detailed reservation and regulation policies were in the Rufiji mangroves. Parry (1962) lists several of these German Reserves, many of which have identical areas and boundaries today, e.g. Nou, Marang and Northern Highlands Reserves, all in Arusha District. German maps show the presence of minor settlements in some reserves, but Lundgren (1978), describing secondary forest in a few forest areas, says it is likely that most reserves were uninhabited. Whilst the Germans established many trial plots and arboreta of exotic plant species (e.g. at Amani: Greenway, 1934), they established very few plantations, totalling less than 10 km².

The onset of the First World War in 1914 led to renewed encroachment of forest land by Tanzanian peoples escaping enlistment and crop loss. Forest resources would also have been cut into for emergency civil and military use.

Forest and land use in colonial times: the British period

Hamilton (1984) states that the Forestry Department of Uganda started in 1898, Honore (1962) tells us that the Kenya Forest Department came into being in 1902; that of Tanganyika started in 1919 with the appointment of a Conservator to administer the reserves of the former German colony. Parry (1962) says that some private German estates with much forest on the Usambaras were converted to forest reserves. Rodgers & Homewood (1982a) list the present-day East Usambara forest reserves which were German reserves or estates.

The pattern of land acquisition in Tanzania was probably similar to that which occurred in Uganda, where it has been well documented by Hamilton:

> The process of acquiring land by the Forest Department was gradual and to a degree unsystematic, and it was not until the 1940s that the boundaries of the forest estate, more or less as they now stand, became established. All larger

blocks of forest were declared Crown Land, coming under Government control . . .
>
> (Hamilton, 1984)

By Independence in 1961, Tanzania had reserved between 121 300 km² (Maagi, Mkude & Mlowe, 1979) and 114 000 km² (Sangster, 1962), including some 9500 km² of natural closed forests, and the rest woodland. By 1979 the figures had reached 133 500 km² or 15% of Tanzania's land surface. By comparison, in Kenya some 16 600 km² was declared as forest reserve by 1961 (Honore, 1962), a similar figure to that of today, covering 3% of Kenya's land surface (Lusigi, 1982).

Early policies regarding acquisition of forest land stressed the role of protection forestry. Principal objectives of forestry stressed the 'maintenance and improvement of climatic and physical conditions of the country' and the need to 'conserve and regulate water supplies by the protection of the catchments' (Honore, 1962). The importance of forest cover in maintaining climatic stability was emphasised in the ruling that tea estates in the Usambaras should maintain some 40% of their area as forest so as to perpetuate mist and occult precipitation (Hamilton & Mwasha, 1990).

Hamilton (1984) describes the growth and changes of forest policy in Uganda; this description is undoubtedly relevant to what happened in Kenya and Tanzania as well. A report in 1929 stressed the role of forestry in climate and water regulation and recommended much greater protection and major afforestation; Uganda then had some 11% natural moist thicket and forest! This report was followed by an official forest policy statement in 1948, which stressed two objectives:

(a) To reserve in perpetuity, for the benefit of the present inhabitants of Uganda, and of posterity, sufficient land to maintain climatic conditions suitable for agriculture, to preserve water supplies, to provide forest produce and to maintain soil stability.

(b) To manage this forest estate to obtain the best returns on its capital value and the expenses of management, in so far as such returns are consistent with the primary aims set out above.

This policy statement downgraded earlier priorities of protective forestry in favour of pressing demands of agriculture, with the declaration that 'forest cover should be limited to the minimum area which will achieve the primary aims of management' (Hamilton, 1984).

Similar policy changes took place at this time in Tanganyika as well (probably due to the post-war economic slump). The Secretary of State Forestry Advisor said, 'Protection, whilst still most important, had been over-stressed. Production should get more attention.' The Forest Department expected export revenue to increase as a consequence (Government of Tanganyika, 1946).

By the end of 1950s, with Independence coming, forest policy grew more complicated, and Tanganyika again gave precedence to protection. The new forest policy, first proposed in 1953 and enacted in 1961, stated objectives as:

(a) To demarcate and reserve in perpetuity, for the benefit of present and future inhabitants of the country, sufficient forested land or land capable of afforestation to preserve or improve local climates and water supplies, stabilize land which is liable to deterioration, and provide a sustained yield of forest produce of all kinds for internal use and export.

(b) To manage forest estate and all forest growth on public land so as to obtain the best financial returns on capital value and cost of management in so far as such returns are consistent with the primary aims above.

(c) To encourage and assist the practice of forestry by local Government bodies and by private enterprise.

(d) To undertake and promote research and education in all branches of forestry and to build up by example and teaching a real understanding among the people of the country, of the value of forests and forestry to them and their descendants.

The Forest Department Annual Report containing this policy statement went on to say that in the past preoccupation had been with paragraph (b), i.e. financial gain (Government of Tanganyika, 1959). It is noteworthy that export value of forest products had gone from UK £120 000 in 1946 to £580 000 in 1959.

Forest management activities were directed towards immediate economic returns. The early 20th century saw the great expansion of industrial pulp and timber requirements, and it rapidly became obvious that the natural forest resources, with only a low proportion of valuable species, could not meet this demand. Indeed, commercial non-destructive harvesting of natural forests gave yields too low to pay for their protection (Lundgren, 1975). As a result both Kenya and Tanganyika started to replace logged natural forests with exotic softwood plantations; large areas of forest on Mounts Meru and Kilimanjaro, and in the West Usambaras were so replanted. In Tanganyika this policy was described by Sangster (1962) as 'the development of productive forests follows lines well known in other countries of the Commonwealth ... replacement of local timbers by plantations of exotic trees, either on the same site or elsewhere ... '. Hamilton & Mwasha (1990) present a detailed review of forest development in the Usambaras in the British period.

However, another set of priorities was developing. The early 1900s saw the beginnings of an international movement to conserve the aesthetic and biological values of natural resources, not merely their economic values. This was first formally stated at a London Convention in 1901. This movement, which led to the creation of 'National Parks' in areas of exceptional natural beauty and richness (for example, Albert Park in the then Belgian Congo, 1925; Kruger Park in South Africa, 1926), achieved recognition in colonial Africa. The London 'Faunal' Conference of 1933, which advocated an increased reservation of land for the conservation of important biological resources, extended these actions.

Foresters had long seen the need for some form of natural reference and protection system for seed sources and the maintenance of refuges for pest-controlling birds and insects, etc. This need was met by a system of small nature reserves or preservation plots in major forest areas in several countries including Kenya and Uganda but not Tanganyika. These plots, each averaging much less than 2 km², were insignificant beside the huge parks and reserves created for large mammal conservation. Tsavo National Park in

Kenya, 20 000 km² in 1948; and Serengeti National Park in Tanzania, 25 000 km² in 1951, were the first two parks in East Africa. Game reserves in Tanganyika, which specifically forbade agriculture, settlement, forest exploitation and, up to 1964, any form of hunting, also covered over 100 000 km² by 1961.

At Independence legal categories of land use for conservation purposes in Kenya and Tanzania included the following:

National Park, no human rights or utilisation, enacted by Parliament.

Game Reserve (or National Reserve in Kenya), no human rights or utilisation, enacted by Ministerial gazettement.

Game Controlled Area/Hunting Block, all forms of utilisation allowed, including settlement and cultivation, but shooting controlled.

Forest Reserve, no settlement or human rights, but varied possibilities of exploitation from complete protection to clear felling and exotic plantation, enacted by Ministerial gazettement. Kenya incorporated a category of 'Nature Reserve' to give greater protection to specific areas in the Forest Reserve Estate.

National Parks and Game Reserves were created largely for faunal conservation (especially for large animals). Controlled Areas regulated offtake of wildlife with commercial values, and Forest Reserves served a variety of functions from protecting watersheds to regulating commercial offtake of timber. The function of conserving unique and threatened biological components had not yet, however, been officially mentioned. *The Tanganyika Handbook* of 1958 (Government of Tanganyika, 1958b), which stresses 'game preservation', makes no mention of the conservation of forest values. Sir Julian Huxley, in a sponsored series of essays on African development, did foresee the need for forest parks to conserve forest values:

> First and foremost the alpine zone of the great mountains, Kilimanjaro, Kenya and Rwenzori, and perhaps Mount Meru, should be made

National Parks, *together with a considerable section of the forest zone.* [my emphasis]

(Huxley, 1932)

The landmark 'Arusha Conference' of the International Union for the Conservation of Nature and Natural Resources (IUCN) in 1961, which lauded the achievements of outgoing colonial regimes and laid much of the groundwork for later wildlife conservation (e.g. the origin of the famed Arusha Manifesto in which President Nyerere stated the nation's commitment to the preservation of wildlife), makes absolutely no mention of the biological values of East Africa's forests (IUCN, 1963).

Forests and land use practices from Independence to 1985

Both Kenya and Tanzania have increased their respective areas of forest reserves in the 20 years since Independence. In Tanzania, some 13 700 km² of new reserves were formed and some 605 km² were revoked in the face of great demand for agricultural land, which did unfortunately include a large area of 120 km² of valuable natural forest in the West Usambaras (Lundgren & Lundgren, 1982). Detailed information on East Usambara forest development is given by Hamilton & Mwasha (1990).

Major changes of forest policy resulted in a great decrease in the volume of hardwoods exported, the increased importance of industrial plantations, the beginnings of a zonation system for natural forests into production and protection forest areas and, belatedly, much greater attention to social forestry and the needs of common people. The village forestry programme of Tanzania (Mnzava, 1980) is an important example of Government's evolving forest policies.

In Tanzania forest policy has, in theory, given equal attention to protecting and managing the reserved forests in order to achieve perpetual wood production and to secure ecological and hydrological stability of watersheds (Lundgren & Lundgren, 1982). In practice, however, protection forestry has received much less attention owing to inadequate staffing and financial resources (Nordic Review Mission, 1979). The

Table 14.1. *The distribution and area of closed protection forest in Tanzania*

Region	Reserve area (km^2)[a]	Total forest area (km^2)[b]	Principal forest area
Arusha[c]	1939	2795	Ngorongoro[d], Monduli, Hanang, Meru
Coast	19	1772	Pugu, Pande, Kichi, Wami
Dodoma	98		Mpwapwa
Iringa	2768	2423	West Uzungwa, Image, Njombe
Kilimanjaro[c]	848	949	Kilimanjaro[d], Pare
Mbeya	404	686	Southern Highlands
Morogoro[c]	1282	2273	Ulugurus, Mahenge, East Uzungwa, Ukaguru, Nguru
Mwanza	7	390	Rubondo[d]
Rukwa	8	18	
Tanga[c]	519	1793	East and West Usambaras, Coastal Hills
Total	7892	13 099	

Note: [a]Data from Maagi, Mkude & Mlowe (1979).
[b]Data from Rodgers, Mziray & Shishira (1985). Discrepancies are due to inclusion of national park forest, non-reserved forest and forest reserve area used for production.
[c]Regions with Regional Forest Catchment Offices, reporting direct to National Ministry.
[d]Forest areas with higher conservation status as park or conservation area.

new forest policy in Tanzania, expected to be approved and implemented shortly, will give greater emphasis to watershed management and protection (see below (p. 291) and Government of Tanzania, 1986).

Part of this reduced protection was attributable to the fragmentation of a consolidated national Forestry Department in 1969 into smaller regional units under the effective control of regional development authorities, most of whom were more interested in rapid production, than in longer term conservation policies. Recently, reserves designated as protection for catchment in major mountain forest regions have again been brought under an increasing measure of national control, e.g. the Usambara and Uluguru mountains. Catchment forest areas are detailed for Tanzania in Table 14.1.

In Kenya, the Forestry Act of 1974 lists four major objectives, giving priority to

the protection of the forest estate to maintain climatic and physical conditions of the country and regulate water supplies and to conserve the soil by prevention of desiccation and soil movement caused by water and wind.

Other objectives make short mention of the need for

provision of timber and forest products; to provide recreation and the preservation of wildlife (the first such official mention) and to provide employment.

(Lusigi, 1982)

These policies of Kenya and Tanzania contrast sharply with those of Uganda. A 1973 policy developed by Canadian aid consultants and adopted by the Government authority is described by Hamilton (1984) as:

(a) To ensure a supply of raw materials for direct use or processing by industry now and in the future.
(b) To provide employment.
(c) To capture returns to the nation from the natural forest resource resulting from the utilisation of the remaining natural high forest areas.

The report then mentions secondary objectives such as protection of water catchments, soil, wildlife and amenity.

It is of interest that the amount of emphasis

Table 14.2. *Relationship between the amount of forest cover and the importance of catchment forestry in East Africa*

Country	Importance given to catchment conservation	Land with over 800 mm precipitation (%)	Natural closed forest cover (%)
Kenya	Much emphasis	15	2.3
Tanzania	Moderate emphasis	51	2.5
Uganda	Little emphasis	78	4.6

Source: From Griffiths, 1962; Lusigi, 1982; Hamilton, 1984; Rodgers *et al.*, 1985a.

given to water catchment protection forestry, in each East African country varies inversely with the proportion of land with high rainfall and good forest cover (Table 14.2).

Forest Departments in East Africa have been slow to make adequate provision in terms of finance, personnel and equipment for the fundamental protection of the forests.

> Such a low profile attitude has led to protection forestry having an inadequate role in development planning and must have contributed to an aura of indifference among producers, administrators and local leaders.
>
> (Lusigi, 1982)

Rodgers, Hall & Mwasumbi (1983) document a case of five separate small forest reserves of closed forest on the wet slopes of the Ulugurus all being controlled by a single forest guard. The largest natural closed forest reserve in Tanzania, West Kilombero Scarp, has a single guard. This contrasts greatly with nearby game reserves which have protection staff in the hundreds. Two departments in the same Ministry, forestry and wildlife, have developed different concepts of the problems and solutions of resource protection.

Forests and land use practices since 1985

The last decade has seen the start of forest protection for biological as well as soil water and timber resources. Kenya has three mountain national parks, all with considerable areas of forest and all valuable water catchments: the Aberdares, Mount Elgon and Mount Kenya. In Tanzania, Arusha National Park has a consider-

able forest area on Mount Meru, but Kilimanjaro National Park still has only a cosmetic forest fraction. All of these five parks have been created primarily for scenic, large mammal and tourist purposes. The forests are included almost as a secondary role. In 1986 Tanzania implemented park status for the Mahale Mountains specifically for forest values – water, fauna (especially chimpanzee) and flora. The Uzungwa National Park was gazetted in 1992, the first to be protected for biodiversity, and the first component of the 'Eastern Arc' to be given realistic protection.

The late 1980s have seen the realisation at government level of the necessity of a natural forest cover for maintaining water, soil and climatic resources in both Kenya and Tanzania. The nations' agriculture is in many ways dependent on these resources. At the same time governments, at both political and administrative levels, have become aware of the rapidly increasing pace of forest loss and land degradation in the closed forest mountainous areas and in the semi-arid regions of the country. After a decade of ambivalence over the conservation/catchment or commercial timber values of the East Usambara Mountains including the closure of FINNIDA supported logging operations, a government sponsored report (Finnmap-Silvestre, 1988) has firmly stated the maintenance of intact forest cover and catchment capability to be the main two objectives of forest management.

Tanzania

In Tanzania these realisations of the importance of good forest cover have led to the development of two important documents, which themselves have stimulated great attention to forest issues. They are discussed in some detail:

A revised Forest Policy Document, 'FORPOL'

A Tanzanian Forest Action Plan, 'TFAP'

FORPOL summarises the shortcomings of the earlier 1953 policy as follows.

1. Highly generalised statements, not giving objective guidelines at implementation level.
2. No emphasis on interdependence with other sectors.
3. The need for community forestry and the importance of wood as rural fuel is not emphasised.
4. The role of forestry as a rural employer is not emphasised and forest legislation as a whole has been negative to people's participation.

FORPOL then gives the following policy outline and principal objectives:

General Forest Development

To demarcate and reserve in perpetuity for the benefit of the nation, forested land or land capable of afforestation to preserve or improve local climates and water balance (supply), stabilise land which is liable to deterioration, and provide sustained yield of forest produce for local and export market.

To manage and develop state forests, state controlled forests (public lands), village and private forests in order to perpetuate the values and benefits of the same.

To set aside forested areas for scientific, recreation and amenity purposes.

Development and Management of Natural Forests

To preserve, develop and manage both public and state owned natural forests.

To ensure that catchment forests are properly preserved and managed in perpetuity to ensure a sustained supply of water, conservation of soil, flora and fauna and amelioration of climate.

Afforestation and Reforestation

To ensure that urban and rural communities and various authorities establish and manage forest woodlots and plantations for fuelwood and charcoal, including other wood energy sources.

To establish sufficient forest plantations in suitable areas to create enough forest produce for the home and export markets.

To encourage and advise the integration of forestry with related sectors in order to capitalise on the role of forests in the development of the national economy.

Extension Services

To advance knowledge and understanding (general public awareness) among the people of this country on the values and benefits of forestry to them and to their descendants. Plus several provisions for industry, research and employment.

FORPOL goes on to state the necessity to have a greatly strengthened and centralised forest administration. This issue is, however, controversial, and probably will not be fully accepted. Reforms of the way the three-tier administration (national, regional and district levels) functions are a probable alternative.

The **TFAP** takes over from the policy document in spelling out the implementation of the policy. TFAP states:

> The overriding strategic goal of the Tanzania Forest Action Plan is to enhance the forestry contribution to sustainable use and development of the country's land resources.
>
> (Government of Tanzania, 1989)

It stresses that the forest sector's overt contribution to development has been largely limited to unplanned release of land for other activities and in providing fuelwood. The perceived contribution to GDP has been small, resulting in tiny budget allocations. The visible outcome has been rapid deforestation and land degradation. TFAP builds on government awareness that forestry has a clear role in soil, water and climate stability, to stress the sectoral linkages to agriculture, livestock, energy, water, tourism, etc. TFAP takes the following as key issues.

1. The enhancement of forest sector contribution to national economic development, especially in the key linkages to agriculture and water.
2. The involvement of local people in forest based rural development activities.
3. The conservation of ecosystem and biological values.

TFAP has generated a great deal of preliminary survey, discussion and reporting; For example Bensted-Smith & Msangi (1989) discuss inputs for a strategy for biological value conservation. Principles behind this strategy are:

> To enable key people (managers, decision makers and local users of the forests) involved in forest conservation to develop the knowledge, capability and motivation to use the forests sustainably and to develop education and motivation amongst local people to conserve forest resources.

The conservation strategy has eight components.

To finalise policies, and guidelines to implement policies, that allow conservation and sustainable use of natural use (the premise here is that FORPOL should be a dynamic document, constantly under review to improve the end results).

To promote conservation awareness about forest resources and their benefits.

To collect scientific information and stimulate research on forest ecosystems.

To ensure information flow to the field forest manager.

To develop management plans for the sustainable use of forests.

To establish a network of nature reserves.

To increase capabilities and motivations of managers and forest users.

To harmonise forestry with other land uses on non-reserved land.

Given the economic conditions in Tanzania today, implementation of TFAP will be de-pendent on foreign economic assistance. The deterioration in Tanzania's economy in the 1980s has led to greatly reduced efficiencies of forest staff in the field. A typical monthly wage of 2200 shs does not buy food for a family. Uniforms, machetes, notebooks, travel allowance payments, transport, funds for labourer assistance, etc. are not available. There is a minimum of supervision from senior staff, and responsibilities and duties are not clearly spelled out. Several Forest Guards and Foresters interviewed in the Usambaras in 1989 clearly stated they were no longer able to patrol.

Today we are in a situation where government is cognisant of the need to conserve and manage efficiently. It is no longer a question of conservationists trying to persuade an unconvinced government. Government is convinced. What conservationists need to do now is to provide detailed implementation plans and help persuade donor agencies to fund them.

Kenya

Like Tanzania, Kenya has experienced a change in governmental attitudes to forest conservation in the past five years. An immediate example comes from two Presidential Directives, one banning the conversion of indigenous forest for plantations (1984), and one banning the cutting of indigenous trees (1988). The introduction to the World Bank forestry review (1988) stresses the urgency of the forest crisis:

> The forestry sub-sector is facing serious problems. If they are not addressed soon, the economic costs incurred could be great. Critical watershed protection forests are in danger of being lost to encroachment or damaged irreparably, threatening downstream agriculture . . . ; human populations . . . , hydroelectric schemes . . . , and future generations through the loss of invaluable genepools of flora and fauna . . .
>
> (World Bank, 1988)

The lack of rain-fed agricultural land and the rapidly growing national population (4% p.a.) point to excessive pressure on forest land. This is real and is already resulting in extensive encroachment on legal and open forest land:

Encroachment, observed in all forests, is especially severe in the Mau forests and Machakos. In Mau forests, those under Narok Country Council are particularly depleted. Frequently the areas under greatest pressure are those outside gazetted forests, and therefore not under effective control of the Forest Department.

The World Bank (1988) and Kokwaro (1988) present a more detailed analysis of forest loss in Kakamega. Young (1984) pointed out that up to 1984 Kenya lost more natural forest to plantation by Forest Department than to agricultural encroachment!

Options to reduce this land hunger are limited, and include ideas such as President Moi's tea belt buffer zone (Spears, 1987), improving agricultural productivity and developing other income generating activities. The last few years have seen more writing, more discussion and more research on forest conservation issues than the two-and-a-half decades since Independence. This is a positive step to understanding that there is a problem and that the problem is a major land use and development issue, not simply a local protection problem.

The Kenya Forest Department has recently entered into a Memorandum of Understanding with the Kenya Wildlife Services to carry out joint management of the major natural forest areas of conservation significance. This will greatly increase protection capability and strengthen management planning and monitoring inputs. Kenya forestry in general is being supported by a World Bank coordinated assistance programme totalling US $ 60 million. This contains a specific component for natural forests and a biodiversity inventory.

The forest resource today

Where is the forest resource?

The distribution and extent of natural closed forest cover in eastern Africa has been quantified for Kenya by Doute, Ochanda & Epp (1981) and Beentje (1990a), and for Tanzania by Rodgers, Mziray & Shishira (1985b), Polhill (1988) and

Table 14.3. *Distribution and area of closed forest in Kenya (km^2)*

1. Tropical rain forest			230
2. Montane forest (Moister)	(a)	*Ocotea*	415
	(b)	*Aningeria–Strombosia*	215
	(c)	*Albizia–Neoboutonia*	160
	(d)	*Newtonia*	7
	(e)	*Croton sylvaticus*	110
	(f)	*Podocarpus latifolius*	683
	(g)	*Diospyros–Olea*	240
	(h)	*Craibia–Cola*	10
		Sub-total	1840
3. Montane forest (Drier)	(a)	*Croton–Brachylaena*	32
	(b)	*Croton megalocarpus*	16
	(c)	*Podocarpus falcatus*	170
	(d)	*Juniperus–Nuxia–Podo*	1250
	(e)	*Juniperus–Rapanea–Hagenia*	25
	(f)	*Juniperus–Olea*	830
		Sub-total	2323
4. Riverine and groundwater			25
5. Coastal			350
6. Uncategorised and Uncertain			560
7. Mangroves			530
Grand total			5858

Source: After Beentje, 1990.

Lovett (1990). Both countries have a total of some 2–3% natural forest cover, concentrated on areas of higher moisture along rivers, on east-facing mountain blocks and near the coast. In both countries there is a considerable spread of altitude and precipitation within forested areas. This is reflected in the great ecological and taxonomic diversity within the forests. The physical factors affecting forest distribution and forest type are discussed elsewhere in this volume, and generalised accounts given in, for example, Lind & Morrison (1974) and Hamilton (1989). An overview for eastern Africa is given in the *Atlas of African Rain Forests* (IUCN/CMC, 1992). Estimates of forest cover are given in Tables 14.1 and 14.3.

Present closed forest cover in Kenya and Tanzania is less than 3%, Potential forest cover is very much higher for both countries, perhaps up to 20% in Tanzania and 10% in Kenya. Shortfalls result from past clearing and burning for cultivation. Myers (1979) suggests that forest area in Kenya was four times greater at the beginning of the century.

A further factor that emerges from these studies is the extreme fragmentation of the forest cover. In Tanzania the largest blocks of natural closed forest are: Shume–Magamba 80 km^2 (West Usambaras), Hundu–Kisiwani 110 km^2 (East Usambaras), Luhomero 150 km^2 (Uzungwas) and the rings of forest on Mounts Kilimanjaro and Meru. The *Atlas of Tanzania* (Government of Tanzania, 1976) shows a network of over 220 small separate closed forest reserves, indicative of a fragmented resource. Evidence in this book and in this chapter suggests that the resource is not fragmented owing to physical or biotic habitat factors, but to the historical impact of man's clearing and burning.

Much of the forest cover is reserved in both Kenya and Tanzania, but there are major exceptions, some of great ecological and biological importance. The Taita Hills in Kenya are an example of major forest values still ungazetted. Most coastal thicket and forest is not reserved, including large areas of tall dry coastal evergreen forest on the Kichi Hills in Rufiji and Kilwa Districts of Tanzania (Vollesen, 1980; Rodgers *et al.*, 1985b; Burgess *et al.*, 1990), which do have endemic tree taxa such as *Pteleopsis apetala* Vollesen and *Tessmannia densiflora* Harms. Kisiju Forest, on the coast south of Dar es Salaam, may have Tanzania's last surviving fragments of gum copal (*Hymenaea verrucosa* Gaertn.) forest (Hawthorne, 1984; Rodgers, Hall & Mwasumbi 1985a; Rodgers, 1992). Many of the surviving riverine forests of Tanzania are not reserved. The highly battered and wafer-thin forest along the Great Ruaha River is a good example, where the evergreen forest has acted as a migratory pathway through extremely arid *Euphorbia–Commiphora* scrub for colobus monkeys, birds and plant taxa, as well as providing past protection to an important water source. Planning for the Uzungwa National Park led to the inclusion of dry forest and woodland communities north to the Great Ruaha River (J. Boshe, personal communication 1989). Most riverine forest along the Umba River bordering Mikomazi Game Reserve has been cleared since 1965, leading to local extinction of colobus monkeys.

This book is largely about forest resources, but it must not be forgotten that these forests contain important animal communities. Table 14.4 documents this resource for the mammals in the eastern African forests (see also Kingdon & Howell, Chapter 11). The birds are discussed by Stuart *et al.* (Chapter 10). A very readable account of animal and plant life in East Africa's fragmented forests is Kingdon's (1990) book, *Island Africa*.

Rodgers, Owen & Homewood (1982) stress the relative low diversity of the Eastern Arc forests for mammals and suggest that this is attributable to the distance and isolation of the forests from the principal refugia of Zaïre–Uganda. Fragmentation and isolation has, however, led to much genetic differentiation at species and subspecies levels. The biological consequences of fragmentation, and management to reduce such impact (see Saunders, Hobbs & Margules, 1991) are discussed in later sections (pp. 303–5).

The type of forest cover

East Africa's forests, which have been formed under a great variety of physical factors, are diverse in composition and structure; this diversity has been described in detail elsewhere in this volume (e.g. Lovett, Chapter 4; Hoffman, Chapter 6; Kingdon & Howell, Chapter 11; Stuart *et al.* Chapter 10). Rodgers *et al.* (1985b) quantify Tanzania's forest cover in terms of three altitude classes and two rainfall regimes. Summaries of broad forest types may be found in White (1983), Hedberg & Hedberg (1968) and Lind & Morrison (1974). Beentje (1990a) describes Kenyan forests in detail.

Monitoring trends in forest cover

Neither Kenya nor Tanzania has had detailed forest monitoring programmes in the past, and there is thus no accurate record of the rate and pattern of deforestation. Some estimated data are given for Uganda by Hamilton (1984). Estimates and spatial presentation of forest cover by Doute

Table 14.4. *Larger mammal species associated with eastern African forests*

Primates

| Black-and-white | *Colobus guereza* | Northern and high altitudes |
| colobus monkey | *Colobus angolensis* | Southern and low altitudes |

Red colobus monkey	*Procolobus rufomitratus**	Tana River
	Procolobus gordonorum*	Uzungwa Mts
	Procolobus kirkii*	Zanzibar

| Mangabey | *Cercocebus galeritus galeritus** | Tana River |
| | *Cercocebus galeritus sanjei** | Uzungwa Mts |

| Blue monkey | *Cercopithecus mitis* | Widespread |

| Galago | *Galago zanzibaricus* | Widespread, coastal |
| | *Galago senegalensis* | Widespread, elsewhere |

Bovidae

Ader's duiker	*Cephalophus adersi**	Zanzibar, South coastal Kenya
Abbot's duiker	*Cephalophus spadix**	Kilimanjaro, Uzungwa
Red duiker	*Cephalophus harveyi* and *Cephalophus natalensis*	Widespread, many races
Blue duiker	*Cephalophus monticola*	Widespread, many races
Suni	*Neotragus moschatus*	Widespread
Bushback	*Tragelophus scriptus*	Widespread
Buffalo	*Syncerus caffer*	Major mountain forests

Squirrels

	Funisciurus lucifer	Block mts
	Funisciurus palliatus	Kenya
	Helios rufobrachium	Lowland forests
	Anomalurus derbianus	Block mts

Others

Leopard	*Panthera pardus*	Widespread
Elephant	*Loxodonta africana*	Kilimanjaro, Uzungwa
Hyrax	*Dendrohyrax validus*	Patchy
	Dendrohyrax arboreus	Patchy

Note: *Indicates a species of threatened or endangered status.
Distribution and conservation information as follows: Colobines, Rodgers (1981); Primates and mammals in general, Kingdon (1982); Bovidae, Rodgers & Swai (1988).

et al. (1981) and Rodgers *et al.* (1985b) are based on interpretation of LandSat imagery; both studies were designed to give baseline data for future monitoring.

On a smaller scale, for individual forests and regions, there are estimates of rates of deforestation. In Tanzania, air photograph interpretation has showed a loss of forest area for several individual reserves. In the period 1955–79 Magombera FR, a groundwater lowland forest of value for primates and endemic plant taxa,

decreased from 15.1 to 10.5 km² (Rodgers, Homewood & Hall, 1979). Rau FR near Moshi decreased from 26 km² of lowland evergreen forest to 6 km² (Rodgers, 1983). Non-coral rag forest on Mafia has been totally destroyed (Rodgers *et al.*, 1988). Larger scale studies in the Usambaras show estimated forest cover reduction in this century of some 60% (Stuart, 1983). Rich lowland forest in Pangani District, well collected by Faulkner, has been greatly destroyed (W. D. Hawthorne, personal communication). Lake

Lutamba, the site of so many plant species type collections near Lindi, has been clear felled (R. Wingfield, personal communication). Rondo Plateau, mentioned in Polhill (1968) has been heavily logged; less than 5 km² of good natural forest remains, and this is under threat (K. M. Howell, personal communication). Scattered forest and thicket patches in the north of Serengeti National Park have declined as a result of fire and elephant (Sinclair & Norton-Griffiths, 1979). Chapters in Hamilton & Bensted-Smith (1990) illustrate the pattern of deforestation in the East Usambaras.

In Kenya the pattern is similar, and LandSat findings are summarised by Epp (1984) who concludes that for many forests, if the present rate of deforestation continues, forest cover will be totally depleted by AD 2020. A Kenya Forest Working Group was established in the mid-1970s to monitor forest change, but unfortunately soon disbanded.

Barnes (1990) presents a disturbing analysis of probable rates of deforestation in the future, based on the relationships between population growth, forest area and past extent of forest clearing. His estimates for Kenya and Tanzania, based on FAO (1981) baseline data, are:

Year:	1980	2000	2020	2040
Kenya				
Area of forest (km²)	6900	3752	1199	44
Area per million people	415	121	23	1
Tanzania				
Area of forest (km²)	14 400	9120	4345	1100
Area per million people	777	287	87	15

Barnes (1990) also stresses that this loss will not be compensated for by new afforestation. He uses FAO data to expect a deforestation loss of 13 310 km² from 37 African countries from 1981 to 1986, whilst new planting would be 1260 km²!

The concept of forest value

The problem of different perceptions

The major problem in discussing forest values is that different people and organisations may hold very different, and sometimes incompatible, views on what is of value in a forest. Essentially, there are three polarised viewpoints. At one extreme, the individual peasant farmer living at the periphery of a forest sees the forest as a source of land, food and fuel, as well as a refuge for crop-eating pests. At another extreme, local government may see the forest as a source of revenue and employment in terms of timber; and national governments may see forests as major water-producing areas for national development as well as suppliers of timber and other products. Finally, an increasing number of local and international organisations are ever more vociferously stressing the less immediately tangible values of natural forest systems, as environmental buffers and gene pools.

These different value systems have led to conflicting objectives of land use. One source of conflict is the degree to which local individual interests can be served – fuelwood, poles, food plants, hunting, religious values, etc. – whilst maintaining government objectives. Another conflict is at government level, concerning the degree to which commercial exploitation and management may be permitted, as opposed to policies of full protection which better serve catchment needs. Recently there have been major attempts to defuse such land policy conflicts by approaches such as the World Conservation Strategy (IUCN, 1980), although these are yet to be seriously considered in East Africa. Tanzania plans to develop a National Conservation Strategy in the coming years, and a major component of that and the TFAP will address forest land use conflicts.

The different user demands have been summarised as the individual or subsistence requirements of rural populations as opposed to the larger scale economic exploitation of resources by governments, which again contrasts with the less tangible goals of species or biological value conservation which are often based on moral, aesthetic and scientific grounds rather than economic arguments.

Economic values themselves may include real quantifiable assets, available now, such as timber; and resources whose value is less easily quantified, such as clean water. In addition there are undoubted potential economic values in forest systems, such as chemicals, foodstuffs, etc. which have yet to be developed and so again are difficult to quantify. Experience suggests that where such products are developed commercially, they rarely benefit the local people or the developing nation (Prescott-Allen, 1984).

This section discusses each set of forest values individually.

Resources of value to people

Some cultural groups in Africa derive their total livelihood from forests (see, for example, Turnbull, 1961) but there are no such people in eastern Africa. Agriculturalists, either purely or as agro-pastoralists, depend on cleared land for most of their income and subsistence. Pastoralists may use forest glades or edges for grazing, but this is not important for forest areas considered here.

The great majority of East Africa's people live away from forests; it is probably only those living within a 10 km radius of forest who make direct *demands* on forest land and its products. Some products exploited by local people may eventually be marketed over a much larger distance, e.g. dug-out canoes, honey, medicines, fruits and alcohol. To such forest edge villagers the forest represents potential values as well as real resources. The biggest potential is land, and land of initial high fertility. Forest land if not reserved may be cleared or, if reserved, encroached on illegally or obtained by successful political lobbying. Public land in well-watered mountain areas is in short supply, pressures on forest land are high.

Most of the natural forests do not offer a great deal of formal employment opportunity. Pit-sawing is usually done by specialists (Wahehe people from Iringa Region), not by locals. Regular forestry practice does not require much labour in non-plantation areas (especially with inadequate inputs to protection and silviculture), and forest industries are largely centred in urban zones away from the producing areas. Plantation development will require considerable labour input, but this may well necessitate migrant workers and not

provide benefit to local people. This is discussed in detail in IIED/IRA (1992) for a proposed CDC (Commonwealth Development Corporation) teak plantation in Kilombero Region, Tanzania.

Forests offer fuelwood and building poles for traditional houses, both products which are not yet in abundant supply from agro-forestry or village woodlot schemes. Evidence is given below to illustrate the scale of wood demands on forest areas. Details for an Usambara community are given in Fleuret & Fleuret (1978).

Few people hunt, but forests offer the possibility of hunting. Vegetables, honey and beeswax, and medicinal values are of greater importance, and are harvested by larger numbers of people, (e.g. Fleuret, 1979, 1980; Mtotomwema, 1982). The increasing scarcity of drugs in Tanzania led to renewed interest in traditional, usually forest based, medicines (W.A.R., personal observations, 1985). A review of medicinal plants is given by Kokwaro (1976). Some natural products could form a base for small-scale village industries, e.g. oil from *Allanblackia* fruits (Lovett, 1983).

Forests offer refuge for religious or traditional rites, and shade, rest and relaxation for local people. These are resource values which are intangible and difficult to quantify; they are not necessarily distinct from the intangibles valued by 'Western' society, but as their use is not regulated, they are often thought of as non-compatible with our ideas of conservation. It is a sad reflection of our understanding of local people's use of forests that there has been no quantitative study of forest resource use by village communities anywhere in eastern Africa. Studies on *kayas* in coastal Kenya (Oxford University, 1981; Robertson, 1984) are a partial exception. Growing evidence from other tropical forested regions suggests that 'minor' forest products can have a resource value equal to or greater than that of major timber values, *and* minor product use is more sustainable and more likely to maintain biodiversity (e.g. De Beer & McDermott, 1989).

Finally, urban centres, by demands for fuelwood, etc., can impose considerable *indirect* demand on resources, where people are cutting for sale to distant locations.

Resources of value to government

Water

The value on a resource varies inversely with the abundance of that resource. This is clearly seen in the case of water, perhaps the most fundamental of all forest resources, and one which is becoming increasingly scarce in absolute terms, as well as in relation to development needs. Rapp, Berry & Temple (1973) document changing patterns of river flow, leading to less dry season flow and increasing floods and siltation in the rains. Most major cities in Tanzania have experienced water shortfalls in the past decade. Dry season hydro-power generation suffers because of inadequate reservoir levels (W.A.R., personal observations).

The renewed interest in forest watersheds by Tanzania's central government, including the development of new catchment forest policy and the establishment of separate offices and staff cadres for catchment protection, shows govern-mental concern. A thesis by a senior government forest officer investigating catchment processes in the Usambaras (Mbwana, 1988) is indicative of this interest, which has continued to a recent broader view of montane forest catchments (Mbwana, 1990).

Timber

Timber is still a major economic resource and an important industry in many regions of Kenya and Tanzania. Both countries did not export indigenous hardwoods for many years although sawn softwood, plantation teak, pulp and charcoal are growing export items. Tanzania has exported some products, for example *Brachylaena* parquet floorboards. *Brachylaena* is a tree of coastal closed forest and is considered threatened by ill-planned exploitation. Natural forest timbers feed local demands for construction and furniture. Whilst 20 years ago relatively few species were in de-mand, many more are being profitably exploited today.

Very few of the major hardwood species are being successfully grown in plantation, *Cephalos-phaera* is one exception. More attention is being given to non-indigenous species: *Tectona, Maesop-sis, Cedrella*. Silvicultural operations to enhance

natural regeneration have been developed for a few species only, e.g. *Ocotea usambarensis*, but much more silvicultural research is needed (Mugasha, 1982).

Minor forest produce

East Africa does not have a major tradition of exploitation of forest resources such as fruit, seeds, leaves, gums and resins, as, for example are harvested from the forests of India and South-East Asia (Krishnamurthy, 1983) although sandalwood oil is a growing export (FAO, 1982a). Once important commodities such as gum copal are no longer harvested because of competition from economically superior synthetics (Rodgers, 1992).

There are potential minor resources such as oils from the fruit of *Allanblackia*, horticultural varieties of *Saintpaulia* (African violet), and potential genetic values in the several species of *Coffea* (wild coffee) (see Lovett, 1985 for a detailed review). They are likely to benefit local populations or organisations outside Tanzania rather than the Government of Tanzania. The high cost of production in Tanzania, and the past virtual government monopoly on forest produce exploitation, means it is unlikely that such prod-ucts will achieve economic importance in the foreseeable future.

Resources of value to conservationists

The international conservation movements, of which IUCN/WWF have been the most obvious in forest conservation in East Africa (IUCN/WWF, 1982), see tropical forest communities as 'the most diverse and complex ecosystem on earth – a virtual powerhouse of evolution – containing 40% of the living species on earth' (Myers, 1984). The forests have value in the many species they contain, several of which are rare and of restricted distribution, and several threatened with imminent extinction. These values are not necessarily related to the real or potential econ-omic worth of the species – it is deeper than that – and are composed of a moral, aesthetic and scien-tific desire for all such species to survive. The tremendous diversity of species and species linkage within a forest community adds to that value. Forest is valued for its complexity and diversity.

Forest conservationists also see forests as environmental buffers, playing some role in regulating climates and pollutants and in maintaining water and soil systems. Conservationists frequently publicise the enormous potential wealth of forest systems in their mass of unique genetic resources: there are plants and products for the chemical, food, pharmaceutical, cosmetic and energy industries. Several books and papers illustrate this potential resource (e.g. Myers, 1979, 1984; Guppy, 1984).

One of the crucial aspects of the species and diversity value arguments is that these values can rapidly be lost by excessive or ill-planned exploitation. Strict biological conservation and exploitation are usually thought to be incompatible.

On the surface it may seem as if different interest groups have radically different viewpoints on the values of a forest, and that a forest cannot continue to satisfy all resource requirements for perpetuity.

Successful conservation, however, does consist of trying to accommodate all resource requirements whilst maintaining the 'capital' balance of the forest system. The last decade has shown that conservation must be more than putting up fences and saying 'hands off'. Effective conservation may involve developing alternatives, the improvement of land and land productivity and a sharing of sustainable forest resources by a system of core and buffer zonation. We have yet to see this in East Africa, although concepts are well developed in India (Panwar, 1984) and have begun in Rwanda (W.A.R., personal observations; and see papers in Proceedings Workshop on Afromontane forests; and in Brown, 1992). Ideas relevant to the East African situation are discussed later in this chapter (p. 309).

The pressures on the natural forest resources of eastern Africa today

An immediate vision of the scale of pressures on East Africa's forest resources comes from the statistics that in Kenya, forests occupy over one quarter of the 12% of the country that is arable under existing climate and technologies. There is a rapidly growing population (4.1% p.a.), of which 88% are dependent on land for their livelihood, and there is virtually no spare potentially arable land. As forest land is already insufficient for water and siltation regulation, a critical problem is inevitable (Mwagiru, 1982).

Similar problems exist in Tanzania. In the West Usambara Mountains, human populations have increased from 15 000 in 1900 to 286 000 in 1978 (Lundgren, 1980). Odner (1971) stated that the human population on Mount Kilimanjaro grew from 50 000 in 1899, to 125 000 in the 1920s, to 500 000 in 1967; a tenfold growth in 70 years. In the Uluguru Mountains, river flows have dried up in the past 20 years as a result of deforestation and poor agricultural practices (Rapp *et al.*, 1973). It is instructive to examine some of these pressures in more detail.

The human population explosion and forest encroachment

Population growth rates in East Africa are high, but rates are heterogenous within each country. An overview of population problems in Africa is given by Goliber (1985). Rates in the agriculturally fertile and productive mountain districts are higher than the national average (Government of Tanzania, 1979). Kurji (1980) analysed human density changes around protected areas; he showed that wards adjacent to Kilimanjaro National Park averaged 3.1% p.a., whilst those around the extensive Selous Game Reserve in south coastal Tanzania averaged 0.4%. Mlay (1982) looked at populations around Mount Meru in detail; despite increasing emigration, the population was still growing by 3.3% p.a. Over half the rural population was at densities above 200 per km^2 and one ward had a density of 357 per km^2. Mlay considered that this population growth had led to reduced plot size, land fragmentation, eradication of fallow, overstocking, soil erosion and a decline in agricultural productivity.

Lundgren & Lundgren (1982) discussed the consequences of population size in the West Usambaras, where human density is close to 100 per km². Severe soil erosion and land degradation with loss of fertility and the failure of large-scale rehabilitation schemes (e.g. Watson, 1972) led to major pressures on forest. In 1963, politicians yielded to this pressure and 120 km² of natural forest reserve in the West Usambaras were degazetted. Poor land management and increasing cultivation on steep slopes have since caused severe erosion of this virgin land in less than 20 years.

Brown (1981, and in Rodgers & Homewood, 1982a) described the agricultural practices of the small and well-watered East Usambara Mountains as extremely poor and non-productive forms of subsistence shifting cultivation. An IUCN/CDC (1985) report on the area showed a 48% population increase from 1967 to 1978 with village population densities as high as 421 people per km². Many people are immigrants whose agricultural practices take little account of soil fragility. This continued agricultural expansion, now over 50% of the area, together with high intensity logging on steep slopes (see below) is stated to be leading to 'a great loss of forest, impaired catchment regulatory efficiency, soil erosion and seriously depleted gene pools'.

Forest plantations

Whilst plantations are rapidly expanding, policy now is generally not to replace existing natural forest but to plant on degraded and otherwise unused lands. There are exceptions, however, and recently the Kenya Forest Department had plans for plantations in parts of Kakamega and Arabuko–Sokoke natural forests, both areas of exceptional biological values and which had each been recommended for UNESCO Biosphere Reserve Status. The increasing demand for timber and the reduced level of remaining resources in natural forests will no doubt lead to increased pressures to 'improve' natural forest in years to come. A report on production forestry in Tanzania (Poyry, 1980) gave the following projections for peeler and saw-logs:

Year:	1985	1990	2000	2010
Demand:	685	805	1220	1750
Potential natural forest supply:	455	415	360	350

(Figures are in thousand cubic metres of timber. Note that new inventories suggest the potential supply is below these figures).

In Tanzania there are no new major plantations planned within natural forests, but small-scale plantations have been developed in the past ten years. For example, Rau Forest near Moshi, the only major locality for the massive leguminous tree, *Oxystigma msoo*, has small plantations of exotic and native timbers (Rodgers, 1983), Kimboza Forest in the Uluguru foothills has several patches of *Cedrella*, now regenerating freely (Rodgers *et al.*, 1983), and Ngezi Forest on Pemba has many small (<10 ha) patches of several exotic species (Rodgers *et al.*, 1986). In the East Usambaras parts of Kwamkoro Forest Reserve (one of the largest blocks of natural forest) have been underplanted with *Cephalosphaera* (indigenous) and *Maesopsis* (an exotic, which is freely regenerating). *Maesopsis* encroachment is now thought to be one of the major problems of natural forest management in the Usambaras (Hamilton & Bensted-Smith, 1990).

Proposals from CDC to plant 10 000 ha of teak in Kilombero in southern Tanzania are thought to benefit conservation by reducing pressure on natural forest hardwoods, as well as contributing to export earnings and local industry (IIED/IRA, 1992).

There is now increasing concern that the level of afforestation for fuelwood is totally inadequate to meet growing demand. Struhsaker (1987) predicts that current and planned reafforestation programme will supply less than 5% of fuelwood needs in Uganda. Pohjonen & Pukkala (1990) analyse fuelwood plantations in Ethiopia. They estimate that 1 million ha are needed to prevent continued loss of natural forests. Today there are 310 000 ha of varied plantation. If populations reach 70 million by AD 2000 and villagers are to switch from cow dung fuel to wood (so improving soils) then 3–4 million ha are needed!

Logging

Within Tanzania, natural forests have long been termed as catchment or production forests as a generalised classification for internal Forest Division management purposes. More recently, a detailed zonation format has been adopted by the Forestry Division, and may soon be in actual field use. Catchment forests are those considered of prime importance for watershed conservation; they should not be exploited, and are managed by staff of central government. Production forests are of lesser catchment value, and are exploited for timber and other resources, usually under the control of regional governments. The classification has no legal basis, may cut across reserve boundaries, and may be altered at any time. Maagi *et al.* (1979) state that the catchment to production forest ratio to be about 45:55. In theory, catchment forest should either not be logged, or logged lightly for selected species. Within production forests trees should not be felled within prescribed distances of streams and river banks, nor on slopes greater than specified inclinations. In practice, however, these restrictions are not enforced, and there is some uncertainty as to what constitutes protection or production forest. Poyry (1980) stated that 120 km² of closed forest in East Usambara are of commercial value and should be logged. This report says, 'Under present practices closed forests are gradually deteriorating or disappearing despite partial under-use.' It also stresses that most accessible forest is overexploited and the amount of valuable timber is dwindling. Many statistical reviews show a decrease in the harvesting of industrial timber, e.g. FAO (1982a) state that the volume of non-coniferous round-wood decreased by 20%, and saw-logs and veneers by 50% from 1970 to 1980. FAO (1982b) show a decrease in saw-logs of 65% from 1967–70 to 1975–8. They qualify this, however, by suggesting that these statistics are for recorded removals which are significantly smaller than actual removals. The Nordic Review Mission (1979) recommends that all Usambara natural forest should be catchment.

Several observers described the severe logging pressures on the East Usambara forests, e.g. IUCN/CDC (1985), Stuart (1983), and Hamilton & Bensted-Smith (1990). The national and international outrage at the scale of land degradation involved in the FINNIDA-aided mechanical logging project in the Usambaras has served to bring forest conservation issues to public and government attention. Government intervened to stop all mechanical logging in the East Usambaras in 1985 but only after the head of the foreign aid organisation stated in the world press that 'logging would continue in these forests despite widespread allegations of environmental damage' (Hellenius, 1985). A fuller history of how a foreign aid–industrial company consortium can break forestry laws and cause such loss is given in Hamilton & Bensted-Smith (1990). Logging was largely concentrated in the public, non-reserved forests and one major effect of the huge level of canopy clearance (from almost 100% cover to less than 20%) has been the inflow of people to plant maize. Those lands will not bear forest again. Mechanisation was too heavy, roads were ruined, access tracks were badly sited, skidding techniques led to accelerated erosion, trees were felled along drainage lines, streams were blocked and no care was taken of growing stock. Finally, there was a great lack of monitoring or documentation of exploitation and forest status. Government has rightly declared, 'never again!'.

Sangster (1962) stated that some three quarters of hardwood timber extraction was taken by sawmillers and the rest by hand-sawyers, J. F. Redhead (personal communication, 1982) suggested more than half was pit-sawn; much of this would be illegal extraction. There is little information on pit-sawing in the forestry literature. As it does not involve extraction roads and machinery damage, it is often thought that it has little environmental impact, e.g. by Kinoti & White (1981) who investigated pit-sawing economy and production in the Pare Mountains. Hall & Rodgers (1986) recorded measurements of pit-saw sites in Kimboza and showed considerable damage, a single site averaging 790 m² clear felled and trampled, with many (30) smaller trees cut for props and rollers. In addition are the problems associated with small labour camps inside the forest for up to a month. Further details are given in Lovett (1985). Rodgers *et al.* (1979) described canopy damage caused by heavy logging pressure in Magombera Forest Reserve. Of nearly 200 canopy segments examined, over 60% had severe exploitation by a mixture of mechanical and hand-sawn logging.

Kelsey & Langton (1984) describe pit-sawing taking place in Arabuko–Sokoke coastal forest. This felling was inside a nature reserve and occurred after a Presidential decree prohibiting indigenous tree cutting. Similar illegal extraction is taking place in the forests along the Tana River (Marsh, 1978).

Of concern is the realisation that since the mechanical logging ban in the Usambaras, there has been a massive increase in pit-saw exploitation, largely by immigrant Wahehe, but funded by local businessmen. Most exploitation was in the public forest land and is illegal. This is aided by the poor control exercised by local District Council forest staff. In 1990, however, 'we have discovered more incidences of illicit pit-sawing in so-called Catchment Forest Reserve than in the public lands. In all cases authorization to harvest has been provided in some form, and "poetic license" given to actual harvesting levels' (Wardell, 1990 in litt.). Planks presumably feed the market left after mechanised logging stopped (D. A. Wardell & I. Mwasha, personal communication, 1989). In Matundu Forest Reserve in the Uzungwa Mountains mechanised logging was stopped in 1990 and pit-sawing licences given to deal with 'badly fallen logs'. Local comment suggests a great deal more timber is coming out than allowed for under these cleaning-up operations.

Throughout the tropics there is now renewed interest in developing greater sustainability of logging evergreen forests, both by improved silvicultural systems (Bruening, 1989; Howard, 1990) and by better logging management (e.g. Poore, 1988). However, little of this initiative filters down to field operations in eastern Africa as natural forest research and silvicultural capabilities are limited and there is virtually no operational control at field level.

Pole and fuelwood extraction

Rough estimates from eastern Africa suggest that over 90% of all wood harvesting is for fuelwood (e.g. FAO, 1982a, b; Openshaw, 1984). Whilst this is biased in favour of woodlands with lower timber values, it is still indicative of the patterns of use of closed forest. The question of rules governing fuelwood extraction is difficult; theoretically, all collection and extraction requires a permit for which a fee should be charged. At a district level people talk of 'customary rights for minor produce' but these are not officially recognised nationally. In practice there is little control or monitoring of extraction, especially when this involves dry wood. Field staff numbers are totally inadequate to allow any checking of such extraction. Cutting of firewood may be discouraged in some places, but in others the practice is rampant. In Zanzibar, the collection of firewood is much more closely controlled, as the supply to townships from coral rag forest forms a considerable commercial undertaking (Swai, 1983). On the mainland, forests close to major settlements are heavily disturbed, charcoal pits being common in Rau Forest near Moshi (Rodgers, 1983) and Pugu and Pande Forests near Dar es Salaam (Howell, 1981).

There is a lack of quantitative studies on the magnitude and effects of fuelwood extraction on the forests themselves. Pressures are concentrated around the periphery of forest, and cutting may be severe enough to allow grass to develop and fires to enter and prevent forest regrowth as, for example, in Pande Forest near Dar es Salaam.

Pole extraction for house construction has been documented for lowland Tanzania forests by Hall & Rodgers (1986). Extraction involves the cutting of straight stems, often the saplings of canopy trees. Pressures near the forest periphery may be high enough to remove up to 80% of all stems of favoured species, with obvious implications on canopy renewal.

Hunting and traditional product extraction

Whilst there is a growing body of data on the general values of forests for food and medicinal products (see above, p. 297), there is little quantitative information of the amounts extracted (Fleuret, 1979, 1980; Mtotomwema, 1982). Further, there are few natural forests protected well enough to serve as reference centres against which exploitation elsewhere can be measured.

It is known that hunting and disturbance factors have caused local extinctions from small forests: Howell (1981) lists several vertebrate species from Pugu Forest, and Rodgers (1981) mentions extinction of *Colobus angolensis* from coastal forest patches. The same must be true for more preferred meat sources such as duiker, suni, bushbuck and bushpig. Buffalo are now absent from Kigogo FR in the West Uzungwa Mountains. Traps, pits and snares are commonplace in East African forests.

The biological basis for forest conservation

Literature discussing the scientific background to forest conservation, has greatly proliferated in the past decade (UNESCO/UNEP/FAO, 1978 to, for example, Saunders *et al.*, 1987), although it has not yet been synthesised in a format relevant to the East African situation. Literature deals either with a single resource, such as water (e.g. Pereira, 1973), or a complex, such as genetic resources (e.g. Roche & Dourojeanni, 1984), or theoretical aspects, such as the maintenance of species numbers in reserves of varied size and shape. This last concept stems from island biogeography theory and has recently been summarised for the USA by Harris (1984).

This section briefly reviews concepts of importance in planning forest conservation in eastern Africa. Topics are treated under the broad headings of watersheds and hydrology, protected area design, and sustainable utilisation of resources.

Watersheds and hydrology

Scientists and conservationists have long known of the link between dry season water flow and forest cover in eastern Africa. The maintenance of a closed canopy to reduce erosivity of raindrop impact, and maintenance of surface ground cover to slow surface runoff, have been major tenets of land use planning theory. The watershed experiments in different catchments in East Africa summarised by Pereira (1973) and the empirical data from field situations in eastern Tanzania in Rapp

et al. (1973) show the importance of forest cover and the magnitude of the problems where cover is reduced. More recent studies such as those of Dunne (1979) in Kenya, and Lundgren & Lundgren (1979) in Tanzania, emphasise the role of forests in maintaining soil and water processes. These studies compare forest and agriculture; few studies compare forest types. L. Lundgren (1980) stressed that different forest communities have different soil and water flow parameters. B. Lundgren (1978) suggested that natural forest was superior to plantation forest in this respect. Pocs (1980) documented the hydrological importance of the dense bryophyte biomass in mist-prone wet montane forest, a characteristic of old growth natural forest. In general, however, the effect of forest operations (silvicultural tending, thinning, felling, extraction) on catchment properties has been little studied, although general guidelines for improving hydrological parameters are given by FAO (1976). Bruen (1990) and Mbwana (1988) have looked at hydrological characteristics of the Sigi River catchment in the East Usambaras, attempting to relate water flow regimes to past excessive logging. Incomplete and inadequate data sets did not permit a firm conclusion. Local comment, however, is that climate is warmer; there is less mist and fewer rainy days. People remark on springs drying up, larger peak floods and heavy water siltation. These observations on reduced number of rainy days, hence less cloud cover, more radiation input and worsening evaporation: precipitation ratios (E/P) have been documented elsewhere, e.g. Meher-Homji (1988) in India. More and more local communities are relating decreased forest cover to reduced spring flow. Deforestation is a result of less care of past strictly controlled village catchment tree cover, such as in the Pare Mountains (Archaeology Department, University of Dar es Salaam, personal communication, 1989). With excessive canopy opening, runoff and sediment yield increase dramatically; this is obvious. But questions remain unanswered, such as: 'What degree of exploitation is permissible before catchment ability declines? Is dense secondary regrowth as good as primary forest? How can timber be extracted in a way which will minimise water loss?' The short reply must be that maximum water retention will occur with the least disturbed primary forest; but, as yet, 'In Tanzania, watershed management is given very little attention' (Temu, 1984).

The design of forest protected areas

The concept of an optimum design for protected areas arose from island biogeographical theory and from the genetics of small populations. How big an area, and how many individuals of a species, are needed to ensure long-term viability of target populations, are questions addressed by these ideas. Design, though, is meaningless unless there is capability to implement design on the ground. The present pattern of fragmentation is largely related to a past and present lack of effective land use planning. Short-term 'local' economic interests frequently dominate over longer term 'national' conservation interests.

The location of the protected area may be important. Many authors have stressed the advantages of conserving areas considered to be species refuges or refugia (e.g. Diamond & Hamilton, 1981; Grubb, 1982; Prance, 1982). Refugia are 'Centres of Richness and Endemism' (Diamond, 1985) and have higher than normal species diversities and numbers of endemics. This book is largely concerned with the East Tanzania/Kenya refuge. Refugia are areas of persistent climate, little subject to long-term patterns of change. You get more species conserved for longer time periods per square kilometre (or per dollar invested) in refugia than elsewhere (Myers, 1982).

The question of how big an area to conserve is central to conservation planning. Initial theory suggested the larger the better, and that one large area was better than several small ones (e.g. May, 1975; Miller, 1978). Other authors (e.g. Simberloff & Abele, 1976) stated that a number of smaller reserves would conserve more species than one large area. Certainly there is much empirical evidence to show that when reserve size is reduced species will become locally extinct, and that such species loss is proportional to area reduction. A large number of good sized reserves

which cover habitat as well as species diversities would be optimal. This is academic in the case of eastern Africa as forest reserves are already highly fragmented and of small size. Recent study in the Usambaras has revealed the impoverished nature of the avifauna of small isolated forest patches (Newmark, 1991a).

Reserves should be big enough to contain populations of viable size. 'How big is a viable population?' and 'how big an area does such a population require?' are unanswered questions for African forest taxa; but see Collar & Stuart (1985) for practical ideas for forest birds. Population geneticists believe 250 individuals may be necessary to avoid problems of loss of genetic heterogeneity (e.g. Wilcox, 1984; but see the review of Roche & Dourojeanni, 1984, and works in Frankel & Soulé, 1981; Soulé, 1986 and Tsingalia, 1990). Lovejoy (1984) and Lovejoy & Oren (1981) suggested that a reserve should be big enough 'to maintain the characteristic diversity of its component communities', and gave a rule of thumb that the required minimum area is twice the space needed by a viable population of the top carnivore. For leopard 500 km^2 may be needed, which is obviously not possible in eastern African forests! Luckily leopard (and other top carnivores – some larger raptors) can range outside forest cover, but several other lesser carnivores and most herbivores cannot.

The short answers to questions of how big an area is needed to conserve genetic diversity must be 'as many big areas as possible'.

Small conservation areas may be more effective at retaining species if they are close to larger areas which can act as a source of immigration or recolonisation. Linkages in the form of corridors or buffer zones, a good example of which is a dendritic riverine forest strip (e.g. Harris, 1984), are important to maintain genetic transfer between forest patches and allow recolonisation. The corridor concept figured prominently in the planning of a protected area network in the forests of India and elsewhere in the world (Rodgers & Panwar, 1988; Hobbs, Saunders & Hussey, 1990). Newmark (1991a) discusses the biology and importance of corridors for bird movement in the Usambara Mountains. Helliwell (1976) and

Pickett & Thompson (1978) discuss ideas of reserve design in the light of resource fragmentation. Janzen (1983) emphasises the dangers of small patches of pristine forest being colonised by pioneer species from peripheral buffer zones of secondary forest. A recent symposium volume from Australia emphasises the continuing role of remnant forest vegetation patches in conserving plants and smaller animals (Saunders *et al.*, 1987). However, they stress the need for informed and continued management efforts.

Harris (1984) summarised the above by saying that the 'effective' size of forest patches is determined by (i) actual size, (ii) distance between patches and (iii) the hostility of the intervening matrix. All three constraints can be lessened by effective management.

The ecological consequences of forest fragmentation have been the subject of much recent study (e.g. Burgess & Sharpe, 1981; Saunders *et al.*, 1991). Hill (1985) makes the distinction between edge and interior forest types and says that forest patches below 8 ha in extent probably will not survive as primary forest (they certainly will not hold their vertebrate populations, as shown by Lovejoy (1986) in South America). Ranney, Bruner & Levenson (1981) liken edge communities to highly disturbed forests, and suggest edges are usually over 20 m wide and will not support a climax community. Whitcomb (1981) stated that 'edge'-dominated forest will have a large number of generalist edge species but a very small subset of habitat specialist forest-dwelling species. These authors go on to say that forest tracts that are sufficiently large to preserve populations of most plant species may be inadequate as reserves for bird species. The maintenance of forest bird communities requires forests of hundreds or even thousands of hectares and the management of these forests should be aimed at minimising disturbances of the forest interior. Mathiae & Stearns (1981) show that excessive fragmentation leads to a remnant fauna of 'opportunist and generalist' species; much of the 'specialist' original fauna will disappear.

The effect of internal patch dynamics and conservation values has been little studied in African forest systems. The effects of shifting cultivation,

of clearings from logging operations and of general overexploitation of fuelwood, pole and tree resources are to create a series of internal gaps or edges. Whilst small-scale gaps from natural treefall and minor perturbations are considered to play a major role in the patterns of community dynamics and maintenance of diversity (Mabberley, 1983), larger gaps may cause more lasting change towards secondary forest conditions (Skorupa & Kasenene, 1984).

Although some species may achieve higher densities in secondary forest, the overall levels of species diversity will be reduced, and conservation goals of maintaining climax forest systems will be lost. Some ecological principles for tropical forest conservation have been put forward by Ng (1983), based on work in Malaysia. Major findings are that forest exploitation is likely to lead to reduction in height and biomass of succeeding forest cover, resulting in less niche space for species occupancy. Patterns of reproduction of most forest species dictate that conserved forests must be large, undisturbed and numerous to be fully effective. The effects of logging on the conservation values of forest have been examined in a major review by FAO (1982c): 'Selective logging of superior trees may cause genetic erosion and damage to residual vegetation. Seed trees may not survive isolation shock.' Simplification of forest to enhance timber values by selective cutting and enrichment planting may reduce overall resilience to pests and pathogens (e.g. Way, 1977; Whitehead, 1982). Defective trees, usually removed, may be ecologically valuable. This finding has led to a rapidly growing literature from the USA on the great conservation importance of such 'snags', often removed as over-mature individuals: see, for example, Thomas (1979). Such old growth may be important in maintaining *Rhizobium* populations. Logging may make adjacent stands more vulnerable to windthrow, fire, species intrusion, illegal felling and encroachment.

The sustainable utilisation of forest resources

Whilst most conservationists are convinced of the need to integrate genetic resources protection with increased investment and considerations to adjacent rural communities, there is little hard experience to show how this may be done. Joint or participatory management approaches with local people are showing considerable success: see below. The *World Conservation Strategy* states that

> Tropical forests are an important renewable resource, acting as a reservoir of genetic diversity, *yielding a continuous supply of products if managed sustainably* [my stress] . . .
> (IUCN, 1980)

The Man and the Biosphere Programme of UNESCO (see von Droste, 1984) has a mandate to integrate conservation and human needs, through its Biosphere Reserves. McNeely (1984) stated that biosphere reserves have yet to succeed with this goal. An African example of these problems is that of Tai Forest National Park, Ivory Coast (Roth, 1986).

Biology has yet to provide us with solutions to problems of how much resource exploitation is permissible before conservation values decrease. We may be at fault for not phrasing precise questions to be answered: what is the effect on this species, or on overall diversity, or on future timber species girth increment, or on water flow. We are certainly at fault for the virtual lack of research effort addressing these problems. The same pattern of forest use will probably affect different values in different ways.

Johns (1985) has demonstrated that the logging of primary forest in Malaysia need not lead to a decrease in numbers of all primates; some thrive in secondary regrowth, others such as orang-utan decline. Shelton (1985) reviews the subject of logging intensity on wildlife. Skorupa & Kasenene (1984) discuss logging intensity on natural tree mortality and show mechanised harvesting to be especially destructive.

A more recent set of ideas on designing protected areas focuses on the need to integrate the park or reserve into the surrounding land use pattern. Protected areas cannot function as isolated fortresses or sterile no-go areas; they have to contribute to sustainable development of adjacent rural communities (see, for example, McNeely & MacKinnon, 1989; McNeely, 1990). However, as Lucas (1990) warns, there is still a need to ensure that core areas are properly protected. The outer buffer zone is where resource sharing should take place (Oldfield, 1987; Sayer, 1991).

The long-term ecological effects of the massive increase of mechanised logging in the East Usambara Mountains have not been fully documented. There are several general statements among contributions to Hamilton & Bensted-Smith (1990). Some results are clear, however. On public land, heavy felling in primary forest has allowed local cultivators to plant maize in the huge gaps. Forest will not regrow. Gaps in reserves are invaded by 'weeds', including *Cardamom* spp. and other Zingiberaceae, grasses, and many climbers which form a dense ground mat with little woody regeneration except the alien *Macsopsis*. Diversity is reduced, stratification is reduced and light and temperatures all rise. K. M. Howell (personal communication, 1989) recorded a savanna species of frog in such a gap – the first collection for the Usambaras. Smaller gaps following limited pit-sawing do have a variety of early succession trees including *Newtonia buchananii*, but also *Maesopsis*.

The incompatibility of different requirements from forests means that management compromise must take place. Improved land use planning policies, and practices which ascribe different intensities of exploitation to different forests is one form of compromise. Management objectives for each forest which provide block-based internal zonation, and the protection of primary forest by more intensively used buffers, are examples of how land use practice may work. These ideas are discussed more fully in the next section (p. 309). However, efficient land use planning is dependent on accurate data sets. Methodologies for rapid survey of forest resources, forest values and the people pressures on these resources are described by, for example, Synott (1977) and Hall (1986).

The conservation and management of eastern Africa's forests and forest resources

Conservation strategies

Previous sections have described the history of forest resources in eastern Africa and discussed the ways in which different people value and exploit these resources. The problems facing conservation have been catalogued and some of the biological principles governing conservation discussed. The stage is now set for the development of conservation and management strategies for the forests. Three strategies will be discussed: those dealing with people, planning, and management inputs. The chapter finishes with discussion on where conservation inputs are needed within the forests of East Africa.

Ultimately a successful conservation strategy must have the support of the people at national and local levels.

> Conservation of natural ecosystems can only be achieved by a total commitment of society ... Conservation cannot come with excessive exploitation, or via forced preservation.
> (Lugo & Brown, 1984).

These sections therefore stress the human dimension of conservation in some detail.

Conservation is a broad-based goal, which must be broken down into component objectives, each of which can be aimed at by management action. Care must be taken in framing these objectives, giving attention to both human and biological requirements. Conservation will not come by luck or goodwill: it must be strenuously planned for and managed towards.

Improving management practice is the third topic of debate. At present natural catchment forests are little managed and overexploited. This can change.

The present conservation position

The total inadequacy of the existing conservation status is evident from the fact that there is only one National Park covering the forests of eastern Tanzania (Uzungwa – gazetted in February 1992!), and only one in the eastern/coastal forests of Kenya, the Shimba Hills NP, and that with only a minimal coverage of forest values and a major objective to conserve sable and roan antelope in savanna woodland. In 1985 the Commission for National Parks and Protected Areas (CNPPA) of IUCN declared the Shimba Hills as

an environment in danger! Note that Kilimanjaro and Meru NPs in Tanzania, and Mount Kenya, Mount Elgon and The Aberdares NPs are on volcanic mountain impoverished forests, not really included in this definition of eastern Africa. Tanzania has two Game Reserves covering some of the less important coastal thicket communities, Saadani and Selous Game Reserves. Arabuko–Sokoke in Kenya is now of National Nature Reserve status, and further conservation areas exist on the Tana River. The critically important forest resources of Taita Hills in Kenya, and the Usambara, Uluguru, Nguru and southern Uzungwa Mountains are all of Forest Reserve status, a category which permits considerable resource exploitation and manipulation, and which to date has little investment in protection. Biosphere Reserve status has long been considered for the Usambaras, but as yet there is no meaningful proposal to implement such suggestions.

Calls for improved conservation often fail as no detailed suggestion as to objectives or boundaries is given, which could then be more easily followed up. Frequently conservation categories are proposed which do not exist in law, and therefore cannot be implemented.

The suggestion of a 'nature reserve' status for Amani–Sigi–Kwamkoro forests of the East Usambaras is an example of this latter point (Finnmap–Silvestre, 1988). Similar discussion on 'forest parks' delayed conservation action in Uganda. However, recently several forest reserves of great conservation significance (e.g. Rwenzori and Impenetrable) were declared as National Parks, much against the wishes of the Forest Department, which said Parks had no funds or expertise to use them. Howard (1991) gives a detailed review of conservation values and planning. There is no category of Forest Park or Nature Reserve in Tanzania, and modification of the existing legislation is a time-consuming process. There may be concern on the part of forest administrations that declaring a forest area as a National Park means they lose all control. This is not necessarily true. National Park status is a legal category of land use. There is nothing to prevent the Forest Division administering the area under the National Parks legislation as a Forest National Park!

Kenya has a complicated set of protected area categories (IUCN, 1987). National Parks provide the highest level of conservation status. National Reserve may allow compatible land uses other than conservation, such as grazing. These two categories are administered by the Ministry of Tourism and Wildlife. The National Park category does have some forest area: the Aberdares, Mount Elgon, Mount Kenya; although National Reserves also have forest areas of conservation importance: Marsabit, Tana River Primate Reserve, Shimba Hills. The Forest Department has a category of 'Nature Reserves' which give a higher conservation status to areas within the Forest Reserve estate. Examples are Arabuko–Sokoke, Mau and Nandi.

This multiplicity of category can lead to confusion. Kakamega Forest, of extreme biological value, has been a Forest Reserve (230 km²) since 1933. Part became a Nature Reserve (12 km²) in 1967. A National Park of 97 km² was proclaimed in 1983, but was not legally gazetted, and a National Reserve order was passed in 1985. The area is still heavily exploited (Kokwaro, 1988). Some forest areas of great importance such as the Taita Hills are still unprotected (Beentje, 1988a) although yet another category, 'Game Sanctuary', supposedly safeguards the non-forest wildlife at lower altitude.

The existing conservation status is of great concern on three distinct counts. First, there is inadequate legal protection. Secondly, there is inadequate field protection and management of the existing protected areas, mainly of Forest Reserve status. There are too few field staff to prevent illegal exploitation and to regulate legal exploitation. Today, their conditions of service in Tanzania do not allow efficient work input. Management inputs are rarely planned, biodiversity conservation is rarely a dominant objective, and resource status is not monitored. Research into natural forest ecology and the wildlife of forests is minimal. Resources are deteriorating. As the following sections illustrate, recovery is possible, but it will require considerable investment of staff, finance and expertise. Thirdly, there is the question of management knowledge for what is increasingly realised is a highly complex system involving ecological and socioeconomic components. Traditional forest management texts do not deal with people problems. There are no forest management textbooks for East African forest situations. A manual of basic forest practice, which will help fill this gap, is under preparation (J. Holmes, personal communication, 1991). General texts may offer some help, but rarely are their guidelines of sufficient detail to offer exact advice or suggestion for the young forest officer. Poore & Sayer's (1987) booklet on managing tropical moist forest is an example; the management of catchment forest gets one page. Buffer zone management is discussed by Oldfield (1987); here the case histories are of interest, but again at a general level of applicability. There is a need for much more documentation on management possibilities for the East African closed forest system.

The need for monitoring and inventory of forest resources

Conservation planning requires knowledge of the distribution and numerical status of the resources. 'Where is this species located, how numerous is it, and is the population status changing,' are necessary pieces of information. The inverse information is also important: 'This forest has the following species, at these densities and their status is declining.'

Unfortunately we do not possess this basic information for most of our natural forest resources. Recent surveys in the Usambara, Uluguru and Uzungwa Mountains have all led to the discovery of new taxa and new records of large and conspicuous species (see, for example, several chapters in this volume). The East Usambaras may have the richest forests in terms of species diversity in all tropical Africa. Forest cover is fragmented into 18 separate reserves. There is no accurate vegetation map of the area, no detailed description of any forest block and no complete plant species list for any one area. White (in Rodgers & Homewood, 1982a) lists some 30 tree species as near endemics for the Usambaras, and

Hedberg (1979) includes smaller plants, but there is little knowledge of their distribution pattern (see Hamilton & Bensted-Smith, 1990, for preliminary data on major species). Past attempts to suggest key forest areas for plant species conservation in the Usambara failed owing to an inadequate data base (W.A.R., personal observations). Finnmap–Silvestre (1988) map density distributions for 11 commercial tree species in East Usambara and roughly map the collective concentration of endemic and near endemic tree species; this is a useful start. Hamilton & Bensted-Smith (1990) extend the data base considerably. Beentje (1988b) gives detailed distributions for some 90 Kenyan forest tree species.

The solution is obvious. A much greater investment into simple inventory and description of the forest resource, with equal interest in biological and commercial values, is required. Recent floristic and ecological surveys of areas such as Pugu (Hawthorne, 1984), Mazumbai (Hall, 1985) and Kimboza (Rodgers et al., 1983) describe realistic methodologies for plants. Surveys such as those of Stuart (1981) for birds, Kielland (1983) for Lepidoptera and Rodgers (1981) for colobus monkeys illustrate what can be done for animals.

Eltringham (1980), in a comprehensive review of wildlife (not forest!) research needs for Tanzania, stressed the need for greatly increased inputs into applied and biological values of natural forests and their wildlife. He advocated an Usambara Wildlife Research Centre, with studies on people, forest and animal ecology. The importance of environmental monitoring and audit (a full evaluation) is increasingly recognised. A thorough review is given by Howlett (1991).

The last few years, however, have seen an increase in the quantum of research into East Africa's natural forests. Recent studies, by Kigomo et al. (1990) into regeneration patterns of the Nairobi *Brachylaena* forest and Tsingalia (1990) into Kakamega gap recolonisation, are good Kenya examples.

Deciding land use objectives and zonation

Successful resource conservation and management is dependent on the formulation of realistic objectives; objectives both for the resource itself and the peripheral human populations. These objectives and the ensuing prescriptions should be contained within properly prepared management plans (MacKinnon et al., 1986). Planning must take into account the human component and their needs of forest based resources; see Saharia (1985) for a discussion on the Indian experience.

The inherent incompatibility between strict conservation needs and resource exploitation can be overcome by spatial separation of land use practices, through zoning. This was initially attempted by a separation into catchment and production forest reserves, but zonation should go much further. Zonation is not a new concept for East African conservation, possibly being first suggested by Lambrecht (1966) and more recently elaborated by Lusigi (1985). It has been attempted on a larger scale in the Ngorongoro Conservation Area and in Amboseli National Park, but scarcely tried within a forest environment. Howard (1991) describes a detailed zonation plan for Ugandan Forests. The Ugandan Forest Department has proposed that at least 20% of all forest would be non-exploited forest reserve, a further 30% would be buffer zone with light exploitation and 50% would be for commercial use.

In 1991 Tanzanian foresters approved a zonation plan giving four principal zones. These are as follows:

Management Zonation System for Catchment Forest Reserves in Tanzania.

Catchment Zone.

Aims: The Catchment zone exists to protect the natural catchment values of the forest, and to prevent erosion. The zone is defined by the criteria of slope and hydrology: all slopes >40% and areas 50 metres either side of streams and upper catchments by watershed ridges are so included.

Management: There will be no exploitation or disturbance to natural vegetation.

Biodiversity Zone.

Aims: The zone is to protect areas of high biodiversity and migratory routes [this may be difficult to define in practice: what is 'high' biodiversity? W.A.R.].

Management: there will be no exploitation or disturbance of natural forest or wildlife.

Amenity Zone.

Aims: This zone is to allow the use of unusual topographic or natural features for education, recreation and research.

Management: this will depend on specific objectives. There will be no commercial exploitation. Visitor facilities will be developed.

Productive Zone.

Aims: This zone is to allow the sustainable production of timber and other forest products, in areas of the reserves not included in the other zones.

Management: Exploitation will be permitted according to rules of selective harvesting. There will be no mechanical logging. Exploited areas will be regenerated as per standard practices.

One component of zonation theory is the core–buffer concept, where a biologically important core zone is given extra protection by the presence of a peripheral buffer zone where resource exploitation is permitted. The buffer absorbs people pressure, and management should be directed towards maintaining that capacity to absorb pressure by, for example, plantations of fuel, fodder, timber trees, etc. The buffer zone has two roles: one is this social buffer, to allow people access to resources; the other is extension buffering, allowing core zone wildlife populations to spill over into larger areas, possibly acting as a gene flow corridor between two core zones (Saharia, 1985).

The plantation of native and exotic fuelwood trees around Nyungwe Forest in Rwanda is a good example of a peripheral buffer zone. We proposed the same for the eastern margins of an Uzungwa Mountains National Park (Rodgers & Hall, 1985). Weber (1987) discusses possible socioecological inputs into the management of mountain forests of great biological values. Buffer zones could be of Forest Reserve status, internally

managed for either more production or more catchment objectives.

Areas considered to be of extreme importance for biological resource conservation should be managed as 'core zones' free from disturbance, and given the legal basis which prevents disturbance, i.e. National Park status. Eastern Africa is still only slowly moving from an early idea that Parks are places for visitors to look at animals!

In Tanzania today there is increasing realisation of the essential role of forest cover in maintaining water flow, and the understanding that overuse of forest resources can lead to watershed degradation, and even to large-scale landslides, mudflows, and flooding and siltation problems, such as has recently happened in Mbeya Region. One answer is to ban all forms of exploitation from the forests (J. Holmes, personal communication, 1989). In practice this will be difficult to implement unless adequate alternative forest resources are provided in plantations and village or homestead woodlots. Careful forest mapping and planning should be able to demarcate buffer zones where legitimate exploitation can be regulated and key core zones or fragile sites where all exploitation should be stopped. Sayer (1991), Oldfield (1987), Howard (1991) and Wood (1990) discuss buffer zones in forested regions in different parts of the tropics. A number of recent initiatives in India and Southeast Asia have shown the great success that can be achieved in increasing buffer zone productivity by joint management or participatory management by local people (Chambers & Leach, 1990; Poffenberger 1990a, b).

Improved management of the natural forests

FAO (1982b) stated that only a small proportion of Tanzania's forests is intensively managed, with in-date working plans. These were the commercially productive forests and plantations. Regular staff establishments of unworked natural forests do not permit adequate management, nor even adequate protection. A first step in improved management must be the provision of sufficient numbers of field staff, supervisory staff and facilities to protect the resource from illegal entry, encroachment and exploitation. This will need the construction of living quarters and patrol camps, the availability of funds for porterage and tentage and the development of a patrolling ethic amongst all staff cadres. Financial resources must be available to pay standard patrol allowances. Again, joint management strategies between government and local people may reduce the need to maintain expensive regulatory control by Government.

Hall (1983) has laid down guidelines for the management of strict nature reserves in forest. Major management activities include patrolling and prevention of intrusion, and reduction of external threats, e.g. fire and exotics. Sinclair (1983) discusses the role of Protected Areas as baseline study sites and suggests that management be greatly limited.

Few guidelines for practical management of forested watersheds exist for the East African situation. FAO (1985) gives an overview of watershed management needs and states

> effective watershed management is rarely found in mountainous watershed of developing countries; the development of appropriate practices is hampered by a lack of relevant training, insufficient applied research, and limited demonstration projects.

Effective watersheds, especially on steep slopes, need total protection from exploitation which can be achieved by effective patrolling.

The concept of zonation of forest and land use objectives allows an intensification of forest management. Whilst a forest designated as a full nature reserve should have the minimum of manipulative management, other objectives demand more intensive inputs. Forest areas designated as priority watersheds may require ameliorative management involving reduction of erosion by gully-plugging, check-dams, the enhancement of natural regeneration and perhaps the planting of selected species on understocked areas.

Silvicultural practices for developing timber values are known for some species. Whilst these practices may cause a loss of biological diversity and even reduced water retention capability, they can be applied to forest areas designated for tim-

ber production. Subsidiary regulations protecting stream banks and spring sources should be enforced. Timber extraction can be optimised for profit or for reduced environmental impact. The scarcity of forest resources in eastern Africa suggests the importance of the latter criterion.

Fire plays a major role in reducing closed forest values and in destroying forest itself. Fire prevention must be a major function of management. Techniques of fire breaks, back burning, shelterbelt planting, etc. are well known; they require implementation. Fire fighting capability requires strengthening; this will require the cooperation of local people.

These ideas are discussed below. Management is categorised into direct practices, including logging regulation, which immediately affect the resource; and indirect practices which affect the resource by modifying pressure by local people. Saunders *et al.* (1991) describe two distinct sets of management action for fragmented forests. These are the management of the natural system or the internal dynamics of the area; and the management of the external influences on the natural system. For large areas the emphasis should be on the former, whereas for smaller remnants the emphasis should be on the latter. Traditional reserve management has stopped at the reserve borders; but the fluxes of water, nutrients and organisms do not. Management therefore has to place the reserve firmly within the context of the surrounding landscape. This has important implications for managers as it forces them to interact with peripheral land users.

Finally, the idea of restoring degraded forest land is discussed. Restoration ecology is a growing field of study, and one of importance for many eastern Africa forest areas (see Jordan, Peters & Allen, 1988; Stanley-Price, 1989; Usher, 1991).

Direct management practices

These practices include the reduction of illegal exploitation. In buffer zones exploitation can be legalised on a sustained yield basis. The village system of government in Tanzania gives local people a great ability to monitor and regulate their own activities (see Bell, 1985; Poffenberger, 1990a, b, for a discussion on the need to involve local communities in conservation regulation). The TFAP stresses the need for greater involvement of people in forest management practice. In core protection zones for biodiversity, exploitation should cease. Such prevention needs adequate field patrolling, the ability to apprehend and fine culprits (which may require enhancement of the law, or greater support by local judiciary) and conservation education at the village level. It must be remembered that fuelwood cutting is not a luxury for many rural populations, it is essential – although much goes to make 'pombe', a local alcohol. The provision of alternative fuelwood resources must take place simultaneously with improved protection. Fuel efficient 'stills' would be a useful development!

Fire prevention and fire fighting ability must be increased. Preventive measures such as fire lines must be maintained. Fireline clearing and boundary demarcation provide a monetary wage income to the poor and so have social benefit. It is a cost that governments have to bear. Strip plantings of *Agave* can act as live permanent fire lines requiring little maintenance and also act as legal or management boundaries. *Tripsacum laxum* and *Acacia melanoxylon* also are useful fireline species (D. A. Wardell, *in litt.*, 1990).

The removal of exotics should be a long-term goal for the more important conservation areas. The problem varies from place to place, and may only amount to weed infestation around disturbance sites. Some plantation species self-seed (*Cedrella*, wattles, *Maesopsis*) and management prescriptions could consider their removal. Hamilton & Bensted-Smith (1990) give a detailed account of the *Maesopsis* problem in the East Usambaras. D. A. Wardell (personal communication) cautions that removal of such a successful pioneer as *Maesopsis*, which has already developed huge soil seed banks, will be all but impossible. Possible success needs the development (and subsidy) of a ready market for the mediocre timber of *Maesopsis*.

Improving logging practice

All logging, no matter how low intensity or selective will have some effects on forest processes.

These effects may be attributable to the physical pressures of extraction (roads, compaction, scarring, erosion and clearing) or to the human pressures involved with extraction (fire, hunting, plant exploitation, disturbance, canopy opening) or to more complicated biological changes such as genetic erosion. The report by FAO (1982c), *Environmental Impact of Forestry*, illustrates the concern increasingly expressed about logging impacts. Similar comments are given in, for example, Ng (1983) and Whitmore (1984).

Ideally each individual forest management unit, compartment and reserve, will have its own objectives for management, designed to maintain the specific values of that unit. There is a need to revert to the old 'compartment' basis of forest management, zonation should be at compartment level and compartments ideally should be microwatersheds. Objectives and values for compartments could be maintenance of water catchment or naturalness, or a specific species or community, or timber or fuelwood, or a combination. Logging rules then should cater to these objectives, varying from no exploitation, to selection felling and minor tending, to fuller exploitation procedures. Slope steepness, precipitation, stream density, canopy and ground layer covers, preferred species densities, etc. all affect the intensity of logging. Hall (1988) brings out the need to develop a set of management strategies for catchment forests.

The practice of sanitary felling – the removal of dead, dying, diseased, bent or over-mature individuals – has no beneficial role in conservation management. Indeed the removal of older trees can materially reduce species diversity by removal of epiphytes, lianes, fungi, nesting holes, insect food for insectivores, etc. A growing literature in the USA acknowledges the benefits of old trees ('snags') in the forest, and designs silvicultural systems to maintain overall biodiversity (e.g. Thomas, 1979; Hoover & Wills, 1987). This has been discussed with relevance to tropical forests in India (Rodgers, 1991).

Indirect management practices

This will include more intensive buffer zone management allowing exploitation, and the provision of alternative fuelwood supplies as peripheral buffers or village woodlots. Ecologically oriented sustainable development or 'ecodevelopment' practices to improve agricultural productivity at village level will reduce land pressure in the long run. Agro-forestry, stall feeding, improved crop varieties, etc. are all major possibilities. Education remains the key behind all these proposals and educational programmes should be carefully designed at community level, appealing to schoolchildren, women, and elders of each tribal unit. These ideas were elaborated in a demonstration conservation proposal now being implemented by IUCN/EEC in the East Usambaras (Wardell, 1990) and elsewhere in Tanzania. These are discussed in a later section.

A second set of indirect inputs is into the development of industrial hardwood plantations to remove pressure on the limited natural forest areas. This policy is slowly being introduced in Tanga Region, where teak is being promoted as an alternative to Usambara timber, and in the Kilombero valley of Morogoro Region. Tanzania teak has a ready export market to India, provides a valuable saw and peeler log and the thinnings serve as poles.

Forest restoration

The restoration of degraded forest land may be an important component of conservation practice. Whilst such land may take centuries to redevelop even a part of past levels of diversity, such restoration will have many more immediate values, such as improved catchment capability and soil stabilisation. Reafforested or restored land can act as corridors for genetic transfer and partial habitat for many forest species. The new forest resources will allow some form of resource availability for local villagers, alleviating pressure on more important natural forests.

Reafforestation measures can create fire barriers or, by removing grasslands, reduce fire risk. Village plantations will keep human populations out of the natural forest, a policy used to good effect on the periphery of the Nyungwe Forest in South Rwanda. Whether gaps in natural forest should be planted is more debatable; if natural regeneration can cope, then artificial planting should be avoided. Exotics should not be considered, multi-species and multi-age plantations of natural species should be attempted. Gaps in forest cover, giving grass and shrub communities, may have a major importance for wildlife and patterns of species diversity. The Ngorongoro mountain forests for example have major elephant, rhino, buffalo and bushbuck populations dependent on the frequent grassy clearings. Unless there is some compelling reason to create tree cover (fire pressure, erosion, major catchment) then gaps should be maintained. Monitoring, however, should ensure that gaps are not spreading.

In the long term, species reintroduction could be considered, for example *Polyceratocarpus* (Annonaceae) to the East Usambaras where it has not been collected for years, from the Uzungwas, where it has recently been discovered. Examples could be considered for animal species as well.

Buffer zone, people, education and politics

Buffers for people

Conservation oriented forest management cannot stop at the edge of the Forest Reserve; indeed, the most pressing and difficult task ahead is to take management actions and benefits to the people outside the core conservation areas. Sumardja (1984), writing of Indonesia, states: 'how to combine goals of nature conservation with the material needs of the people, especially those adjacent to the reserve is the biggest question mark for managers.' People's interests and objectives of reserves must be brought closer together, and this can only be done in a well-managed buffer zone separating core resources from settlement.

This of course is a key component of the 'Man and Biosphere' (MAB) programme of UNESCO, where

Protected areas should be seen as part of an integrated system of action for environmental care in a region. They should not be isolated islands. Mutual linkages between peoples & resources must be recognized and taken into account. The buffer zone is a major tool of such a system.

(Thelen & Child, 1984)

They believe that buffer zones should be providers of goods and services to a multiple land use development, and that it is a lack of policy and management capability that has prevented such development to date. Whilst MAB reserves have been created in East Africa (Serengeti, Ngorongoro, Manyara, Selous, Kulal), it is only the latter in the deserts of North Kenya that has functioned as a planning, research and development oriented unit. Suggestions for the whole of the Usambara Mountains to be a Biosphere Reserve have not been followed up. There will need to be a much greater level of understanding of the role of Biosphere Reserves if the concept is to succeed. The case of Tai Forest in Ivory Coast illustrates the potential beneficial role of Biosphere Projects (Dosso, Guillaumet & Hadley, 1981).

These concepts of buffer zone development are not merely part of MAB; they are a cornerstone of the World Conservation Strategy's Theme 16 for Tropical Forests, wherein the forests of East Africa figure as a continental priority. The Tropical Forestry Action Plan of FAO also sees ecodevelopment in rural areas as of major importance for long-term conservation.

But these are all words; how do these translate into action? Oldfield (1987) and Sayer (1991) provide field examples. Shepherd (1991) discusses both policy and practice of communal management in Africa. First, there is a need for a national conservation strategy which recognises the need for such development and associated educational inputs. Development will not take place overnight, there will be much need for experimentation of approaches, much trial and error. There will be a great need for extension expertise and funding. People must see benefit from conservation, in the form of some improved supply of requirements, principally water and fuel. Education should show both the need to conserve and how conservation can produce benefits. Joint participatory management must evolve.

The past few years have shown that conservation of forest resources based on legal protection, a policeman 'keep out' strategy, has not worked, and cannot work in the face of a growing resource-dependent rural people and an inadequate resource base. Such policies may in some cases have exacerbated the conservation situation. More and more resource conservationists are accepting that without the involvement of local communities, and without addressing the underlying problems of subsistence agricultural production, encroachment on the forest resource will continue. Tanzania has several integrated forest conservation-sustainable development projects under way; in Iringa District, parts of Morogoro District, West Usambaras and East Usambaras. Most inputs are planned. Initial lessons are very clear, however. Input has to be long-term: the acceptance of input is slow by people used to the stick not the carrot approach. People participation in management and in planning the management is the key, along with an assured stake in the alternative resources on development. Wardell (1990) gives an overview of the East Usambara experience.

Conservation education

The long-term conservation of Eastern Africa's natural resources is totally dependent on persuading local communities to reduce their pressures of direct exploitation. This could in theory be done by greater use of force – more patrolling, stiffer penalties, etc. – but this strategy is not proving effective in the conservation of wildlife populations (e.g. rhino and elephant in eastern and central Africa: Bell, 1985). Potentially more viable strategies include education inputs to persuade people to reduce exploitation to sustainable levels and show them how to develop agriculture along ecologically appropriate lines (the goal of ecodevelopment) and the provision of alternative resources such as village woodlots.

These latter aspects are themselves dependent on their acceptance by local people, again a function of education inputs. Forest conservation education programmes have many potential components, ranging from compulsory school curriculum inputs at primary, secondary and professional or vocational levels, to filmstrip, radio and political slogan messages, to specifically designed extension and demonstration educational inputs by trained extension workers. Demonstration plots are a major component of successful extension and education programmes. Only recently have professional wildlife courses at tertiary levels incorporated elements of forest ecology, forest wildlife and catchment planning into their syllabuses. The degree courses in Wildlife Ecology at the University of Dar es Salaam devoted lecture and field tour time to forest study (Rodgers, 1984). Conservation and natural forest components were largely neglected in the initial syllabuses of the university's forestry degree programme, emphasis being placed on wood production from pine plantations.

Children's conservation education in East Africa was pioneered by Kenya's Wildlife Clubs and then followed by Malihai in Tanzania. Neither have concentrated on forest issues, although IUCN funded a brief survey of Usambara educational needs with Malihai assistance. The Ministry of Natural Resources in Tanzania has undertaken some extension work, and a model programme of ecodevelopment with considerable educational elements has been started in East Usambara with EEC funding. These efforts, however, are mere tokens; the need is enormous and continually growing. Government and non-governmental organisations alike must devote greater attention to all forms of environmental education.

Politics and economics

Our abilities to implement and maintain conservation strategies are often determined by political and economic forces rather than biology and technical expertise. Conservation requires land, forested land, often fertile; such land is in short supply and in high demand by rural populations. It is difficult for politicians, especially those from forest areas, to argue against such demands. Conservation costs money, and taking forests out of productive use (timber and fuelwood) reduces economic inputs to local and national exchequers. It is difficult for many administrators to see value in increased conservation. In this section I wish to consider just three issues of political and economic concern, though there are many more. The role of the 'Western developed nations', the con-

cept of monetary value on resources, and the problem of 'development projects' warrant increased attention. It has taken the developed world many decades to develop the now powerful environmental lobbies to argue for conservation. But it will still be decades before such forces are operative in eastern Africa, notwithstanding some limited successes by Kenya's Wildlife Clubs. Much concern for Africa's natural resources (and elsewhere in the tropics) comes from foreigners, from Europe and North America. Too much pressure or vociferous appeal can be counterproductive: 'Now they have destroyed all their forests, who are they to tell us what to do with ours?' If the West wishes Africa to conserve its natural resources, then it must help subsidise the costs involved in such conservation. Frequently it is the foreigners who reap the benefits of such conservation, if not first-hand as tourists, then at a distance via television and coffee-table books. Tanzanians rarely get such opportunities.

A second topic of concern is pricing of natural resources. Timber is expensive; fuelwood is free, but costs effort; water is free. Fuel and water are essentials, and are resources in dwindling supply. They might be better husbanded if there were some monetary value attached to their production. Forests could be judged in cumecs of water, not cubic metres of wood. People could pay a tiny sum to illustrate that there is a value, and that the product requires care.

East Africa's mountain forests have escaped much of the land use conflict over hydroelectric reservoirs and forest loss, common in much of the tropics. Lowland forests, especially those along rivers, can be at risk. Hughes (1984) discusses the ecological problems of the Tana floodplain forests following river impoundment. Forests are at risk from other development activities, however. The scale of degradation in the East Usambaras from what was envisaged as long-term sustainable logging exploitation is a case in point. Hosier (1988) discusses the economic costs of forest degradation in eastern Africa. Railway development in Magombera (Rodgers et al., 1979), roads in the Ulugurus to service radio towers and in the Uzungwas to service pulp mills, are two immediate concerns. Impact assessments of development projects require the forest values to be known. Usually we are not in a position to document potential impact.

Planning for a forest protected area network in eastern Africa

Conservation planning embraces a larger framework than the management of existing forest areas, and has to consider regional, national and even continental issues. Planning must decide how many areas of what size should be incorporated into the national protected area network.

These questions can be answered only by an awareness of the level of diversity within the resource to be conserved. If all was uniform, then one large area could be sufficient; two, perhaps, to allow for catastrophic loss. We know there is great diversity of both botanical and animal resources, but there is still little hard information on the distribution of distinct forest communities. Surveys at continental level make broad separations within eastern Africa, for example, by White (1983) into:

Coastal Margins, the Mangroves (outside the scope of this book).

Coastal Forests and Thickets.

Block Mountain Intermediate level Forests.

Block Mountain Higher level Forests.

But detailed biogeographical studies within eastern Africa recognise a very much greater level of diversity, at community and species levels; many of these studies are summarised in this volume (Lovett, Chapter 4; Hawthorne, Chapter 5; Scharff, Chapter 7; Stuart et al., Chapter 10; Kingdon & Howell, Chapter 11). Mackinnon & Mackinnon (1986) discuss conservation requirements for all African nations against a continental level biogeographic framework. Rodgers (in IUCN, 1992) extends this for the forested regions of east and northeast Africa. Appendix 14.1 summarises conservation needs from these discussions of diversity.

Many of these areas are not new suggestions: many have been proposed as conservation sites

long ago. Hedberg & Hedberg's (1968) volume on conserving plant resources of Africa has detailed proposals for Kenya (Lucas, 1968) and Tanzania (Polhill, 1968). Rodgers & Homewood (1982a) discussed broad conservation needs in the Usambaras, which have been further elaborated by Stuart (1983) and Hamilton & Bensted-Smith (1990). Rodgers & Homewood (1982b) recommended the creation of Uzungwa Mountains National Park, a proposal approved for funding by IUCN and implementation by Tanzania. Hawthorne (1984, Chapter 5) described conservation needs for the coastal belt, north of the Rufiji River.

The IUCN Primate Specialist Group recognises the extreme importance of the Tana River Forests, Jozani on Zanzibar and the Uzungwas, at a continental level for primate species (Oates, 1985). The IUCN Antelope Specialist Group draws attention to two species of the eastern African forests, Ader's duiker and Abbot's duiker, both considered as threatened, both without adequate habitat under effective protection (East, 1988). The Red Data Book on Birds for Africa (Collar & Stuart, 1985) considers seven sites of exceptional importance in eastern Africa: Tana, Arabuko, Usambara, Nguru–Ukaguru, Uluguru, Uzungwa, Pugu and others in Mozambique. Details are given by Stuart et al. (Chapter 10).

Many of these suggestions for Tanzania have been collated in the planning materials towards the Tanzanian Forest Action Plan (TFAP: Bensted-Smith & Msangi, 1988). Suggestions for conservation now need implementation.

Conclusion

I started to write this chapter in 1985. I first rewrote the conclusion in September 1989 after I revisited Tanzania and saw how much had changed. I again rewrote the conclusion and amended parts of the text after larger visits to both Kenya and Tanzania in late 1990 and in 1991, when it was obvious governments and agencies had new perceptions of the forest sector. I have been struck by these enormous differences in attitudes and circumstances. Some change leads to optimism for the future, some does not. This section is largely about Tanzania but much, I am sure, is appropriate to Kenya as well.

Firstly is the change in 'knowledge'. We know much more about the biological values of the forests of eastern Africa. We have developed methodologies for survey and study. There is in Tanzania a body of expertise in the forestry division, the universities and, a welcome addition, in the NGO conservation societies. This expertise still has a core of expatriates but it is spreading. The Wildlife Conservation Society of Tanzania, through its magazine *Miombo*, addresses many issues of forest wildlife and forest conservation. The biggest gap, and this is crucial, is in forest botany. There is a dearth of identification skills. This skill is essential if we are to understand the forest, as Hall (1984) pointed out in a symposium relating biology to development in Tanzania. Tanzania's two most knowledgeable field and herbarium taxonomists retire in the coming few years. There is no one to replace them.

Secondly this knowledge has been directed to conservation issues and Government has become aware of the importance of forest conservation. No longer is there the feeling of fighting an uninterested bureaucracy. Conservation inputs and proposals have been welcomed. The Forest Policy (FORPOL) is changing. The Tanzania Forest Action Plan (TFAP) recognises the importance of conserving the Eastern Arc forests and the coastal forests for catchment and biological values. The role of people in conservation is acknowledged. The will to conserve is now there.

But, and this is the last point I wish to make, the means to conserve have become sadly depleted, for two reasons. First is the economic situation and the grave fact that field staff salaries do not permit a full day's work. The infrastructure to support work has collapsed. Staff need incentives in order to work; they are willing, given the support mechanisms. Secondly, the changes in governmental administration have affected conservation activity. Twenty years of experimenting with decentralisation and a reduction of hierarchical control has resulted in less discipline, less supervision and less commitment to long-term programmes. The huge increase in illegal forest

use and the almost total lack of prosecution indicate the collapse of effective conservation.

This, we learn, may change. FORPOL suggested that forestry should revert to a centralised Ministerial system with direct chains of command to all field staff. This has been made a major plank of a large forest aid project (World Bank/ODA/FAO, 1990) but it has not been accepted. Total change is unlikely, but advantageous changes in the way that district and national forest cadres operate are expected. Conservation-orientated management and aggressive policies of forest reservation and encroachment rationalisation can lead to long-term success. This will require expensive financing. Tanzania is aware of the economic linkages between forest maintenance and the more productive sectors of agriculture, energy, and water. Tanzania seeks to invest more into the forestry sector, but has limited internal funds. Forest conservation will need foreign financing. In August 1989 I was associated with a FINNIDA/IUCN mission to consider funding forestry inputs for the East Usambaras. We estimated needs at over US $2 million for the 3–5 year project period which is now under way.

In September 1990 I participated in a Tanzania Forestry Action Plan Seminar to prepare guidelines for anticipated donor support to the Eastern Arc Block Mountain Forests. Two sets of results are encouraging. First, there is great donor interest; Denmark, EEC, Finland, Germany, IUCN, Norway and Sweden indicated future financial support. Secondly, much is already taking place on small and large scales: Germany in the West Usambaras, EEC and now Finland in the East Usambaras, Denmark in Iringa, Sweden in part of Morogoro, Norway to overall catchment forest support and new interest in the important coastal forests. Some of this input involves Tanzanian NGOs, such as the Tanzania Wildlife Conservation Society. Some is directed at District, or Regional, or National level governmental organisation. Much is broad based, aiming at sustainable development of people as well as sustainable conservation.

Some aspects give cause for concern. There is a need for donor project coordination to prevent wasteful overlap and competition. Tanzania does not yet have a natural forest section within a Ministry large enough to achieve coordination and direction of donor effect. The distinction between administrative units of the Central Ministry, Region, District and Local Government forestry sector hampers all forms of forest management. There is difficult coordination between forestry and agricultural sectors of Government, although potentially achievable through district government.

Forests still provide important timber resources and there is a great deal of illegal trade in pit-sawing, some with collusion of forest officials. Despite a partial and temporary moratorium on pit-sawing in the East Usambaras, there was no sign of field effort to reduce this massive activity. The answers to these problems will not be easy in Tanzania's difficult economic climate. But the answers must depend on local people support and a motivated forest department. Foreign aid can help stimulate such support and motivation; it cannot replace it.

Kenya, too, is going through similar internal political debate on the need for forest conservation. There is some level of donor support to the natural forest sector (ODA of UK) and much more to rural development and rural reafforestation programmes which will directly and indirectly affect the natural forests. Critics of the developing Kenya Forest Action Plan as alluded to in World Bank (1988) reports, point to the great difficulty in implementing conservation, and preventing encroachment and overuse, in the face of huge population growth and consequent land hunger. Arguments for biodiversity will not work, arguments for water catchment might.

Both countries still have an inadequate forest protected area network. There are still forest lands outside the forest reserve estate. There are still many areas of extreme biological importance which have inadequate protection in the forest reserve network. While forests are degrading there is interminable debate on categories and administrative power. The first steps to conservation are obvious, legal gazettement at the highest levels. Then comes improved management capability to well-formulated plans.

But conservation is a human problem not a biological problem. Conservation will not succeed unless human needs are catered for and adequate alternative resources provided. Forestry has little expertise in human development and must join forces with other agencies to develop on integrated sustainable land use in the forested areas. We need urgent action to find out 'how?'.

Acknowledgements

John Hall, Katherine Homewood, Jack Holmes and Andrew Wardell discussed many of the issues raised in this chapter and commented on earlier drafts of the manuscript. Many people contributed ideas, information and literature. I thank in particular Alex Boswell, Neil and Liz Baker, Gil Child, Colin Congdon, Henry Fosbrooke, Geoff and Vicki Fox, William Hawthorne, Kim Howell, Jan Kielland, Jon Lovett, E. J. Mlowe, I. Mwasha and Simon Stuart. Saidi Mbwana, who has headed the Forest Management section of the Tanzania Forest Division for many years has been my companion on several forest fact-finding surveys. I learned much practical forestry with him. Jon Lovett and Sam Wasser helped 'tidy up' early manuscript drafts. Messrs Rajesh Thapa and R. N. Tyagi typed several drafts. To all of them, my thanks.

References

ALLAN, W. (1965). *The African Husbandman.* Edinburgh: Oliver & Boyd.

ALPERS, E. A. (1975). *Ivory and Slaves in East Central Africa.* Nairobi: Heinemann Educational Books.

ANON. (1902). Forestry in German East Africa. *Indian Forester* **28**, 372–4.

BARNES, R. F. W. (1990). Deforestation trends in tropical Africa. *African Journal of Ecology* **28**(3), 161–73.

BEENTJE, H. J. (1988a). An ecological and floristic study of the forests of the Taita Hills, Kenya. *Utafiti (Occasional Papers of the National Museums of Kenya)* **1**(2), 23–66.

BEENTJE, H. J. (1988b). Atlas of rare trees of Kenya. *Utafiti* **1**(3), 71–123.

BEENTJE, H. J. (1990a). The Forests of Kenya. *Mitteilungen aus dem Institut für allgemeine Botanik in Hamburg* **23A**, 265–8.

BEENTJE, H. J. (1990b). Botanical assessment of Ngezi Forest, Pemba Island, Zanzibar. Report to the FINNIDA Forest Project, Zanzibar, Tanzania.

BELL, R. H. V. (1985). Traditional use of wildlife resources in protected areas. In *Conservation and Wildlife Management in Africa*, ed. R. H. V. Bell and E. McShane-Caluzi. Washington, DC: US Peace Corps.

BENSTED-SMITH, R. & MSANGI, T. H. (1989). Material for the Tanzania Forest Action Plan: Ecosystems and Communities for Conservation. Dar es Salaam: Ministry of Lands, Natural Resources and Tourism.

BROWN, L. H. (1981). The conservation of forest islands in areas of high human density. *African Journal of Ecology* 19, 27–32.

BROWN, M. (1992). Buffer zone management in Africa. In Proceedings of a workshop in Queen Elizabeth National Park, Uganda, October 1990. Washington, DC: PVO/NGO/NRMS Project.

BRUEN, M. (1990). Hydrological considerations for development in the East Usambara Mountains. In *Forest Conservation in the Eastern Usambara Mountains, Tanzania*, ed. A. C. Hamilton and R. Bensted-Smith, pp. 117–40. Tropical Forest Programme. Gland, Switzerland: International Union for Conservation of Nature and Natural Resources.

BRUENING, E. F. (1989). Use and misuse of tropical rain forests. In *Tropical Rain Forest Ecosystems. Ecosystems of the World 14B*, ed. H. Lieth and M. J. A. Werger, pp. 611–36. Amsterdam: Elsevier.

BURGESS, N. D., MWASUMBI, L., KINGDON, J. & DOGGETT, A. (1990). Distribution and Status of Tanzanian Coastal Forests. MS. Royal Society for the Protection of Birds, Sandy, UK.

CHAMBERS, R. & LEACH, M. (1990). Trees as savings and security for the rural poor. *Unasylva* 161 (41), 39–52.

CLARKE, J. D. (ed.) (1969). *Kalambo Falls Pre-historic Site*, Vol. 1. Cambridge: Cambridge University Press.

COLLAR, N. J. & STUART S. N. (1985). *Threatened Birds of Africa and Related Islands*. The IUCN Red Data Book, Part 1. Cambridge: International Council for Bird Preservation and International Union for Conservation of Nature and Natural Resources.

DALE, I. R. (1954). Forest spread and climatic change in Uganda during the Christian era. *Empire Forest Review* 33, 23–9.

DAVIDSON, B. (1961). *Old Africa Rediscovered*. London: Gollancz.

DE BEER, J. H. & MCDERMOTT, M. J. (1989). *The Economic Value of Non-Timber Forest Products in South East Asia*. Amsterdam: Netherlands IUCN Committee.

DIAMOND, A. W. (1985). The selection of critical areas and current conservation efforts in tropical forest birds. In *Conservation of Tropical Forest Birds*, ed. A. W. Diamond and T. E. Lovejoy, pp. 33–48. ICBP Technical Publication No. 4. Cambridge: ICBP.

DIAMOND, A. W. & HAMILTON, A. C. (1981). The distribution of forest passerine birds and quaternary climatic change in Tropical Africa. *Journal of Zoology (London)* 191, 379–402.

DOBSON E. B. (1955). Comparitive land tenure of ten Tanganyika tribes. *Tanganyika Notes and Records* 38, 31–8.

DOSSO, H., GUILLAUMET, J. L. & HADLEY, M. (1981). The Tai project: land use problems in a tropical rain forest. *Ambio* 10, 120–5.

DOUTE, R., OCHANDA, N. & EPP, H. (1981). *Forest Cover Mapping in Kenya Using Remote Sensing Techniques*. Technical Report Series 30, KREMU, Ministry of Environment and Natural Resources, Nairobi.

DUNNE, T. (1979). Sediment yield and land use in tropical catchments. *Journal of Hydrology* 42, 281–300.

EAST, R. (1988). *Antelopes, Global Survey and Regional Action Plans*. Part I. East and North East Africa. Gland: IUCN.

EICHHORN, A. (1911–1923). *Beiträge zur Kenntnis der Washambaa*. Nachhinter lassener Aufzeichnungen von A. Karaset I–IV. Baessler Archiv 1, 3, 7, 8.

ELTRINGHAM, S. K. (1980). *Recommendations for a Comprehensive Wildlife Research Programme for Tanzania*. Gland: UNEP/IUCN report.

EPP, H. (1984). Monitoring forest cover change in Kenya. In *Endangered Resources for Development*, pp. 74–81. Workshop Proceedings, National Environment Secretariat, Nairobi.

FAO (1976). *Hydrological Techniques for Upstream Conservation*. FAO Conservation Guide 2. Rome: Food and Agriculture Organization.

FAO (1981). *Environmental Impact of Forestry*. FAO Conservation Guide 7. Rome: FAO.

FAO (1982a). *Africa: Regional Tables of Production 1970–1980. Trade and Consumption of Forest Products. Supplement to Yearbook of Forest Products.* Rome: FAO.

FAO (1982b). *Tropical Forest Resources: Africa, Tanzania*, pp. 491–504. Rome: Forest Resources Division, FAO.

FAO (1985). *Strategies, Approaches and Systems in Integrated Watershed Management*. Rome: FAO.

FARLER, A. (1879). Mlinga Mission Reports. *See* Hamilton, A. C. (1990).

FAYAD, V. C. & FAYAD, C. C. (1980). An ecological survey of the Nguruman Range. Mimeo. Nairobi, Kenya.

FEIERMAN, S. (1970). The Shambaa Kingdom: a history. PhD thesis, North-Western University, Evanstown, Illinois, USA.

FINNMAP-SILVESTRE (1988). East Usambara Mountains, Amani Forest Inventory and Management Plan Project (3 vols). Helsinki: FINNIDA.

FLEURET, A. (1979). The role of wild foliage plants in the diet, a case study of Lushoto, Tanzania. *Ecology, Food and Nutrition* 8, 87–93.

FLEURET, A. (1980). Non food uses of plants in Usambara. *Economic Botany* 34(4), 320–33.

FLEURET, P. & FLEURET, A. (1978). Fuelwood use in a peasant community: a Tanzanian case study. *Journal of Developing Areas* 12, 315–22.

FOSBROOKE, H. A. & SASSOON, H. (1965). Archaeological remains on Kilimanjaro. *Tanzania Notes and Records* 64, 62–4.

FRANKEL, O. H. & SOULÉ, M. E. (1981). *Conservation and Evolution*. Cambridge: Cambridge University Press.

GOLIBER, T. J. (1985). Sub-Saharan Africa: population pressures and development. *Population Bulletin* 40, 1–47.

GOVERNMENT OF TANGANYIKA (1946). *Annual Report of the Forest Department*. Dar es Salaam.

GOVERNMENT OF TANGANYIKA (1958a). *Annual Report of the Forest Department*. Dar es Salaam.

GOVERNMENT OF TANGANYIKA (1958b). *The Tanganyika Handbook*. Dar es Salaam: Government Printer.

GOVERNMENT OF TANGANYIKA (1959). *Annual Report of the Forest Department*. Dar es Salaam.

GOVERNMENT OF TANZANIA (1976). *Atlas of Tanzania*, 2nd edn. Dar es Salaam: Surveys and Mapping Division.

GOVERNMENT OF TANZANIA (1979). *Report of Population Census, 1978*. Dar es Salaam: Bureau of Statistics.

GOVERNMENT OF TANZANIA (1986). *Draft Forest Policy Document*. Dar es Salaam: Ministry of Lands, Natural Resources and Tourism.

GOVERNMENT OF TANZANIA (1989). *Tanzania Forest Action Plan*. Dar es Salaam: Ministry of Lands, Natural Resources and Tourism.

GREENWAY, P. J. (1934). Amani Plantations. MS. East African Agriculture and Forestry Organisation, Nairobi.

GREENWAY, P. J. (1955). Ecological observations on an extinct East African volcanic mountain. *Journal of Ecology* 43, 544–63.

GRIFFITHS, J. F. K. (1962). The climate of East Africa. In *The Natural Resources of East Africa*, ed. E. W. Russell, pp. 77–87. Nairobi: East African Literature Bureau.

GRUBB, P. (1982). Refuges and dispersal in the speciation of African forest mammals. In *Biological Diversification in the Tropics*, ed. G. T. Prance, pp. 537–52. New York: Columbia University Press.

GUPPY, N. (1984). Tropical deforestation: a global view. *Foreign Affairs* (Spring 1984), pp. 928–65.

HALL, J. B. (1983). Positive management for strict natural reserves: reviewing effectiveness. *Forest Ecology and Management* 7, 57–66.

HALL, J. B. (1984). Ecological and taxonomic knowledge and natural resource utilisation in Tanzania: a survey of position, problems and prospects. In *Proceedings of a Symposium on the Role of Biology in Development*, University of Dar es Salaam, September 1983, ed. J. Middleton, A. Nikundiwe, F. Banyikwa and J. R. Mainoya, pp. 22–9.

HALL, J. B. (1985). *Mazumbai Forest, Tanzania: Report on Large Tree Survey 1981–1984*. Report to Department of Forestry, Tanzania: University of Dar es Salaam.

HALL, J. B. (1986). Luhomero Massif, Iringa Region, Tanzania. Reconnaissance Vegetation Survey in August 1985. Mimeo report. Department of Forestry. University College of North Wales, Bangor.

HALL, J. B. (1988). The Conservation Role of Catchment Forest. Paper presented at the Seminar on the Management of Tropical Rain Forest, Belize.

HALL, J. B. & RODGERS, W. A. (1986). Pole cutting pressure in Tanzania forests. *Journal of Forest Ecology and Management* 14, 133–40.

HAMILTON, A. C. (1974). Distribution patterns of forest trees in Uganda and their historical significance. *Vegetatio* 29, 21–35.

HAMILTON, A. C. (1982). *Environmental History of East Africa: A Study of the Quaternary*. London: Academic Press.

HAMILTON, A. C. (1984). *Deforestation in Uganda.* Nairobi: Oxford University Press.

HAMILTON, A. C. (1989). African forests. In *Tropical Rain Forest Ecosystems. Ecosystems of the World,* Vol. 14B, ed. H. Lieth and M. J. A. Werger, pp. 155–82. New York: Elsevier.

HAMILTON, A. C. (1990). History of resource use and utilisation and management: the precolonial period. In *Forest Conservation in the Eastern Usambara Mountains, Tanzania,* ed. A. C. Hamilton and R. Bensted-Smith, pp. 35–8. Tropical Forestry Programme. Gland: IUCN.

HAMILTON, A. C. & BENSTED-SMITH, R. (ed.) (1990). *Forest Conservation in the East Usambara Mountains, Tanzania.* Tropical Forestry Programme. Gland: IUCN.

HAMILTON, A. C. & MWASHA, I. V. (1990). History of resource utilisation and management: The German rule, British rule and post independence. In *Forest Conservation in the East Usambara Mountains, Tanzania,* pp. 39–56. Tropical Forestry Programme. Gland: IUCN.

HARRIS, L. D. (1984). *The Fragmented Forest.* Chicago: University of Chicago Press.

HAWTHORNE, W. D. (1984). Ecological and biogeographical patterns in the coastal forests of East Africa. DPhil thesis, University of Oxford.

HEDBERG, I. (1979). Possibilities and needs for conservation of plant species and vegetation in Africa. In *Systematic Botany, Plant Utilisation and Biosphere Conservation,* ed. I. Hedberg, pp. 80–105. Stockholm, Sweden: Alanqvist & Wiksell International.

HEDBERG, I. & HEDBERG, O. (ed.) (1968). Conservation of vegetation in Africa south of the Sahara. *Acta Phytogeographica Suecica* 54, 1–320.

HELLENIUS, K. (1985). Future for East Usambaras. *Oryx* 20, 249–50.

HELLIWELL, D. R. (1976). The extent and location of nature conservation areas. *Environmental Conservation* 3, 255–8.

HILL, D. B. (1985). Forest fragmentation and its implications in central New York. *Forest Ecology and Management* 12(2), 113–28.

HITCHINS, D. E. (1907). Report on the Forests of Kenya. *Colonial Reports,* Misc. No. 41.

HOBBS, R. J., SAUNDERS, D. A. & HUSSEY, B. M. T. (ed.) (1990). *Nature Conservation – the Role of Corridors.* Sydney, Australia: Surrey–Beatty.

HONORE, E. J. (1962). Forestry in East Africa: Kenya. In *The Natural Resources of East Africa,* ed. E. W. Russell, pp. 117–21. Nairobi: East African Literature Bureau.

HOOVER, R. L. & WILLS, D. L. (ed.) (1987). *Managing Forested Lands for Wildlife.* Denver, Colorado: Colorado Division of Wildlife.

HOSIER, R. H. (1988). The Economics of deforestation in eastern Africa. *Economic Geography* 64(2), 121–36.

HOWARD, P. (1991). *Nature Conservation in Uganda's Tropical Forest Reserves.* IUCN Tropical Forest Programme Books, No. 3. Gland: IUCN.

HOWARD, W. J. (1990). Forest conservation and sustainable hardwood logging. *Mitteilungen aus dem Institut für allgemeine Botanik in Hamburg* 23A, 31–7.

HOWELL, K. M. (1981). Pugu Forest Reserve: biological values and development. *African Journal of Forestry* 19, 73–81.

HOWLETT, D. (1991). *Environmental Management: the Content of Social and Environmental Studies for Plantation Management, Impact Assessment and Environmental Audits.* Objective 9. Tree Plantation Review. London: Shell International and World Wide Fund for Nature.

HUGHES, F. M. R. (1984). A comment on the impact of development schemes on the floodplain forests of the Tana River of Kenya. *Geographical Journal* 150(2), 230–44.

HUXLEY, J. (1932). *Africa View.* London: Chatto and Windus.

IIED/IRA (1992). *An Assessment of the Environmental Impact of the Kilombero Valley Hardwood Project, Tanzania.* International Institute for Environment and Development, London; and Institute of Resource Assessment, University of Dar es Salaam, Tanzania.

IUCN (1963). *Conservation of Nature and Natural Resources in Modern African States.* Morges (Gland), Switzerland: International Union for Conservation of Nature and Natural Resources.

IUCN (1980). *World Conservation Strategy: Living Resource Conservation for Sustainable Development.* Gland: IUCN/UNEP/WWF.

IUCN/WWF (1982). *Tropical Forests and Primates Programme, East Africa.* Documentation. Gland: IUCN.

IUCN/CDC (1985). *Agricultural Development and Environmental Conservation in the East Usambara Mountains, Tanzania.* Mission Report of Conservation for Development Center. Gland: IUCN.

IUCN (1987). *IUCN Directory of Afrotropical Protected Areas.* Gland: IUCN.

IUCN (1992). *An Atlas of African Rain Forests* Gland: IUCN and London: MacMillan.

IVERSEN, S. T. (1991). The Usambara Mountains, NE Tanzania: phytogeography of the vascular plant flora. *Acta Universitatis Upsaliensis, Symbolae Botanicae Upsaliensis XXIX (3), 1–234.*

JANZEN, D. H. (1983). No park is an island: increase in interference from outside as park size decreases. *Oikos* 41, 402–10.

JOHNS, A. D. (1985). Selective logging and wildlife conservation in tropical rain forest: problems and recommendations. *Biological Conservation* 31, 355–75.

JORDAN, W. R., PETERS, R. L. & ALLEN, E. B. (1988). Ecological restoration as a strategy for conserving biological diversity. *Environmental Management* 12(1), 55–72.

KELSEY, M. G. & LANGTON, T. E. S. (1984). *The Conservation of the Arabuko–Sokoke Forest, Kenya.* ICBP Study Report No. 4. Cambridge: ICBP.

KIELLAND, J. (1983). A report on the butterfly fauna in the montane and lowland forests of eastern Tanzania. Unpublished Manuscript.

KIGOMO, B. N., SAVILL, P. S. & WOODELL, F. R. (1990). Forest composition and its regeneration dynamics: a case study of a semi-deciduous tropical forest in Kenya. *African Journal of Ecology* 28(3), 174–288.

KINGDON, J. (1982). *East African Mammals: An Atlas of Evolution in Africa*, Vol. III, Parts C and D. London: Academic Press.

KINGDON, J. (1990). *Island Africa: The Evolution of Africa's Rare Animals and Plants.* London: Collins.

KINOTI, W. E. & WHITE, M. G. (1981). *Pitsawing in the Pare Mountains of Tanzania.* Record Number 21, Division of Forestry, University of Dar es Salaam.

KJEKSHUS, H. (1977). *Ecology Control and Development in Eastern Africa.* Nairobi: Heinemann Educational Books.

KOKWARO, J. O. (1976). *Medicinal Plants of East Africa.* Nairobi: East African Literature Bureau.

KOKWARO, J. O. (1988). Conservation status of the Kakamega Forest in Kenya. *Monogr. Syst. Bot. Missouri Botanical Garden* 25, 471–89.

KRISHNAMURTHY, A. V. R. G. (1983). *Forests and Wildlife in India.* Madras: COSTED, Indian Institute of Technology.

KURJI, F. (1980). *Human Population Densities and Their Changes Around the Major Conservation Areas of Tanzania.* BRALUP Research Paper 51, University of Dar es Salaam.

LAMBRECHT, F. L. (1966). Some principles of tsetse control and land use with emphasis on wildlife husbandry. *East African Wildlife Journal* 4, 89–98.

LIND, E. M. & MORRISON, M. E. S. (1974). *East African Vegetation.* London: Longman.

LOVEJOY, T. E. (1984). Biosphere resources. The size question. In *Conservation, Science and Society*, pp. 146–51. Paris: UNESCO/UNEP.

LOVEJOY T. E. (1986). Sizing the Safety Factor. *IUCN Bulletin* 17(1–3), 21–2.

LOVEJOY, T. E. & OREN, D. (1981). The minimum critical size of ecosystems. In *Forest Island Dynamics in Man Dominated Landscapes*, ed. R. L. Burgess and D. M. Sharpe, pp. 7–13. Ecological Studies 41. New York: Springer-Verlag.

LOVETT, J. (1983). *Allanblackia stuhlmannii* and its potential as a basis for soap production in Tanzania. Mimeo. Department of Chemical Engineering, University of Dar es Salaam.

LOVETT, J. (1985). *An overview of the moist forests of Tanzania.* Research Monograph, Utafiti. Tanzania National Scientific Research Council, Dar es Salaam.

LOVETT, J. C. (1990). Classification and status of the moist forest of Tanzania. *Mitteilungen aus dem Institut für allgemeine Botanik in Hamburg* 23A, 287–300.

LUCAS, G. L. (1968). Kenya. *Conservation of Vegetation in Africa South of the Sahara*, ed. I. Hedberg and O. Hedberg, pp. 150–63. Stockholm: Acta Phytogeographica Suecica, Vol. 54.

LUCAS, P. H. C. (1990). Parks and sustainable development: a global perspective. *Parks* 1(1), 3–7.

LUGO, A. & BROWN, S. (1984). Conserving tropical rain forest ecosystems: assembling coverage and assigning priorities. In *Conservation, Science and Society*, pp. 37–43. Paris: UNESCO/UNEP.

LUNDGREN, B. (ed.) (1975). *Land Use in Kenya and Tanzania.* Stockholm: Royal College of Forestry.

LUNDGREN, B. (1978). Soil Conditions and Nutrient cycling under Natural and Plantation Forests in Tanzania Highlands. Department of Forest Soils, Swedish University of Agricultural Sciences, Sci. 31, Uppsala. 426pp.

LUNDGREN, B. & LUNDGREN, L. (1972). Comparison of some soil properties in one forest and two grassland ecosystems on Mount Meru, Tanzania. *Geografiska Annaler* 54A (3–4), 227–40.

LUNDGREN, B. & LUNDGREN, L. (1982). Socio-economic effects and constraints in forest management: Tanzania. In *Socio-economic Effects* and *Constraints in Tropical Forest Management*, ed. E. G. Hallsworth, pp. 43–52. London: Wiley.

LUNDGREN, L. & LUNDGREN, B. (1979). Rainfall, interception and evaporation in the Mazumbai Forest Reserve, West Usambara Mts, Tanzania. *Geografiska Annaler* 61A (3–4), 157–8.

LUNDGREN, L. (1980). Comparison of surface run off and soil loss from run off plots in forest and small scale agriculture in the Usambara Mountains, Tanzania. University of Stockholm, Department of Physical Geography. STOU–Ng. 38. 118pp.

LUSIGI, W. J. (1982). Socio-economic effects and constraints in forest management in Kenya. In *Socio-economic Effects and Constraints in Tropical Forest Management*, ed. E. G. Hallsworth, pp. 123–30. London: Wiley.

LUSIGI, W. (1985). Future directions for the Afrotropical realm. In *National Parks, Conservation and Development*, ed. J. A. McNeely and K. R. Miller, pp. 137–46. Washington, DC: Smithsonian Press.

MAAGI, Z. G. N., MKUDE, M. J. & MLOWE, E. J. (1979). *The Forest Area of Tanzania Mainland*. Forest Resource Series No. 34. Ministry of Natural Resources, Tanzania.

MABBERLEY, D. J. (1983). *Tropical Rain Forest Ecology*. Glasgow: Blackie.

MACKINNON, J. & MACKINNON, K. (1986). *Review of the Protected Areas System in the Afrotropical Realm*. Gland: IUCN.

MACKINNON, J., MACKINNON, K., CHILD, G. & THORSELL, J. (1986). *Managing Protected Areas in the Tropics*. Gland: IUCN.

MAREALLE, T. L. M. (1949). The Wachagga of Kilimanjaro. *Tanganyika Notes and Records* 32, 57–64.

MARSH, C. W. (1978). Problems of primate conservation in a patchy environment along the lower Tana River, Kenya. In *Recent Advances in Primatology*, Vol. 2. *Conservation*, ed. D. Chivers and W. Lane-Petter, pp. 85–6. London: Academic Press.

MARTIN, G. H. G. (1983). Bushmeat in Nigeria as a natural resource with environmental implications. *Environmental Conservation* 10(2), 125–33.

MATHIAE, P. E. & STEARNS, F. (1981). Mammals in forest islands in South Eastern Wisconsin. In *Forest Island Dynamics in Man Dominated Landscapes*, ed. R. L. Burgess and D. M. Sharpe, pp. 55–6. Ecological Studies 41. New York: Springer-Verlag.

MATZKE, G. M. (1976). The development of the

Selous Game Reserve. *Tanzania Notes and Records* 79–80, 37–48.

MAY, R. M. (1975). Island biogeography and the design of wildlife preserves. *Nature* 257, 177–8.

MBWANA, S. B. (1988). Land use cover changes and water yield in the Sigi River catchment, East Usambaras, Tanzania. MSc thesis, University of Wales, Bangor.

MBWANA, S. B. (1990). Management of the catchment forests with emphasis to the Eastern Arc Mountains. *Coordination Seminar on Tanzania Forestry Action Plan*, 5–7 September 1990. FINNIDA/Forestry Division, Dar es Salaam.

MCNEELY, J. A. (1984). Biosphere reserves and human ecosystems. In *Conservation, Science and Society*, pp. 492–9. Paris: UNESCO/UNEP.

MCNEELY, J. A. (1990). The future of national parks. *Environment* 32(1), 16–20, 36–40.

MCNEELY, J. A. & MACKINNON, J. R. (1989). Protected areas, development and land use in the tropics. *Resource Management and Optimisation* 7(1–4), 189–206.

MEHER-HOMJI, V. M. (1988). Effects of forests on precipitation in India. In *Forests Climate and Hydrology – Regional Impacts*, ed. E. R. C. Reynolds and F. B. Thomson. United Nations University.

MILLER, K. R. (1973). Conservation and Development of Tropical Rain Forest Areas. In *Conservation for Development*, ed. H. I. F. I. Elliot, pp. 259–70. Morges, Switzerland: IUCN.

MILLER, R. I. (1978). Applying island biogeographic theory to an East African reserve. *Environmental Conservation* 5(3), 191–5.

MLAY, W. (1982). Population Pressure in Arumeru District. Report to Arumeru Area Commissioner's Office, Arusha, Tanzania.

MNZAVA, E. M. (1980). Village afforestation, lessons of experience in Tanzania. TF/INT 271 (SWE). Rome: FAO.

MOREAU, R. E. (1935). A synecological study of Usambara, Tanganyika Territory, with particular reference to birds. *Journal of Ecology* 23, 1–43.

MORRISON, M. E. S. & HAMILTON, A. C. (1974). Vegetation and climate in the uplands of South-Western Uganda during the later Pleistocene period. II. Forest clearance and other vegetational changes in the Rukife Highlands during the past 8000 years. *Journal of Ecology* 62, 1–31.

MTOTOMWEMA, K. (1982). Conservation of lowland forests in Tanzania as sources of food for humans and a reservoir of Medicinal Plant Germplasm. Paper presented at Workshop on Forest

Conservation in Tanzania, Tanga, February 1982. Tanzania National Scientific Research Council, Dar es Salaam.

MTURI, F. A. (1991). The feeding ecology and behaviour of the Zanzibar red colobus (*Colobus badius kirkii*). PhD thesis, University of Dar es Salaam.

MUGASHA, A. G. (1982). Conservation of remnants of Tropical Rain Forests in Usambara, with emphasis on Silvicultural Research. Paper presented at Workshop on Forest Conservation in Tanzania, Tanga, February 1982. Tanzania National Scientific Research Council, Dar es Salaam.

MWAGIRU, W. (1982). The forest resource in Kenya in relation to other natural resources. In *Socio-economic Effects and Constraints in Tropical Forest Management*, ed. E. G. Hallsworth, pp. 131–41. Chichester: Wiley.

MYERS, N. (1979). *The Sinking Ark*. Oxford: Pergamon Press.

MYERS, N. (1982). Forest refuges and conservation in Africa. In *Biological Diversification in the Tropics*, ed. G. T. Prance, pp. 658–72. New York: Columbia University Press.

MYERS, N. (1984). *The Primary Source: Tropical Forests and Our Future*. New York: W. W. Norton & Co.

NEWMARK, W. D. (1991a). Tropical forest fragmentation and the local extinction of understory birds in the Eastern Usambara Mountains, Tanzania. *Conservation Biology* 5, 67–78.

NEWMARK, W. D. (ed.) (1991b). *The Conservation of Mount Kilimanjaro*. IUCN Tropical Forest Series. Gland: IUCN.

NG, F. S. P. (1983). Ecological principles of tropical lowland rain forest conservation. In *Tropical Rain Forest: Ecology and Management*, ed. S. L. Sutton, T. C. Whitmore and A. C. Chadwick. Oxford: Blackwell.

NORDIC REVIEW MISSION (1979). *The Forestry Sector of Tanzania*. Ministry of Natural Resources, Dar es Salaam.

OATES, J. (1985). *Action Plan for African Primate Conservation, 1986–90*. IUCN/SSC Primate Specialist Group. Gland: IUCN.

ODNER, K. (1971). A preliminary report on an archeological survey on the slopes of Mount Kilimanjaro. *Azania* 6, 131–50.

OLDFIELD, S. (1987). *Buffer Zone Management in Tropical Moist Forests: Case Studies and Guidelines*.

IUCN Tropical Forest Series. Gland: IUCN.

OPENSHAW, K. (1984). Use and management of indigenous woody plant species for energy. In *Endangered Resources for Development*, pp. 100–9. Nairobi, Kenya: National Environment Secretariat.

OXFORD UNIVERSITY (1981). Kenya – an Ethnobotanical Perspective. Botany School, University of Oxford.

PANWAR, H. S. (1984). What to do when you have succeeded, Project Tiger ten years later. In *National Parks, Conservation and Development*, ed. J. A. McNeely and K. R. Miller, pp. 183–9. Washington, DC: Smithsonian Institution Press.

PARRY, M. S. (1962). Progress in the protection of stream-source areas in Tangayika. *East African Agricultural and Forestry Journal* 43 (Special Issue), 104–6.

PEREIRA, H. C. (1973). *Land Use and Water Resources*. Cambridge: Cambridge University Press.

PICKETT, S. T. A. & THOMPSON, J. N. (1978). Patch dynamics and the design of nature reserves. *Biological Conservation* 13, 27–37.

POCS, T. (1976a). Bioclimatic studies in the Uluguru mountains (Tanzania, East Africa). II. Correlations between orography, climate and vegetation. *Acta Botanica Academiae Scientiarum Hungaricae* 22, 163–83.

POCS, T. (1976b). Vegetation mapping in the Uluguru mountains (Tanzania, East Africa). *Boissiera* 24, 477–98.

POCS, T. (1980). The epiphytic biomass and its effect on the water balance of two rain forest types in the Uluguru Mountains. *Acta Botanica Academiae Scientarum Hungaricae* 26(1–2), 143–67.

POFFENBERGER, M. (ed.) (1990a). *Keepers of the Forest, Land Management Alternatives in South-east Asia*. Connecticut, USA: Kumarian Press.

POFFENBERGER, M. (1990b). *Joint Management of Forest Lands*. New Delhi: Ford Foundation.

POHJONEN, V. & PUKKALA, T. (1990). *Eucalyptus globulus* in Ethiopia. *Forest Ecology and Management* 36, 19–31.

POLHILL, R. M. (1968). Tanzania. In *Conservation of Vegetation in Africa South of the Sahara*, ed. I. Hedberg and O. Hedberg, pp. 166–78. Stockholm: Acta Phytogeographica Suecica, Vol. 54.

POLHILL, R. M. (1988). East Africa (Kenya, Tanzania, Uganda). In *Floristic Inventory of Tropical Countries*, ed. D. G. Cambell and H. D. Hammond, pp. 218–31. New York: Science Publication Department, New York Botanical Gardens.

POORE, D. (1988). *Natural Forest Management for Sustainable Timber Production*. London: IIED.

POORE, D. & SAYER, J. (1987). *The Management of Tropical Moist Forest Lands: Ecological Guidelines*. IUCN Tropical Forest Series. Gland: IUCN.

POYRY, J. (1980). Long Term Development Plan for Wood Industries in Tanzania. J. Poyry Report, 1980-2008, to Tanzania Wood Industries Corporation, Dar es Salaam.

PRANCE G. T. (ed.). (1982). *Biological Diversification in the Tropics*. New York: Columbia University Press.

PRESCOTT-ALLEN, R. (1984). *Genes from the Wild*. London: Earthscan.

PROCEEDINGS FROM THE FIRST INTERNATIONAL WORKSHOP FOR THE CONSERVATION AND MANAGEMENT OF AFRO-MONTANE FORESTS. Cyangugu, Rwanda. (1991). Wildlife Conservation International, New York; and ORTPN Rwanda.

RANNEY, J. W., BRUNER, M. C. & LEVENSON, J. B. (1981). Importance of edge in the structure and dynamics of forest islands. In *Forest Island Dynamics in Man Dominated Landscapes*, ed. R. C. Burgess and D. M. Sharpe, Ecological Studies 41. New York: Springer-Verlag.

RAPP, A., BERRY, L. & TEMPLE, P. (ed.) (1973). Studies of soil erosion and sedimentation in Tanzania. *Geografiska Annaler* 54A (3-4), 105-379.

ROBERTSON, S. A. (1984). The status of Kaya forests. In *Endangered Resources for Developments*. The proceedings of a workshop on the management of plant communities in Kenya. National Environment Secretariat, Nairobi.

ROCHE, L. & DOUROJEANNI, M. J. (1984). *A guide to In Situ Conservation of Genetic Resources of Tropical Woody Species*. Rome: Forest Resources Division, FAO.

RODGERS, W. A. (1975). Past Wangindo settlements in the eastern Selous Game Reserve. *Tanzania Notes and Records* 77/78, 21-6.

RODGERS, W. A. (1981). The distribution and conservation status of Colobus Monkeys in Tanzania. *Primates* 22(1), 33-45.

RODGERS, W. A. (1983). A note on the distribution and conservation of *Oxystigma msoo* Harms. *Bulletin du Jardin Botanique National de Belgique* 53, 161-4.

RODGERS, W. A. (1984). The teaching of Wildlife at the University of Dar es Salaam, Tanzania. Rome: African Forestry Commission, FAO.

RODGERS, W. A. (1991). *Managing Forests for Biodiversity in India: A Review of Concepts and Practices*. Report to the Wildlife Institute of India, Dehra Dun, India.

RODGERS, W. A. (1992). A history of the gum copal trade in Tanganyika. *Tanzania Notes and Records* manuscript.

RODGERS, W. A. & HALL, J. B. (1985). Tree planting programs for Mwanihana Forest Reserve Villages. A draft project proposal for EEC village project funding. Forestry Division, Ministry of Natural Resources and Tourism, Dar es Salaam.

RODGERS, W. A., HALL, J. B. & MWASUMBI, L. (1983). Kimboza Forest Reserve. Report to the Ministry of Natural Resources, Tanzania.

RODGERS, W. A., HALL, J. B. & MWASUMBI, L. B. (1985a). The vegetation of three coastal forests. MS, University of Dar es Salaam.

RODGERS, W. A., HALL, J. B., MWASUMBI, L. B., SWAI, I. & VOLLESEN, K. (1986). The conservation status and values of Ngezi Reserve, Pemba Island, Tanzania. Report to the Forest Department, Zanzibar, Tanzania.

RODGERS, W. A. & HOMEWOOD, K. M. (1982a). Species richness and endemism in the Usambara Mountain Forests, Tanzania. *Biological Journal of the Linnean Society* 18, 197-242.

RODGERS, W. A. & HOMEWOOD, K. M. (1982b). Conservation of the Uzungwa Mountains, Tanzania. *Biological Conservation* 24, 285-304.

RODGERS, W. A., HOMEWOOD, K. M. & HALL, J. B. (1979). Magombera Forest Reserve and the Iringa Red Colobus, Tanzania. Report to the Ministry of Natural Resources, Dar es Salaam, Tanzania.

RODGERS, W. A. & LOBO, J. D. (1981). Elephant control and legal ivory exploitation 1920-1976. *Tanzania Notes and Records* 84/85, 25-54.

RODGERS, W. A., MZIRAY, W. & SHISHIRA, E. K. (1985b). The Extent of Forest Cover in Tanzania Using Satellite Imagery. *IRA Research Paper* No. 12. University of Dar es Salaam.

RODGERS, W. A., OWEN, C. & HOMEWOOD, K. M. (1982). Biogeography of African forest mammals. *Journal of Biogeography* 9, 41-54.

RODGERS, W. A. & PANWAR, H. S. (1988). *Planning a Wildlife Protected Network for India* (2 vols). Wildlife Institute of India, Dehra Dun, India.

RODGERS, W. A., STRUHSAKER, T. T. & WEST, C. C. (1984). Observations on the Red Colobus of Mbisi Forest, SW Tanzania. *African Journal of Ecology* 22, 187-94.

RODGERS, W. A. & SWAI, I. (1988). Tanzania. In *Antelopes, Global Survey and Regional Action Plans*,

Part I. East and North East Africa, ed. R. East, Gland: IUCN.

ROTH, H. H. (1986). We all want the trees: Resource Conflict in Tai National Park, Ivory Coast. pp. 122–9.

SAHARIA, V. B. (1985). Human dimensions in wildlife management: the Indian experience. In *National Parks, Conservation and Development*, ed. J. A. McNeely and K. R. Miller, pp. 190–6. Washington, DC: Smithsonian Institution Press.

SANGSTER, R. G. (1962). Forestry in East Africa: Tanganyika. *The Natural Resources of East Africa*, ed. E. W. Russell, pp. 122–5. Nairobi: East African Literature Bureau.

SAUNDERS, D. A., ARNOLD, G. W., BURBIDGE, A. A. & HOPKINS, A. J. M. (ed.) (1987). *Nature Conservation: The Role of Remnants of Native Vegetation*. New South Wales, Australia: CSIRO and Surrey Beatty & Sons.

SAUNDERS, D. A., HOBBS, R. J. & MARGULES, C. R. (1991). Biological consequences of ecosystem fragmentation: a review. *Conservation Biology* 5(1), 18–32.

SAYER, J. A. (1991). Buffer zones in tropical forested regions. Paper Presented at Regional Expert Consultation on Management of Protected Areas in Asia Pacific Region, *Tiger Paper* 18 (4), 10–17; FAO-RAPA, Bangkok, Thailand.

SCHABEL, H. G. (1990). Tanganyika Forestry under German Colonial Administration, 1891–1919. *Forest and Conservation History* 1, 130–41.

SCHMIDT, P. R. (1990). Early exploitation and settlement in the Usambara Mountain. In *Forest Conservation in the East Usambara Mountains, Tanzania*, ed. A. C. Hamilton and R. Bensted-Smith, pp. 75–78. Tropical Forestry Programme. Gland: IUCN.

SHELTON, N. (1985). Logging versus the natural habitat in the survival of tropical forests. *Ambio* 14(1), 39–41.

SHEPHERD, G. (1991). The communal management of forests in the semi-arid and sub-humid regions of Africa: past practice and prospects for the future. *Development Policy Review* 9, 151–76.

SIEBENLIST, T. (1914). *Forest wirtschaft in Deutsch-Ostafrika*. Berlin: Paul Parey Verlag.

SIMBERLOFF, D. S. & ABELE, L. G. (1976). Island biogeography, theory and conservation practice. *Science* 191, 285–6.

SINCLAIR, A. R. E. (1983). Management of conservation areas as ecological baseline controls.

In *Management of Large Animals in African Conservation Areas*, ed. R. N. Owen-Smith, pp. 13–22. Pretoria: Haum.

SINCLAIR, A. R. E. & NORTON-GRIFFITHS, M. (ed.) (1979). *Serengeti: Dynamics of an Ecosystem*. Chicago: University of Chicago Press.

SKORUPA, J. P. & KASENENE, J. M. (1984). Tropical forest management: can rates of natural tree falls help guide us? *Oryx* 18(2), 96–101.

SOPER, R. (1967). Iron age sites in north-east Tanzania. *Azania* 2, 19–36.

SOULÉ, M. E. (ed.) (1986). *Conservation Biology: Science of Scarcity and Diversity*. Sunderland, MA: Sinauer Associates.

SPEARS, J. (1987). Kenya Forestry Sector Review: key issues. Discussion Paper, World Bank, Nairobi and Washington, DC.

STANLEY-PRICE, M. R. (1989). Reconstructing ecosystems. In *Conservation for the Twenty-First Century*, ed. D. Western and M. Pearl, pp. 210–18. Oxford: Oxford University Press.

STUART, S. N. (1981). A comparison of the avifaunas of seven East African Forest Islands. *African Journal of Ecology* 19, 133–51.

STUART, S. (1983). A forest conservation plan for the Usambara Mountains, Chapter 9. PhD Thesis, University of Cambridge.

STRUHSAKER, T. T. (1987). Forestry issues and conservation in Uganda. *Biological Conservation* 39, 209–34.

STRUHSAKER, T. T. (1991). Conservation of the Zanzibar Red Colobus Monkey. Report of a field trip; June–July 1991. Department of Wildlife and Range Sciences, University of Florida.

STRUHSAKER, T. T. & LELAND, L. (1980). Observations on two rare and endangered populations of red colobus monkeys in East Africa: *Colobus badius gordonorum* and *C. b. kirkii*. *African Journal of Ecology* 18, 191–216.

SUMARDJA, E. A. (1984). *In situ* conservation of tropical forest genetic resources, Indonesia's experience. In *Conservation, Science and Society*, pp. 271–5. Paris: UNESCO/UNEP.

SUTTON, J. (1966). The archaeology and early peoples of the Highlands of Kenya and Northern Tanzania. *Azania* 1, 37–57.

SWAI, I. (1983). The conservation status of small Bovidae in Zanzibar. MSc thesis, University of Dar es Salaam.

SYNOTT, T. J. (1977). *Monitoring Tropical Forests: A Review with Special Reference to Africa*. Special

Report of MARC of the Scientific Committee on the Problems of the Environment, University of London.

TEMPLE, P. H. (1972). Soil and water conservation policies in the Uluguru Mountains, Tanzania. *Geografiska Annaler* **54A** (3–4), 110–23.

TEMU, A. B. (1984). Soil and water conservation in forestry. In *Proceedings of the Symposium on The Role of Biology in Development*, pp. 59–62. University of Dar es Salaam, Tanzania.

THELEN, K. D. & CHILD, G. S. (1984). Biosphere reserves and rural development. In *Conservation, Science and Society*, pp. 470–7. Paris: UNESCO/UNEP.

THOMAS, J. W. (1979). *Wildlife Habitat in Managed Forest*. USDA Forest Service Agricultural Handbook. Washington, DC: US Department of Agriculture.

TSINGALIA, M. H. (1990). Habitat disturbance, severity and patterns of abundance in Kakamega Forest, Western Kenya. *African Journal of Ecology* **28**(3), 213–26.

TURNBULL, C. M. (1961). *The Forest People*. London: Chatto and Windus.

UNESCO/UNEP/FAO (1978). *Tropical Forest Ecosystems*. Paris: UNESCO.

USHER, M. B. (1991). Biological invasions into tropical nature reserves. In *Ecology of Biological Invasion in the Tropics*, ed. P. S. Ramakrishnan, pp. 21–34. New Delhi: International Scientific Publications.

VON DROSTE, B. (1984). How UNESCO's Man and the Biosphere Programme is Contributing to Human Welfare. In *Conservation, Science and Society*. Paris: UNESCO/UNEP.

VOLKENS, G. (1897). *Zur Frage der Aufforstung in Deutsch-Ost Africa*. Notizblatt des (Konigl.) Botanischen Gartens and Museums zu Berlin **2**(11), 12–20.

VOLLESEN, K. (1980). An annotated checklist of the vascular plants of the Selous Game Reserve, Tanzania. *Opera Botanica* **59**, 1–71.

WARDELL, D. A. (1990). Community participation in the management and conservation of the forest resources of the East Usambara Mountains. TFAP Coordination Seminar for Eastern Arc Forest Management Programme, Dar es Salaam, September 1990. Ministry of Natural Resources.

WATSON, J. R. (1972). Conservation problems, policies and the origins of the Mlalo Rehabilitation Scheme, Usambara Mountains, Tanzania. *Geografiska Annaler* **54A** (3–4), 221–6.

WAY, M. J. (1977). Pest and disease status in mixed stands versus monocultures, the relevance of ecosystem stability. *18th Symposium of the British Ecological Society*. Oxford: Blackwell.

WEBER, W. A. (1987). Socioecologic factors in the conservation of Afromontane Forest Reserves. *Monographs in Primatology* **9**, 205–29.

WHITCOMB, R. F. (1981). Effects of forest fragmentation on avifauna of the eastern deciduous forests. In *Forest Island Dynamics in Man Dominated Landscapes*, ed. R. L. Burgess and D. M. Sharpe. Ecological Studies 41. New York: Springer-Verlag.

WHITE, F. (1983). *The Vegetation of Africa*. Paris: UNESCO.

WHITEHEAD, D. (1982). Ecological aspects of natural and plantation forests. *Forestry Abstracts* **43**(10), 615–24.

WHITMORE, T. (1984). *Tropical Rain Forests of the Far East*, 2nd edn. Oxford: Clarendon Press.

WILCOX, B. S. (1984). *In Situ Conservation of Genetic Resources: Determinants of Minimum Area Requirements*. Washington, DC: Smithsonian Institution Press.

WOOD, P. J. (1965). The forest glades of Kilimanjaro. *Tanganyika Notes and Records* **64**, 56–61.

WOOD, P. J. (1990). Deflecting pressure on forests: use of agroforestry in establishing buffer zones to protect nature reserves. *Mitteilungen aus dem Institut für allgemeine Botanik in Hamburg* **23A**, 99–107.

WORLD BANK (1988). *Forestry Sector Review: Kenya*, Vol. I. Washington, DC: World Bank.

WORLD BANK/ODA/FAO (1990). *Tanzania: Forest Resources Management Project*. A Report. Dar es Salaam.

YOUNG, R. & FOSBROOKE, H. (1960). *Land and Politics Among the Luguru of Tanganyika*. London: Routledge & Kegan Paul.

YOUNG, T. P. (1984). Kenya's Indigenous Forests. Status Threats and Prospects for Conservation Action. East Africa Office, Nairobi: WWF/IUCN.

Appendix 14.1
Summary of biological values for forest areas of conservation significance

Note that further details of many of these sites are given in the reviews by Lucas (1968), Beentje (1990a) and Polhill (1988) for Kenya; and Polhill (1968, 1988) and Lovett (1985, 1990) for Tanzania. Details of areas gazetted as National Park or National Reserve are given in IUCN, 1987.

Coastal and lowland communities
(A) Islands

1. Jozani Forest, Zanzibar. FR, and potential National Park. 26 km². Swamp forest (*Calophyllum, Anthocleista, Bridelia, Elias, Pandanus*, etc.), surrounded by forest and thicket communities on coral rag (*Manilkara, Mimusops, Terminalia,* Annonaceae, Celastraceae, etc.). Major habitat for endemic red colobus monkey, plus 3 endemic bovid subspecies. Struhsaker & Leland, 1980; Struhsaker, 1991; Mturi, 1991.

2. Ngezi Forest, Pemba. FR, and potential NP. 14 km². Diverse forest communities from almost pure *Antiaris, Odyendea* stands to 40 m height through *Uapaca, Syzygium, Pachystela* mixed forest to swamp forest of *Barringtonia* and *Samadera indica* (the only African locality), to mangrove. Many exceptional distributional and altitudinal records, e.g. *Polyscias*; and linkages to Madagascar. Three Pemba endemic plants. Poorly explored. Rodgers *et al.*, 1986; Beentje 1990a.

3. Coral Rag Forests, Mafia. NO STATUS. Mixture of lowland evergreen forest and coral rag thicket. Better forest already cleared for agriculture. Still extensive mangrove. Need for survey and choice of areas important for conservation interests. Rodgers *et al.*, 1988.

(B) Mainland, Tanzania

1. Kisiju Forest, Coast Region. NO STATUS. <3 km². Dry lowland evergreen forest, dominated by *Hymenaea* (gum copal), merging into *Barringtonia* and mangrove communities. Rapidly being converted for cultivation. Important as example of once widespread forest type of historical commercial value. Rodgers *et al.*, 1983; Hawthorne, 1984; Burgess *et al.*, 1990.

2. Pugu Forest Reserve, Coast Region. <4 km² good forest. Biologically well known (e.g. Howell, 1981; Hawthorne, 1984). Important coast bird locality (Collar & Stuart, 1985), many plant endemics (e.g. *Baphia, Millettia*). Urgent need of improved protection from mining, plantation development, encroachment, and illegal fuel and pole collection. Growing support from Tanzania Wildlife Conservation Society.

3. Pande Forest (now Game) Reserve, Coast Region. <10 km² good forest. Distinctive botanically, several very local endemics, including many Annonaceae. Diverse avifauna, relict small bovid fauna. Poorly described (see Hawthorne, 1984; Collar & Stuart, 1985).

4. Kiono, Kazimzumbwie, Zaraninge Forest Reserves, Coast Region. Key examples of fragmented patches of coastal forests. Hawthorne, 1984; Burgess *et al.*, 1990.

5. Rufiji River Delta, FRs, Coast Region. Mainly mangroves (Forest Dept. 1990), 100 km². Area long considered as a Biosphere Reserve and marine park.

6. Kichi Hills, Coast Region. NO STATUS. Extensive areas of dry closed forest and evergreen thicket. Several endemic plants (e.g. *Pteleopsis*: Vollesen, 1980). Nearby **Matumbi Hills, Coast–Lindi Region**, have deep moist ravines; recent exploration found *Saintpaulia* (*African violet*) populations, and many new and significant bird species records. Burgess *et al.*, 1990.

7. Lindi Area (Baobab forest, Lake Lutamba), Lindi Region. NO STATUS. Lake Lutamba is the type locality for many endemic rare plants. Polhill, 1968.

8. Rondo Plateau FR, Lindi Region. <10 km² natural forest. Poorly described: see Burgess *et al.* (1990) for recent reconnaissance. *Milicia, Khaya* communities, important 'stepping-stone' site for lowland forest bird distributions.

9. Magombero FR (remaining intact areas now in Selous Game Reserve), Morogoro Region. 9 km². Groundwater forest, 300 m a.s.l. *Milicia, Khaya* (greatly disturbed by past logging), *Erythrophleum, Isoberlinia scheffleri*, many endemics (tree, shrub and herb). Avifaunal links to adjacent Mwanihana Forest in Uzungwa Mts. Rodgers *et al.*, 1979; Vollesen, 1980.

10. Kimboza FR, Morogoro Region. 4 km². Relict patch on limestone karst, exceptionally rich botanically, *Parkia, Milicia, Antiaris, Scorodophleus, Sorindeia, Elaeis, Pandanus*. 17 plant endemics, e.g. *Cynometra*, and Lepidoptera and a lizard. Plants, birds and butterflies well documented (Rodgers *et al*, 1983). Under localised pressure for fuel and poles (Hall & Rodgers, 1986). Nearby **Ruvu FR** is similar but drier, and needs survey.

11. Kitulanghalo, Dindili, Pongwe FRs, Morogoro and Coast Regions. 25 km². Isolated hills, with dry closed forest, e.g. *Scorodophleus, Manilkara*, etc. and locally, *Brachylaena*. Urgent need for more survey.

12. Ruaha River, Iringa and Morogoro Regions. Narrow fringing riverine gallery forest, increasingly fragmented. NO STATUS. Varied composition. *Diospyros, Milicia, Parkia, Terminalia, Ficus*, etc. Past migratory pathway for birds, colobines, etc. Better patches urgently need protection.

13. Rau FR, Kilimanjaro Region. 6 km^2 but <1 km^2 good forest. Groundwater forest, main locality for giant leguminous tree *Oxystigma msoo* (Rodgers, 1983). Very heavily disturbed; despite TFAP call for better status, no action yet taken.

(C) Mainland, Kenya

1. Arabuko-Sokoke Nature Reserve and FR, Coast Province. 417 km^2 of FR, 33 km^2 NR. Probable National Park site, but borders are controversial. Largest coastal forest area. *Afzelia, Brachylaena, Brachystegia, Manilkara*, etc. Very important avifaunal locality, with many endemics, e.g. Sokoke owl (Collar & Stuart, 1985). Rare mammals, e.g. golden-backed elephant shrew.

2. Boni FR, Coast Province. 183 km^2 woodland and forest. *Afzelia, Diospyros, Oldfieldia*, etc. More survey needed.

3. Kayas, the Sacred Groves of Coast Province. Scattered relict forest patches from Chale Island to Kaurna near Kilifi. Groves protected by tribal traditions. Many areas on limestone are botanically rich, with some endemics. Community dominated by *Cynometra, Gyrocarpus, Milicia, Sterculia*, etc. (Hawthorne, 1984; Robertson, 1984). Target of a major study conservation initiative of The National Museums of Kenya, the Kayas may become a National Heritage Site.

4. Diani, Jadini, Chale Point Coral Rag Forest Thickets, Coast Province. Tiny fragmented patches. NO STATUS. Adder's duiker at Chale Point. Plant endemics.

5. Shimba Hills NR and FR, Coast Province. 20 km^2 of forest patches in 200 km^2 woodland. Botanical interest; logging, plantation and tourism (around woodland wildlife interests, e.g. sable) disturbances.

6. Mrima Hill FR, Coast Province. <4 km^2 forest. Floristically distinctive, *Antiaris, Newtonia, Sterculia*, with endemics, e.g. two *Uvariodendron* spp. Needs greater protection. Lucas, 1968.

7. Taveta Forest, Coast Province. NO STATUS. Tiny, <3 km^2, patches of groundwater forest, *Albizia, Diospyros, Milicia*, etc.

8. Tana River Forests, Coast Province. Part Game Reserve, part FR. Groundwater forest on dynamic floodplain and delta. Primate values, two endemics

(IUCN, 1987). Endemic poplar. Growing pressure from villages, pastoralists, and upstream hydroschemes (Hughes, 1984). Site for massive World Bank conservation project, proposed under GEF (Global Environment Facility).

Montane and Plateau Communities

(A) Eastern Arc block mountains, Kenya

1. Taita Hills, Coast Region. NO STATUS. <3 km^2 closed forest. Biologically extremely rich, with high density of endemic plant, reptile and bird species. Forest to 2228 m a.s.l., *Albizia, Cola, Macaranga* (*Ocotea and Podocarpus* mainly cut out), *Strombosia* communities. 14 plant endemics including *Coffea, Saintpaulia.* 4 bird endemics. A priority for improved status (Beentje, 1990).

2. Kasigau FR, Coast Region. Isolated mountain forest, <2 km^2, to 1450 m asl. *Newtonia* forest, endemic *Psychotria taitensis.* (Similar to nearby **Ngulia Hills** in Tsavo NP, site of recent discovery of new *Podocarpus* sp.)

(B) Eastern Arc block mountains, Kenya

1. Pare Mountains, Kilimanjaro Region. Few small FRs, mostly on south Pare Mts, closed montane forest from 1700 to 2300 m a.s.l. Poorly documented, but thought to be of minor conservation interest. The gap between the Pares and Mt Kilimanjaro is the barrier between a southern and northern migratory element of the eastern Africa forest fauna.

2. East and West Usambara Mountains, Tanga Region. Several separate FRs, both large and small. The most important forest areas for conservation in East Africa, also of major catchment significance. Values summarised in Rodgers & Homewood, 1982a; Hamilton & Bensted-Smith, 1990. Recent detailed forest description is in Iversen, 1991. The East Usambaras are wetter with forest from <300 to 1000 m asl, *Ocotea, Cephalosphaera.* The West Usambaras are drier with more montane communities from 1000 to 2600 m asl. Forest communities are varied, internally diverse and with exceptional levels of generic and species endemism. There are several bird species of conservation concern. The increasing fragmentation and loss of continuity in forest cover is of great worry, as are the growing pressures from human settlements, tea and coffee estates, legal and illegal felling, etc. The area desperately needs higher conservation status than FR.

3. Nguru and Ukaguru Mountains, Tanga and Morogoro Regions. Separate FRs on wetter east-facing and drier west-facing slopes. Most at higher altitudes, botanically richer lowland forest virtually all cleared. Poorly documented; bird and herp data suggest minor conservation significance, but more survey needed.

4. Uluguru Mountains, Morogoro Region. Three main FRs on upper slopes and ridges of mountain range, 1500 to 2700 m a.s.l. Only the North Uluguru FR explored as closer access, but all survey curtailed in recent years. Good forest, and montane moorland, grassland, swamp communities, of extreme conservation and catchment value. No good biological review exists, (see Pocs, 1976a, b, for physical descriptions and vegetation map). High plant endemism (especially Rubiaceae, e.g. *Lasianthus*) and many significant birds, herps. There is urgent need for higher level conservation status than FR.

5. Uzungwa Mountains. Extensive block mountain range running NE–SW, only recently explored (Rodgers & Homewood, 1982b; Collar & Stuart, 1985; Lovett, 1985). Four distinct areas warrant description:

(a) Sanje Scarp, Morogoro Region. FR and now Uzungwa NP. Rises from 300 to 2250 m a.s.l. on Mwanihana Mt, the greatest continuous forest altitudinal gradient in East Africa. Wet-facing scarp, 2200 mm rain, with luxuriant forest, many endemics, still being discovered. *Cephalosphaeria, Parinari, Albizia* community giving rise, to *Syzygium, Podocarpus*, etc., very diverse. Two endemic primates, many birds and herps.

(b) Southern Uzungwa Scarp, Morogoro Region. FR. Less well known, steep scarp face rising from Kilombero Swamps, some similarity to Sanje. Both red colobus and mangabey endemics.

(c) Luhomero Massif, Iringa Region. FR, and part in Uzungwa NP. Rises to 2660 m a.s.l. Diverse forest, much *Hagenia, Ocotea, Podocarpus*, several endemics. Surrounded by montane grassland, swamp and woodland, with large mammal values.

(d) Mufindi eastern forests, Iringa Region. FR. Drier forest, distinct botanically and in bird life, endemics.

6. Southern Highlands, Mbeya Region. End of the Eastern Arc in Tanzania, confused geologically with block mountains (Porotos), volcanics (Rungwe) and older ranges (Kipengere). FRs. Diverse forest communities, not yet reviewed fully. Important orchid-rich high-altitude grasslands. Whole area under threat from dense human populations; sites of especial conservation significance need identification and protection.

7. Image FR, Iringa Region. Isolated block north of Ruaha River, drier high and medium altitude forest. Biogeographical interest.

8. Mahenge Forests, Morogoro Region. FRs. Small forest patches, isolated south of the Kilombero River. Hardly explored, in urgent need of biological inventory.

9. Njombe Forests, Rovuma Region. FRs. At edge of Eastern Arc, isolated and poorly explored. Generally lower altitudes, away from dense settlement pressures. Need for further survey.

Non-Eastern Arc Forests, Kenya

1. Cherangani Hills, Western Province. FR. 1285 km^2, including plantations. *Olea, Podocarpus, Syzygium* and *Juniperus* forest. Alpine *Erica, Lobelia, Protea* heath. Past locality for bongo. Needs biological review and higher conservation status.

2. Loita Hills, Nguruman Scarp, Southern Kenya. NO STATUS. Fragmented Loita forests total 21 km^2. *Juniperus, Nuxia, Podocarpus*. More extensive Nguruman Forest, 477 km^2 of great catchment value; Podo, Juniper and broad-leaved forest. Fayed & Fayad, 1970.

3. Kakamega–Nandi FR, NR, and National Park designate, Western Kenya. FR is 572 km^2 NR is 12 km^2 of which part is already degraded. Park may be 45 km^2. Easternmost example of Guineo–Congolian flora and fauna, and so of great East African conservation significance, *Celtis, Croton, Prunus* community dominates. Only Kenyan locality for L'Hoest's monkey and 62 bird species (Collar & Stuart, 1985). Kokwaro (1988) discusses deteriorating status. There is an obvious need for much greater investment in conservation, both planning and management.

4. Mt Kenya, Central Province. FRs and NP. Park is basically the afro-alpine heath and moorland above the forest line; the forest communities are mainly of reserve status. Lower edges are under growing pressure. Extensive *Juniperus* stands at altitude on drier NW slopes and *Ocotea* and the bamboo *Arundinaria alpina* on wetter sites. Planning is needed to identify key areas for improved conservation status.

5. Aberdares, Central Province. FRs and NP. Park has good area of Forest as well as moorland and bamboo.

6. Mau Summit, Rift Valley Province. FRs Extensive forest area, much plantation and encroachment. *Aningeria, Drypetes, Strombosia* and *Juniperus, Olea, Podocarpus* communities. Bongo, yellow-backed duiker and giant forest hog are major mammal rarities. Need for greater level of survey and extending NR network.

7. Suk Mts and East Turkana Mts, Rift Valley and Eastern Provinces. FRs on many isolated mountains in Northern Kenya, perhaps 2000 km^2 closed forest:

Loima, Lorosuk, Sepich in Suk; Kulal, Nyiru, Ndoto, Lerochi and Mathews in East Turkana. Dry forest, much *Juniperus*. Need for overview of conservation needs and enhanced status for key sites. Mt Kulal is a Biosphere Site.

8. Marsabit, Eastern Province. NR of 200 ha; needs extension to cover most of the biogeographically interesting *Olea* forest.

9. Mount Elgon, Nyanza Province. NP. Some good forest in the national park, 170 km², which may need extension to protect the lower altitude forest values from growing resource pressures. Joint planning with adjacent Uganda forest conservation agencies would be valuable.

10. Nairobi Forests, Nairobi and Central Provinces. FRs, part of Nairobi NP, and unreserved fragments. Relict patches of *Croton* forest. Need to plan long-term conservation requirements.

Non-Eastern Arc Forests, Tanzania

1. Mount Meru NP, Arusha Region. Diverse forest communities, well protected. No resource description published.

2. Mount Kilimanjaro, Kilimanjaro Region. FR and tiny area of NP. The national park covers the high altitude moorland and afro-alpine areas, but only cosmetic narrow forest corridors along tourist paths. Good forest, but low diversity and limited endemism. Forest area needs further protection (Newmark, 1991b).

3. Ngorongoro Massif, Northern Highlands FR, Arusha Region. Conservation Area. Forest area over 300 km². *Croton* up to *Cassipourea*, *Nuxia* and then *Hagenia*, *Ekebergia* communities, etc. at higher levels. Bamboo and *Juniperus* on drier slopes.

4. Masailand Mts, Arusha Region. FRs. Several isolated mountains rising up to 2700 m a.s.l. from dry bushland, e.g. Loliondo, Elsimingor, Monduli, Longido, Lolkisale, etc. Dry forest including *Olea, Juniperus* communities. All areas under severe and growing threat from fire, grazing and felling. Conservation interests should select key areas and plan long-term protection.

5. Mt Hanang, Singida Region. FR. Isolated old volcano, relict dry forest. Greenway (1955). Urgently needs biological survey.

6. Marang and Nou Forests, Arusha Region. FRs and part of Marang now in Manyara NP.

7. Minziro Forest, West Lake Region. FR. On Uganda border, quite distinct from other Tanzanian forest communities, 250 km². Much Swamp Forest (1200 m asl), with lowland *Podocarpus* (Lind & Morrison, 1974). Recent survey shows many new bird, mammal and butterfly species for Tanzania (N. Baker pers. comm.). Still poorly described and a need for conservation planning.

8. Mbisi Forest, Sumbawanga Region. FR, 26 km². Isolated forest on Ufipa Plateau (2300 m a.s.l.), surrounded by forb-rich grassland. Red colobus population and interesting bird community (Rodgers *et al.*, 1984).

9. Mahale Mountains NP, Kigoma Region. Guineo–Congolian affinities. Many fragmented forest patches in woodland area rising from shores of Lake Tanganyika. Only locality for several plant, butterfly and bird species. Rich primate fauna. IUCN, 1987.

10. Itigi Thickets, Singida Region. NO STATUS. Extensive areas of closed thicket (<8 m tall) on hardpan soils. Many endemic plant species in an unusual and distinctive community. A large and representative sample should be protected as a matter of urgency.

Index

Page numbers in *italics* refer to figures and tables.